VOLUME ONE HUNDRED AND THREE

ADVANCES IN FOOD AND NUTRITION RESEARCH

Nano/micro-Plastics Toxicity on Food Quality and Food Safety

VOLUME ONE HUNDRED AND THREE

ADVANCES IN FOOD AND NUTRITION RESEARCH

Nano/micro-Plastics Toxicity on Food Quality and Food Safety

Edited by

FATIH ÖZOGUL

Department of Seafood Processing Technology, Faculty of Fisheries, Cukurova University, Adana, Turkey

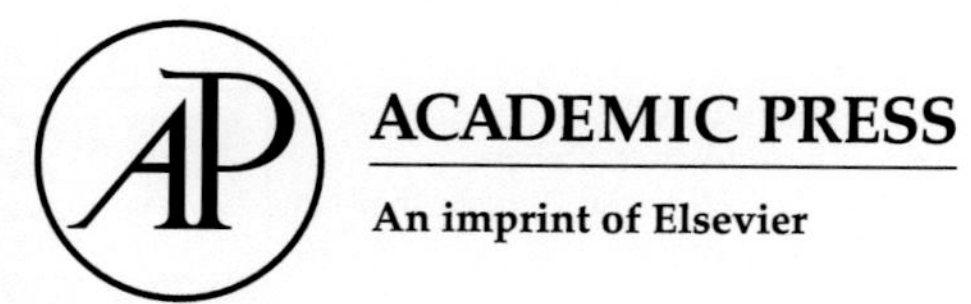

ACADEMIC PRESS
An imprint of Elsevier

Academic Press is an imprint of Elsevier
50 Hampshire Street, 5th Floor, Cambridge, MA 02139, United States
525 B Street, Suite 1650, San Diego, CA 92101, United States
The Boulevard, Langford Lane, Kidlington, Oxford OX5 1GB, United Kingdom
125 London Wall, London, EC2Y 5AS, United Kingdom

First edition 2023

ISBN: 978-0-323-98835-3
ISSN: 1043-4526

For information on all Academic Press publications
visit our website at https://www.elsevier.com/books-and-journals

Publisher: Zoe Kruze
Acquisitions Editor: Mariana L. Kuhl
Developmental Editor: Jhon Michael Peñano
Production Project Manager: Vijayaraj Purushothaman
Cover Designer: Matthew Limbert

Typeset by STRAIVE, India

Contents

Contributors

Abderrahmane Aït-Kaddour
Université Clermont Auvergne, INRAE, VetAgro Sup, UMRF, Lempdes, France

Tigran Garrievich Ambartsumov
Food Technology and Engineering Department, North Caucasus Federal University, Stavropol, Russia

M. Anjaly Shanker
Department of Agriculture and Environmental Sciences, National Institute of Food Technology Entrepreneurship and Management (NIFTEM), Sonepat, Haryana, India

Nur Alim Bahmid
Research Center for Food Technology and Processing, National Research and Innovation Agency (BRIN), Yogyakarta, Indonesia

Sneh Punia Bangar
Department of Food, Nutrition and Packaging Sciences, Clemson University, Clemson, SC, United States

Andrey Vladimirovich Blinov
Food Technology and Engineering Department, North Caucasus Federal University, Stavropol, Russia

Gioacchino Bono
Institute for Biological Resources and Marine Biotechnologies, National Research Council (IRBIM-CNR), Mazara del Vallo, TP; Dipartimento di Scienze e Tecnologie Biologiche, Chimiche e Farmaceutiche (STEBICEF), Università Di Palermo, Palermo, Italy

Suman Chaudhary
Department of Veterinary Physiology and Biochemistry, Lala Lajpat Rai University of Veterinary and Animal Sciences, Hisar, Haryana, India

Vandana Chaudhary
Department of Dairy Technology, College of Dairy Science and Technology, Lala Lajpat Rai University of Veterinary and Animal Sciences, Hisar, Haryana, India

Tatiana Chenet
Department of Environmental and Prevention Sciences, University of Ferrara, Ferrara, Italy

Nariman El Abed
Laboratory of Protein Engineering and Bioactive Molecules (LIP-MB), National Institute of Applied Sciences and Technology (INSAT), University of Carthage, Tunis, Tunisia

Scott W. Fowler
School of Marine and Atmospheric Sciences, Stony Brook University, Stony Brook, NY, United States; Institute Bobby, Cap d'Ail, France

Alexey Alekseevich Gvozdenko
Food Technology and Engineering Department, North Caucasus Federal University, Stavropol, Russia

Abdo Hassoun
Sustainable AgriFoodtech Innovation & Research (SAFIR), Arras, France; Syrian Academic Expertise (SAE), Gaziantep, Turkey

B.K.K.K. Jinadasa
Analytical Chemistry Laboratory, National Aquatic Resources Research & Development Agency (NARA), Colombo; Department of Food Science and Technology (DFST), Faculty of Livestock, Fisheries & Nutrition (FLFN), Wayamba University of Sri Lanka, Makandura, Gonawila, Sri Lanka

Sipper Khan
Institute of Agricultural Engineering Tropics and Subtropics Group, University of Hohenheim, Stuttgart, Germany

Anandu Chandra Khanashyam
Department of Food Science and Technology, Kasetsart University, Ladyao, Chatuchak, Bangkok, Thailand

Andrey Ashotovich Nagdalian
Food Technology and Engineering Department, North Caucasus Federal University, Stavropol; Saint Petersburg State Agrarian University, St Petersburg, Russia

Asad Nawaz
College of Civil and Transportation Engineering; Shenzhen Key Laboratory of Marine Microbiome Engineering, Institute for Advanced Study; Institute for Innovative Development of Food Industry, Shenzhen University, Shenzhen, China

Nilesh Prakash Nirmal
Institute of Nutrition, Mahidol University, Nakhon Pathom, Thailand

Natalya Pavlovna Oboturova
Food Technology and Engineering Department, North Caucasus Federal University, Stavropol, Russia

Fatih Özogul
Department of Seafood Processing Technology, Faculty of Fisheries, Cukurova University, Adana, Turkey

Luisa Pasti
Department of Environmental and Prevention Sciences, University of Ferrara, Ferrara, Italy

Girija Gajanan Phadke
Network for Fish Quality Management & Sustainable Fishing (NETFISH), The Marine Products Export Development Authority (MPEDA), Navi Mumbai, Maharashtra, India

Héctor J. Pula
Fish Nutrition and Feeding Research Group, Faculty of Science, University of Granada, Granada; Aula del Mar Cei-Mar of the University of Granada, Faculty of Sciences, Granada, Spain

Nikheel Bhojraj Rathod
Department of Post Harvest Management of Meat, Poultry and Fish, Post Graduate Institute of Post Harvest Technology & Management, Dr. Balasaheb Sawant Konkan Krishi Vidyapeeth, Roha, Raigad, Maharashtra, India

João Miguel Rocha
LEPABE – Laboratory for Process Engineering, Environment, Biotechnology and Energy, Faculty of Engineering, University of Porto; ALiCE – Associate Laboratory in Chemical Engineering, Faculty of Engineering, University of Porto, Porto, Portugal

Celia Rodríguez-Pérez
Department of Nutrition and Food Science, Faculty of Health Sciences, University of Granada (Melilla Campus), Melilla; Biomedical Research Centre, Institute of Nutrition and Food Technology (INYTA) 'José Mataix', University of Granada; Instituto de Investigación Biosanitaria ibs.GRANADA, Granada, Spain

Polina Rusanova
Institute for Biological Resources and Marine Biotechnologies, National Research Council (IRBIM-CNR), Mazara del Vallo, TP; Department of Biological, Geological and Environmental Sciences (BiGeA) – Marine Biology and Fisheries Laboratory of Fano (PU), University of Bologna (BO), Bologna, Italy

Miguel Sáenz de Rodrigáñez
Department of Physiology, Faculty of Health Sciences, University of Granada (Melilla Campus), Melilla, Spain

Sayed Hashim Mahmood Salman
Department of Science, Arabian Pearl Gulf Private School, Bilad Al Qadeem, Bahrain

Aysha Sameen
National Institute of Food Science and Technology, University of Agriculture, Faisalabad, Pakistan

Garima Kanwar Shekhawat
Department of Microbiology, School of Life Sciences, Central University of Rajasthan, Jaipur, India

Shahida Anusha Siddiqui
Technical University of Munich Campus Straubing for Biotechnology and Sustainability, Straubing; German Institute of Food Technologies (DIL e.V.), Quakenbrück, Germany

Slim Smaoui
Laboratory of Microbial Biotechnology and Engineering Enzymes (LMBEE), Center of Biotechnology of Sfax (CBS), University of Sfax, Sfax, Tunisia

Tayyaba Tariq
National Institute of Food Science and Technology, University of Agriculture, Faisalabad, Pakistan

Mishela Temkov
Department of Food Technology and Biotechnology, Faculty of Technology and Metallurgy, University Ss. Cyril and Methodius, Skopje, RN, Macedonia

Neha Thakur
Department of Livestock Products Technology, Lala Lajpat Rai University of Veterinary and Animal Sciences, Hisar, Haryana, India

Saif Uddin
Environment and Life Sciences Research Center, Kuwait Institute for Scientific Research (KISR), Kuwait City, Kuwait

Elena Velickova Nikova
Department of Food Technology and Biotechnology, Faculty of Technology and Metallurgy, University Ss. Cyril and Methodius, Skopje, RN, Macedonia

Noman Walayat
College of Food Science and Technology, Zhejiang University of Technology, Hangzhou, China

K.A. Martin Xavier
Department of Post-Harvest Technology, Fishery Resource Harvest and Postharvest Management Division, ICAR-Central Institute of Fisheries Education, Mumbai, Maharashtra, India

Preface

This volume of *Advances in Food and Nutrition Research* will be of potential interest to a wide range of audiences with different backgrounds, from academicians and researchers to individuals in the food industry (food engineers, food processing developers/managers in food packaging and preservation, food safety, food quality and control, etc.) to all persons in general with an interest in all these fields, who are united by a common denominator that is nano-/microplastics toxicity in food quality and food safety research.

Plastics are decomposed into smaller particles, i.e., microplastics (MPs) and nanoplastics (NPs), which have become a serious global safety concern owing to their overuse in many products and applications and their insufficient management, resulting in possible leakage into the environment and eventually into food webs, thus potentially affecting human health. A considerable mass of plastics is released into the marine environment annually through different human activities that involve the use of industrial, agricultural, medical, pharmaceutical, and daily care products.

This volume comprises 10 chapters that provide in-depth knowledge on the recent developments in nano-/microplastics (NPs/MPs) toxicity in food quality and food safety research. The topics covered in the volume include the sources, occurrence, and migration of NPs/MPs in terrestrial/marine environments, the mechanisms for the release of MPs from packaging into foods, the impacts of MPs on the quality of food products, their toxicological impacts, seafood safety, complications associated with the presence of NPs/MPs in the environment, and methods for the quantification of NPs/MPs as well as the existing regulations and requirements for plastics.

Chapter 1 describes the sources and occurrence of NPs/MPs in marine environments, their classification based on the properties influencing associated hazards, and their impact on the quality and safety of aquatic food products as well as the existing regulations and requirements for a robust framework on plastics. Chapter 2 focuses on the occurrence of plastic constituents in food packed with different types of plastic packaging, films, and coatings and how the type of food and packaging may influence the extent of contamination. The main types of contaminant phenomena and types of migration phenomena and factors are comprehensively discussed, along with the regulations for the use of plastic food packaging. In addition, migration components related to packaging polymers and packaging additives are individually reviewed in terms of their chemical structure and adverse effects on foodstuffs, health, migration factors, and the regulated residual values of such components. Chapter 3 discusses the recent developments with regard to the presence of MPs in different marine food chains, their translocation and accumulation potential, MPs as a critical vector for pollutant transfer, toxicological impact, cycling in the marine environment, and seafood safety as well as concerns and challenges related to MPs. Chapter 4 deals with the sources, complications, and toxicity associated with the presence of NPs/MPs in the environment along with evidence of trophic transfer and quantification methods. Chapter 5 mainly focuses on the techniques for the identification and quantification of NPs/MPs in different food matrices (mostly seafood). Chapter 6 provides a high-level overview of the potential risks of marine NPs/MPs for seafood safety and human health. Chapter 7 highlights the most recent evidence on the occurrence of NPs/MPs in wild and farmed edible species and the potential impacts of exposure to NPs/MPs on human health. The chapter also describes the identification of NP/MP particles in human biological samples, including the standardization of methods for their collection, characterization, and analysis. Chapter 8 discusses commercial plastic food packaging materials, the mechanisms for the release of MPs from such packaging materials into foods, and the factors influencing this release, e.g., high temperature, ultraviolet radiation, and bacteria. In addition, the potential threats and negative effects on human health are highlighted. Finally, future trends that aim to reduce the migration of MPs by enhancing public awareness and improving waste management are summarized. Chapter 9 reviews the potential risks and toxicological effects of both MPs and NPs on human health. The main points of distribution of various toxicants in the food chain are established. The impacts of some of the main sources of NPs/MPs on the human body are also emphasized. The processes of the entry and accumulation of NPs/MPs are described and the mechanism of accumulation that occurs inside the body is briefly explained. Chapter 10 discusses the challenges encountered in controlling NPs/MPs and improved technologies such as density separation, continuous flow centrifugation, oil extraction protocol, and electrostatic separation that are used to extract and quantify them. Besides the control measures, practical alternatives to MPs can be developed such as core–shell powder, mineral powder, and biobased food packaging systems such as edible films and coatings, developed using various nanotechnological tools. Finally, the existing and ideal situations with regard to global regulations are compared and the key research areas are highlighted.

I am especially grateful for the superb contributions made by remarkable food scientists and engineers from all over the world. This volume presents the combined efforts of 42 professionals with a variety of backgrounds and expertise from 18 countries, including Germany, France, India, Portugal, Spain, Sri Lanka, Thailand, Tunisia, Turkey, and the United States. I am also thankful for the high-quality editorial assistance provided by Mr. Jhon Michael Peñano (developmental editor) and the production staff at Elsevier for making the publication of this book possible. Finally, I acknowledge the kind encouragement from Prof. Fidel Toldra, editor of the book series *Advances in Food and Nutrition Research.*

Fatih Özogul
Department of Seafood Processing Technology
Cukurova University, Adana, Turkey

CHAPTER ONE

Impacts of nano/micro-plastics on safety and quality of aquatic food products

Nikheel Bhojraj Rathod[a,*], K.A. Martin Xavier[b], Fatih Özogul[c], and Girija Gajanan Phadke[d]

[a]Department of Post Harvest Management of Meat, Poultry and Fish, Post Graduate Institute of Post Harvest Technology & Management, Dr. Balasaheb Sawant Konkan Krishi Vidyapeeth, Roha, Raigad, Maharashtra, India
[b]Department of Post-Harvest Technology, Fishery Resource Harvest and Postharvest Management Division, ICAR-Central Institute of Fisheries Education, Mumbai, Maharashtra, India
[c]Department of Seafood Processing Technology, Faculty of Fisheries, Cukurova University, Adana, Turkey
[d]Network for Fish Quality Management & Sustainable Fishing (NETFISH), The Marine Products Export Development Authority (MPEDA), Navi Mumbai, Maharashtra, India
*Corresponding author: e-mail address: nikheelrathod310587@gmail.com

Contents

Abstract

The spread of nano/microplastics (N/MPs) pollution has gained importance due to the associated health concerns. Marine environment including fishes, mussels, seaweed and crustaceans are largely exposed to these potential threats. N/MPs are associated with plastic, additives, contaminants and microbial growth, which are transmitted to higher trophic levels. Foods from aquatic origin are known to promote health and have gained immense importance. Recently, aquatic foods are traced to transmit the nano/microplastic and the persistent organic pollutant poising hazard to humans. However,

Advances in Food and Nutrition Research, Volume 103
ISSN 1043-4526
https://doi.org/10.1016/bs.afnr.2022.07.001

microplastic ingestion, translocation and bioaccumulation of the contaminant have impacts on animal health. The level of pollution depends upon the pollution in the zone of growth for aquatic organisms. Consumption of contaminated aquatic food affects the health by transferring the microplastic and chemicals. This chapter describes the sources and occurrence of N/MPs in marine environment, detailed classification of N/MPs based on the properties influencing associated hazard. Additionally, occurrence of N/MPs and their impact on quality and safety in aquatic food products are discussed. Lastly, existing regulations and requirements of a robust framework of N/MPs are reviewed.

Abbreviations

AFP aquatic food products
APC polycarbonate
BIS Bureau of Indian Standards
BPA bisphenol
CP cellophane
DBP dibutyl phthalate
DDT dichloro diphenyl trichloroethane
DEHP di (2-ethyl-hexyl) phthalate
DEP diethyl phthalate
ECHA European Chemicals Agency
FAO Food and Agriculture Organization
FDA Food and Drug Administration
FSSAI Food Safety and Standards Authority of India
GESAMP Joint Group of Experts on the Scientific Aspects of Marine Environmental Protection
GI gastrointestinal
GMPL Global Partnership on Marine Litter
HDPE high-density polyethylene
LDPE low-density polyethylene
MP microplastic
N/MPs nano/microplastics
NP nanoplastic
OECD Organization for Economic Co-operation and Development
PA polyamide
PAE phthalic acid esters/phthalates
PAH polycyclic aromatic hydrocarbon
PAR polyarylether
PBDE polybrominated diphenyl ether
PBT polybutylene terephthalate
PCB polychlorinated biphenyl
PE polyethylene
PET polyethylene terephthalate
POP persistent organic pollutants
PP polypropylene
PR polystrene

PS polystyrene
PU polyurethane
PVC polyvinyl chloride
UNEP United Nations Environmental Programme
UV ultraviolet

1. Introduction

Plastic is regarded as a convenience product due to its durability, having applications across all segments from packaging, infrastructure, food, automobile, electronics, etc., with huge production, has gained them ubiquitous status (Galafassi et al., 2021; Jia, Evans, & van der Linden, 2019). However, lower recycling aspects (non-biodegradable nature) and huge production tune, leading to its release in the environment causing pollution. Around 10% of plastic produced is estimated to end up in the marine water environment; plastic is around 80% of the pollutants found in water bodies (Dehaut, Hermabessiere, & Duflos, 2019; Laglbauer et al., 2014). The pollution caused by plastics has reached alarming levels in aquatic foods, with an estimate to be more than fish by weight by 2050 (MacArthur, 2017; Simon & Schulte, 2017). The improper management of plastic waste from landfills, polluted rivers, aquatic activities, etc., is adding huge quantities of plastics in the marine ecosystem. Making oceans a largest reservoir of plastics, with an estimate of plastic waste around 23 million metric tons (Borrelle et al., 2020; Hale, Seeley, La Guardia, Mai, & Zeng, 2020; Lebreton et al., 2017; Park et al., 2020).

Plastic are known to breakdown/fragment/degrade due to several biotic and abiotic process into particles in size ranging <5–1 mm/<1 mm – 1 μm/<1 μm forming microplastic (MP)/Mini-microplastic/nanoplastic (NP) (EFSA, 2016; Kershaw, Turra, Galgani, et al., 2019). Micro/nanoplastics (M/NPs) are categorized based on their origin into primary and secondary particles. M/NPs could be a mixture of different shapes from fragments, fibers, spherical, granular, pellets, flakes or beads. Manufacturing of plastic involves incorporation of several harmful/toxic chemicals improving the quality having significant impacts on the surrounding environment, by releasing them (especially aquatic ecosystem) (EFSA, 2016; Park et al., 2020; Wright, Thompson, & Galloway, 2013; Zettler, Mincer, & Amaral-Zettler, 2013). In addition, large surface area of M/NPs causes concentration or interaction of different kinds of pollutants such as persistent organic pollutants, polycyclic aromatic hydrocarbons, organochlorine pesticides and polychlorinated biphenyls, etc. Some studies have also reported growth of pathogenic microorganisms on M/NPs (EFSA, 2016; Wright, Rowe, Thompson, & Galloway, 2013; Wright, Thompson, et al., 2013; Zettler et al., 2013). Lack of analytical development in detection, characterization and quantification of M/NPs has led to underreporting of actual presence and translocation (Kershaw et al., 2019).

All the M/NPs and the associated chemicals pollute the aquatic environment and transfer across the food web through bioaccumulation ending up on our table. M/NPs has been reported in the majority of aquatic food products (AFPs) moving across all the trophic levels of the marine environment (Gündogdu et al., 2022; Lusher, Hollman, & Mendoza-Hill, 2017; Tanaka & Takada, 2016). The consumption of M/NPs is either intentional (resemblance to food) or unintentional (feeding behavior) in the aquatic food system. It is not only the M/NPs but also the associated pollutant (released or absorbed) or pathogen that further increases the risk to the consumer.

AFP is a myriad of nutrition with established health benefits. They are a cheap source of proteins for large populations worldwide (Rathod, Ranveer, Benjakul, et al., 2021). The pollution of water bodies by M/NPs has raised a concern about the safety and quality of AFP. M/NPs are ingested and having size >1.5 μm could enter the circulation and further accumulate (EFSA, 2016). Usually M/NPs ingested are held in the digestive tract, which is not usually consumed. Loss in the quality of AFP was reported due to physical stress induced by M/NPs ingestion (Lusher, 2015). In exceptional cases where whole fish are dried or used as such for consumption, while the tract portion is used for fish meal. This is used as animal feed which is further used for human consumption. Apart from this, the toxicant accumulated or transmitted by M/NPs further lowers the nutritional status of AFP. Several cases of growth of pathogenic microorganisms on M/NPs are reported (Amaral-Zettler et al., 2015). Hence, AFP could be regarded as a source for transmission of M/NPs, associated organic chemical and pathogenic microorganism to humans.

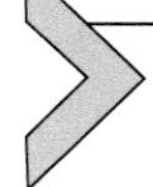

2. Occurrence of nano/microplastics and associated hazards

2.1 Sources of nano/microplastics in water-bodies

Recently, there is an increased demand for goods (clothing, storage, transportation, packaging, construction, a host of consumer goods, etc.) formed using plastics have made them indispensable material of today's life. Durability of plastic, which was the advantage in their application, has turned into the main threat in aquatic systems. Risk is reported to increase with increase in the migration of plastic to the ocean. Plastic particles floating on the surface of oceans are estimated to be 5.25 trillion weighing approximately 269,000 metric tons. Classified on the basis of size about 4.85 trillion particles were found to be microplastic (<4.75 mm) (Eriksen et al., 2014). Additionally, on yearly basis it is estimated that around 4.8 to 12.7 million metric tons of waste plastic enter as fresh into oceans (Jambeck et al., 2015), and there have been emerging environmental and health issues raised in regards to the plastics materials being disposed of in the ocean water around the world. The occurrence of plastic as pollutants in marine environments are linked in majority from inland and nearby urban areas (Cable et al., 2017; Hajbane & Pattiaratchi, 2017; Wang, Ndungu, Li, & Wang, 2017). Due to resemblance between microplastic and food, marine organisms (zooplankton, invertebrates, fish, bivalves, birds, and cetaceans) incidentally ingest them as mistaken for food (Cole et al., 2013; Ferreira, Fonte, Soares, Carvalho, & Guilhermino, 2016; Lusher, 2015).

Microplastics pollution in aquatic ecosystems is contributed mostly from the plastics being disposed of into the river system which ultimately make their way to the ocean water and thus degrade within a given time period and contribute to microplastics pollution in the marine environment. The rivers are known to contribute about 70–80% of plastic to marine ecosystems (Bowmer & Kershaw, 2010). The other source is also contributed by the fishing industries from wear and tear of vessels having plastic materials in its construction in some form or other as well as lost fishing gear materials. The discarded or abandoned or lost material from the fishing activity such as fishing gear contribute as a major source of plastic waste (nanofibers) in aquatic systems (FAO, 2016). The synthetic fibers reported in the fish are similar to plastic used in making fishing gears (Lusher, Mchugh, & Thompson, 2013). Significantly higher proportion of microplastic concentration was observed in digestive tract of fishes caught from the area near to wastewater treatment plant in comparison with freshwater zones in Argentina (Pazos, Maiztegui, Colautti, Paracampo, & Gómez, 2017). The presence of microplastics is observed to be directly proportional with population density, also the higher population density is associated with higher proportion (count and mass) of microplastics in river basins. Yonkos, Friedel, Perez-Reyes, Ghosal, and Arthur (2014) concluded that significant relationship is observed between presence of microplastic based on a relationship determined for four estuarine rivers that flow into the Chesapeake Bay. Hence, it can be clearly evident that population density and urbanization are significantly correlated to microplastics concentrations.

M/NPs are found to be distributed globally all over the world's oceans. Microplastics are defined as any plastic particles smaller than 5 mm size as accepted globally (Arthur, Baker, & Bamford, 2009; Kershaw & Rochman, 2015). Nanoplastics are the plastic particles ranging between 1 and 100 nm/0.001–0.1 μm. Currently, no legislation exists for the presence of microplastics and nanoplastics as contaminants in food. Considering the small dimensions of microplastic, they have propensity making it possible for the microplastic to translocate in the circulatory system through the digestive tract, where they may reside for relatively long periods of time (Browne, Dissanayake, Galloway, Lowe, & Thompson, 2008; Collard et al., 2017; Van Cauwenberghe, Claessens, Vandegehuchte, & Janssen, 2015). Considering the high levels of microplastic pollution in marine environment, many investigations have proposed several modeling studies for identifying the source, distribution and accumulation of microplastic in marine environments (Clark et al., 2016; Eriksen et al., 2014; Van Sebille et al., 2015). Among the marine habitats of oceans, seas, beaches, water surface, column and seafloor are reported to contain microplastic (Lusher, 2015). The properties of its small size and low-density MPs in the oceans are readily transported across a wide area and distance by ocean tidal current (Cole, Lindeque, Halsband, & Galloway, 2011; Eriksen et al., 2013).

A recent study has provided evidence that the Mediterranean Sea has relatively higher MPs abundance in comparison to other water bodies (Suaria et al., 2016). MPs have been reported in deep-sea habitats (Van Cauwenberghe, Vanreusel, Mees, & Janssen, 2013), oceanic gyres and accumulation zones, especially in the North Pacific and North Atlantic Ocean (Eriksen et al., 2013). MPs are found to be bioavailable in many marine organisms owing to their smaller sizes, which is very similar to that of plankton species (Wright, Rowe, et al., 2013).

2.2 Classification of nano/microplastics

Microplastics have been categorized into different groups based on their shapes and sizes by different researchers as reported in our previous review (Gündogdu et al., 2022). Based on the size, the microplastics have been classified as primary (few μm to 5 mm) usually found in personal care products (i.e., facial cleansers, toothpaste or shower gels), textile (i.e., stockings, faux leather, fur or suits) and medical applications (i.e., vectors for drugs). They are purposefully engineered or manufactured for industrial or domestic use including virgin or pristine plastic pellets. Besides, the large amounts of primary sized MPs are generated by abrasion and laundering of synthetic textiles makes them the main sources of generation of this type of MPs. While the secondary microplastics are those plastics materials which are generated by the withering of larger plastics materials due to mechanical and oxidative processes occurring in the environment (Hidalgo-Ruz, Gutow, Thompson, & Thiel, 2012). Several studies in the literature have suggested that beaches are the source of secondary microplastics (Andrady, 2011; Kataoka & Hinata, 2015; Kataoka, Nihei, Kudou, & Hinata, 2019). The other classification of microplastics is based on its shapes, colors, and sizes. Several authors have classified microplastics based on different shapes, colors and sizes and there has not been a fixed classification for microplastics so far. Based on its shapes microplastics had been classified types as fragments, fibers (filaments and lines), foams, beads, and pellets (Free et al., 2014; Karami, Golieskardi, Ho, Larat, & Salamatinia, 2017; Lusher et al., 2017).

The color and size categorization of microplastics are generally based on the source and location, and it varies with various researchers and their study area. Abbasi et al. (2018) has classified MPs colors into five categories (white-transparent, yellow-orange, red-pink, blue-green and black-gray) and sizes into five groups (<100, 100–250, 250–500, 500–1000, and >1000 μm) as found in fish tissues of Persian Gulf. In another study of microplastics in tissues of some commercial fish species by Akhbarizadeh, Moore, and Keshavarzi (2018), the MPs have been classified into six categories based on their size (<100, 100–250, 250–500, 500–1000, 1000–5000, and >5000 μm).

Microplastics have been rising as imminent environmental pollutants in recent years due to mass production of plastics materials worldwide coupled with inappropriate waste management, thus resulting in serious threat to the environment and its inhabitants. Evidence from several experiments has indicated that microplastics have successfully moved across different trophic levels (Bouwmeester, Hollman, & Peters, 2015; Farrell & Nelson, 2013). Serious physiological impact on aquatic creature health leading to obstruction in the gut and food depletion was reported as a consequence of microplastic ingestion into the organism body (Wright, Rowe, et al., 2013). The presence of several harmful additives/monomers used for producing plastic polymers could possibly leach into the digestive tract reducing survival, feeding, immunity or antioxidant capacity (Browne, Niven, Galloway, Rowland, & Thompson, 2013).

The study on microplastics pollution had already been globally reported from the Pacific Ocean (Day & Shaw, 1987; Yamashita & Tanimura, 2007), Atlantic Ocean (do Sul, Costa, Barletta, & Cysneiros, 2013; Thompson et al., 2004), and the Arctic Ocean (Obbard et al., 2014) where it is uninhabited by the humans, and study report have given a positive result for the presence of microplastics. Microplastics contamination in the future will be increasing due to the breakdown of the present stock piles of plastic materials already present in the environment along with the production of new plastics materials in the coming days.

2.3 Hazards associated with nano/microplastics

Microplastics exhibit wide variation in their characterization based on physical (size, weight, shape and color) and chemical (composition) properties (Hidalgo-Ruz et al., 2012). The behavior of the particles influences, i.e., their dispersion, accumulation of contaminants (chemicals, microorganisms), their toxicity and bioavailability (Lambert & Wagner, 2018). There are several physical and physiological threats associated with plastics materials. Virgin microplastics do not cause any imminent harm to fishes after dietary exposure (Jovanović, 2017) although some of the additives, colors, and fillers used in the production of plastics polymers are harmful to the animal body. These additives have several health impacts as it can alter the hormonal balance of the living organism and the additives like polychlorinated biphenyls (PCBs), a carcinogenic compound. Deleterious impacts of plastic monomers along with chemical additives used for manufacturing of plastic have been reported in fish tissues. Plastic monomers such as ethylene terephthalate and ethylene are not considered as hazards, while styrene is known to exhibit hormonal (estrogen) resembling activity (Yang, Yaniger, Jordan, Klein, & Bittner, 2011). In addition, vinyl chloride is reported to have genotoxic and mutagenic effects (Brandt-Rauf et al., 2012; Giri, 1995). Some additives (bisphenol A, nonylphenols, and phthalates) included in plastic to enhance their mechanical properties namely, flame retardants, are known to disrupt the functioning of endocrine activity (Bang et al., 2012; Diamanti-Kandarakis et al., 2009; Halden, 2010; Hermabessiere et al., 2017). Microplastics are known to exhibit higher surface area to volume ratio, with potential to absorb and leach large quantities of harmful chemicals/toxic compounds. Persistent toxic substances such as naturally occurring chemicals (polycyclic aromatic hydrocarbons-PAH), pesticides (dichloro diphenyl trichloroethane-DDT) and synthetic toxicants (polychlorinated biphenyl) are known to accumulate on microplastics. Inclusion of heavy metals in manufacturing of plastic to act as stabilizers, antioxidants, dyes, or antifouling agents are released in the aquatic ecosystem as well.

Recently, it has been reported about the ability of microplastic to support the formation of biofilm forming microorganisms (Rao, 2019). In polluted aquatic environments, microplastics are the potential surfaces for bacterial biofilm formation. The bacterial strains like Vibrios and Bacillus sp. are reported in weathered microplastics (Kirstein et al., 2016). MPs contaminate water and have detrimental effects on seafood and products like sea salts that humans use daily (Iñiguez, Conesa, & Fullana, 2017). As a result, the predominant route of exposure for humans exposed to MPs is through marine foods and salts.

Plastics are made by polymerizing repeated units of monomers, together with additional components like chemical additives, generally by addition or condensation processes (Lithner, Larsson, & Dave, 2011). Ethylene polymers such as low-density polyethylene (LDPE) and high-density polyethylene (HDPE) are examples of plastics. Polyvinyl chloride (PVC) is a vinyl chloride polymer, while polypropylene (PP) is a propylene polymer, polyethylene terephthalate (PET) is an ethylene terephthalate polymer, polystyrene is a styrene polymer and nylon is a polyamide. Chemicals are added to polymers altering their physical and mechanical characteristics, making processing more accessible and increased mechanical performance. Inclusion of additives changes the structure of polymers, modifying their sorptive capacity of the plastic (Endo, Yuyama, & Takada, 2013). Antioxidants and additives such as phthalates, bisphenol A, bisphenone, brominated flame retardants, polybrominated diphenyl ethers, nonylphenols, phthalates, triclosan, and organotin are the common additives included and traced from the microplastics (Hermabessiere et al., 2017). Antioxidants, colors, flame retardants, fillers, plasticizers, preservatives and UV stabilizers, such as bisphenol A, phthalates and triclosan, are additional chemical additives added to improve the desired quality of the final product. (Rao, 2019). Some of the plastic monomers (ethylene and ethylene terephthalate) have no health risk, while styrene and vinyl chloride disturbs hormonal glands, genotoxic and mutagenic effects. Besides, additives such as phthalates and nonylphenols exhibit endocrine disruption capacity.

Table 1 Major endocrine disruptor compounds associated with micro/nanoplastics.

Additives	Function	Uses	Effects on human health	Reference
BPA—Bisphenol A	An antioxidant or as a plasticizer in other polymers (PP, PE and PVC)	Many plastics, as well as in the lining of canned food containers	Endocrine disruptors	Rani et al. (2015)
Phthalic acid esters (PAE) or Phthalates	Plasticizers to soften plastic mainly in polyvinyl chloride	Toys, vinyl flooring and wall covering, cleansing and cosmetics (personal care products)	Endocrine disruptors	Grindler et al. (2018), Qian, Shao, Ying, Huang, and Hua (2020)
Nonylphenol	Antioxidant and plasticizer in some plastics	Paints, pesticides, detergents and personal care products	Endocrine disruptors	Hermabessiere et al. (2017)

Phthalates are phenolic acid esters with a broad polarity range. They are used as plasticizers in polymers, particularly in PVC. Di (2-Ethylhexyl) phthalate (DEHP) is commonly used as a plasticizer in diagnostic instruments, cosmetics, personal care products, and furniture. DEHP and other phthalates are endocrine disruptors (Table 1) that have been linked to kidney, reproductive, cardiac, and neurological toxicity in organisms (Rowdhwal & Chen, 2018).

Bisphenol A (BPA) found in polycarbonate (PC) and epoxy resin plastics are well-known for their endocrine-disrupting ability. It constitutes 30–65% of the monomeric units by volume (Hermabessiere et al., 2017). Commonly BPA is used for their plasticizing and antioxidant activity in polymers such as polyethylene, polypropylene and polyvinyl chloride. The disadvantages of BPA like other additives are leaking of additives into food and beverage packaging (Rani et al., 2015).

Flame retardants (poly-brominated) are chemicals that function to slow down the combustion and disperse flame by releasing bromine radicals. Nonylphenol are added as antioxidants with intent to delay oxidation, responsible for aging of polymers. They are recovered from microplastic in the form of octylphenol, nonylphenol and BPA. They are abundantly applied in the form of nonylphenol surfactants (especially nonylphenol ethoxylates, NPEs) through cleaners, detergents, paints and cosmetics (United Nations Environment Programme (UNEP), 2018).

Aluminium, iron, lead, copper, manganese, silver, and zinc were all absorbed by virgin polyethylene (Ashton, Holmes, & Turner, 2010). High concentration of PAH (2.4 mg/g) and PCB (0.1 mg/g) in plastic pellets from Chinese beaches was found by Zhang, Gong, Lv, Xiong, and Wu (2015). Heavy metals, zinc (Zn) and copper (Cu) were sourced to be leached from antifouling paint and adsorbed on virgin polystyrene (PS) beads, and old polyvinyl chloride (PVC) recovered from seawater. Heavy metals discharged into water from the antifouling paint were found to be absorbed on microplastics. Cu concentration on PVC fragments was substantially higher than in PS, attributed to higher polarity and large surface area of PVC. Except for Zn on PS, Cu and Zn concentrations on PVC and PS increased significantly during the course of experiment. As a result, Brennecke, Duarte, Paiva, Caçador, and Canning-Clode (2016) reported a strong connection between microplastic and heavy metals, suggesting consequences on marine life and the environment. Table 2 provides detailed information on heavy metals associated with microplastic polymers.

Microplastics are possible sites for the production of bacterial biofilms. Microorganisms can form a biofilm on floating plastics in as little as 7 days. Human diseases and spoilage bacteria could be present in the biofilm microbiome. *V. alginolyticus*, *V. coralliilyticus*, *V. fluvialis*, *V. harveyi*, *V. parahaemolyticus*, *V. splendidus* and pathogenic *Vibrio* species were found on microplastics (Kirstein et al., 2016). Biofilms can also serve as rafts, assisting in transferring microorganisms to remote locations.

Microplastics have proven to leak chemicals and provide a surface for chemical sorption. The sorption of persistent organic pollutants (POPs) into microplastics tends to concentrate the chemicals, and this process has been associated with increases in microplastic toxicity and environmental risk. POPs are toxic compounds made by humans that do not degrade naturally. POPs include PCBs, organochlorine, and flame retardants. Due to lipid-soluble nature of POP, ingested POPs are likely to get accumulated in fat tissues. POP-containing microplastics might potentially make contact with an organism's external and internal surfaces. POPs are known to cause adverse effects in humans by causing bioaccumulation in tissues and organs, decreasing the immunity and impairing the reproductive ability (Endo et al., 2013). POPs interact with nonpolar particles in the environment, such as microplastics, as a subgroup of organic pollutants that are frequently hydrophobic (Hartmann et al., 2017). Compared to seawater, the quantities of organic compounds absorbed by microplastics could be six times higher (Wright, Rowe, et al., 2013).

Table 2 Heavy metals associated with different microplastic polymers.

Plastics	Additives	Heavy metal associated	Adverse effects	Reference
PVC	Heat stabilizers, UV stabilizers and inorganic pigments	Cadmium (Cd)	Bone related disorders in postmenopausal women; Changes in metabolism of calcium, phosphorus and bone. Promote lipid oxidation, carcinogenesis, apoptosis and DNA methylation	Hahladakis, Velis, Weber, Iacovidou, and Purnell (2018)
PU - Polyurethane	Biocides	Mercury (Hg)	Mutagenic and carcinogenic	Andrady (2015)
PBT, PE, PS & PP	Flame retardants	Bromine (Br)	Apoptosis and genotoxic	Nusair, Almasaleekh, Abder-Rahman, and Alkhatatbeh (2019)
PVC, LDPE and polyesters	Biocides	Arsenic (As)	Congenital disabilities and Carcinogenic, gastrointestinal damage; death	Ashton et al. (2010)
PVC and all types of plastics, where red pigments are used	Heat stabilizers, UV stabilizers and inorganic pigments	Lead (Pb)	Anemia (less Hb); hypertension; miscarriages; hamper nervous Systems; infertility; induce stress	Brennecke et al. (2016), Massos and Turner (2017)

3. Properties of nano/microplastics

In general properties of nano/microplastics are classified based on physical and chemical properties of polymers.

3.1 Physical properties of nano/microplastics

3.1.1 Particle size

Particle size is an essential factor to consider while evaluating interaction among particles and organisms (Montes-Burgos et al., 2010). Weathered MP particle size distributions will be polydisperse, i.e., increase in concentration with decrease in size distribution (Lambert & Wagner, 2018). As environmental plastics weather and degrade, they break and disintegrate, creating particles with different size distribution and diverse spectrum of particle morphology. The particle uptake by an organism is determined based upon the feeding and digestion system of the organism.

3.1.2 Particle shape

Particle shape of polymeric particles determines the interaction with the biological system (Wright, Rowe, et al., 2013). Effects of particle size on amphipod *Hyalella azteca* were reported. Findings suggested polypropylene (PP) fibers were more hazardous in comparison to beads. In case of mortality and hatching inhibition, zinc oxide nano-sticks were more harmful than nanospheres on zebrafish embryos (Hua, Vijver, Richardson, Ahmad, & Peijnenburg, 2014). Study concluded that particle shape, i.e., particles with irregular shape and needle-like shape is found to have more adherence to the surface of the organism.

3.1.3 Surface area

Surface area is a significant metric since it rises with reduced particle size; hence, material having nanoscale can have higher impacts (Van Hoecke, De Schamphelaere, Van der Meeren, Lcucas, & Janssen, 2008). Although surface area is not commonly reported in MP research, it can be determined by measuring spherical equivalent diameter; however, it can cause overestimation.

3.1.4 Polymer crystallinity

Crystallinity highlights the ordered and dense organization of polymer chains, an important polymer feature. This influences physical qualities like density and permeability, which impacts hydration and swelling ability. The crystallinity of MPs will alter over time as they degrade (Chen, Bei, & Wang, 2000). The shrinkage in size of MP by preferential degradation of the amorphous region increases the overall crystallinity. This will result in the development of crystallites, which may be more poisonous than original material. Due to change in crystallinity the material differs significantly from their original counterparts. The change in the crystallinity impacts physical and chemical properties such as surface area, particle shape, particle size, density, leaching of additives, adsorption of pollutants, etc., which are related to their ingestion, and hampering outcomes/performance.

3.2 Chemical properties of nano/microplastics

3.2.1 Polymer types and additives

The discharge of chemicals as solvents and catalysts, as well as additives such as antioxidants, colors, plasticizers and residual monomers introduced during processing, may induce plastic-associated toxicity (Andrady, 2015; Muncke, 2009). The toxic effects of monomers and additives used in the manufacture of specific kinds of plastics are well established as:

- **i.** High quantities of additives used in the manufacturing and dioxins released during burning are regarded as dangerous polymers (Rossi & Lent, 2006).
- **ii.** Polycarbonate, which is made from bisphenol A, is an endocrine disruptive substance.
- **iii.** Polyacrylonitriles and acrylonitrile butadiene styrene, are regarded as carcinogenic by global agencies.

The leaching monomers and additives are influenced by physical (pore diameter, molecular size) quality of monomer and the additive used (Göpferich, 1996). However, the rate of chemical leaching posing potential hazards is dependent on the concentration of chemicals in plastic, portioning coefficient, age and level of degradation. For instance, an aged MP may have a higher degree of crystallinity, which may result in a reduced leaching.

3.2.2 Surface chemistry

Surface chemistry of plastics is altered based on degradation process and quality. Light and oxidative degradation processes will alter the plastic surface by forming new functional groups via interactions with OH radicals, oxygen, nitrogen oxides, and other photogenerated radicals (Chandra & Rustgi, 1998). An increase in chemical processes leads the surface of a plastic to fracture, allowing more degradation processes to proceed (Lambert & Wagner, 2016). These actions may degrade the plastic surface, producing further microscopic particle release when ingested, increase release of chemicals, and increase gut retention durations through the production of more angular shaped particles, distinguishing environmental MPs from main micro-beads.

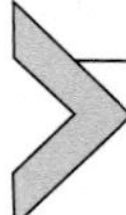

4. Occurrence of nano/microplastics in aquatic food products and health concerns

Microplastic pollution has spread throughout the marine ecosystem because of anthropogenic activities. Micro- and nanoplastics are becoming commonplace in the environment. They are reported to be present in air, water, and sediments, as well as in terrestrial and aquatic species. Their primary sources range from single-use plastic products, fishing equipment to garments, cosmetic products, farming tools, and waste generated from various industries such as paints, tyres, and many others. The contamination of the aquatic domain is particularly important in urban areas, and is caused through both territory (effluent, industrial sites, and other anthropogenic impacts) and marine causes (navigation, fishing, and oil platforms) (Bouwmeester et al., 2015). They can make their way past wastewater treatment systems that are not built to keep them. As a response, they concentrate at higher population concentrations (Browne et al., 2011), while the dispersion is controlled by air and water movements such as sea ripples and wind and density of the particle (Engler, 2012). Oceans accumulate plastic as a primary and final reservoir, exposing the existing flora and fauna as a consequence (Li, Tse, & Fok, 2016). A brief description on occurrence of microplastics in aquatic food is presented in Table 3.

Microplastics have been reported to be consumed via different processes by many kinds of wildlife, including fresh/marine fish and shellfish, resulting in widespread contamination (Smith, Love, Rochman, & Neff, 2018). Entry of microplastics into human beings is mostly through food items contaminated with microplastics primarily from shellfish sources contributing to nearly one fourth of the total items (Prata, da Costa, Lopes, Duarte, & Rocha-Santos, 2020). There are limited studies and literature available regarding the occurrence of microplastic in food. However, occurrence of microplastics has been observed in a variety of fish and seafood items. Almost all marine aquatic organisms have been reported to have microplastics (Hantoro, Löhr, Van Belleghem, Widianarko, & Ragas, 2019).

Table 3 Occurrence of microplastics in aquatic food products.

Aquatic food	Occurrence of microplastics	Reference
Farmed mussels and wild caught mussels	Higher microplastic concentration in farmed mussels (178 nos.) compared to that in wild mussels (126 nos.)	Van Cauwenberghe and Janssen (2014)
Commercial and wild caught fish in Indonesia markets and USA markets	28% of fish processed in Indonesia contained microplastics, 25% of commercial fish ingested with microplastics	Rochman et al. (2015)
Dried fish (gutted and whole)	Presence of microplastics in 4 species out of 30 with nearly half of the particles identified as plastic polymers	Karami et al. (2017)
Canned sardines and sprats from 13 countries	20% of samples were contaminated with 1 to 3 particles	Karami et al. (2018b)
Commercial fish from Portuguese, 26 samples from 7 locations	Occurrence of microplastics in 17 species out of 26 species studied	Neves, Sobral, Ferreira, and Pereira (2015)
Commercial mussel samples from Supermarkets of Belgian country and wild mussels from Belgian groynes	0.37 particles/g	De Witte et al. (2014)
9 bivalve samples from Chinese market	2–10 particle/g	Li, Yang, Li, Jabeen, and Shi (2015)
Oysters from USA (*Crassostrea gigas*)	2 particles/sample	Rochman et al. (2015)
Commercial marine fish species from South west coast of India	Presence of microplastics in 15 samples from 70 samples studied	Robin et al. (2020)
Commercial squid and crab samples from Kerala fish markets	Edible tissues of 18% of Indian squid, and 13.3% of blue crab	Daniel, Ashraf, Thomas, and Thomson (2021)
Fenneropenaeus indicus commercial white shrimp sample	0.39 microplastics/shrimp	Daniel, Ashraf, and Thomas (2020)
Canned fish	Occurrence of microplastics	Akhbarizadeh et al. (2020)

Table 3 Occurrence of microplastics in aquatic food products.—cont'd

Aquatic food	Occurrence of microplastics	Reference
Stuffed mussels	Occurrence of microplastics due to processing	Gündoğdu, Çevik, and Ataş (2020)
Fish meal	Occurrence of microplastics in fishmeal as a main source	Duis and Coors (2016)
Whole dried fish with gut and intact head	Important source of microplastic contamination with possibilities of transfer to humans	Karami et al. (2017), Mistri et al. (2022)
Dry fish across Asian countries	Presence of PE, PP, PS, PET and PVC in commercial dry fish samples	Piyawardhana et al. (2022)
Processed Japanese seaweed product: nori (*Pyropia* spp.)	Occurrence of microplastics	Li, Feng, Zhang, Ma, and Shi (2020)

Studies on microplastics and their health concerns are still underway due to the presence of additives in plastics. Studies on potential harm caused by microplastic is of great concern (Bouwmeester et al., 2015; Vethaak & Legler, 2021). Biota and microplastics frequently interact through ingestion. Shellfish and other aquatic organisms ingesting microplastics are major concerns for exposure of human beings. Major issue is associated with shellfish and other marine creatures consumed with intact GI tract, which retains and are of special concern (Smith et al., 2018). Microplastics and nanoplastics are widely spread in both freshwater and marine environments, posing a serious threat to aquatic life right from planktons to fish (Tanaka & Takada, 2016). Additionally, trophic transfer and mechanism of biomagnification allow these microplastics to reach humans. There is concern about physical and chemical toxicity, apart from plastic material chemicals used in manufacturing and that sorb from the environment increases the hazard. Microplastic toxicity and epidemiological evidence is accumulating. Bivalves such as mussels show a wide range of microplastic size ingestion from 6000 μm to 5 mm with a range of microplastic types. Ingestion of microplastics occurs in fish at different levels of trophic including phytoplankton, zooplankton, carnivores and filter feeding organisms like bivalves. Aquatic plants including seaweeds which are popular worldwide also exhibit microplastic contamination for smaller size (Sundbæk et al., 2018). Microplastics occurs in a wide range of aquatic foods products right from nori (Li et al., 2020) to by-products from fish such as fish meal (Gündoğdu, Eroldoğan, Evliyaoğlu, Turchini, & Wu, 2021). Other than direct contamination of seafood by microplastics in the aquatic environment, seafood processing and packaging used may also lead to contamination of seafood by microplastics.

Occurrence of microplastic in seafood is mainly dependent on feeding habits of fish and season. Fish that prefer selective diets may eat plastic rarely whereas fish that prefer a wide range of food may ingest more quantities of microplastics (Renzi et al., 2019). *Scomberomorus* species exhibit seasonal variation in ingested microplastic amounts with higher intake of microplastics in October when compared to that in March (Barboza, Vethaak, Lavorante, Lundebye, & Guilhermino, 2018). In bivalves, higher concentrations of microplastics have been observed due to their filter feeding habit (Bouwmeester et al., 2015). Oysters and mussels excrete ingested microplastics however, scallops keep larger and longer microplastics (Santillo, Miller, & Johnston, 2017). Occurrence of microplastics in crustaceans is also related to their activities and feeding habits. Larger crustaceans such as crabs and lobsters are exposed to microplastics due to the presence of microplastics in their feed (Bouwmeester et al., 2015). Report on presence of microplastics in marine biota exhibited 69% higher proportion of microplastics in 2016 as compared to those in 1977. Approximately 25% of the species ingested microplastic debris in nature with occurrence of microplastics in digestive tract, body and circulatory system (Lusher et al., 2017; Murray & Cowie, 2011; Van Cauwenberghe & Janssen, 2014). Presence of microplastics in commercial samples of fishmeal is of great concern. Potential contamination of fishmeal with microplastics need further evaluation studies in details which are lacking at present (Thiele, Hudson, Russell, Saluveer, & Sidaoui-Haddad, 2021). As discussed in above sections the whole fishes are utilized for formulation of fish meal, which in-turn are used as a protein source for formulation of fish feed. The impacts of microplastic on human health are summarized in Table 4.

Contamination of food and, in particular, the marine environment with nano/microplastic particles has become a major source in recent years. Pollution of the environment and the consequent seafood microplastic contamination poses a possible risk to consumers. Proportion of microplastic and nanoplastic that may enter into the gastrointestinal system of fish during its lifecycle is quite high and it will be more and more in future

Table 4 Impact of microplastics on human health.

System impacted	Microplastic	Reference
Respiratory system	Impact of polystyrene nanoparticles on human cell lines at different concentrations showed increased toxicity at higher concentrations in positively charged particles	Paget et al. (2015)
Digestive system	Effect of nanoparticles of polystyrene on functioning of human lung cells and observed internalization by non-specific phagocytosis	Xu et al. (2019)
	Death of cells due to higher proportions of polystyrene nanoparticles	Bexiga et al. (2011)
	Enhanced In-vitro cytotoxicity level	Stock et al. (2019)
Nervous system	Cytotoxicity assay of microplastics (PE and PS) on human health	Schirinzi et al. (2017)
Placental barrier	Microplastics may cross placental barrier in humans by perfusion model	Grafmueller et al. (2015)

(Jovanović, 2017). Consumption of fish without proper gutting and evisceration or in whole form is of great concern for exposure of human beings to hazards associated with micro and nanoplastics (Smith et al., 2018). Ingested micro and nanoplastics are excreted through viscera after ingestion of microplastics in oral mode (EFSA, 2016). Pieces of plastic sizing $<150\,\mu m$ in diameter (by definition, the tiniest microplastics and all nanoplastics) can pass the gastrointestinal epithelium and cause systemic absorption. Microplastics in fish may contain additives, contaminants and pathogens of drug resistant nature, which can enter through ingestion, inhalation and dermal contact in vertebrates and invertebrates. These may lead to various gut related issues and concerns such as impaired gut epithelial tissues, due to immune cell recruitment, barrier function, secretion of cytokines and oxidative stress. If microplastics/nanoplastics are ingested, it enters the GI tract by the mechanism of endocytosis by M type of cells and moves to tissues through paracellular transportation (Cox et al., 2019).

Diversified microbes and their composition may lead to dysbiosis of gut and immune-toxic gut due to oxidative stress, action of enzymes and immune response (Hirt & Body-Malapel, 2020). Microplastics can include up to 4% additives on average, and they can absorb pollutants. Microplastics in seafood, based on a conservative assessment, would have a minor impact on overall exposure to additives or pollutants. Physicochemical properties and potential damage induced by micro and nano plastics has impacts on human health (Hahladakis et al., 2018; Wilkinson, Hooda, Barker, Barton, & Swinden, 2017). However, the kinds of particles and types of pollutants exhibit physical and chemical toxicity in humans. Physical effects could include increased chronic inflammation, increased lipid peroxidation, cellular damage, and shape toxicity, among others. Moreover, because the immune system is unable to remove the plastic particles, inflammatory processes and an elevated chance of neoplasia may occur. In case of chemical consequences, microplastics are known to transfer a variety of chemicals, including compounds introduced during the manufacturing process and environmental contaminants such as hazardous metals and polychlorinated biphenyls. Bisphenol A (BPA) and phthalates, among the plastic additives, are endocrine disruptors that can cause carcinogenic and neurotoxic effects in animals and humans. Scientific data is increasingly pointing to different routes of microplastic exposure through food items, including findings that microplastics are prevalent in marine fish species making microplastics a growing seafood safety issue. Concentrations to which humans are exposed affect human health. Aquatic food products consumed as a whole pose a particular concern for exposure to microplastic in humans. There are various factors on which microplastic impacts on human health are dependent, which includes; toxicity level, type of polymer, size of polymer, its chemical structure and its hydrophobic nature (Smith et al., 2018). There exists a research gap in quantification of microplastic exposure to humans through food (Lusher et al., 2017). According to researchers, shellfish consumers in Europe consume around 11.000 plastic particles per year (Karami et al., 2018, 2017; Yang et al., 2015). Exposure to higher proportions of poly chlorinated biphenyls (PCBs) in fish may contribute to development of detrimental effects including changes in hormonal level, affected immune system, cancer development and thyroid (Faroon & Olson, 2000). There are several reports on inflammation, necrosis and decreased immune responses due to ingestion of microplastics (Wright & Kelly, 2017). Toxicity of additives in microplastics leads to transfer of hazardous compounds affecting biological systems. For a human risk assessment, toxicology and toxicokinetic data for both microplastics and nano-plastics are insufficient. Quality assurance should also be implemented and demonstrated. More research is needed for estimating chemical concentrations affecting human health in relation to microplastics. Analytical methodologies for microplastics and nanoplastics should be further developed, improved and standardized to determine the exact presence, identification, and quantification in food (EFSA, 2016). Seafood which is contaminated with microplastics and nanoplastics may cause severe health impacts on elderly population, pregnant women and children (Lusher et al., 2017). Studies in animals showed that exposure of animals to micro and nano plastics lead to impaired oxidative and intestinal function and act as potential agents for further adverse effects (Thiele et al., 2021).

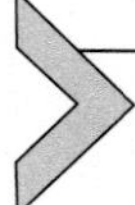

5. Impacts of nano/microplastics on safety and quality of aquatic food products

Considering the importance of aquatic foods and derived by products have been highlighted relating to bioactivity and health benefits (Ozogul et al., 2021; Phadke et al., 2021). Increase in consumer demand for foods processed minimally and preserved naturally, several novel processing technologies have been widely being evaluated for processing and preservation (Hassoun et al., 2022; Rathod, Kulawik, Ozogul, Ozogul, & Bekhit, 2022; Rathod, Phadke, et al., 2021; Rathod, Ranveer, Benjakul, et al., 2021; Rathod, Ranveer, Bhagwat, et al., 2021).

The M/NPs has been known to move across the human food chain, especially aquatic food chains. The recent study has confirmed its presence in human placenta justifying the ubiquitous nature and our inability to exclude M/NPs (Ragusa et al., 2021). However, the amount of M/NPs consumed is dependent upon type of AFP consumed or the feeding trophic level or consumption type, i.e., whole fish or degutted ones (Li, Ma, Zhang, & Shi, 2021; Vázquez-Rowe, Ita-Nagy, & Kahhat, 2021). Recently, the chemical and biological hazards associated with M/NPs occurrence in foods affecting the food safety has been highlighted (EFSA, 2016; Walkinshaw, Lindeque, Thompson, Tolhurst, & Cole, 2020). Zooplankton, the food for fish ingested polystyrene beads between 1.7 and 30.3 μm size which would further be transmitted to fishes indicating trophic transfer of M/NPs (Cole et al., 2013). Microbeads less than 10 μm were ingested by all marine invertebrates (bivalves, crustaceans and deposit feeders). However, their accumulation was found to be mostly based on concentration of M/NPs and feeding habits (Setälä, Norkko, & Lehtiniemi, 2016). Trophic transfer of M/NPs in crabs through mussels was also reported by Farrell and Nelson (2013). Table 5 specifies the studies suggesting the impact of occurrence of M/NP on AFP safety.

Table 5 Impact of presence of microplastic in aquatic seafood.

Product evaluated	Microplastic reported	Result	Reference
Shrimp—*Litopenaeus vannamei*, *Pleoticus mueller*, *Fenneropenaeus indicus*	Fibers, fragments and spheres. They ranged between 13 and 7050 items per shrimp	Highlighted the role of shrimp as a route to transmit microplastics in humans	Curren, Leaw, Lim, and Leong (2020)
Red belly tilapia, common carp	Polyethylene, polypropylene, ethylene-propylene copolymer, polyethylene terephthalate. 0.2 to 27.4 items/individual	Feeding habit (omnivores) with bottom feeding habit increased plastic ingestion with seasonal variation	Zheng et al. (2019)
Wild crucians (*Carassius auratus*)	Polypropylene, polyethylene. Sized <0.5 mm with 0–18 items per individual	Sewage water, human activity and fishing activity material dumped in lake contributed to microplastic	Yuan, Liu, Wang, Di, and Wang (2019)
Brown shrimp (*Crangon*, Linnaeus 1758)	63% of evaluated samples were contaminated with microplastic, 1.23 item/sample were observed. Size: 200–1000 μm	Higher feeding rate were responsible for more microplastic estimation. Particle greater than 20 μm size do not translocate	Devriese et al. (2015)
Tilapia (*Oreochromis niloticus*)	Polyamide, polyester and synthetic and natural cellulose	Microplastic presence and accumulation is related to physiological characteristics of fish	Martinez-Tavera et al. (2021)
Mollusks	Microfibers, fragments, film and pellets. Polyethylene, polyethylene terephthalate and nylon. About 3.7 to 17.7 items/individual were found	Mollusks from Persian Gulf exhibited higher risk of transmitting microplastic to humans. The filter feeding habit led to ingestion of microplastic	Naji, Nuri, and Vethaak (2018)
Mussels (*Mytilus galloprovincialis*)	6.2 to 7.2 items/g were observed, filament were between 750 and 6000 μm size	Cooking lowered the microplastic levels by around 14% in comparison to raw mussels	Renzi, Guerranti, and Blašković (2018)

Grouper (*Epinephelus* spp.)	Films and fishing thread. Belonging to polypropylene and polyethylene	The level of microplastic in habitats was directly related to their occurrence in fishes	Baalkhuyur et al. (2018)
Blue panchax (*Aplocheilus* sp.)	75% samples contained microplastic in the form of fragments, fibers and foam (1.97 particle/sample). Size: 300–500 µm	The small size led to consumption of microplastic with possibilities of transmission of toxic chemicals	Cordova, Riani, and Shiomoto (2020)
Freshwater and wild fishes	Microplastic in the form of flakes and fibers. Among polymers: polypropylene-polyethylene copolymer, polyethylene and polyethylene terephthalate	Effluents, local harbors and vessels contributed to microplastic. Sampling method has impacts on microplastic yields	Zhang et al. (2021)
Perch (*Perca fluviatilis*, Lin. 1758)	86% of sampled species contained microplastic at 1.24 particle/sample. They were fragments and films	Linear relation was observed between feeding activity and presence of microplastic	Galafassi et al. (2021)
Nile tilapia and catfish	75% of fishes were contaminated with fibers, films and fragments	Microplastic pollution in Nile and its potential to transfer to human population were reported	Khan, Siddique, Shereen, et al. (2020)
Oreochromis niloticus, *Prochilodus magdalenae* and *Pimelodus grosskopfii*	The samples were contaminated with microplastic (22%). Cellophane, polyethylene terephthalate, polyester and polyethylene were prominent	Occurrence of microplastic was dependent upon growing environment (natural or cultured)	Garcia, Suárez, Li, and Rotchell (2021)

Recent literature updated the relationship of M/NPs with different chemicals used during manufacturing (BPA, DBP, DEP, PAH, PBDEs, etc.), bioaccumulation of contaminants from water bodies (PAHs, heavy metals, pesticides, etc.), microbial contamination (*Vibrio* spp., *Escherichia coli*, *Bacillus cereus* and *Aeromonas salmonicida*) and production of toxins (diarrhetic and paralytic shellfish poisoning) could possibly accumulate in aquatic food products having severe implications on human health. The chemicals used for the manufacturing of plastics such as BPA, DEHP, NP, phthalates, heavy metals, flame retardant, etc., are not safe for human consumption anyhow (Campanale, Massarelli, Savino, Locaputo, & Uricchio, 2020). The proportions of contaminants present on PP resin pellets from the coast of Japan were about 10^5–10^6 higher than sediment indicating extreme risk to marine organisms and environment (Mato et al., 2001). Considering the toxic potential of BPA, a recent study by Meli, Monnolo, Annunziata, Pirozzi, and Ferrante (2020) linked defects in the reproductive system, oxidative stress and diseases in humans to exposure to BPA. Plasticizer use (DEHP) was recommended for assessment of toxicity data due to their potential hazard to humans by migrating to humans through medical devices was reviewed by Den Braver-Sewradj, Piersma, and Hessel (2020). Considering the constant exposure of a negligible dose of toxic chemicals interferes with the endocrine system that is usually neglected could further exhibit damage (Muncke, 2021). Pathogenic microorganisms such as *Escherichia coli*, *Bacillus cereus*, Vibrio spp., have been found on plastic debris (Barboza et al., 2018; Kirstein et al., 2016). The association of pathogenic microorganisms can further increase the risk of spreading diseases.

Persistence of M/NPs and associated pollutants in edible parts of shrimps, crab and squid as an emerging issue in food safety was reported by Daniel et al. (2021). They are poised to transmit hazardous compounds to consumers, especially humans. Squid (7.7 particles/kg) had maximum M/NP followed by crab (3.2 particles/kg) present in the edible part. Collard et al. (2017) proved the translocation of M/NPs to organs of commercial fishes, which are further used for direct or indirect consumption. Apart from chemicals included during manufacturing, M/NPs absorbs several hazardous chemical toxicants (brominated flame retardant, PAHs, PCBs, etc.) from water bodies that pose a great heat concern for humans (Hantoro et al., 2019). Some earlier findings by Yoo, Doshi, and Mitragotri (2011) and Wright and Kelly (2017) have demonstrated the ability of M/NPs $>1.5\,\mu m$ could penetrate capillaries and be transmitted across epithelial tissues.

Majority of edible items from marine environments such as seafood (shell fish or fresh fish) and sea salt have been contaminated with microplastic has been reported (Wang, Zhao, & Xing, 2021). The detailed description on ambiguity of consumption/ingestion of M/NPs has been detailed by Li et al. (2021). Authors reported variation in consumption of M/NPs from 2602 items/capita per year to 27,825 items/capita per year depending upon consumption part (whole fish, edible fillet, etc.), type (fish, shell fish, etc.) and consumption amount (Danopoulos, Jenner, Twiddy, & Rotchell, 2020; FAO, 2019; Senathirajah et al., 2021). Considering the above mentioned categories, this wide variation was found making it difficult to ascertain the harmful effects on human health (Li et al., 2021). However, the possibility of contamination by chemical pollutant or microorganism further ingestion by the marine organisms could transmit the risk to end consumers (Cole et al., 2011). Recently quantification of presence of M/NPs in bivalves, clams, crabs and fish from Portugal were investigated (Vital et al., 2021). Authors reported that mussels were most contaminated with M/NPs (86%) and attributed the increased levels of contamination to traveler, fishing gear disposal and disposing sewage and effluent water in marine water bodies. Pelagic and demersal fishes caught from Mumbai coast contained 6.74–9.12 and 5.62–6.6 items/fish, respectively (Gurjar et al., 2022). The M/NPs was of 7 colors consisting fibers, fragments, pellets and films; sizing below 250 μm. Authors confirmed trophic transfer of M/NPs through interlinked food chains. Exposure to M/NPs in humans through edible seafood from the Mediterranean region was assessed to evaluate risk (Ferrante et al., 2022). Among evaluated six types, all were contaminated irrespective of fish, farmed or wild caught and mussels. The biggest diameter of M/NPs was found to be 2.5 μm and estimated daily intake among the evaluated seafood ranged from 23.7 to 25.50 μm.

Further in the case of processed canned sardines, tuna and mackerel were reported to contain M/NPs intended for human consumption (Akhbarizadeh et al., 2020; Karami et al., 2018). Authors reported that about 80% of tuna and mackerel samples were contaminated by M/NPs with at least one particle, while sardines contained between 1 and 3 particles. They suggested the improper handling, processing, contact materials, additives added (salt) and the canning process resulted in migration of M/NPs in the flesh. Popular and frequently consumed traditionally stuffed mussels were found to contain on average 0.6 item M/NP/mussels sold in the Turkish market (Gündoğdu et al., 2020). The characterization using μ-Raman analysis identified the polymers to be polyethylene and polypropylene. Commercial industrially processed edible seaweed nori was also found to be contaminated with M/NPs (Li et al., 2020). It contained 0.9–3.0 items/g on a dry weight basis. The fractions found were of large size between 1 and 5 mm, mainly constituted of PP, PE and poly copolymers. Authors suggested the polluted water body and processing condition influenced the content of M/NPs. Piyawardhana et al. (2022) reported the occurrence of M/NP in dried fish from Asian countries. The entire sample was contaminated with M/NPs, of which 80% were fibers. Spectroscopy revealed the samples to be PE, PET, PS, PVC and PP. The highest proportions of M/NP, i.e., 1.92/g of dried fish were found in *Etrumeus micropus* samples obtained from Japan. Sea cucumber (*Apostichopus japonicus*), a delicacy, was reported to be contaminated by M/NPs (Mohsen, Lin, Liu, & Yang, 2022). The M/NPs were transmitted from the contaminated water body through the outer surface, however after shifting to clean water, the sea cucumber retained M/NPs till the 60 days period evaluated. About 86% of samples were found to be contaminated (0–15 M/NP/animal and 0–2 M/NP/g). M/NPs samples were microfiber, fragment, film and beads. Infrared spectroscopy identified samples to be rayon, polyester, chlorinated polyethylene. The presence of M/NPs in the gut of fishes has to be taken into consideration for development of processing interventions to eliminate M/NPs in processed seafood as suggested by Gurjar et al. (2022). Occurrence of M/NPs in commercial shrimps as direct exposure to humans was evaluated (Curren et al., 2020). Commercial shrimps (Pacific white leg shrimp, Argentine red shrimp and Indian white shrimp) were evaluated, M/NPs in the form of film were found at greater extent followed by fragments and spheres. The higher exposure to M/NPs in humans is expected due to contaminated shrimp as they are consumed as such without removal of gut.

Sea salt was evaluated as a source of M/NPs and could act as an indicator for determining pollution in water bodies (Kim, Lee, Kim, & Kim, 2018). Different types of salts (sea salt, rock salt, lake salt) evaluated contained M/NPs between 0 and 13,629 numbers/kg. Especially sea salt obtained from Asian countries had the highest M/NPs indicative of highest pollution. Characterization and toxicity evaluation of sea salt from India was done by Sivagami et al. (2021). The commercial brands evaluated contained less than 700 M/NP per kg of salt samples. The particle size of the sample ranged between 5.2 nM and 3.8 μM, identified as fragments, fibers and pellets and made up of CP, PR, PA, PAR. While the toxicity study suggested,

M/NP induced apoptosis in HET cells. Among the eight commercial Indian sea salts evaluated, all of them were found to contain M/NPs. Moreover, the classification revealed fibers and fragments consisting of 80% of them were less than 2000 and 500 μm size, respectively (Seth & Shriwastav, 2018). Authors also suggested using sand filters as remediation for excluding M/NPs. They also found that sand filters could eliminate 85% and 90% of M/NPs by weight and number basis.

6. Regulations related with nano/microplastics

Nutritional authorities encourage people to consume more seafood as they have an enormous amount of essential nutrients; however, the incidence of microplastics in seafood and seafood environments presents several questions among the researchers and consumers. In 2012, Global Partnership on Marine Litter (GPML) was created to reduce and manage marine litter to protect human health and the environment. Joint Group of Experts on the Scientific Aspects of Marine Environmental Protection (GESAMP) gives technical advice to UN Nations on marine pollution; in that GESAMP working group 40 deals with the source, fate, and effects of microplastics in the marine environment. Table 6 provides an overview about legal regulation for usage of single use plastics in different countries.

Initiatives by international organizations pertinent to marine debris include UNEP (United Nations Environmental Programme) with resolution in 2014 to note the effects of marine litter, which includes plastics and their effects on human health. Several OECD (Organization for Economic Co-operation and Development) nations have enacted bans and phase-outs on certain chemicals or goods, and regulatory restrictions that apply to use of specific cosmetics. The Netherlands declared a ban on usage of microbeads in cosmetic products in 2014 and became the pioneer in the world. In 2017, the European Commission requested ECHA (European Chemicals Agency) to evaluate the scientific evidence for taking regulatory action at the European level on the intentional production of microplastics. Followed by the US government, which approved the Microbead—Free Waters Act in 2015. Later in January 2018, the UK banned usage of microbeads in the rinse off cosmetics and personal care products (*Environmental Protection (Microbeads) (England) Regulations, 2017*). Further, in 2019, ECHA suggested a wide-ranging restriction on microplastics in the products placed on the EU market in order to avoid their release to the environment.

Table 6 Legal regulations on single use plastics and microplastic by different countries.

Country	Prohibition	Detailed description	Legal regulations
Canada	Cannot manufacture, import, or sell	Microbeads-containing toiletries must not be manufactured or imported. Any toiletry containing microbeads shall not be sold	Microbeads in Toiletries regulations, 2017
France	Cannot sell	The introduction of exfoliating and rinsing off cosmetics containing microplastics into the market is prohibited	Reclaiming Biodiversity, Nature and Landscapes, 2016
Italy	Cannot sell	It is illegal to sell cosmetic rinse solutions containing microplastics that have an exfoliating or detergent function	General Budget Law 2018 Law No:205
Republic of Korea	Cannot manufacture/sell	Plastic raw materials should not be utilized in cosmetics, and there should be regulations on the usage of cosmetics	Regulations on safety standards for cosmetics, 2017
New Zealand	Cannot manufacture/sell	In New Zealand, it is illegal to manufacture a prohibited wash-off product	Waste Minimisation (Microbeads) Regulations, 2017
UK England, Scotland, Wales, N. Ireland	Cannot sell or manufacture	A person who manufactures/supplies microbeads as an ingredient in any rinse-off personal care product is guilty of a crime	The Environmental Protection (Microbeads) (England) Regulations 2017 The Environmental Protection (Microbeads) (Scotland) Regulations 2018 The Environmental Protection (Microbeads) (Wales) Regulations 2018 The Environmental Protection (Microbeads) (Northern Ireland) Regulations 2018
US	Cannot sell or manufacture	A person who manufactures/supplies microbeads as an ingredient in any rinse-off personal care product is guilty of a crime	Micro bead-Free Waters Act of 2015
Sweden	Cannot sell	It is illegal to put microplastics containing cosmetic products on the market	Microplastics, in cosmetic products (2018)
India	Cannot use	BIS classified microbeads as unsafe to use in cosmetic products	BIS (2017)

Source: United Nations Environment Programme (UNEP). (2018). *Legal limits on single-use plastics and microplastics: A global review of national laws and regulations* (pp. 1–144). World Resources Institute. https://www.unep.org/resources/publication/legal-limits-single-use-plastics-and-microplastics-global-review-national

Recently, China has announced plans regarding the ban of microbeads production for its usage in cosmetics during December 31, 2020 and the prohibition of sales of existing stock from December 31, 2022. In India, Bureau of Indian Standards (BIS) has classified microbeads as unsafe to be used in cosmetic products. BIS in collaboration with food safety and standards authority of India (FSSAI) are working on microplastics contamination followed by the report of Orb Media, a US-based non-profit company on microplastics in drinking water. Microplastics on average contain 4% of additives that are harmful (Campanale et al., 2020).

BPA, phthalates, and several brominated flame retardants used in household items and food packaging have all been found to be endocrine disruptors that can harm human health if consumed or inhaled (De Coster & van Larebeke, 2012; Diamanti-Kandarakis et al., 2009). One of the main additives is the BPA (Bisphenol A), an estrogen-mimicking substance that is banned in many countries. In 2012, FDA banned BPA in baby bottles and sippy cups due to the risk of BPA leaching which poses a health risk. In 2015, BIS India also banned the use of BPA in feeding bottles.

7. Future trends

In view of tremendous work being focused on research about microplastic, every day newer claims about their existence in the human system are being reported (Leslie et al., 2022; Ragusa et al., 2021). Recent studies demonstrated the uptakes of microplastic in the human system with higher uptake on blood over organ deposition. There is a gap between the possible metabolism of uptake, deposition, excretion and impact on the immune system. Considering the high level of pollution in water bodies and processing involving usage of machines with possibilities of migration, more studies are required to evaluate the influences on microplastic. Formulation of hazard analysis and industry standards for evaluation of microplastic in food samples requires robust legislation.

Although human exposure to microplastic by aquatic food is low, considering the alarming level of contamination of water bodies by microplastic, several global food safety regulations like European Food Safety Authority, United States Environmental Protection Agency, FAO and WHO have performed detailed assessment addressing the problem.

8. Conclusion

Considering the ubiquitous nature of plastic (micro/nanoplastics), they are contaminating the food and water bodies affecting human health. Therefore, sources, types and contaminants from micro/nanoplastics and their impact (short and long term) should be evaluated. Recent studies have demonstrated the presence of micro/nanoplastics in human blood and fetuses exhibiting their systemic exposure in humans. The ability of micro/nanoplastics to be deposited in organs is very likely. However, the toxicity and adverse effects induced by micro/nanoplastics or associated contaminants need to be identified with exact mechanisms and limits should be established. Finally, the ingested micro/nanoplastics are known to interact with the human immune system with significant concern. The results reported so far are inconclusive regarding the risk proposed by consumption of seafood contaminated by micro/nanoplastics. In general, there is a strong need to understand the risk posed by micro/nanoplastics in seafood and their possibilities to transmit to humans and pose risk to human health. It is also the need of the day to determine the limits for micro/nanoplastics contamination and term the micro/nanoplastics and associated toxicants as marine pollutants.

References

Abbasi, S., Soltani, N., Keshavarzi, B., Moore, F., Turner, A., & Hassanaghaei, M. (2018). Microplastics in different tissues of fish and prawn from the Musa estuary, Persian Gulf. *Chemosphere, 205*, 80–87.

Akhbarizadeh, R., Dobaradaran, S., Nabipour, I., Tajbakhsh, S., Darabi, A. H., & Spitz, J. (2020). Abundance, composition, and potential intake of microplastics in canned fish. *Marine Pollution Bulletin, 160*, 111633. https://doi.org/10.1016/j.marpolbul.2020.111633.

Akhbarizadeh, R., Moore, F., & Keshavarzi, B. (2018). Investigating a probable relationship between microplastics and potentially toxic elements in fish muscles from northeast of Persian Gulf. *Environmental Pollution, 232*, 154–163.

Amaral-Zettler, L. A., Zettler, E. R., Slikas, B., Boyd, G. D., Melvin, D. W., Morrall, C. E., et al. (2015). The biogeography of the Plastisphere: Implications for policy. *Frontiers in Ecology and the Environment, 13*(10), 541–546.

Andrady, A. L. (2011). Microplastics in the marine environment. *Marine Pollution Bulletin, 62*(8), 1596–1605.

Andrady, A. L. (2015). *Plastics and environmental sustainability*. John Wiley & Sons.

Arthur, C., Baker, J. E., & Bamford, H. A. (2009). *Proceedings of the international research workshop on the occurrence, effects, and fate of microplastic marine debris, September 9–11, 2008*. Tacoma, WA, USA: University of Washington Tacoma.

Ashton, K., Holmes, L., & Turner, A. (2010). Association of metals with plastic production pellets in the marine environment. *Marine Pollution Bulletin, 60*(11), 2050–2055.

Baalkhuyur, F. M., Dohaish, E.-J. A. B., Elhalwagy, M. E., Alikunhi, N. M., AlSuwailem, A. M., Røstad, A., et al. (2018). Microplastic in the gastrointestinal tract of fishes along the Saudi Arabian Red Sea coast. *Marine Pollution Bulletin, 131*, 407–415.

Bang, D. Y., Kyung, M., Kim, M. J., Jung, B. Y., Cho, M. C., Choi, S. M., et al. (2012). Human risk assessment of endocrine-disrupting chemicals derived from plastic food containers. *Comprehensive Reviews in Food Science and Food Safety, 11*(5), 453–470.

Barboza, L. G. A., Vethaak, A. D., Lavorante, B. R., Lundebye, A.-K., & Guilhermino, L. (2018). Marine microplastic debris: An emerging issue for food security, food safety and human health. *Marine Pollution Bulletin, 133*, 336–348.

Bexiga, M. G., Varela, J. A., Wang, F., Fenaroli, F., Salvati, A., Lynch, I., et al. (2011). Cationic nanoparticles induce caspase 3-, 7-and 9-mediated cytotoxicity in a human astrocytoma cell line. *Nanotoxicology, 5*(4), 557–567.

Borrelle, S. B., Ringma, J., Law, K. L., Monnahan, C. C., Lebreton, L., McGivern, A., et al. (2020). Predicted growth in plastic waste exceeds efforts to mitigate plastic pollution. *Science, 369*(6510), 1515–1518. https://doi.org/10.1126/science.aba3656.

Bouwmeester, H., Hollman, P. C., & Peters, R. J. (2015). Potential health impact of environmentally released micro-and nanoplastics in the human food production chain: Experiences from nanotoxicology. *Environmental Science & Technology, 49*(15), 8932–8947.

Bowmer, T., & Kershaw, P. (2010). *Proceedings of the GESAMP international workshop on microplastic particles as a vector in transporting persistent, bio-accumulating and toxic substances in the ocean, 28-30th June 2010, UNESCO-IOC, Paris*. GESAMP.

Brandt-Rauf, P. W., Li, Y., Long, C., Monaco, R., Kovvali, G., & Marion, M.-J. (2012). Plastics and carcinogenesis: The example of vinyl chloride. *Journal of Carcinogenesis, 11*(5). https://doi.org/10.4103/1477-3163.93700.

Brennecke, D., Duarte, B., Paiva, F., Caçador, I., & Canning-Clode, J. (2016). Microplastics as vector for heavy metal contamination from the marine environment. *Estuarine, Coastal and Shelf Science, 178*, 189–195.

Browne, M. A., Crump, P., Niven, S. J., Teuten, E., Tonkin, A., Galloway, T., et al. (2011). Accumulation of microplastic on shorelines woldwide: Sources and sinks. *Environmental Science & Technology, 45*(21), 9175–9179. https://doi.org/10.1021/es201811s.

Browne, M. A., Dissanayake, A., Galloway, T. S., Lowe, D. M., & Thompson, R. C. (2008). Ingested microscopic plastic translocates to the circulatory system of the mussel, Mytilus edulis (L.). *Environmental Science & Technology, 42*(13), 5026–5031.

Browne, M. A., Niven, S. J., Galloway, T. S., Rowland, S. J., & Thompson, R. C. (2013). Microplastic moves pollutants and additives to worms, reducing functions linked to health and biodiversity. *Current Biology, 23*(23), 2388–2392.

Cable, R. N., Beletsky, D., Beletsky, R., Wigginton, K., Locke, B. W., & Duhaime, M. B. (2017). Distribution and modeled transport of plastic pollution in the Great Lakes, the world's largest freshwater resource. *Frontiers in Environmental Science, 5*, 45.

Campanale, C., Massarelli, C., Savino, I., Locaputo, V., & Uricchio, V. F. (2020). A detailed review study on potential effects of microplastics and additives of concern on human health. *International Journal of Environmental Research and Public Health, 17*(4), 1212.

Chandra, R., & Rustgi, R. (1998). Biodegradable polymers. *Progress in Polymer Science, 23*(7), 1273–1335.

Chen, D., Bei, J., & Wang, S. (2000). Polycaprolactone microparticles and their biodegradation. *Polymer Degradation and Stability, 67*(3), 455–459.

Clark, J. R., Cole, M., Lindeque, P. K., Fileman, E., Blackford, J., Lewis, C., et al. (2016). Marine microplastic debris: A targeted plan for understanding and quantifying interactions with marine life. *Frontiers in Ecology and the Environment, 14*(6), 317–324.

Cole, M., Lindeque, P., Fileman, E., Halsband, C., Goodhead, R., Moger, J., et al. (2013). Microplastic ingestion by zooplankton. *Environmental Science & Technology*, *47*(12), 6646–6655.

Cole, M., Lindeque, P., Halsband, C., & Galloway, T. S. (2011). Microplastics as contaminants in the marine environment: A review. *Marine Pollution Bulletin*, *62*(12), 2588–2597.

Collard, F., Gilbert, B., Compère, P., Eppe, G., Das, K., Jauniaux, T., et al. (2017). Microplastics in livers of European anchovies (Engraulis encrasicolus, L.). *Environmental Pollution*, *229*, 1000–1005.

Cordova, M. R., Riani, E., & Shiomoto, A. (2020). Microplastics ingestion by blue panchax fish (Aplocheilus sp.) from Ciliwung Estuary, Jakarta, Indonesia. *Marine Pollution Bulletin*, *161*, 111763.

Cox, K. D., Covernton, G. A., Davies, H. L., Dower, J. F., Juanes, F., & Dudas, S. E. (2019). Human consumption of microplastics. *Environmental Science & Technology*, *53*(12), 7068–7074.

Curren, E., Leaw, C. P., Lim, P. T., & Leong, S. C. Y. (2020). Evidence of marine microplastics in commercially harvested seafood. *Frontiers in Bioengineering and Biotechnology*, *8*(562760). https://doi.org/10.3389/fbioe.2020.562760.

Daniel, D. B., Ashraf, P. M., & Thomas, S. N. (2020). Abundance, characteristics and seasonal variation of microplastics in Indian white shrimps (Fenneropenaeus indicus) from coastal waters off Cochin, Kerala, India. *Science of the Total Environment*, *737*, 139839. https://doi.org/10.1016/j.scitotenv.2020.139839.

Daniel, D. B., Ashraf, P. M., Thomas, S. N., & Thomson, K. T. (2021). Microplastics in the edible tissues of shellfishes sold for human consumption. *Chemosphere*, *264*, 128554. https://doi.org/10.1016/j.chemosphere.2020.128554.

Danopoulos, E., Jenner, L. C., Twiddy, M., & Rotchell, J. M. (2020). Microplastic contamination of seafood intended for human consumption: A systematic review and meta-analysis. *Environmental Health Perspectives*, *128*(12), 126002.

Day, R. H., & Shaw, D. G. (1987). Patterns in the abundance of pelagic plastic and tar in the North Pacific Ocean, 1976–1985. *Marine Pollution Bulletin*, *18*(6), 311–316.

De Coster, S., & van Larebeke, N. (2012). Endocrine-disrupting chemicals: Associated disorders and mechanisms of action. *Journal of Environmental and Public Health*, *2012*, 1–52. https://doi.org/10.1155/2012/713696.

De Witte, B., Devriese, L., Bekaert, K., Hoffman, S., Vandermeersch, G., Cooreman, K., et al. (2014). Quality assessment of the blue mussel (Mytilus edulis): Comparison between commercial and wild types. *Marine Pollution Bulletin*, *85*(1), 146–155.

Dehaut, A., Hermabessiere, L., & Duflos, G. (2019). Current frontiers and recommendations for the study of microplastics in seafood. *Trends in Analytical Chemistry*, *116*, 346–359.

Den Braver-Sewradj, S. P., Piersma, A., & Hessel, E. V. (2020). An update on the hazard of and exposure to diethyl hexyl phthalate (DEHP) alternatives used in medical devices. *Critical Reviews in Toxicology*, *50*(8), 650–672.

Devriese, L. I., Van der Meulen, M. D., Maes, T., Bekaert, K., Paul-Pont, I., Frère, L., et al. (2015). Microplastic contamination in brown shrimp (Crangon crangon, Linnaeus 1758) from coastal waters of the Southern North Sea and Channel area. *Marine Pollution Bulletin*, *98*(1–2), 179–187.

Diamanti-Kandarakis, E., Bourguignon, J.-P., Giudice, L. C., Hauser, R., Prins, G. S., Soto, A. M., et al. (2009). Endocrine-disrupting chemicals: An Endocrine Society scientific statement. *Endocrine Reviews*, *30*(4), 293–342.

do Sul, J. A. I., Costa, M. F., Barletta, M., & Cysneiros, F. J. A. (2013). Pelagic microplastics around an archipelago of the Equatorial Atlantic. *Marine Pollution Bulletin*, *75*(1–2), 305–309.

Duis, K., & Coors, A. (2016). Microplastics in the aquatic and terrestrial environment: Sources (with a specific focus on personal care products), fate and effects. *Environmental Sciences Europe*, *28*(1), 2. https://doi.org/10.1186/s12302-015-0069-y.
EFSA. (2016). Presence of microplastics and nanoplastics in food, with particular focus on seafood. *EFSA Journal*, *14*(6), e04501.
Endo, S., Yuyama, M., & Takada, H. (2013). Desorption kinetics of hydrophobic organic contaminants from marine plastic pellets. *Marine Pollution Bulletin*, *74*(1), 125–131.
Engler, R. E. (2012). The complex interaction between marine debris and toxic chemicals in the Ocean. *Environmental Science & Technology*, *46*(22), 12302–12315. https://doi.org/10.1021/es3027105.
Environmental Protection (Microbeads) (England) Regulations. (2017). https://www.legislation.gov.uk/uksi/2017/1312/made/data.pdf.
Eriksen, M., Lebreton, L. C., Carson, H. S., Thiel, M., Moore, C. J., Borerro, J. C., et al. (2014). Plastic pollution in the world's oceans: More than 5 trillion plastic pieces weighing over 250,000 tons afloat at sea. *PLoS One*, *9*(12), e111913.
Eriksen, M., Maximenko, N., Thiel, M., Cummins, A., Lattin, G., Wilson, S., et al. (2013). Plastic pollution in the South Pacific subtropical gyre. *Marine Pollution Bulletin*, *68*(1–2), 71–76.
FAO. (2016). The state of world fisheries and aquaculture 2016. In *Contributing to food security and nutrition for all* (p. 200). Publications of Food and Agriculture Organization of the United Nations Rome.
FAO. (2019). *Fishery and aquaculture statistics 2016*. Rome: FAO.
Faroon, O., & Olson, J., N. (2000). *Toxicological profile for polychlorinated biphenyls (PCBs)*. Agency for Toxic Substances and Disease Registry, United States. https://stacks.cdc.gov/view/cdc/6480.
Farrell, P., & Nelson, K. (2013). Trophic level transfer of microplastic: Mytilus edulis (L.) to Carcinus maenas (L.). *Environmental Pollution*, *177*, 1–3.
Ferrante, M., Pietro, Z., Allegui, C., Maria, F., Antonio, C., Pulvirenti, E., et al. (2022). Microplastics in fillets of Mediterranean seafood. A risk assessment study. *Environmental Research*, *204*, 112247.
Ferreira, P., Fonte, E., Soares, M. E., Carvalho, F., & Guilhermino, L. (2016). Effects of multi-stressors on juveniles of the marine fish Pomatoschistus microps: Gold nanoparticles, microplastics and temperature. *Aquatic Toxicology*, *170*, 89–103.
Free, C. M., Jensen, O. P., Mason, S. A., Eriksen, M., Williamson, N. J., & Boldgiv, B. (2014). High-levels of microplastic pollution in a large, remote, mountain lake. *Marine Pollution Bulletin*, *85*(1), 156–163.
Galafassi, S., Sighicelli, M., Pusceddu, A., Bettinetti, R., Cau, A., Temperini, M. E., et al. (2021). Microplastic pollution in perch (Perca fluviatilis, Linnaeus 1758) from Italian south-alpine lakes. *Environmental Pollution*, *288*, 117782.
Garcia, A. G., Suárez, D. C., Li, J., & Rotchell, J. M. (2021). A comparison of microplastic contamination in freshwater fish from natural and farmed sources. *Environmental Science and Pollution Research*, *28*(12), 14488–14497.
Giri, A. (1995). Genetic toxicology of vinyl chloride—A review. *Mutation Research/Reviews in Genetic Toxicology*, *339*(1), 1–14.
Göpferich, A. (1996). Mechanisms of polymer degradation and erosion. In D. F. Williams (Ed.), *The biomaterials: Silver jubilee compendium* (pp. 117–128). Elsevier.
Grafmueller, S., Manser, P., Diener, L., Diener, P.-A., Maeder-Althaus, X., Maurizi, L., et al. (2015). Bidirectional transfer study of polystyrene nanoparticles across the placental barrier in an ex vivo human placental perfusion model. *Environmental Health Perspectives*, *123*(12), 1280–1286.

Grindler, N., Vanderlinden, L., Karthikraj, R., Kannan, K., Teal, S., Polotsky, A., et al. (2018). Exposure to phthalate, an endocrine disrupting chemical, alters the first trimester placental methylome and transcriptome in women. *Scientific Reports*, *8*(1), 1–9.
Gündoğdu, S., Çevik, C., & Ataş, N. T. (2020). Stuffed with microplastics: Microplastic occurrence in traditional stuffed mussels sold in the Turkish market. *Food Bioscience*, *37*, 100715. https://doi.org/10.1016/j.fbio.2020.100715.
Gündoğdu, S., Eroldoğan, O., Evliyaoğlu, E., Turchini, G. M., & Wu, X. (2021). Fish out, plastic in: Global pattern of plastics in commercial fishmeal. *Aquaculture*, *534*, 736316.
Gündogdu, S., Rathod, N., Hassoun, A., Jamroz, E., Kulawik, P., Gokbulut, C., et al. (2022). The impact of nano/micro-plastics toxicity on seafood quality and human health: Facts and gaps. *Critical Reviews in Food Science and Nutrition*, 1–19. https://doi.org/10.1080/10408398.2022.2033684.
Gurjar, U. R., Xavier, K. M., Shukla, S. P., Jaiswar, A. K., Deshmukhe, G., & Nayak, B. B. (2022). Microplastic pollution in coastal ecosystem off Mumbai coast, India. *Chemosphere*, *288*, 132484.
Hahladakis, J. N., Velis, C. A., Weber, R., Iacovidou, E., & Purnell, P. (2018). An overview of chemical additives present in plastics: Migration, release, fate and environmental impact during their use, disposal and recycling. *Journal of Hazardous Materials*, *344*, 179–199. https://doi.org/10.1016/j.jhazmat.2017.10.014.
Hajbane, S., & Pattiaratchi, C. B. (2017). Plastic pollution patterns in offshore, nearshore and estuarine waters: A case study from Perth, Western Australia. *Frontiers in Marine Science*, *4*, 63.
Halden, R. U. (2010). Plastics and health risks. *Annual Review of Public Health*, *31*, 179–194.
Hale, R. C., Seeley, M. E., La Guardia, M. J., Mai, L., & Zeng, E. Y. (2020). A global perspective on microplastics. *Journal of Geophysical Research: Oceans*, *125*(1), e2018JC014719.
Hantoro, I., Löhr, A. J., Van Belleghem, F. G. A. J., Widianarko, B., & Ragas, A. M. J. (2019). Microplastics in coastal areas and seafood: Implications for food safety. *Food Additives & Contaminants: Part A*, *36*(5), 674–711. https://doi.org/10.1080/19440049.2019.1585581.
Hartmann, N. B., Rist, S., Bodin, J., Jensen, L. H., Schmidt, S. N., Mayer, P., et al. (2017). Microplastics as vectors for environmental contaminants: Exploring sorption, desorption, and transfer to biota. *Integrated Environmental Assessment and Management*, *13*(3), 488–493.
Hassoun, A., Siddiqui, S. A., Smaoui, S., Ucak, İ., Arshad, R. N., Garcia-Oliveira, P., et al. (2022). Seafood processing, preservation, and analytical techniques in the age of industry 4.0. *Applied Sciences*, *12*(3), 1703. https://doi.org/10.3390/app12031703.
Hermabessiere, L., Dehaut, A., Paul-Pont, I., Lacroix, C., Jezequel, R., Soudant, P., et al. (2017). Occurrence and effects of plastic additives on marine environments and organisms: A review. *Chemosphere*, *182*, 781–793.
Hidalgo-Ruz, V., Gutow, L., Thompson, R. C., & Thiel, M. (2012). Microplastics in the marine environment: A review of the methods used for identification and quantification. *Environmental Science & Technology*, *46*(6), 3060–3075.
Hirt, N., & Body-Malapel, M. (2020). Immunotoxicity and intestinal effects of nano-and microplastics: A review of the literature. *Particle and Fibre Toxicology*, *17*(1), 1–22.
Hua, J., Vijver, M. G., Richardson, M. K., Ahmad, F., & Peijnenburg, W. J. (2014). Particle-specific toxic effects of differently shaped zinc oxide nanoparticles to zebrafish embryos (Danio rerio). *Environmental Toxicology and Chemistry*, *33*(12), 2859–2868.
Iñiguez, M. E., Conesa, J. A., & Fullana, A. (2017). Microplastics in Spanish table salt. *Scientific Reports*, 7(1), 1–7.
Jambeck, J. R., Geyer, R., Wilcox, C., Siegler, T. R., Perryman, M., Andrady, A., et al. (2015). Plastic waste inputs from land into the ocean. *Science*, *347*(6223), 768–771.

Jia, L., Evans, S., & van der Linden, S. (2019). Motivating actions to mitigate plastic pollution. *Nature Communications*, *10*(1), 1–3.

Jovanović, B. (2017). Ingestion of microplastics by fish and its potential consequences from a physical perspective. *Integrated Environmental Assessment and Management*, *13*(3), 510–515.

Karami, A., Golieskardi, A., Choo, C. K., Larat, V., Karbalaei, S., & Salamatinia, B. (2018). Microplastic and mesoplastic contamination in canned sardines and sprats. *Science of the Total Environment*, *612*, 1380–1386.

Karami, A., Golieskardi, A., Ho, Y. B., Larat, V., & Salamatinia, B. (2017). Microplastics in eviscerated flesh and excised organs of dried fish. *Scientific Reports*, 7(1), 1–9.

Kataoka, T., & Hinata, H. (2015). Evaluation of beach cleanup effects using linear system analysis. *Marine Pollution Bulletin*, *91*(1), 73–81.

Kataoka, T., Nihei, Y., Kudou, K., & Hinata, H. (2019). Assessment of the sources and inflow processes of microplastics in the river environments of Japan. *Environmental Pollution*, *244*, 958–965.

Kershaw, P., & Rochman, C. (2015). Sources, fate and effects of microplastics in the marine environment: Part 2 of a global assessment. In *Reports and studies-IMO/FAO/Unesco-IOC/WMO/IAEA/UN/UNEP joint group of experts on the scientific aspects of marine environmental protection (GESAMP) Eng No. 93*.

Kershaw, P., Turra, A., Galgani, F., et al. (2019). *Guidelines for the monitoring and assessment of plastic litter and microplastics in the ocean*. London, UK: GESAMP Joint Group of Experts on the Scientific Aspects of Marine Environmental Protection.

Khan, S., Siddique, R., Shereen, M. A., et al. (2020). The emergence of a novel coronavirus (SARS-CoV-2), their biology and therapeutic options (published online ahead of print, 2020 Mar 11). *Journal of Clinical Microbiology*, *58*(5), e00187–20. https://doi.org/10.1128/JCM.00187-20.

Kim, J.-S., Lee, H.-J., Kim, S.-K., & Kim, H.-J. (2018). Global pattern of microplastics (MPs) in commercial food-grade salts: Sea salt as an indicator of seawater MP pollution. *Environmental Science & Technology*, *52*(21), 12819–12828.

Kirstein, I. V., Kirmizi, S., Wichels, A., Garin-Fernandez, A., Erler, R., Löder, M., et al. (2016). Dangerous hitchhikers? Evidence for potentially pathogenic Vibrio spp. On microplastic particles. *Marine Environmental Research*, *120*, 1–8.

Laglbauer, B. J., Franco-Santos, R. M., Andreu-Cazenave, M., Brunelli, L., Papadatou, M., Palatinus, A., et al. (2014). Macrodebris and microplastics from beaches in Slovenia. *Marine Pollution Bulletin*, *89*(1–2), 356–366.

Lambert, S., & Wagner, M. (2016). Characterisation of nanoplastics during the degradation of polystyrene. *Chemosphere*, *145*, 265–268.

Lambert, S., & Wagner, M. (2018). Microplastics are contaminants of emerging concern in freshwater environments: An overview. In M Wagner, & S Lambert (Eds.), *Freshwater microplastics*. Springer.

Lebreton, L., Van Der Zwet, J., Damsteeg, J.-W., Slat, B., Andrady, A., & Reisser, J. (2017). River plastic emissions to the world's oceans. *Nature Communications*, *8*(1), 1–10.

Leslie, H. A., van Velzen, M. J. M., Brandsma, S. H., Vethaak, D., Garcia-Vallejo, J. J., & Lamoree, M. H. (2022). Discovery and quantification of plastic particle pollution in human blood. *Environment International*, *163*, 107199. https://doi.org/10.1016/j.envint.2022.107199.

Li, Q., Feng, Z., Zhang, T., Ma, C., & Shi, H. (2020). Microplastics in the commercial seaweed nori. *Journal of Hazardous Materials*, *388*, 122060.

Li, Q., Ma, C., Zhang, Q., & Shi, H. (2021). Microplastics in shellfish and implications for food safety. *Current Opinion in Food Science*, *40*, 192–197. https://doi.org/10.1016/j.cofs.2021.04.017.

Li, W. C., Tse, H. F., & Fok, L. (2016). Plastic waste in the marine environment: A review of sources, occurrence and effects. *Science of the Total Environment, 566–567*, 333–349. https://doi.org/10.1016/j.scitotenv.2016.05.084.
Li, J., Yang, D., Li, L., Jabeen, K., & Shi, H. (2015). Microplastics in commercial bivalves from China. *Environmental Pollution, 207*, 190–195.
Lithner, D., Larsson, Å., & Dave, G. (2011). Environmental and health hazard ranking and assessment of plastic polymers based on chemical composition. *Science of the Total Environment, 409*(18), 3309–3324.
Lusher, A. (2015). Microplastics in the marine environment: Distribution, interactions and effects. In M. Bergmann, L. Gutow, & M. Klages (Eds.), *Marine anthropogenic litter* (pp. 245–307). Cham: Springer.
Lusher, A., Hollman, P. C. H., & Mendoza-Hill, J. (2017). *Microplastics in fisheries and aquaculture: Status of knowledge on their occurrence and implications for aquatic organisms and food safety*. Food and Agriculture Organization of the United Nations.
Lusher, A. L., Mchugh, M., & Thompson, R. C. (2013). Occurrence of microplastics in the gastrointestinal tract of pelagic and demersal fish from the English Channel. *Marine Pollution Bulletin, 67*(1–2), 94–99.
MacArthur, E. (2017). *The new plastics economy: Rethinking the future of plastics & catalysing action* (p. 68). Ellen MacArthur Foundation.
Martinez-Tavera, E., Duarte-Moro, A., Sujitha, S., Rodriguez-Espinosa, P., Rosano-Ortega, G., & Expósito, N. (2021). Microplastics and metal burdens in freshwater tilapia (Oreochromis niloticus) of a metropolitan reservoir in Central Mexico: Potential threats for human health. *Chemosphere, 266*, 128968.
Massos, A., & Turner, A. (2017). Cadmium, lead and bromine in beached microplastics. *Environmental Pollution, 227*, 139–145.
Mato, Y., Isobe, T., Takada, H., Kanehiro, H., Ohtake, C., & Kaminuma, T. (2001). Plastic resin pellets as a transport medium for toxic chemicals in the marine environment. *Environmental Science & Technology, 35*(2), 318–324.
Meli, R., Monnolo, A., Annunziata, C., Pirozzi, C., & Ferrante, M. C. (2020). Oxidative stress and BPA toxicity: An antioxidant approach for male and female reproductive dysfunction. *Antioxidants, 9*(5), 405.
Mistri, M., Sfriso, A. A., Casoni, E., Nicoli, M., Vaccaro, C., & Munari, C. (2022). Microplastic accumulation in commercial fish from the Adriatic Sea. *Marine Pollution Bulletin, 174*, 113279.
Mohsen, M., Lin, C., Liu, S., & Yang, H. (2022). Existence of microplastics in the edible part of the sea cucumber Apostichopus japonicus. *Chemosphere, 287*, 132062.
Montes-Burgos, I., Walczyk, D., Hole, P., Smith, J., Lynch, I., & Dawson, K. (2010). Characterisation of nanoparticle size and state prior to nanotoxicological studies. *Journal of Nanoparticle Research, 12*(1), 47–53.
Muncke, J. (2009). Exposure to endocrine disrupting compounds via the food chain: Is packaging a relevant source? *Science of the Total Environment, 407*(16), 4549–4559.
Muncke, J. (2021). Tackling the toxics in plastics packaging. *PLoS Biology, 19*(3), e3000961.
Murray, F., & Cowie, P. R. (2011). Plastic contamination in the decapod crustacean Nephrops norvegicus (Linnaeus, 1758). *Marine Pollution Bulletin, 62*(6), 1207–1217.
Naji, A., Nuri, M., & Vethaak, A. D. (2018). Microplastics contamination in molluscs from the northern part of the Persian Gulf. *Environmental Pollution, 235*, 113–120.
Neves, D., Sobral, P., Ferreira, J. L., & Pereira, T. (2015). Ingestion of microplastics by commercial fish off the Portuguese coast. *Marine Pollution Bulletin, 101*(1), 119–126.
Nusair, S. D., Almasaleekh, M. J., Abder-Rahman, H., & Alkhatatbeh, M. (2019). Environmental exposure of humans to bromide in the Dead Sea area: Measurement of genotoxicy and apoptosis biomarkers. *Mutation Research/Genetic Toxicology and Environmental Mutagenesis, 837*, 34–41.

Obbard, R. W., Sadri, S., Wong, Y. Q., Khitun, A. A., Baker, I., & Thompson, R. C. (2014). Global warming releases microplastic legacy frozen in Arctic Sea ice. *Earth's Future*, *2*(6), 315–320.

Ozogul, F., Cagalj, M., Šimat, V., Ozogul, Y., Tkaczewska, J., Hassoun, A., et al. (2021). Recent developments in valorisation of bioactive ingredients in discard/seafood processing by-products. *Trends in Food Science & Technology*, *116*, 559–582. https://doi.org/10.1016/j.tifs.2021.08.007.

Paget, V., Dekali, S., Kortulewski, T., Grall, R., Gamez, C., Blazy, K., et al. (2015). Specific uptake and genotoxicity induced by polystyrene nanobeads with distinct surface chemistry on human lung epithelial cells and macrophages. *PLoS One*, *10*(4), e0123297.

Park, T.-J., Lee, S.-H., Lee, M.-S., Lee, J.-K., Park, J.-H., & Zoh, K.-D. (2020). Distributions of microplastics in surface water, fish, and sediment in the vicinity of a sewage treatment plant. *Water*, *12*(12), 3333.

Pazos, R. S., Maiztegui, T., Colautti, D. C., Paracampo, A. H., & Gómez, N. (2017). Microplastics in gut contents of coastal freshwater fish from Río de la Plata estuary. *Marine Pollution Bulletin*, *122*(1–2), 85–90.

Phadke, G. G., Rathod, N. B., Ozogul, F., Elavarasan, K., Karthikeyan, M., Shin, K.-H., et al. (2021). Exploiting of secondary raw materials from fish processing industry as a source of bioactive peptide-rich protein hydrolysates. *Marine Drugs*, *19*(9), 480. https://doi.org/10.3390/md19090480.

Piyawardhana, N., Weerathunga, V., Chen, H.-S., Guo, L., Huang, P.-J., Ranatunga, R., et al. (2022). Occurrence of microplastics in commercial marine dried fish in Asian countries. *Journal of Hazardous Materials*, *423*, 127093.

Prata, J. C., da Costa, J. P., Lopes, I., Duarte, A. C., & Rocha-Santos, T. (2020). Environmental exposure to microplastics: An overview on possible human health effects. *Science of the Total Environment*, *702*, 134455.

Qian, Y., Shao, H., Ying, X., Huang, W., & Hua, Y. (2020). The endocrine disruption of prenatal phthalate exposure in mother and offspring. *Frontiers in Public Health*, *8*, 366.

Ragusa, A., Svelato, A., Santacroce, C., Catalano, P., Notarstefano, V., Carnevali, O., et al. (2021). Plasticenta: First evidence of microplastics in human placenta. *Environment International*, *146*, 106274.

Rani, M., Shim, W. J., Han, G. M., Jang, M., Al-Odaini, N. A., Song, Y. K., et al. (2015). Qualitative analysis of additives in plastic marine debris and its new products. *Archives of Environmental Contamination and Toxicology*, *69*(3), 352–366.

Rao, B. M. (2019). Microplastics in the aquatic environment: Implications for post-harvest fish quality. *Indian Journal of Fisheries*, *66*, 142–152.

Rathod, N. B., Kulawik, P., Ozogul, Y., Ozogul, F., & Bekhit, A. E.-D. A. (2022). Recent developments in non-thermal processing for seafood and seafood products: Cold plasma, pulsed electric field and high hydrostatic pressure. *International Journal of Food Science & Technology*, *57*(2), 774–790.

Rathod, N. B., Phadke, G. G., Tabanelli, G., Mane, A., Ranveer, R. C., Pagarkar, A., et al. (2021). Recent advances in bio-preservatives impacts of lactic acid bacteria and their metabolites on aquatic food products. *Food Bioscience*, *44*, 101440.

Rathod, N. B., Ranveer, R. C., Benjakul, S., Kim, S.-K., Pagarkar, A. U., Patange, S., et al. (2021). Recent developments of natural antimicrobials and antioxidants on fish and fishery food products. *Comprehensive Reviews in Food Science and Food Safety*, *20*, 4182–4210.

Rathod, N. B., Ranveer, R. C., Bhagwat, P. K., Ozogul, F., Benjakul, S., Pillai, S., et al. (2021). Cold plasma for the preservation of aquatic food products: An overview. *Comprehensive Reviews in Food Science and Food Safety*, *20*(5), 4407–4425. https://doi.org/10.1111/1541-4337.12815.

Renzi, M., Guerranti, C., & Blašković, A. (2018). Microplastic contents from maricultured and natural mussels. *Marine Pollution Bulletin*, *131*, 248–251.

Renzi, M., Specchiulli, A., Blašković, A., Manzo, C., Mancinelli, G., & Cilenti, L. (2019). Marine litter in stomach content of small pelagic fishes from the Adriatic Sea: Sardines (Sardina pilchardus) and anchovies (Engraulis encrasicolus). *Environmental Science and Pollution Research*, *26*(3), 2771–2781.

Robin, R., Karthik, R., Purvaja, R., Ganguly, D., Anandavelu, I., Mugilarasan, M., et al. (2020). Holistic assessment of microplastics in various coastal environmental matrices, southwest coast of India. *Science of the Total Environment*, *703*, 134947.

Rochman, C. M., Tahir, A., Williams, S. L., Baxa, D. V., Lam, R., Miller, J. T., et al. (2015). Anthropogenic debris in seafood: Plastic debris and fibers from textiles in fish and bivalves sold for human consumption. *Scientific Reports*, *5*(1), 1–10.

Rossi, M., & Lent, T. (2006). *Creating safe & healthy spaces: Selecting materials that support healing*. The Centre for Health Design. Designing The, 21.

Rowdhwal, S. S. S., & Chen, J. (2018). Toxic effects of di-2-ethylhexyl phthalate: An overview. *BioMed Research International*, 1750368. https://doi.org/10.1155/2018/1750368. https://www.hindawi.com/journals/bmri/2018/1750368/.

Santillo, D., Miller, K., & Johnston, P. (2017). Microplastics as contaminants in commercially important seafood species. *Integrated Environmental Assessment and Management*, *13*(3), 516–521.

Schirinzi, G. F., Pérez-Pomeda, I., Sanchís, J., Rossini, C., Farré, M., & Barceló, D. (2017). Cytotoxic effects of commonly used nanomaterials and microplastics on cerebral and epithelial human cells. *Environmental Research*, *159*, 579–587.

Senathirajah, K., Attwood, S., Bhagwat, G., Carbery, M., Wilson, S., & Palanisami, T. (2021). Estimation of the mass of microplastics ingested–A pivotal first step towards human health risk assessment. *Journal of Hazardous Materials*, *404*, 124004.

Setälä, O., Norkko, J., & Lehtiniemi, M. (2016). Feeding type affects microplastic ingestion in a coastal invertebrate community. *Marine Pollution Bulletin*, *102*(1), 95–101.

Seth, C. K., & Shriwastav, A. (2018). Contamination of Indian sea salts with microplastics and a potential prevention strategy. *Environmental Science and Pollution Research*, *25*(30), 30122–30131.

Simon, N., & Schulte, M. L. (2017). *Stopping global plastic pollution: The case for an international convention*. Großbeeren: ARNOLD Group.

Sivagami, M., Selvambigai, M., Devan, U., Velangani, A. A. J., Karmegam, N., Biruntha, M., et al. (2021). Extraction of microplastics from commonly used sea salts in India and their toxicological evaluation. *Chemosphere*, *263*, 128181.

Smith, M., Love, D. C., Rochman, C. M., & Neff, R. A. (2018). Microplastics in seafood and the implications for human health. *Current Environmental Health Reports*, *5*(3), 375–386.

Stock, V., Böhmert, L., Lisicki, E., Block, R., Cara-Carmona, J., Pack, L. K., et al. (2019). Uptake and effects of orally ingested polystyrene microplastic particles in vitro and in vivo. *Archives of Toxicology*, *93*(7), 1817–1833.

Suaria, G., Avio, C. G., Mineo, A., Lattin, G. L., Magaldi, M. G., Belmonte, G., et al. (2016). The Mediterranean plastic soup: Synthetic polymers in Mediterranean surface waters. *Scientific Reports*, *6*(1), 1–10.

Sundbæk, K. B., Koch, I. D. W., Villaro, C. G., Rasmussen, N. S., Holdt, S. L., & Hartmann, N. B. (2018). Sorption of fluorescent polystyrene microplastic particles to edible seaweed Fucus vesiculosus. *Journal of Applied Phycology*, *30*(5), 2923–2927.

Tanaka, K., & Takada, H. (2016). Microplastic fragments and microbeads in digestive tracts of planktivorous fish from urban coastal waters. *Scientific Reports*, *6*(1), 1–8.

Thiele, C. J., Hudson, M. D., Russell, A. E., Saluveer, M., & Sidaoui-Haddad, G. (2021). Microplastics in fish and fishmeal: An emerging environmental challenge? *Scientific Reports*, *11*(1), 1–12.

Thompson, R. C., Olsen, Y., Mitchell, R. P., Davis, A., Rowland, S. J., John, A. W., et al. (2004). Lost at sea: Where is all the plastic? *Science*, *304*(5672). https://doi.org/10.1126/science.1094559. https://www.science.org/doi/full/10.1126/science.1094559.

United Nations Environment Programme (UNEP). (2018). *Legal limits on single-use plastics and microplastics: A global review of national laws and regulations* (pp. 1–144). World Resources Institute. https://www.unep.org/resources/publication/legal-limits-single-use-plastics-and-microplastics-global-review-national.

Van Cauwenberghe, L., Claessens, M., Vandegehuchte, M. B., & Janssen, C. R. (2015). Microplastics are taken up by mussels (Mytilus edulis) and lugworms (Arenicola marina) living in natural habitats. *Environmental Pollution*, *199*, 10–17.

Van Cauwenberghe, L., & Janssen, C. R. (2014). Microplastics in bivalves cultured for human consumption. *Environmental Pollution*, *193*, 65–70.

Van Cauwenberghe, L., Vanreusel, A., Mees, J., & Janssen, C. R. (2013). Microplastic pollution in deep-sea sediments. *Environmental Pollution*, *182*, 495–499.

Van Hoecke, K., De Schamphelaere, K. A., Van der Meeren, P., Lcucas, S., & Janssen, C. R. (2008). Ecotoxicity of silica nanoparticles to the green alga Pseudokirchneriella subcapitata: Importance of surface area. *Environmental Toxicology and Chemistry: An International Journal*, *27*(9), 1948–1957.

Van Sebille, E., Wilcox, C., Lebreton, L., Maximenko, N., Hardesty, B. D., Van Franeker, J. A., et al. (2015). A global inventory of small floating plastic debris. *Environmental Research Letters*, *10*(12), 124006.

Vázquez-Rowe, I., Ita-Nagy, D., & Kahhat, R. (2021). Microplastics in fisheries and aquaculture: Implications to food sustainability and safety. *Current Opinion in Green and Sustainable Chemistry*, *29*, 100464.

Vethaak, A. D., & Legler, J. (2021). Microplastics and human health. *Science*, *371*(6530), 672–674.

Vital, S., Cardoso, C., Avio, C., Pittura, L., Regoli, F., & Bebianno, M. (2021). Do microplastic contaminated seafood consumption pose a potential risk to human health? *Marine Pollution Bulletin*, *171*, 112769.

Walkinshaw, C., Lindeque, P. K., Thompson, R., Tolhurst, T., & Cole, M. (2020). Microplastics and seafood: Lower trophic organisms at highest risk of contamination. *Ecotoxicology and Environmental Safety*, *190*, 110066.

Wang, W., Ndungu, A. W., Li, Z., & Wang, J. (2017). Microplastics pollution in inland freshwaters of China: A case study in urban surface waters of Wuhan, China. *Science of the Total Environment*, *575*, 1369–1374.

Wang, C., Zhao, J., & Xing, B. (2021). Environmental source, fate, and toxicity of microplastics. *Journal of Hazardous Materials*, *407*, 124357.

Wilkinson, J., Hooda, P. S., Barker, J., Barton, S., & Swinden, J. (2017). Occurrence, fate and transformation of emerging contaminants in water: An overarching review of the field. *Environmental Pollution*, *231*, 954–970.

Wright, S. L., & Kelly, F. J. (2017). Plastic and human health: A micro issue? *Environmental Science & Technology*, *51*(12), 6634–6647.

Wright, S. L., Rowe, D., Thompson, R. C., & Galloway, T. S. (2013). Microplastic ingestion decreases energy reserves in marine worms. *Current Biology*, *23*(23), R1031–R1033.

Wright, S. L., Thompson, R. C., & Galloway, T. S. (2013). The physical impacts of microplastics on marine organisms: A review. *Environmental Pollution*, *178*, 483–492.

Xu, M., Halimu, G., Zhang, Q., Song, Y., Fu, X., Li, Y., et al. (2019). Internalization and toxicity: A preliminary study of effects of nanoplastic particles on human lung epithelial cell. *Science of the Total Environment*, *694*, 133794.

Yamashita, R., & Tanimura, A. (2007). Floating plastic in the Kuroshio current area, western North Pacific Ocean. *Marine Pollution Bulletin*, *4*(54), 485–488.

Yang, D., Shi, H., Li, L., Li, J., Jabeen, K., & Kolandhasamy, P. (2015). Microplastic pollution in table salts from China. *Environmental Science & Technology*, *49*(22), 13622–13627.

Yang, C. Z., Yaniger, S. I., Jordan, V. C., Klein, D. J., & Bittner, G. D. (2011). Most plastic products release estrogenic chemicals: A potential health problem that can be solved. *Environmental Health Perspectives*, *119*(7), 989–996.

Yonkos, L. T., Friedel, E. A., Perez-Reyes, A. C., Ghosal, S., & Arthur, C. D. (2014). Microplastics in four estuarine rivers in the Chesapeake Bay, USA. *Environmental Science & Technology*, *48*(24), 14195–14202.

Yoo, J.-W., Doshi, N., & Mitragotri, S. (2011). Adaptive micro and nanoparticles: Temporal control over carrier properties to facilitate drug delivery. *Advanced Drug Delivery Reviews*, *63*(14–15), 1247–1256.

Yuan, W., Liu, X., Wang, W., Di, M., & Wang, J. (2019). Microplastic abundance, distribution and composition in water, sediments, and wild fish from Poyang Lake, China. *Ecotoxicology and Environmental Safety*, *170*, 180–187.

Zettler, E. R., Mincer, T. J., & Amaral-Zettler, L. A. (2013). Life in the "plastisphere": Microbial communities on plastic marine debris. *Environmental Science & Technology*, *47*(13), 7137–7146.

Zhang, K., Gong, W., Lv, J., Xiong, X., & Wu, C. (2015). Accumulation of floating microplastics behind the Three Gorges Dam. *Environmental Pollution*, *204*, 117–123.

Zhang, L., Xie, Y., Zhong, S., Liu, J., Qin, Y., & Gao, P. (2021). Microplastics in freshwater and wild fishes from Lijiang River in Guangxi, Southwest China. *Science of the Total Environment*, *755*, 142428.

Zheng, K., Fan, Y., Zhu, Z., Chen, G., Tang, C., & Peng, X. (2019). Occurrence and species-specific distribution of plastic debris in wild freshwater fish from the Pearl River catchment, China. *Environmental Toxicology and Chemistry*, *38*(7), 1504–1513.

CHAPTER TWO

Occurrence of *meso*/micro/nano plastics and plastic additives in food from food packaging

Elena Velickova Nikova[a], Mishela Temkov[a], and João Miguel Rocha[b,c,*]

[a]Department of Food Technology and Biotechnology, Faculty of Technology and Metallurgy, University Ss. Cyril and Methodius, Skopje, RN, Macedonia
[b]LEPABE – Laboratory for Process Engineering, Environment, Biotechnology and Energy, Faculty of Engineering, University of Porto, Porto, Portugal
[c]ALiCE – Associate Laboratory in Chemical Engineering, Faculty of Engineering, University of Porto, Porto, Portugal
*Corresponding author: e-mail address: jmfrocha@fe.up.pt

Contents

Abstract

This chapter focuses on the occurrence of plastic constituents in food due to the contact with different types of plastic packaging, films and coatings. The type of mechanisms occurring during the contamination of food by different packaging materials are described, as well as how the type of food and packaging may influences the extent of contamination. The main types of contaminants phenomena are considered and comprehensively discussed, along with the regulations in force for the use of plastic food packaging. In addition, the types of migration phenomena and factors that may influence such migration are comprehensively highlighted. Moreover, migration

Advances in Food and Nutrition Research, Volume 103
ISSN 1043-4526
https://doi.org/10.1016/bs.afnr.2022.08.001

components related to the packaging polymers (monomers and oligomers) and the packaging additives are individually discussed in terms of chemical structure, adverse effects on foodstuffs, health, migration factors, as well as regulated residual values of such components.

Abbreviations

ABS	acrylonitrile butadiene-styrene
ANVISA	National Health Surveillance Agency
ASTM	American Society for Testing Materials
BADGE	BPA-diglycidyl ether
BHA	butylated hydroxyanisole
BHDB	bishydroxydeoxybenzoin
BHT	butylated hydroxytoluene
BPA	bisphenol A
CBA	chemical blowing agents
CO	carbon monoxide
CO_2	carbon dioxide
DMT	dimethyl terephthalate
DP	degree of polymerization
EDC	endocrine disruptor chemical
EFSA	European Food Safety Authority
EPS	expanded polystyrene
ESBO	epoxidized soybean oil
FCAs	food contact articles
FCMs	food contact materials
FDA	Federal Food and Drugs Administration
GC–MS	gas chromatography with mass spectrometry
HALS	sterically hindered amines
HDPE	high-density polyethylene
HPLC-CLND	high-performance liquid chromatography-chemiluminescent nitrogen detection
HPLC-FLD	high-performance liquid chromatography with fluorescence detection
HPLC-UV	high-performance liquid chromatography with ultra-violet detection
HS-SPME-GC–MS	headspace coupled gas chromatography and mass spectrometry
ISO	International Organization for Standardization
JRC	Joint Research Centre
LC/MS/MS	liquid chromatography-tandem mass spectrometry
LDPE	low-density polyethylene
MDI	4,4′-methylenebisphenyl diisocyanate
Mercosur	common market of the south
N_2	nitrogen gas
NHC	National Health Commission
NIAS	non-intentionally added substances
OM	overall migration
PA	polyamide
PA-6	polyamide-6

PA-66	polyamide-66
PE	polyethylene
PET	polyethylene terephthalate
PP	polypropylene
PS	polystyrene
PVA	polyvinylacetate
PVC	Polyvinyl chloride
PVdC	polyvvinylidene chloride
RCF	regenerated cellulose film
SAN	styrene-acrylonitrile
SM	specific migration
SML	specific migration limit
TDA	toluene-2,4-diamine
TDI	2,4- or 2,6-toluene diisocyanate
TDI	total daily intake
T_g	glass transition temperature
TNPP	tris-nonylphenyl phosphite
TPA	terephthalic acid
TTC	threshold of toxicological concern
UV	ultra-violet
VA	vinyl acetate
VC	vinyl chloride
VdC	vinylidene chloride
WEEE	waste electrical and electronic equipment
WHO	World Health Organization
WTO	World Trade Organization

1. Introduction

Basic functions of food packaging encompasses containment, protection, preservation, presentation, brand communication, environmental responsibility and benchmarking. When considering a material suitable for food packaging, it should meet all the required specifications. The production of plastic materials is in constant expansion since their invention in 1940–50's, which was considered by some as the greatest invention of the last millennium. Since the beginning of their introduction in food packaging, these materials rapidly replaced the existing wood, metal and glass materials having in mind many of their advantages. The characteristics of the plastics brought several benefits to the production plants, products and economy, namely their easiness of shaping (sheets and structures) due to their characteristics of flowing and molding, chemical inertness, transparency, barrier and heat sealing, as well as their specific weight, low price and economic distribution.

In Europe, the total plastic demand in 2018 was 51.2 million metric tons, out of which 20.5 tons were used in general packaging, whereas 8.2 million metric tons belong to the food sector. Globally, the total amount of plastic production in 2020 was 367 million metric tons (Statista, December 2021). Considering the production needs according to the type of polymer, then the demand in food sector mainly concerns to the types of thermoplastic polypropylene (PP, 21%), low-density polyethylene (LDPE, 18%), polyvinyl chloride (PVC, 17%), high-density polyethylene (HDPE, 15%), polystyrene (PS) and expanded polystyrene (EPS, 8%), and polyethylene terephthalate (PET, 7%). However, packed food comes in direct interaction with the inner surface of the plastic film or container, where there are increased chances for absorption or reaction of the aforementioned material with food goods. Usually, polymeric long-chain molecules forming the plastic material are non-reactive with food, but many other components of lower molecular size used as coatings, additives and slip agents in the plastic production process may and most likely will migrate into the food that human consume. Therefore, all additional chemical compounds used in the production of plastics must be previously approved for direct contact with food.

Many countries have different safety regulations concerning the contact of plastic with food. In USA, this type of regulation (21CFR177) is originated from the United States (US) Federal Food and Drugs Administration (FDA). In European Union (EU), the safety of food products from potential hazardous materials coming from plastic packaging is regulated by the Council of the European Union Directive 89/109/EEC, which is related to general requirements for all plastic materials and plastic articles coming in contact with foodstuffs. These regulations monitor the production in accordance with good laboratory practice and certify the product safety regarding public health or organolepticly unacceptable quality changes of the product. The definitions and migration limits for specific substances being exposed to foodstuff are included in EU Directive 90/128 EEC, and following amendment 2001/62/EC. Usually, migration of any substance from flat surface into food is expressed as weight released per unit area or as weight of the migrating substance released per weight affected food product. The latter expression usually comes in place when migration of substances occurs from caps, gaskets or stoppers. In another Directive

82/711/EEC, migration test simulants are specified as following: water (Simulant A), 3% w/v acetic acid (Simulant B), 15% v/v ethanol (Simulant C), and rectified olive oil (Simulant D) for aqueous, acetic, alcoholic or fatty food, respectively. The migrating substances, extracted by these tests are quantified by several analytical procedures such as gas-liquid chromatography, mass spectrometry, infrared analysis, electronic nose, etc., as well as by sensory organoleptic assessment executed by well-trained panelists.

Packaging made from plastic is not only from polymeric chains, but also from other substances called additives, typically added in tiny quantities during processing to ameliorate some of the packaging properties or to facilitate the production and handling. Such compounds are very responsible ones for many useful properties and recognizable functions of plastic packaging that otherwise the packaging itself would not display.

The majority of these additives are low-molecular weight components that can easily permeate into the food matrix. This group of low-molecular weight additives includes the stabilizers, antioxidants, light stabilizers, slip agents, antistatic agents and optical brighteners. Among the components related to the polymer itself that can migrate into food matrix comprises the unreacted monomers/oligomers or catalysts, cross-linkers and emulsifiers.

However, a small fraction of other additives is added in slightly higher concentrations, which is linked with their function in the packaging. These types of additives include fillers, used for volume and weight increase, or plasticizers, to help improving the processability of the material, its elasticity or flexibility. The low-molecular weight of these constituents, whether they are polymeric or non-polymeric, makes them very mobile and easily relocated from one to another media. As a result, it is very likely that the packaged food in plastic sachets/containers will be contaminated with potentially hazardous substances and pass afterwards to the consumers if the directives and guidelines for plastic materials that come into direct contact with foodstuffs are not fully followed.

2. Legislation for plastic food packaging

Plastic food packaging is ubiquitous and transversal to all societies because most of the food is packaged in different types of plastic materials. Plastics provide protection against physical damages, soiling and microbial spoilage. After the boom of the plastic packaging over the last half of the 20th century, the authorities and population in general became aware that besides the numerous benefits, the use of plastics for packaging food materials could also have negative side effects. The use of several food contact materials (FCMs) and food contact articles (FCAs) is steadily rising and, therefore, it is necessary to proceed with continuous risk assessments of the plastic constituents.

The set of strict rules concerning the safety and quality of FCMs and FCAs from countries is very important. Usually, the countries have their own legislation directives to follow although there are also several authorities with their principles that are recognized and approved worldwide.

In European Union, the European Food Safety Authority (EFSA) is responsible for proposing scientific advices to the European Commission (EC) for authorizing substances to be used for plastics. In turn, the European Commission creates rules and regulations for food contact packaging. In Europe, the most important document is the Regulation (EU) N° 10/2011 on plastic materials in contact with food, which contains a positive list of additives, monomers, starting substances and macromolecules obtained from microbial fermentation of different substrates (EC, 2004b, 2011, 2020). In addition, the most common directives are: Directive 2002/72/EEC for plastic materials and articles; Directive 90/128/EEC for plastic monomers; Directive 82/711/EEC for basic rules for migration tests; Directive 85/572/EEC for list of simulants/foodstuffs; Directive 80/766/EEC for VC in PVC; Directive 81/432/EEC for method of analysis for vinyl chloride released into foodstuffs; Directive 78/142/EEC for limits of vinyl chloride monomer; Directive 80/590/EEC for determining symbols; Directive 83/329/EEC for regenerated cellulose film (RCF); Directive 84/500/EEC for ceramic articles; Directive 93/11/EEC for nitrosamines in elastomers and rubber; and Directive 2001/61/EEC for epoxy derivatives.

In the United States, the safety of food contact packaging materials is governed by the Food and Drug Administration, through their Code of Federal Regulations Title 21, Chapter I, Subchapter B, 177.1010 to 177.2910 on Indirect Food Additives-Polymers and sections 174.5 on general provisions applicable to indirect food additives and 174.6 on threshold of regulation for substances used in food-contact articles (FDA, 2010, 2022).

The Common Market of the South (Mercosur)/Groupo Mercado Común (GMC), a trade alliance that represents multiple countries in Latin America, follows the resolution GMC 39/19 to control the additives that are allowed for plastics and polymeric coatings for use in FCMs since

July 2019. The National Health Surveillance Agency (ANVISA) from Brazil officially published (3rd December 2019) the resolution RDC N° 326, establishing the positive list of additives for the preparation of plastics and polymeric coatings in contact with food and other measures. Such a resolution integrates the GMC/MERCOSUR N° 39/19 to the National legal system. Food contact plastics are also controlled by GMC Resolution N° 02/12 (positive list of monomers and polymers), 56/92 (general provisions and overall migration limits) and 32/10 (framework test conditions) (Kato, & Conte-Junior, 2021).

Japan announced (on 9 August 2019) its Positive List System for food-contact plastics [WTO Notification G/TBT/N/JPN/630) to the World Trade Organization (WTO)]. Conversely, in China, their own National authority, the China Food Safety Law has been regulating FCMs since 2009. The most vital Chinese food safety standard for food contact additives is GB 9685-2016 on National Food Safety Standard: Standard for Use of Additives in Food Contact Materials and Articles. National Health Commission (NHC) of the People's Republic of China has issued 13 announcements until August 2019, including the approvals of 90 additives used for FCMs and articles (38 of them are food contact additives with an extended scope or using quantity).

In India, the Central Food Technological Research Institute, Mysore, drafted the IS:9845–1998 for "*Determination of overall migration of constituents of plastics materials and articles intended to come in contact with foodstuffs – method of analysis (second revision)*", which is now into force in the country to be followed for global migration of plastics constituents for their food grade quality (Raj & Matche, 2012).

Several interactions may exist between food and plastic packaging that cause all sorts of effects on human health. Contamination of foodstuff as a result of migration of plastic additives as well as their degraded products is an important issue for health legislation due to the potential risks on health (Muncke et al., 2017), especially for those substances that exhibit certain toxicity. The Threshold of Toxicological Concern (TTC) concept defines human exposure thresholds to substances with unidentified toxicity and identified structure, and can be used to assess materials concerning their potential toxicity when the exposure is superficial. In fact, the TTC concept is an applied risk assessment tool that determines human exposure levels to substances considered to pose no significant risk to human health. Nowadays, TTC is recognized to be used in EU and USA (Canellas, Vera, & Nerín, 2017; EFSA, 2012).

The TTC concept uses a classification scheme, originally proposed by Cramer and Ford (1978), as a priority-setting tool and as a mean of making expert judgments on food chemical risk assessment more transparent and reproducible. For the proposed criteria, there are three structural classes.

Class I includes compounds with simple chemical structures. These compounds have efficient metabolic mechanisms, thus implying a low degree of oral toxicity. Accordingly, class I includes: normal constituents of the body (excluding hormones); simply-branched, acyclic aliphatic hydrocarbons; common carbohydrates; common terpenes; and compounds that are sulfonate or sulfamate salts, without any free primary amines. Their TTC exposure limit (μg/person/day) is 1800.

Class II includes compounds with structures less innocuous than Class I compounds but, simultaneously. It does not comprise structural structures implying toxicity found in compounds from Class III. Accordingly, class II includes common constituents of food; compounds containing no functional groups other than alcohol, aldehyde, side-chain ketone, acid, ester or sodium, potassium or calcium sulfonate or sulfamate, or acyclic acetal or ketal and are either a monocycloalkanone or a bicyclic substance with or without a ring ketone. Their TTC exposure limit (μg/person/day) is 540.

Class III includes compounds with chemical structures that allow no robust initial presumption of safety or may even imply substantial toxicity or have reactive functional groups. Hence, class III includes structures having chemical elements other than carbon, hydrogen, oxygen, nitrogen or divalent sulfur; certain benzene derivatives; certain heterocyclic substances; aliphatic substances containing more than three types of functional groups. Their TTC exposure limit (μg/person/day) is 90 (EFSA, 2012).

Following the legislation, being familiar with the main composition and global extractable amount of plastic constituents may allow prediction of migration phenomena, which can be helpful to the producers of these materials as well as for quality control laboratories. Time and financial efforts may also be avoided if studies are undergone in the assessment of laminates comprehending layers of recycling material(s) with unidentified impurities—and which can migrate through the virgin plastic layer (functional barrier) into the contact with foodstuff. For this purpose, there are varieties of migrating tests that can be used. The choice of the proper food simulant and testing conditions (time versus temperature) is governed by the type of foodstuff and conditions for the use of food products. Currently, the food products have been classified into seven major groups: (1) aqueous food, (2) all aqueous and acidic foods; (3) alcoholic foods; (4) fats/oils and fatty foods;

(5) alcoholic and acidic foods; (6) fatty and aqueous foods; and (7) all fatty and acidic foods. For testing these food products, there are four types of food simulants. They are A—distilled water suitable for groups 1 and 6; B—3% acetic acid suitable for groups 2 and 7; C—8, 10 and 50% ethanol suitable for groups 3 and 5; and D—*n*-heptane or substitute of olive oil (isooctane and 95% ethanol) suitable for groups 4, 5, 6 and 7 (Raj & Matche, 2012).

3. Migration of plastic materials

3.1 Type of migration

The mass transfer of different chemicals from the plastic food packaging materials to food is generally described as migration. The migration process usually proceeds in four steps: (I) diffusion of chemical substances trough the polymer; (II) desorption of molecules from the polymer surface; (III) sorption of compounds at the plastic-food interface; and (IV) absorption of substances in food (Fig. 1) (Cruz, Rico, & Vieira, 2019; Hahladakis, Velis, Weber, Iacovidou, & Purnell, 2018).

Eight types of migration have been identified so far (Fig. 2).

(1) **According to number of migrants**: The migration occurrence can be overall or specific migration. Overall migration (OM) gives the summation of all compounds migrating from the plastic food contact materials into food, while the specific migration (SM) represents the individual substances or group of substances that migrate in the food (Hoppe, de Voogt, & Franz, 2016; Shin et al., 2021).

(2) **Related to food nature**: Depending on the type of food packaged (liquid, aqueous, fatty, alcoholic, solid, etc.), migration can be classified as non-migrating, volatile and leaching. Non-migrating system actually exhibits insignificant migration of the high-molecular weight polymers. The volatile migration can be found in solid food packaging, where there is an extremely low contact between the food and the polymer package and it usually happens in three phases: diffusion or evaporation of the migrating component; desorption; and adsorption of the migrant in the food. Finally, the leaching migration which to occur needs a contact between food and package. For example, it is the case of mass transfer of compounds to liquid or semisolid foods from daily plastic packaging materials upon direct shared interaction (Cruz et al., 2019).

(3) **Based on diffusion coefficient**: This migration type can be grouped into three categories. First one is the diffusion coefficient approaches zero and a minimal migration potential exists. Second one is the

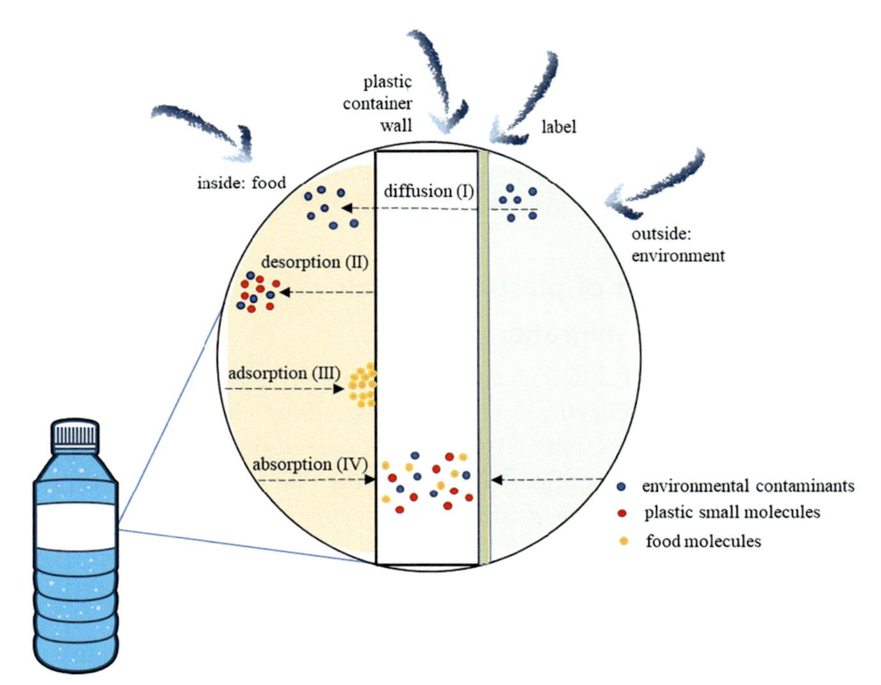

Fig. 1 Steps of the migration of plastic food packaging materials in food and plastic materials.

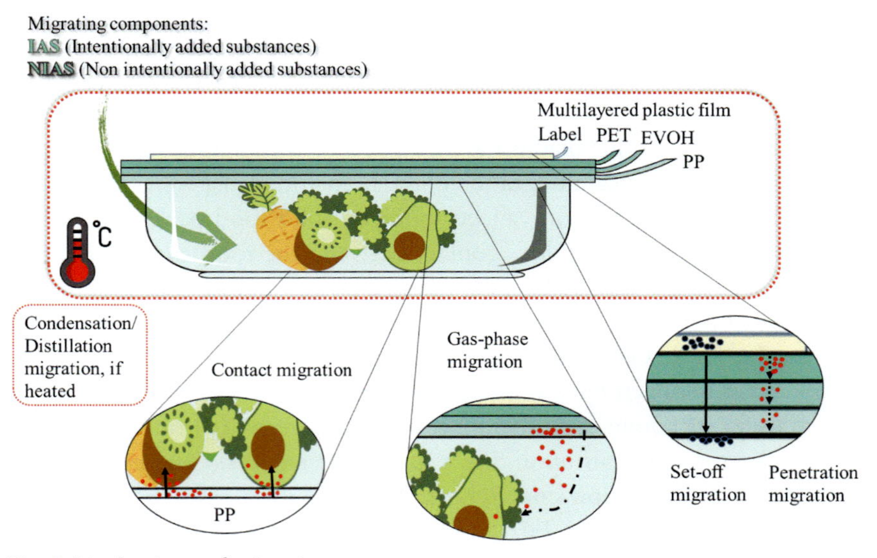

Fig. 2 Mechanisms of migration.

diffusion coefficient that has a constant value and exhibits no influence in the food component or time of storage and third one is the diffusion of a compound remains irrelevant except if the food is in straight interaction with the packaging material (Alamri et al., 2021).

(4) Contact migration. This migration process of the chemicals occurs only upon contact from the packaging to the food, like as the transfer of monomers and plasticizers from the plastic bottles, trays, pouches or wrappings to the foods (Barnes, Sinclair, & Watson, 2007).

(5) Gas-phase migration. Gas-phase migration is actually permeation, and the mass transfer of the substances occurs through the gas medium. Usually, the substances permeate from the external coating or printed layer of the package to the inner layer of the packaging material (Barnes et al., 2007).

(6) Condensation/distillation migration. The food heat treatments, such as blanching, boiling or sterilization are used to increase shelf-life and stability of food, but the transfer of substances may happen during these processes trough the food pouches or trays. In fact, heated aqueous food releases moisture from steam or distillation, which causes migration of volatile components from the packaging to the food and vice-versa (Alamri et al., 2021).

(7) Penetration migration. The penetration migration describes the migration of substances from the external layer to the inner layer trough the polymer material itself. When reaching the inner layer of the package migration continues towards the food through contact or gas-phase migration (Barnes et al., 2007).

(8) Set-off migration. This migration describes the mass transfer of inks, varnishes and coatings printed on the external side of the plastic packaging to the inner side of the packaging films by reeling (for instance, winding printed wrappers into a reel). This migration can be easily noticeable or not depending on the migrant. Compounds transferred by set-off migration can migrate further through gas-phase migration or direct interaction and can cause contamination of the wrapped and packaged food (Alamri et al., 2021).

The net flows of different chemical substances into and out of the polymer matrix, in case of a diffusion-controlled transport, are determined by the concentration gradients of reactants. The local steepness of gradients depends on the transport rates, in addition to the size, porosity and intra-particle reactant diffusivity. Transport properties, principally diffusion coefficients, make available a direct measure of molecular mobility, a central factor for the mass-transfer of substances in food. Diffusion coefficients under various conditions are required to model the diffusion kinetics of the system. The most common methods for obtaining diffusion coefficient are the non-steady state method and the diaphragm-diffusion chamber method. In the first method, the diffusion rate is measured while the substance diffuses into or out of polymer-matrix in a well-stirred solution. The second method is reliable for measuring diffusion through polymer packaging with uniform thickness (Velickova, Kuzmanova, & Winkelhausen, 2011).

Usually the mass-transfer of substances during migration follows the Fick's laws of diffusion. In order to quantify the rate at which a diffusion process occurs, the terms diffusivity or diffusion coefficient (expressed in m^2/s) are used. Diffusion in homogeneous media is based upon the postulation that the rate of transfer (R) of a migrant passing perpendicularly through the unit area of a section is proportional to the concentration gradient between the two sides of the packaging (Fick's first law):

$$R = \partial C/\partial t = -D(C)dC/dX \quad (1)$$

where D(C) is the diffusion coefficient in m^2/s. Generally, D is a function of the local diffusant concentration, C (mol/m^3), t is the time (s) and X is the thickness of the material (m) (Tehrany & Desobry, 2004).

Small substances like monomers and residual solvents with molecular weight in the range between 200 and 2000 g/mol migrate quickly and the steady state diffusion process points out no changes in concentration over time. However, in reality the majority of the interactions between the packaging and food are controlled by non-steady state conditions. The bigger molecules with high molecular weight have slow migration rate and obey the second Fick's law, where migration is governed by the diffusion and the partitioning between the food and the polymer (Ferrara, Bertoldo, Scoponi, & Ciardelli, 2001; Hahladakis et al., 2018; Hansen, Nilsson, Lithner, & Lassen, 2013). The rate of molecular diffusion shown mathematically by Fick's second law is:

$$dC_p/dt = D\left(d^2C_p/dx^2\right) \quad (2)$$

where C_p is the concentration (mg/g) of migrant in packaging material, D is the diffusion coefficient (cm^2/s), t is time of diffusion (s) and x is the distance (cm) between food and packaging material. Mathematical models used for describing these phenomena are constantly changing, developing and improving, thus research relies on measuring migration/contamination from packaging chemicals to better and more accurately present diffusion processes (Silva, Freire, García, Franz, & Losada, 2007). For example, in cases when migration is completely diffusion controlled, the next type of equations may be used:

$$M_t = 2C_o\rho \left(D_p t/\pi\right) \tag{3}$$

where: M_t, is the total migrant from the polymer in time, t (s); C_o is the initial migrant concentration in the polymer (mg/g); ρ is the polymer density (g/cm^3); D_p is the diffusion coefficient of the migrant in the polymer (cm^2/s), and t is the package lifetime in (s). All parameters in Eq. (3) are usually known from literature data or easily measured, excepting for the diffusion coefficient, which needs to be measured with kinetic experiments.

Migration from the plastic material into foodstuffs through diffusion is defined by two parameters. The first is the diffusion coefficient, which measures the rate at which the diffusion process occurs. Diffusion coefficient is a parameter that is used to measure the migration of the chemicals in the polymer matrix or in the foods. It is dependent on several important factors: temperature, initial concentration of migrant in polymer matrix, type of the polymer and physicochemical properties of the food that is in interaction with the packaging, especially its physical state (Maia et al., 2016).

The second parameter is the partition coefficient, $K_{P/F}$. Partition coefficient gives the ratio of the concentration of the migrant in the polymeric material (C_P) and the concentration in the food system (C_F) at equilibrium ($K_{P/F}=C_P/C_F$). When the migrant concentration is higher in the polymer than in the food system, then $K_{P/F} > 1$. Higher values for the partition coefficient, $K_{P/F}$, are preferable for food safety because the migration is limited. Factors that influence partition coefficient value are: lipophilicity or hydrophobicity state of foodstuff and polymer; polarity or unpolarity state of foodstuff and polymer; molecular weight of migrant; steric hindrance of migrant molecules in polymer matrix; type of migrant interaction with polymer; variation of polymer thickness; and homogeneously of migrant distribution within the polymer matrix (Maia et al., 2016). When the migrating substance has comparable polarity with foodstuff, then migration rate is faster. However, lower migration levels and higher value of $K_{P/F}$ means that their affinity to polymer matrix is higher than the food. The higher value of

$K_{P/F}$ for some substances is due to their relatively high-molecular weight, strong interaction with the polymer and steric hindrance (Yao, Hong-fei, Li-jia, & Shi-chun, 2013). Usually for migrants with high solubility in food, the $K_{P/F}$ can be considered approximately 1 and for migrants which are poorly soluble in food the value of $K_{P/F}$ will be around 1000, meaning that their transportation to food is low (Maia et al., 2016).

3.2 Factors and phenomena affecting migration

Several factors and phenomena can affect the migration, which are (a) contact time and temperature, (b) contact type (i.e., direct or indirect contact) of packaging materials with foodstuff, (c) packaging material itself and its properties, such as thickness and permeability (gaseous or aqueous), (d) initial concentration of migrating substance in the packaging material and its characteristics such as structure, molecular size and polarity, (e) nature of the food and its pH; as well as (f) the ratio of the surface area of the packaging to the volume of the food product (Tehrany & Desobry, 2004).

The rate and extent of the migration of different substances from the polymer package are strongly dependent on the temperature of food during storage and duration of the contact. Higher temperatures enable faster migration rates because the equilibrium is more rapidly reached between the packaging headspace and the food. In addition, several researchers proved that mass transfer of substances is proportional to the square root of the duration of contact between food and packaging material (Alamri et al., 2021; Arvanitoyannis & Bosnea, 2004). Other researchers have examined the influence of the type of contact (i.e., direct or indirect) between the food and the packaging on migration levels. They proved that direct contact enables faster migration rate, while with the indirect contact there is a much slower migration owing to the presence of gas between the food and the package (Anderson & Castle, 2003). The nature of packaging material can also influence the migration of a substance. Thickness and plasticization of the polymer material affect the migration of additives. Thicker packaging slows migration whereas thinner packaging enhances it. The nature of the migrants is also important for the migration extent and rate. The highly volatile substances migrate at a greater pace while compounds with relatively higher molecular weights exhibit lower migration rates. The microstructure of the migrating substance also affects its migration extension. More precisely, the configuration of migrating molecules (e.g., spherical versus branched and with or without side chains) affects migration in a different way; for example, branched molecules have lower migration rates. Finally, the nature and composition of the food is also important. Foods with surplus fats exhibit high levels of migration. Thus, various food simulants are already approved by different authorities in EU and USA to study the influence of food nature on plastic migration (Alamri et al., 2021).

A full comprehension of the factors influencing the migration is well suited with improving quality control by determining the variables with greatest influence. The better evaluation of chemical migration from package to food would be helpful in limiting and controlling food contamination as well as in improving food safety. However, the phenomena that can effectively affect the process of migration are mainly the molecular energy, free volume theory, diffusion and partition coefficient, swelling, crystallization and presence of nanomaterials in the polymer package (Anderson & Castle, 2003; Barnes et al., 2007; Cruz et al., 2019; Zabihzadeh Khajavi, Mohammadi, Ahmadi, Farhoodi, & Yousefi, 2019).

The free volume theory postulates that in order for chemical migration to happen, there must be free space in polymer network for molecular circulation and diffusion through the polymer. Based on this rule, the polymers matrix with more porosity (more voids and free volume) have higher chances for migration of chemical substances. Besides enough space, chemical substances also need certain amount of energy to migrate. According to the molecular energy rule, the migration pick-up pace is higher at higher temperatures due to the increased mobility and activity of the polymer chain on one hand, and on the other hand, the increased activity of the chemicals that allows them to overcome attraction from neighboring molecules and results in their separation from the polymer network and further release into the voids (Cheng, 2011).

Swelling is also an important phenomenon occurring in polymer matrices. Mainly it is associated with the reverse migration, i.e., migration of the packaged food into the packaging. This phenomenon can cause irreversible changes in the polymer matrix regarding its structure and morphology and can inflict volume expansion. These reactions are triggered by the solvents that come from the food, located between the polymer chains. During their interaction, free volume is created which increases the migration rate (Grassi, Colombo, & Lapasin, 2001; Zabihzadeh Khajavi et al., 2019).

Regarding crystallization, it is known that polymers are built up from crystalline and amorphous regions. The crystal regions have a well-organized structure to which strong polymer chains are attached. On the contrary, amorphous regions have irregularly arranged structure with weak bonds between the polymer chains. They also have high molecular activity, which allows formation of extra free volume and voids where the chemicals easily migrate from the packaging to food (Yao et al., 2013).

Currently many polymer materials for food packaging have been modified by adding nanoparticles, fibers and macro fillers to the polymers creating composites. The addition of the nanoparticles in the polymers causes structural changes of the polymers and can provide enhanced characteristics. When the migration properties of nanocomposites and the effect of nanoparticles on the migration of other chemicals into food were examined, it was concluded that the nanoparticles affected the overall and specific migration of chemicals. Food packaging based on nanotechnology can provide novel and pioneering solutions to increase safety, e.g., in the reduction of any critical contact of package with food matrices, inhibiting migration of chemicals into foodstuffs (Chaudhry et al., 2008; Lorite et al., 2017).

The existence of nanoparticles in polymer matrix can slow down the rate of packaging contaminant migration into food in two ways. The first one is via creation of a tortuous path around the particles, forcing the migrating compound to travel a longer path to migrate into packed food. The second one is based on crystallinity. The stoichiometric ratio between the crystalline and amorphous phases of the polymers exhibits the crystallinity level in percentage. Crystals are stiffer than amorphous regions (impermeable) and thus could inhibit the migration. Therefore, the polymers with higher crystallinity level have enhanced barrier properties (Zabihzadeh Khajavi et al., 2019).

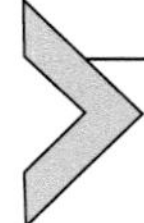

4. Migrating components

4.1 Polymer-bound migrating components (monomers and polymers)

Monomers in plastic packaging can occur as non-reacted species or molecules, which have not been incorporated in the polymeric chain. These molecules are lower in molecular size and highly reactive in comparison with oligomers, thus their migration more feasible. Each monomer has different levels of toxicity, and the residual levels in the packing is restricted and regulated (Lau & Wong, 2000).

4.1.1 Styrene

Styrene monomer can migrate from polystyrene packaging materials, whether in rigid or expanded form. Many adverse health effects are reported in humans when exposed to styrene vapors including eyes, nose, throat and skin irritation, toxic effect on the liver, neurological damage on the central nervous system, etc. The most dangerous problem in human health is the conversion of styrene monomer to styrene epoxide in liver microsomes and the release of the peroxide radical (Khaksar & Ghazi-Khansari, 2009). The annual exposure to this component is estimated to vary between 6.7 and 20.2 mg per person. Higher-level occurrence of styrene in food causes off-flavor and food taints. Sensory properties of food can be affected even at low-level concentrations as 200–500 ppb in yoghurt, 500 ppb in homogenized whole milk, and 37–730 ppb in water (Belz, van Gemert, & Nettenbreijer, 1977; Ehret-Henry, Ducruet, Luciani, & Feigenbaum, 1994). Several factors affect the migration of this component. One of them is the temperature, where higher temperature speeds up the release of the monomer. It is especially pronounced when expanded polystyrene disposable cups are used for hot beverages. Paraskevopoulou, Achilias and Paraskevopoulou (2012) studied the migration of styrene monomer from commercial PS or EPS packaging materials into food stimulants (distilled water, ethanol/water and isooctane) at different temperatures taken in accurate time intervals for styrene diffusion coefficient estimation. As expected, as the temperature rises, the rapidity increases as well as the amount of the released monomer as a function of the food stimulant. At the beginning, a swift release is reported reaching a plateau after certain amount of time. Styrene monomer was mostly detected in ethanol solutions rather than in isooctane, while no monomer was detected in distilled water. Higher styrene quantity was observed when EPS cups were used instead of PS, which was confirmed with higher diffusion coefficients.

In another study, the migration of styrene monomer from general purpose polystyrene and high impact polystyrene cups into hot drinks was determined in the course of 10, 30 and 60 min at 20 °C, 60 °C and 100 °C, respectively (Khaksar & Ghazi-Khansari, 2009). The general conclusion was that the migration of styrene monomer from different types of cups into hot drinks did not exceed 0.05% from the overall quantity of styrene present in the cup. Their findings show similar levels of migration from general-purpose polystyrene cups: between 0.61 and 8.15 μg/L in hot tea, between 0.65 and 8.30 μg/L in hot milk, and between 0.71 and 8.65 μg/L in hot cocoa milk. For high impact polystyrene cups known for their non-porosity, strength, crack-resistance and use in low-heat settings, the migration level was estimated between 0.48 and 6.85 μg/L in hot tea, between 0.61 and 7.65 μg/L in hot milk, and between 0.72 and 7.78 μg/L in hot cocoa milk. Styrene leaching in hot beverages depends strongly on the temperature of consumption, time of exposure and the fat content in the food, shown by the differences in leached styrene monomer in tea (aqueous solution) and cocoa milk (containing fats).

Similar conclusion was reported in a study where styrene monomer migration from polystyrene cups in the food systems was examined. A large variety of food systems was employed comprising water, milk with different fat content, cold drinks (fruit juices, fizzy drinks, beer and chocolate drink), hot drinks (tea, coffee, hot chocolate and soups with different fat content), take away food as yoghurt, pudding, ice-cream and jelly along with aqueous food stimulants and olive oil. Test conditions for stimulants were 24 h at 40 °C for cold drinks, and 1 h at 100 °C for hot drinks, while the real food samples were much rigorously exposed. From all examined samples, styrene migration was the highest in food systems with the highest fat content and stored at the highest temperature. The authors found that whole milk had the highest styrene monomer concentration, gradually decreasing in half fat milk and skimmed milk with statistical differences only at storage temperatures of 60 and 100 °C, respectively. Drinking water did not leach much of the styrene from the cups. However, the food stimulant 15% ethanol leached styrene monomer to a comparable level of highest fat containing milk or soup (0.025% of the total styrene in the cup). The highest migration occurred in the food stimulant 100% ethanol (0.37% of the total styrene in the cup) (Tawfik & Huyghebaert, 1998).

Foodstuff containers made of polystyrene foam such as plates, meat trays, egg cartons, cups, bowls were included in a study for determination of styrene migration in food oil as a stimulant, excluding the egg carton that was analyzed in 8% ethanol as food simulant. Normally, food comes into contact with this type of packaging material for a short time because they are all meant to be transported in containers. However, the migration of styrene from thermoformed containers into food stimulant was found to be proportional to the square root of the time of the exposure following the Fick low of diffusion, except for the cup that was extrusion-coated foam. The overall migration increased two fold over a period of 4 days and 3.2 fold over a period of 10 days. The log of diffusion coefficients with the inverse of the absolute temperature of exposure showed a linear relationship (Lickly, Lehr, & Welsh, 1995). While many studies showed only on styrene monomer migration, in the study of Choi, Jitsunari, Asakawa and Sun Lee (2005), the migration kinetics of PS monomers, PS dimers and PS trimers into food stimulants (distilled water (aqueous food) and n-heptane (fatty food)) were analyzed using gas chromatography with flame ionization detector (GC-FID). The research was performed at 10 °C, 24 °C and 40 °C for n-heptane and 40 °C, 60 °C and 90 °C for distilled water during 72 h. During this time, n-heptane completely extracted styrene monomers and oligomers with higher migration rate at higher temperatures, particularly for smaller molecules. In distilled water stimulant, only styrene trimers were identified, with the diffusion coefficient much lower than the one estimated in n-heptane even at the highest temperature. The diffusion coefficient and activation energy of styrene molecules were dependent on their molecular weight.

4.1.2 Bisphenol A

Bisphenol A (BPA) can be used as monomer in polycarbonate plastic containers for food and beverages, microwave containers, along with its non-plastic use as additive in the epoxy resins for internal can coating, plastic toys, water pipes, tableware, cigarette fillers, etc. (Abou Omar et al., 2017). The migration of BPA is mainly facilitated by the pH of the food and the temperature (sterilization of cans or polycarbonate plastics), inducing breakage of the liaison between BPA molecule and polycarbonate plastic/epoxy resin (Hahladakis et al., 2018). Migrated compounds in food (packed in cans or polycarbonate containers) are prone to be ingested by consumers and even low levels can be detected in humans' blood, urine, breast milk, serum, etc. BPA molecule is known as endocrine disruptor chemical (EDC) and prolonged exposure to this molecule can lead to higher risk of prostate and breast cancer, diabetes and cardiovascular diseases. BPA molecule is also classified as reproductive toxicant, showing moderate estrogenic activity that can influence fertility (Abou Omar et al., 2017; Ćwiek-Ludwicka, 2015). Since many researches have proven that the molecule is harmful, BPA was banned from the polycarbonate infant bottle production in EU countries in 2011. In 2015, the EFSA released a document for BPA migration in food and total daily intake (TDI) for all plastic materials along with epoxy resin internal coatings for canned food. The TDI in this EFSA document was lowered from previously established 50 μg/kg of bodyweight/day to 4 μg/kg of bodyweight/day (EFSA, 2015). In addition, in 2020 the usage of this component above the concentration of 0.02% in thermal papers has been banned. However, in the last years many studies were performed evaluating the risk assessment of BPA and reporting its harmful effect on the immune system that can potentially develop allergic lung inflammation. Therefore, EFSA made a decision based on building evidences to prepare a draft for re-evaluation the TDI of BPA and its restriction of its use in the manufacturing that will be submitted to the European Chemical Agency for review in February 2022. The newly TDI value for BPA is set to be 0.04 ng/kg of bodyweight/day, which is about 10^5 times lower than the one previously given in 2015 (EFSA, 2021). Many studies were undertaken to examine the migration capacity of BPA in different food systems. Kubwabo et al. (2009) evaluated the BPA migration from different baby bottles (polycarbonate, non-polycarbonate, reusable and liners) in food stimulants. To simulate migration into aqueous and acidic food, water was used. For alcoholic food, 10% ethanol solution was used, and for fatty food, 50% ethanol solution was utilized. Migration of BPA from polycarbonate bottles was dependent on storage time, storage temperature and food matrix composition, varying from 0.11 mg/L in water for period of 8 h to 2.39 mg/L in 50% ethanol for period of 240 h, whereas other examined plastic containers showed only traces of BPA.

It was reported the effect of repeated use of polycarbonate bottles up to 100 times on BPA extraction in water at different temperatures varying from 40 °C to 100 °C. BPA was transferred into the water in a concentration of 0.03 ppb (40 °C) to 0.13 ppb (95 °C) from very new bottles. Conversely, the BPA migrated concentration from repeated use bottles (6 months) was much higher, mainly 0.18 and 18.47 ppb at 40 °C and 95 °C, respectively, with temperatures above 80 °C sharply increasing the BPA migration (Nam, Seo, & Kim, 2010).

In another study, polycarbonate baby bottles were exposed to simultaneously dishwashing, boiling and brushing to test the migration level of BPA leaching from new and simulated used bottles. The findings showed significant rise of the compound in the samples of water simulant heated at 100 °C for 1 h. New bottles released 0.23 ± 0.12 μg/L of BPA, while the dish washed bottles for 51 and 169 times released 8.4 ± 4 μg/L BPA and 6.7 ± 4 μg/L of BPA, respectively, which was still below the levels of TDI established by EFSA (Brede, Fjeldal, Skjevrak, & Herikstad, 2003). Except polycarbonate baby bottles, studies were undertaken with BPA migrating into olive oil for human's exposure. In Lebanon, 27 olive oil samples were collected and stored in plastic and non-plastic containers. The results showed substantial increase of BPA in the olive oil packed in plastic containers (333 μg/kg) in comparison to those packed in non-plastic container (150 μg/kg), with the estimated humans' exposure of 1.38% of the EFSA TDI prescribed at the time. Prolonged storage time (>1 year) also increased the level of BPA in the oil (452 μg/kg) rather than short-term storage (<1 year) releasing 288 μg/kg of BPA in the olive oil (Abou Omar et al., 2017).

BPA can migrate from epoxy resin used for internal coatings in cans, apart from polycarbonate materials. Goodson, Robin, Summerfield and Cooper (2004) investigated the storage conditions (1, 3 and 9 months at 5 °C, 20 °C and 40 °C), heat treatment (121 °C for 90 min) and can damage on the BPA migration in four food matrixes. The food matrixes were minced beef in gravy (20% fat), canned carrots in brine, vegetable soup (0.3% fat) and evaporated milk (8% fat), while using food simulant (10% ethanol) as a worst case scenario. The results showed no significant increase on BPA migration of food matrixes upon storage for the accelerated shelf-life tests. However, BPA migration in 10% ethanol was relatively higher compared to other food matrixes, which might be due to the solubilization of the can coating in ethanol. Can damaging did not show any effect on the accelerated BPA migration. Similar conclusions were found by Munguia-Lopez and Soto-Valdez (2001) in their research to inspect the effect of thermal processing and storage period on (BPA) and BPA-diglycidyl ether (BADGE) from tuna and jalapeno cans in aqueous food simulant. Heat processing did affect the compounds migration, whereas storage time did not influenced BPA and BADGE from tuna cans. BPA migration can differ whether leaching from beverages (1.0 ng/mL), or from canned food (40.3 ng/mL), suggesting that BPA concentration in food depend on the can type and heat treatment, rather than food type (Geens, Apelbaum, Goeyens, Neels, & Covaci, 2010).

4.1.3 Vinyl chloride

Vinyl chloride (VC) is most commonly used in the production on polyvinylchloride (PVC) material or together with vinyl acetate (VA) or vinylidene chloride (VdC) as *co*-monomers in manufacturing of polyvinylacetate (PVA) and polyvvinylidene chloride (PVdC), respectively (WHO, 2004). Low levels (up to 1 ppm) of unreacted VC monomer can be found in the polymer, which can easily migrate from food packaging into foodstuff, changing its physical and organoleptic characteristics, in addition to the serious negative damages in the consumers' health. The toxicological effect of vinyl chloride monomer is dependent on the dose of exposure. Low doses are metabolized and excreted mainly by urine, while higher doses cannot be metabolized in the lungs. Frequent and long (15 to 20 years) exposures to high doses of vinyl chloride for industrial workers is linked with higher risk of liver, brain, lung and lymphatic cancer. The monomer can cause other sever diseases as acro-osteolysis, thrombocytopenia, liver damage and disorders of the nervous system (ATSDR, 2006). The toxicity of VC monomer was discovered in 1974, when the newly appeared angioscarcoma was associated with inhalation of this monomer. Several scientific articles have been published regarding the toxicity of VC monomer and its linkage with developing of other diseases. That raised the alarm of the authorities to set the maximum allowance of this monomer in the food. When the maximum allowed levels in the 1974 were 1000 ppm in PVC bottles, those were gradually decreased to 250 ppm in 1975, to 30 ppm in 1977 and to 1 ppm in 1980 (Pearson, 1982). Many organizations including Agency for Toxic Substances and Disease Registry, Food and Drug Administration, World Health Organization (WHO) set their limitations to maximum content of VC monomer in food and daily total exposure (Alamri et al., 2021). According to the European Commission Regulation (EU) N° 10/2011 about the plastic materials and articles denoted to come into contact with food, the concentration of VC monomer should not be higher than 1 mg/kg (EC, 2011).

Drinking water can become a concern if the distribution system with PVC pipes are used. According to WHO report on drinking water, VC monomer was occasionally identified in drinking water samples in Germany with a highest concentration of 1.7 μg/L. However, the release of vinyl chloride from plasticized PVC (uPVC) was greater (2.5 μg/L) at elevated temperatures of 45 °C, which represents a concern for warm climate countries. Bottled water contains lower levels of VC that is less than 0.6 μg/L (WHO, 2004). Vinson and Banzer (1981) confirmed the dependence of VC diffusivity on the temperature. Giving the fact that vinyl chloride is volatile; its concentrations can be reduced by boiling the water. After measuring the migration of VC in water from drinking pipes with different residual monomer concentration at different temperatures, they reported 2.153 μg/kg migration at 50 °C, compared to 0.1 μg/kg at 3 °C for the same time of exposure. In the same study, they also predicted releasing of $<$002 mg/kg of vinyl chloride from a pipe containing 3 mg/kg residual VC monomer by mathematic modeling. The "early era" PVC pipes in water distribution systems may require flushing of the dead ends to control the release of vinyl chloride (Beardsley & Adams, 2003). The partition coefficient of the monomer between the polymer and the food system is dependent on the solubility of the migrant in each of their phases, its initial concentration, along with the sorptive potential of the polymer. Unplasticized PVC can retain VC monomer better than the plasticized PVC due to the presence of active sites immobilizing the monomer (Gilbert, Miltz, & Giacin, 1980). Chan, Anselmo, Reynolds, and Worman (1978) proposed a model to estimate the initial VC concentration in the packaging material that will release the monomer within the established limit. Food systems that act as poor solvents, such as water or oil, could not extract more than 20 ppb of vinyl chloride, while to keep the levels of leached monomer low when food systems acting as strong solvents are used, the most important is the package/solvent ratio to be kept below 0.1.

4.1.4 Polyethylene terephthalate oligomer

Polyethylene terephthalate (PET) polymer is produced either from terephthalic acid (TPA) or from dimethyl terephthalate (DMT) by esterification in the presence of ethylene glycol. To enhance the properties, performance, and heat resistance of the plastic material, other components might be added as co-monomers: isophthalic acid, 1,4-cyclohexanedimethanol, diethyl [[3,5-bis(1,1-dimethylethyl)-4-hydroxyphenyl]methyl]phosphonate diethyl [[3,5-bis(1,1-dimethylethyl)-4-hydroxyphenyl]methyl] phosphonate, etc. (De Cort, Godts, & Moreau, 2017). PET has found many uses in food containers, trays, dishes, bottles, films, microwavable containers owing to its heat resistance (partly crystalized), light weight (blow molded three dimensional oriented), mechanical strength (reinforcement with glass fibers), toughness, unbreakable characteristics, etc. Nevertheless, a small fraction of the polymer corresponds to low-molecular PET, PBT oligomers (degree of polymerization, DP, 2–5) which are susceptible to migration into food. These components are appraised as non-intentionally added substances (NIAS) (Alberto Lopes, Tsochatzis, Karasek, Hoekstra, & Emons, 2021). According to EU Regulation N°. 10/2011(EC, 2011), only cyclic butylene terephthalates are added to the positive list under N°.885 although their application as additives is restricted to a concentration of 1% (w/w) in PET, PBT, PC, PS, and PVC envisioned for contact with aqueous, acidic and alcoholic food at room temperature. There are no toxicological data for PET and PBT oligomers, thus there are no migration limits (EC, 2011). Over the years, numerous research efforts were conducted to estimate the level of oligomers desorption in food. Laurence Castle, Mayo, Crews, and Gilbert (1989) studied the total level migration of PET oligomers from roasting bags, PET trays for conventional and microwavable use and "susceptor pads" for microwave browning applications into different foods. The oligomers were detected by gas chromatography with mass spectrometry (GC–MS) after hydrolysis to terephthalic acid and methylation to dimethyl terephthalate. Thermal conditions, time of exposure and nature of food surface significantly affected migration level. The higher amount of migrated PET oligomers occurred during conventional cooking at oven at 204 °C during 30 to 90 min, compared to microwave oven cooking. PET oligomers were found in the level of 2.73 mg/kg in French fries cooked in a microwave oven with the aid of a susceptor pad. Nevertheless, the authors reported very low migration of PET oligomers, almost no detectable from bottles in aqueous beverages and somewhat slightly higher migration in alcoholic spirits. After repeated use of the PET containers, the oligomer migration in olive oil declines from 1.71–1.79 to 0.34–0.35 mg/dm^2. Migration of PET cyclic oligomers from susceptor type food packaging was measured in French fries, popcorn, waffles, pizza and fish sticks, while the food blanks were heat treated in glass containers. The migration of cyclic PET trimer is far greater than the cyclic tetramer and pentamer accounting for the biggest fraction of the migrated substances. The total migrated quantity of PET oligomers in food varied from 0.012 to 7 μg/g. The highest migration was observed in French fries and popcorns cooked in CPET bowl, yet still to a lesser extent than the oligomer migration in corn oil treated in a similar way. These observations were due to the pronounced solubility of the oligomers in oil and higher oil/polymer surface area than food/polymer contact area contributing to increase mass transfer (Begley, Dennison, & Hollifield, 1990; Begley & Hollifield, 1990). In another study, PET oligomers (DP 2–5) were detected by high-performance liquid chromatography with ultra-violet detection (HPLC-UV) method in olive oil leached from PET roasting bags. Total migration of PET oligomers was reported as 2.7 mg/dm^2 under heating it in microwave oven for 7 min at 850 W and 3.5–4.1 mg/dm^2 after cooking regime of 200 °C for 60 min in conventional oven (Lopez-Cervantes, Sanchez-Machado, Simal-Lozano, & Paseiro-Losada, 2003). Potential migration of PET oligomers (DP 2–7) and PBT oligomers (DP 2–5) from coffee capsules made of PBT and PP in water and simulant C (20% ethanol) at 80–85 °C was examined by HPLC-UV/FLD detection. Migration of oligomers was found to be higher in simulant C than water. Total desorption of oligomers from PBT capsules was higher than PP capsules and varied from 63.9 to

76.8 μg/capsule and from 231 to 234 μg/capsule in water and food simulant C, respectively. The authors concluded outstanding TDI of 50 μg/kg food for oligomers with a molecular weight lower than 1000 Da following the Regulation (EU) N° 10/2011 (Alberto Lopes et al., 2021). PET cyclic oligomers were also determined in water, simulant C (20% *v*/v ethanol) and simulant D1 (50% v/v ethanol) from tea bags. Simulant D1 disclosed the presence of all studied oligomers. Water was able to extract only low-molecular mass oligomers, but as the concentration of ethanol increased in food simulants, the content of oligomers with high molecular mass also increased. In that content, the presence of PET cyclic oligomers from the first series was reported as water (9.6–29.5 μg/kg), food simulant C (49.3–167.5 μg/kg), and food simulant D1 (578.7–933.3 μg/kg). As authors discussed, according to the approach if all identified oligomers are non-genotoxic, they can be classified as Cramer class III with a maximum exposure of 90 μg/person/day per individual oligomer, e.g., the threshold is not exceeded if only one cup of tea per day is consumed (Tsochatzis, Alberto Lopes, Kappenstein, Tietz, & Hoekstra, 2020). HPLC, MS, and ^{1}H and ^{13}C NMR techniques were used to identify cyclic oligomers from PET bottles for mineral water and fruit juice. The results showed quantities ranging from 300 to 462 mg/100 g of first series cyclic trimer (Nasser, Lopes, Eberlin, & Monteiro, 2005).

4.1.5 Caprolactam

Caprolactam is the monomer used for manufacturing of polyamide-6 (PA-6) (nylon) by polymerization with PA-66, which however is prepared by polycondensation of hexamethylenediamine and adipic acid (Heimrich, Nickl, Bönsch, & Simat, 2015). If the polymerization reaction is not conducted to completion, residuals of this monomer and low-molecular polymers can migrate into food. Usually commercial PA hold up to 2% caprolactam monomer. Other components that have potential for migration in the food are those formed by degradation and reaction by-products (Araújo, Félix, Manzoli, Padula, & Monteiro, 2008). Since PA-6 material is widely used for production of vacuum and modified atmosphere packaging, cook-in packaging, roasting bags and boil-in-the-bag other than conventional mono- and multilayer films and containers, there is a greater chance to caprolactam monomer and oligomers exposure. Acute exposure or exposure by inhalation can cause irritation of eyes, nose, throat, skin and mucous membranes as well as headaches, confusion and nervous system irritation, respectively. Nevertheless, chronic or long term exposure might cause hand's peeling and eyes, nose, throat and skin irritations (US EPA, created in April 1992, updated in January 2000). Caprolactam vapor is irritating at 56 mg/m^3, whereas the irritation threshold of caprolactam dust is 84 mg/m^3 for the skin, and 61 mg/m^3 for the mucous membranes (EC, 2011). Following the EU regulation 10/2011 (EC, 2011), these components are classified as non-intentionally added substances, for which there is no specific migration limit (SML). However, the same component is placed in-group restrictions of substances with SML of 15 mg/kg of food. The human exposure threshold value for these substances is 1.5 μg/kg bodyweight/day, corresponding to a daily exposure of 90 μg/person (EFSA, 2012). During the process of polymerization when equilibrium is attained, the level of residual caprolactam and cyclic oligomers is high (9–12%), further reduced by vacuum or hot water extraction. Additionally, reformation of these components occurs when the polymer is thermally processed into shapes (Heimrich et al., 2015). Migration of the components relies on several conditions, which are initial concentration of the migrant in the polymer, diffusion coefficient of the migrant through the polymer and partition coefficient of the migrant between polymer and food. Heimrich et al. (2015) noticed 44% increase of caprolactam and 27% increase of cyclic oligomers in the production of extruded PA-6 polymer film from granulates. Moreover, they conducted migration tests for two PA-6 films (23 μm, 33 μm) and polyamide-66 (PA-66, 45 μm) in food stimulants (water, 50% ethanol, 95% ethanol, isooctane, and oil) for different time-temperature contact conditions. The solubility of the oligomers was found to be very high (41–52% of the initial concentration) in aqueous and ethanol solutions due to their polarity. The water contributes to the increase of the diffusion of the migrants caused by swelling of the polymer, especially for dimers and trimers. Nevertheless, higher concentrations of oligomers (DP 6–9) was observed in 50% ethanol. Oil extracted only caprolactam, trimers and tetramers, whereas none was detected in isooctane. At the harsh time-temperature conditions (2 h/100 °C), only 17% of oligomers from the initial concentration were transferred into oil compared to complete extraction in water at the same conditions, thus suggesting different kinetics. By reducing the time of contact to 1 h, lower migration into oil was observed, however the one in water remained the same. The extraction behavior of caprolactam and oligomers from PA-66 film was analogous with the one from PA-6 film, with 60–81% total fraction of initial concentration migrated in aqueous/ethanol simulants and the trimer being the one in the highest concentration. It was reported that equal amounts of cyclic

PA-66 monomer and dimer are migrating into water for 2h at 100 °C and for 10 days at 40 °C, respectively. Furthermore, when the migrated level of caprolactam and cyclic oligomers was evaluated in scalded sausage by high-performance liquid chromatography-chemiluminescent nitrogen detection (HPLC-CLND), it was unfold that 59% of the initial caprolactam amount (4.2 mg/kg) migrated into the sausage during the scalding process, while the boiling water extracted the remaining 41% during the soaking (Heimrich et al., 2015). Begley, Gay, and Hollifield (1995) developed a method for the residual quantity of caprolactam oligomers in nylon baking bag (thickness of 2.5 mm) that remains after their migration into the food stimulant (oil), when cooked for 30 min at 176 °C. The authors reported a percentage of 43% migration of initial concentration of nylon 6/66 oligomers, which represents 15.5 μg/g (ppm) equivalent to 11.9 μg/cm^2. Migration of caprolactam in food stimulants: A simulant (ethanol 10% v/v), B simulant (acetic acid 3% w/v) and simulant ethanol 95% v/v, was also studied from two flexible multilayer packing [PET//PA//CPP] and [PET//Al//PA//CPP] materials at 60 °C for 10 days. Both materials leached variable migrants more in ethanol 95% (v/v) with an overall migration kinetics of 7.5 μg/g than in the other two remaining simulants. The concentration of caprolactam oligomer DP4 was the highest although below the limit established by European Regulation (Ubeda et al., 2017). Song et al. (2018) investigated the caprolactam monomer migration from laboratory produced PA-6 using a twin-screw extruder into food stimulants. The results were 0.982 mg/L for distilled water, 0.851 mg/L for 4% acetic acid, 0.328 mg/L for 20% ethanol and 0.624 mg/L for 50% ethanol and it was not detected for heptane at different temperatures (25 °C, 60 °C, and 95 °C). Higher temperatures can speed up the process of migration or favor plastic degradation producing many components with low-molecular weight that can be easily transferred into food. When the test for caprolactam migration from multilayer films packaging PA-6 for meat and cheese was conducted under 40 °C during 10 days and 100 °C during 30 min in water, 3% acetic acid and olive oil, the higher temperature extracted more amounts of the monomer. In this context, the concentration of caprolactam extracted in water from PA-6 intended for meat treated at 100 °C for 30 min ranged from 0.92 to 1.21 mg/dm^2, compared to 0.89 to 1.22 mg/dm^2 for the one treated at lower temperature for longer time. In 3% acetic acid, the monomer concentration varied from 1.29 to 1.74 mg/dm^2 and 1.13 to 1.62 mg/dm^2, while in the olive oil varied from 1.18 to 1.98 mg/dm^2 and 0.50 to 0.80 mg/dm^2 for less vigorous and high vigorous treatment. For cheese films, caprolactam migration into olive oil was also higher at 40 °C for 10 days (Félix, Manzoli, Padula, & Monteiro, 2014). Specific migration of caprolactam from polyamide/polyethylene (PA/PE) film over a long-term storage period of 40.5 months in the dark was evaluated. Migration tests comprised 10 days at 40 °C extracted in food simulants as 10% ethanol, 50% ethanol, 3% acetic acid, and distilled water. During the investigation time, the film was subjected to harsh conditions (elevated temperatures or ultra-violet radiation) that accelerate the aging process (Funk, Schlettwein, & Leist, 2016; Ritter, Schlettwein, & Leist, 2020). There was no significant difference in the variation of food simulants, possibly due to the complete migration of all initial concentration of caprolactam. Moreover, no significant increase of migration was observed at elevated temperatures (60 °C) while a full desorption in water was observed at 100 °C for 2 h. Bomfim, Zamith, and Abrantes (2011) investigated migration of caprolactam in 95% ethanol. They studied 40 samples intendent to pack food that contains fat such as Bologna sausage ($n=23$), turkey blanquette ($n=3$), poultry breast ($n=2$), pâté ($n=3$), and ham ($n=2$). They found 8 samples for Bologna sausage (35% of samples), 1 sample for turkey blanquette (33% of samples), 2 samples for poultry breast (100%) and 3 samples to contact pâté (100%) exceeded the limit of 15 mg/kg and were not in conformation with EU Regulation (Bomfim, Zamith, & Abrantes, 2011; EC, 2011). Caprolactam migration is slowing down after prolonged storage time due to PA slow post crystallization or the decreasing of glass transition temperature (T_g) affected by polymer water sorption from the air (Ritter et al., 2020). Ultra-violet or heat treatment can induce other migrants that would not be present under regular conditions (Funk et al., 2016). The effect of the processing under high pressure on the leaching of ε-caprolactam from commercial LDPE/PA/LDPE and PET/LDPE/PA/-EVOH/PA/LDPE packaging materials into food stimulants (acid, aqueous and fatty) was evaluated by Marangoni et al. (2020). Time-temperature-pressure conditions were as 10 min/25 °C/600 MPa, 10 min/90 °C/600 MPa, and 10 min/90 °C/0.1 MPa, while test migration conditions were 40 °C/10 days. The treatment under high pressure did not induce higher migration of ε-caprolactam as it was the case with higher temperatures. The highest migration was observed in fatty simulant after processing at 10 min/90 °C/0.1 MPa, which might be due to the absorption of oil by the polymer that will act as plasticizer. In another study, the effect of γ-irradiation on the migration of caprolactam from multilayer PA-6 film in ethanol, methanol, acetone, dichloromethane and water was investigated.

The results showed an effect of an increment in caprolactam levels as the doses of irradiation increased for one brand, while the opposite results for the other brand destined for meat stuff. Nevertheless, cheese PA-6 films showed reduction of caprolactam levels. Authors explained the increased levels as degradation of polymer liberating the caprolactam, while the crosslinking of residual caprolactam with other components induced reduction of its content (Araújo et al., 2008).

4.1.6 Isocyanates

Isocyanates are a mixture of compounds used in manufacturing of polyurethane polymers as well as food packaging adhesives. In polyurethane fabrication, the reaction involves 2,4- or 2,6-toluene diisocyanate (TDI), 4,4′-methylenebisphenyl diisocyanate (MDI) and polyols, which eventually form aromatic primary amines and CO_2 (Mutsuga, Yamaguchi, & Kawamura, 2014). Polyurethane foam is commonly used as absorbent pad for fish and meat, cushioning for fruits or can be used as a coating in pastry bags having a direct contact with food. However, polyurethane adhesives are employed in the manufacturing of laminated films, which possess superior qualities over monolayer films. These adhesives bind the material through crosslink reactions. As adhesives, they are not directly exposed to the food but remain risky components in the food packing that can release potential harmful components (Yan, Hu, Tong, Lei, & Lin, 2020). Most isocyanates used in the manufacture of polymeric materials are designed for exposure to food. Nevertheless, especially TDI and MDI are proven to have toxicological effects. TDI toxicological assets are ascribed to the –N=C=O group, that in aqueous environment forms carbamic acid further decomposing to primary amine and eventually forming urea derivative. It might react with hydroxyl, sulfhydryl, and amine groups from proteins, modifying their normal function, caused by system of reaction at a given time period. The product of TDI hydrolysis, toluene-2,4-diamine (TDA) is a carcinogen that may further decompose or polymerize with TDI. Orally ingested, can retain up to 20% in the body, while inhaled all is absorbed. Acute exposure causes irritation to the eyes, respiratory and gastrointestinal tracts, and produces dermal hypersensitivity reactions. Chronic exposure causes permanent lung dysfunction and asthma induction. Nonetheless, inhalation of MDI can cause asthma, dyspnea, immune disorders as well as nasal and lung lesions, in addition to eczema and skin and eyes irritation when it comes to dermal contact (Kapp, 2014). According to the EU Regulation N° 10/2011 (EC, 2011), twelve isocyanates are permitted to be used as monomers or other starting substances in the production of plastic materials for food packaging, but not as additives or aids in the food packaging plastic fabrication. With this regulation, the residual maximum of those isocyanates in the final product is set to 1.0 mg/kg and no detectable amount should be present in food. Adhesives applied in plastic materials in contrast are not specifically regulated, with a specific directive. They are mentioned as a part of the Framework Regulation (EC) N° 1935/2004 (EC, 2004b), under which must meet the requirements for quality assurance and risk assessment. In Annex XVII to Regulation (EC) N° 1907/2006 from 24 August 2023, isocyanates will be allowed to be used as components in other blends for industrial purpose only in a total concentration less than 0.1% (w/w). This annex anticipates safety trainings on the use of the diisocyanates for the concerned workers (EC, 2020). In 2010, EFSA published a scientific report on the safety evaluation of the substance bis(2,6-diisopropylphenyl)carbodiimide for use in food contact material for long-term storage at room temperature, which can be used as layer behind polyesters or in blends of PET, PGA and PLA. The previously established limit of 0.05 mg/kg food was confirmed (EFSA, 2010). Since isocyanates represent the large group, they can be classified under Cramer I, II and III with the respect to their toxicity. The reported toxicological limit was 1800, 540 and 90 μg per person per day for the classes I, II, and III, respectively (Yan et al., 2020). Isocyanates are very often analyzed by liquid chromatography-tandem mass spectrometry (LC/MS/MS) detection or headspace coupled to gas chromatography and mass spectrometry detection (HS-SPME-GC–MS) for selective and sensitive identification. Ten different isocyanates and their decomposition product amines (11 different) were identified in polyurethane foam and polyurethane coated products. The foam (fruit cushioning) contained 2,4-TDI in the amount of 0.04–0.92 mg/kg and 2,6-TDI in the quantity of 0.02–0.26 mg/kg or in the levels of 0.03–1.15 mg/kg for total isocyanates. In contrast, the degradation products 2,4-TDA and 2,6-TDA were present in much higher quantity, ca. 10–59 and 7.0–26 mg/kg, respectively, with the total amount of amines varying from 14 to 89 mg/kg. As a difference, in polyurethane coated pastry bag 4,4′-MDI was quantified, with the total amount of isocyanates was 0.047 mg/kg, while no amines were detected (Mutsuga et al., 2014). The effect of time against temperature conditions and nanomaterial addition on migration of isocyanates from polyurethane adhesives employed as the middle layer between two LDPE films in isooctane was investigated. For this purpose, the researchers used "solvent free" and "solvent based" adhesives, each of them enriched with graphene or functionalized graphene. They reported higher migration from "solvent free" laminated films than the ones based on solvent. In addition, higher temperature was accountable for the excess of the kinetic energy applied to the target molecule stimulating the migration. In this particular case, the addition of nanomaterials to the adhesives did not cause any significant difference in migration either on target molecule or the nanomaterial itself. From the safety point perspective, many of the desorbed components in isooctane are classified as Cramer III with safety threshold of 0.09 mg/person/day, which was exceeded for the highly toxic materials at 70 °C, 2 h, and 40 °C, 10 d (Yan et al., 2020). In another study, 45 commercial multi-layered films glued with 29 varieties of adhesives (including polyurethane and vinyl) were studied. Films packaging were intended to be exposed to dry, fresh, vacuumed, and pasteurized and deep freeze food. Migration from polyurethane adhesives was tested in dry food stimulant Tenax® and isooctane. Authors reported migration of 55 different compounds from adhesives, four of which classified as Cramer III regarding their toxicity. However, their affinity to migrate in food stimulants was reported as low and below the limit of detection. Conversely, 57% of all identified compounds in the laminates leached in the dry food stimulant, thus confirming the inevitability of regular control of adhesive composition used in food packaging (Aznar et al., 2011). Primary amines can occur in food from colorants or printings on the plastic other than derivatives of isocyanates. In that context, a study was carried out to analyze the migration of amines from 15 different cooking utensils made of polyamide in water and olive oil food stimulants with repeated use. The migration of aniline occurred in both food stimulants above the EU permitted limits, gradually decreasing with repeated use. It was shown that the migration would drop to the established level after 100 h of use at 100 °C (Brede & Skjevrak, 2004). Different ways in reducing the exposure to toxic compounds leaching from polyurethane involve adjusting the technologies to already described synthesis of non-isocyanate polyurethanes. These technologies comprise polyaddition of cyclic carbonates and polyfunctional amines obtaining polyhydroxyurethanes, an environmentally friendly polymeric material (Stachak, Lukaszewska, Hebda, & Pielichowski, 2021).

4.1.7 Acrylonitrile

Acrylonitrile is a component used as a monomer in plastic production, usually obtained by a process called catalytic ammoxidation from ammonia and propylene. Only 15% of its production is used for acrylonitrile butadiene-styrene (ABS) and styrene-acrylonitrile (SAN), while the rest imparts the production of textile fibers, acrylamide, adiponitrile, nitrile rubber, etc. The polymers with high content of nitrile represent outstanding barrier toward gases, a property that is very useful in containers for carbonated drinks. There are several ways for humans to be exposed to this monomer through either wearing textile containing unreacted monomer and ingestion of food stored in a plastic ABS or SAN containers or indirectly via environment. Although the risk assessment report from 2004 written by European Commission Joint Research Centre (JRC) stated that human exposure is negligible. Moreover, the exposure in the production sites was significantly decreased rather than in the past (JRC, 2004). Nevertheless, acrylonitrile is proven to be toxic if ingested, inhaled or skin contacted, due to the release of cyanide. In addition, it shall also be classified as "irritant, irritating to respiratory system and skin, risk of serious damage to eyes". Long-term exposure cause developing of tumors on the nervous system, intestines and mammary glands (JRC, 2004). Therefore, the specific migration limit is set to 0.02 mg acrylonitrile/kg food according the Commission Regulation (EU) N° 10/2011 of 14 January 2011 (EC, 2011) and Commission Directive of 23 February 1990 relating to plastics materials and articles intended to come into contact with foodstuffs (90/128/EEC) (EC, 1990). According to FDA regulations (21CFR177.1020), the content of acrylonitrile in ABS *co*-polymer product shall not exceed 11 ppm (μg/g). The migration limits to food simulants (water, 3% acetic acid and n-heptane) were established as $0.353\,mg/mm^2$ for 8 days at 50 °C (FDA, 2022). Acrylonitrile monomer was a concern even in the second half of the 20 century. In a research work, Brown, Breder, and Mcneal (1978) developed a procedure for acrylonitrile detection using gas chromatography (GC) coupled with nitrogen/phosphorous detector in food stimulants, chiefly water, heptane, 3% acetic acid, 8% ethanol and 50% ethanol. They postulated useful remarks regarding developing the method of detection as such to minimize the loss of the compound, headspace volume not to be minimized, as well as for greater sensitivity (20 ppb or less) hydrogen flow and bead power settings shall be increased. In another study, Gilbert & Startin (1982) quantified the migration of acrylonitrile monomer from ABS films and tubs in butter, soft margarine and shortening following certain storage time. The concentration of acrylonitrile was varying from 0 to 10 mg/kg in ABS tubs and from 0.01 to 0.04 mg/kg in soft margarines. No correlation was found between the initial concentration of the monomer in the tub and its level of migration. Similar postulation was made by Poustková, Poustka, Babĺčka, and Dobláš (2007) that no correlation between the acrylonitrile content in food contact materials (kitchenware and ABS and SAN polymer granulates) and the extent of migration in food simulants (distilled water, 3% acetic acid, 10% ethanol and 95% ethanol) was established. The range of acrylonitrile in the kitchenware ranged from 0.4 to 25.1 mg/kg, whereas it was found to be much superior in the range of 6.2–283.9 mg/kg in polymer granulates. Monomer migration from the kitchen utensils was the highest in 3% acetic acid (5.2 μg/dm^2), whereas from polymer granules the highest value was observed in 95% ethanol (9.6 μg/dm^2). Nevertheless, Lickly, Markham, and Rainey (1991) found a correlation ($R^2 = 0.99$) between the concentration of the monomer in the polymer and the migrated level in the food stimulants under certain exposure conditions. Eight percent of ethanol at 50°C for 24 h did not extract detectable quantities of the monomer, while the water leached acrylonitrile from the polymer at 100°C for 2 h or at 50°C for 10 days. Simultaneous detection of acrylonitrile, 1,3-butadiene, propionitrile, and 4-vinyl-1-cyclohexene in ABS kitchenware was carried out by Ohno & Kawamura (2010). From all the measured utensils, measuring cups, ice-cream baller and measuring spoon contain the highest amount of residual acrylonitrile (50.4, 29.3 and 20.2 μg/g, respectively). In contrast, Chinese spoon, Japanese radish grater and lunch bow container exhibited the lowest residual quantity of acrylonitrile (0.3, 2.2 and 3.1 μg/g, respectively). Besides, acrylonitrile in PS kitchenware was not detected. In certain conditions, i.e., high ethanol concentration ABS polymer can swell, followed by facilitated acrylonitrile migration. Guazzotti, Ebert, Gruner, and Welle (2021) observed the swelling kinetics of ABS material (thickness 350 μm) in ethanol and water in different ratios at varying temperatures as well as in real food (milk, cream and olive oil). It was found that ABS material in contact with 95% ethanol swells up to 20% (w/w) during 15 days at 60 °C, and all migrating substances are rapidly released in the food simulant. However, after 2 days the equilibrium was attained. At 40 °C, the same material in the same simulant swells up to 9% in the first 2 days, which remain constant until the end of the study. At 20 °C, the swelling and the migration were slower. The swelling of the ABS polymer was lower in 50% ethanol, reaching a maximum of 10% (w/w) at 60 °C after 38 days, followed by lower level of migration of all tested compounds, which normally increased with increasing the temperature. The migration of substances was negligible in 10% ethanol and 3% acetic acid. ABS material in isooctane starts to swell after 8 days, reaching equilibrium of 7% after 100 days, with the migration rapidly increased when the swelling starts unrelated to the molecular weight of the migrating compound, following a non-Fikian behavior. Regarding food matrixes, the highest migration of corresponding compounds was observed in milk, ice cream and oil at 40 °C. The swelling at 20 and 40 °C in milk and ice cream after 90 days was 1.0% and 1.9%, respectively, whereas not all other tested food, including drinks at hot fill conditions, increased the weight of the ABS material more than 1%. The food simulants isooctane and 50% ethanol at 20 °C and 40 °C reproduce the migration accurately as in the food matrixes; however, the migration in 95% ethanol was largely misjudged. Table 1 summarizes the migration of the various migrating components used for packaging of different types of food.

4.2 Migrating components connected to the additives

Plastic materials made from basic polymers alone do not have desirable properties for use in everyday applications, which over time become more and more demanding and driven by the consumer's needs. Thus, plastic industry has been constantly innovating by including a variety of chemical substances in the basic polymer to improve its functionality, aging and performances. The employment of additives in plastic formulas led to a rapid increase of the production of polymers and widened the range of applications.

Additives represent a vast group of chemicals that can be categorized in several groups according to their functionality, rather than their chemistry, and each group having individual roles in amelioration of a certain function of the plastic material by tailoring it to the specific use. The most usual additives employed in food packing materials are plasticizers, antioxidants, slip agents, lubricants, pigments, light and heat stabilizers, antifog agents, antistatic agents, acid scavengers, flame retardants, mold release agents, chemical blowing agents, nucleating agents, fillers, etc. As much as they improve the packaging material and its processability and price, the safety concern arises when it comes to considering their contact with food. The problem of potential migration is even more actual as they are substances with low-molecular weight and, in addition, as they are not chemically bound to the polymeric chain (with few exceptions). Table 2 summarizes the most commonly used additives categorized according to their functionality with respect to migration phenomena.

Table 1 Mechanisms and factors affecting the migration of polymers.

Migrating components	Mechanism/type of migration	Factors affecting migration	Type of polymer packaging	Food
Styrene	Leaching	Temperature of consumption, time of exposure, fat content of food	Plates, meat trays, egg cartons, cups, bowls	Aqueous food, milk, meat, soups, coffee, ice-cream, hot chocolate, yoghurt, jelly, pudding, olive oil, fruit juices, fizzy drinks, beer
Bisphenol A	Condensation, distillation, leaching, diffusion, contact migration (solubilization)	pH of food, temperature of food, can type, heat treatment	Can coatings, plastic bags, tableware	Minced beef in gravy, canned carrots in brine, vegetable soup, evaporated milk, tuna cans, jalapeno cans
Vynil chloride	Contact migration, leaching, condensation	Food type, temperature of food	Bottles, drinking pipes	Water and beverages
PET	Desorption, leaching	Thermal condition, time of exposure, nature of food surface	Food containers, trays, dishes, bottles, films, microwaveable containers, coffee capsules	French fries, popcorn, waffles, pizza, fish sticks, olive oil, coffee, tea, mineral water, fruit juice
Caprolactam	Diffusion, leaching	Type of food, solubility, swelling, time-temperature-pressure	Vacuum and modified atmosphere packaging, cook-in packaging, roasting bags, boil-in-the-bag	Bologna sausage, turkey blanquette, poultry breasts, pate, ham, cheese, food simulants A, B, C and D
Isocyanates	Contact migration, leaching	Time/temperature conditions	Food packaging adhesives, coating in pastry bags, absorbent pad, cushioning pad	Fish, meat, fruits, vegetables, pastry, dry food, dry food stimulants, food stimulants
Acrylonitrile	Contact migration, leaching	Temperature, swelling	ABS containers, ABS films and tubs, SAN containers, kitchenware	Carbonated drinks, butter, soft margarine, shortening, milk, cream, olive oil, food simulants

Table 2 Additives used in food packaging materials.

Category	Type	Function/Mechanism	Compound	Polymer	Food/Food simulant	References
Processing aids	Plasticizers	Decreasing viscosity and glass transition temperature	Phthalates	PET	Distilled water isooctane, ethanol 20%, ethanol 50%	Li, Wang, Lin, & Hu (2016)
			Phthalates, sebacates, ESBO	PVC gaskets of lids	Simulant D (olive oil), oily food in a jar	Fankhauser-Noti & Grob (2006)
			Phthalates, adipates, citrates, sebacates, diacetyllauroyl glycerol	PVC tubing, PVC gloves, cap gasket	Beverages, fat and oily food, dairy products, refreshments, fast food, instant food, baby food	Tsumura, Ishimitsu, Kaihara, Yoshii, & Tonogai (2002)
	Lubricants	Reduction of internal friction, reduce the flow resistance of the polymer, influence the rheology of the melt	Paraffin, wax esters, cholesterol esters, fatty acid methyl esters, triacyl glycerols, fatty alcohols, free fatty acids, cholesterol, 1,3-diacyl glycerols, 1,2-diacyl glycerols, monoacyl glycerols and fatty acid amide	Tinplate strips coated with a commercial, epoxy-anhydride	95% Ethanol (4 h at 60 °C) isooctane (2 h at 60 °C)	Schaefer, Kuchler, Simat, & Steinhart (2003)
			Mineral hydrocarbons	Polystyrene cups	Lager, beer, cola, lemon barley water sparkling apple juice, instant chocolate drink, instant coffee with 25% whole milk, coffee with 'creamer', instant tea with skimmed milk, lemon tea and chicken soup, distilled water, 3% acetic acid and 15% ethanol	Castle, Kelly, & Gilbert (1991)
	Slip agents		Oleamide, erucamide, stearamide, stearyl erucamide, oleyl palmitamide	LDPE, PP, PS (crystal), PVC	Water, 3% aqueous acetic acid, 15% aqueous ethanol and olive oil	Cooper & Tice (1995)

Suppression in thermal decomposition	Anti-oxidants	Scavenge the free radical, decomposes the peroxide	Tris-(2,4-di-tert-butylphenyl) phosphite –*Irgafos 168,* octadecyl-3-(3,5-di-tert-butyl-4-hydroxyphenyl) propionate – *Irganox 1076*	Irradiated LLDPE film	Distilled water, acetic acid 4%, ethanol 20%	Jeon, Park, Kwak, Lee, & Park (2007)
			BHA, DBP, BHT, Irganox 1010, Ethanox 330, Irgafos 168, Irganox 1076	LDPE	Distilled water (simulant A)	Dopico-Garcia, Lopez-Vilarino, & Gonzalez-Rodriguez (2003)
			Irgafos 168, Irganox 1010	Micro-wavable PP packaging	Distilled water, 10% ethanol, 99.9% ethanol, 3% acetic acid, 90:10 isooctane/ethanol	Alin & Hakkarainen (2011)
			Irganox 1076	LDPE	Distilled water, acetic acid 3%, ethanol 10%, 95% ethanol, olive oil, isooctane, chocolate, processed cheese, soft cheese, cottage cheese, chicken breast, pork minced meat 5%, pork minced meat 24%, fresh salmon, wheat flour	Beldi, Pastorelli, Franchini, & Simoneau (2012)
			Hostanox SE-2, Irgafos 168, Irganox 1076	PP films	n-Heptane, 95% ethanol	Garde, Catala, Gavara, & Hernandez (2001)
Esthetic enhancers Performance enhancers	Antistatic agents	Prevent the accumulation of static charge on the surface of the film	Nonylphenol	PVC stretch films	Rapeseed oil, Japanese radish, pumpkin, melon pineapple, cooked radish with fried tofu, meat sauce, hamburger steak, croquette, potato salad, minced chicken white meat, minced pork, and minced tuna	Kawamura, Ogawa, & Mutsuga (2017)

Continued

Table 2 Additives used in food packaging materials.—cont'd

Category	Type	Function/Mechanism	Compound	Polymer	Food/Food simulant	References
	Light and UV stabilizers	Scavenge the radical or absorb the UV light instead of the polymer	UV stabilizers, Tinuvin 326, Tinuvin 327, Cyasorb UV 5411	PET	Fatty-food simulants such as olive oil, soybean oil, n-heptane and isooctane	Monteiro, Nerin, & Reyes (1999)
			Tinuvin 234 (2-(2H-benzotriazol-2-yl)-4,6-bis (1-methyl-1-phenylethyl) phenol)	PET	Ethanol/water, isooctane, fractionated coconut oil simulant (Miglyol)	Begley, Biles, Cunningham, & Piringer (2004)
			2-(2H-benzotriazol-2-yl)-4,6-bis(1-methyl-1-phenylethyl) phenol (UV-234) (2′-hydroxy-3′-tert-5′-methylphenl)-5-chlorobenzotriazole (UV-326) and 2-(2′-hydroxy-5′-methylphenyl) (UV-P)]	Virgin PET bottles	Distilled water, acetic acid 3%, ethanol 20%, ethanol 50% and isooctane	Li et al. (2016)
			Chimasorb 8, Tinuvin 326, Tinuvin 328	Retail packaging material	Distilled water (simulant A), 3% acetic acid (simulant B), 10% ethanol (simulant C), oil (simulant D)	Gao, Gu, & Wei (2011)
	Flame retardants	Reduce flammability	Halogenated compounds, phosphorus compounds, metallic oxides and inorganic fillers, nanocomposites, bishydroxydeoxybenzoin	All polymers		
Foams formation	Blowing agents	Create foam or expanded structure by evaporation or sublimation or decomposition reaction				

Change morphology	Nucleating agents	Increase crystallization rate and temperature of semi-crystalline polymers by changing their morphology	Talc, silica, clay, mono- or polycarboxylic acids, sodium benzoate, pigments, ethylene/acrylic ester copolymers, g-cyclodextrin	Polyamide 6, polypropylene, polyethylene terephthalate		Seven, Cogen, & Gilchrist (2016); Singh, Saengerlaub, Abas Wani, & Langowski (2012)
Common additives	Colorants	Improve the color of final product, can affect mechanical properties	Titanium oxide, zinc oxide, zinc sulfide, and lead carbonate, carbon black, iron oxide, chrome yellow; nickel-chrome-titanium yellow; iron oxides; lead chromate, molybdate orange, cadmium orange, iron oxide, combination of chrome/iron oxides, cadmium sulfide/selenide, ultramarine (aluminosilicate with sodium ion and ionic sulfur groups); chrome oxide; cobalt-based mixed oxides, monoazo pigments, polyazo fillers, phthalocyanine, quinacridones, dioxazines, isoindolines, perylenes, flavanthrones, anthraquinones, titanium dioxide-coated mica (muscovite), ferric oxide-coated mica, bismuth oxychloride, aluminum, copper and alloys of copper and zinc (bronze)	Baby bottles, thermometers, kettles	Distilled water (simulant A), 3% acetic acid (simulant B), 10% ethanol (simulant C), 95% ethanol	Drobny (2014), Singh et al. (2012)
	Fillers	Reduce cost, improve rheological and absorption properties and act as colorants;	Calcium carbonate, barium sulfate	All polymers		Hohenberger (2001)

Continued

Table 2 Additives used in food packaging materials.—cont'd

Category	Type	Function/Mechanism	Compound	Polymer	Food/Food simulant	References
		Impart tear and compressive strength, dimensional stability, scratch resistance, and stiffness to compounds	Glass and ceramic beads	Resins and foams		Hohenberger (2001)
		Reinforce polymers, act as antiblocking additive, viscosity regulator and matting agent	Syntethic silica	Films		Hohenberger (2001)
		Reinforce epoxy resins to have high tensile strength, high stiffness and flexural modulus, high creep resistance, impact resistance, and high heat deflection temperatures	Glass fibers	Laminated pouches, microwaveable packaging		Henkel (2020)
		Produce very good mechanical properties at low loadings, scratch resistance, superior barrier properties, enhance fire resistant properties, improve heat distortion performance	Montmorrilonite, smectite	Packaging films, rigid containers		Hohenberger (2001)

4.2.1 Plasticizers

These compounds are incorporated in the polymer base to mend the processability, flexibility and elasticity of the packaging material. Their mechanism of action lies on decreasing the viscosity and glass transition temperature without changing the chemical traits of the base polymer.

There are two major groups of plasticizers, which are internal and external according to their liaisons with the polymer. External plasticizers do not form chemical bonds, but they are physically incorporated in the matrix, whereas internal plasticizers do form chemical bonds and copolymerize together with the polymer (Nerin, Canellas, & Vera, 2016). The migration rate in food depends, as for general compounds, on the molecular weight, concentration of plasticizer (up to 20% are added by the weight of polymer), polarity of the plasticizer (affinity towards to polymer) as well as the fat content in food.

The most commonly used compounds are phthalates (accounting for 80% of total plasticizers), followed by adipates, sebacates, citrates, epoxidized soybean oil (ESBO), trimellitates, maleates, fumarates, succinates, etc. However, many of these have shown toxicological effects with special attention to the phthalates, which have been described as endocrine disruptors. Therefore, the use of plasticizers is regulated by European legislation on plastics materials and articles intended to come into contact with food (EC, 2011), with strictly prescribed low migration limits. Thus, benzyl butyl ester and dibutyl ester of phthalic acid have a SML of 0.3 and 30 mg/kg, respectively.

4.2.2 Lubricants

Lubricants are compounds used in the processing phase of polymer packing production to reduce the flow resistance of the polymer and influence the rheology of the melting point. They are added at levels up to 7–8% by weight. While the polymer is in its molten state, (temperatures above crystalline melting point or glass transition temperatures, for semi-crystalline or amorphous polymers, respectively) and due to its exceptionally high-molecular weight, the viscosity of melt is correspondingly high.

To process the melt, high shear needs to be applied, resulting in flow resistance and heat generation (friction energy). Under the additional generated heat, the degradation of the polymer can only accelerate. By incorporating the lubricants into the polymer base, the internal and external friction is reduced. Additionally, lubricants can act as mold release agents. They are classified as external lubricants that are applied on the surface of the product in order to decrease adhesion between the plastic material and food system, or internal lubricants that are applied in the formulation supposing to increase flexibility for the drawing process (Schaefer et al., 2003).

Most commonly used substances for this purpose include compounds containing long-chain aliphatic acids like waxes, paraffins, fats and oils or partial acyl glycerols and fatty acid amides, as well as their isomers and analogues. The major disadvantage of highly effective lubricants is the change of appearance of the polymer known as clouding. This phenomenon can be corrected by lubricant's derivatization. Such derivatization is reached through the introduction of functional groups, which will make them similar to the polymer that has been modified. Hence, there are constant developments in their production aimed at obtaining high performance technologies. The health risk and use of lubricants is regulated with several directives and the Revision of the European Ecolabel Criteria for Lubricants (Vidal-Abarca Garrido Candela et al., 2018).

4.2.3 Slip additives (internal lubricants)

This group of chemicals mainly belongs to the group of internal lubricants. Initially incorporated in the base polymer formulation and later they are slowly transferred to the surface to inhibit the adhesion between films, reduce the friction and facilitate mold release. Fatty acid amides (erucamide and oleamide) and metallic stearates (zinc stearate) are generally used as slip agents.

Most of the lubricants and slip additives are authorized and listed in Regulation (EU) N° 10/2011 with their specific migration levels. According to the European Commission Regulation (EU) 2016/1416 from 2016 for amending and correcting Regulation (EU) N° 10/2011 on plastic materials and articles intended to come into contact with food (EC, 2016), the use of additives saturated fatty acids C16–18 and hexaesters with dipentaerythritol do not harm human's health and can be added to the list without restrictions.

4.2.4 Antioxidants

Antioxidants are compounds added to the base polymer to enhance its stabilization and delay the aging. The polymer is prone to degradation due to oxidative reactions when exposed to air or UV light. Oxidative reactions are triggered by free radicals often produced by heat, radiation, mechanical shear forces or metal impurities. In the case of polymers used for food packaging, the probability of being exposed to high temperatures is on the high margin (contact with hot food, sterilization and pasteurization, microwave heating). The mechanism of reaction of antioxidants is trough reacting with the free radicals affecting degradation themselves, thus preserving the polymer. These compounds are effective in very low concentrations and can be added below 1%, preferably as early as possible during the production.

Antioxidants can be classified as chemical and natural compounds according to their source of origin. Tinuvin P, Tinuvin 326, Tinuvin 776 DF, Tinuvin 234, Chimasorb 81, Irganox 1076, Irganox 1330, Irganox 1010, Irganox168 and Irganox P-EPQ are the most common synthetic antioxidants. However, the natural antioxidants that have been used so far are tocopherols, ascorbates, carotenoids, vitamin A, phytochemicals, variety of metals (e.g., selenium), as well as food antioxidants including butylated hydroxyanisole (BHA), butylated hydroxytoluene (BHT) (Arvanitoyannis & Bosnea, 2004). Phenols are regarded as primary antioxidants, whereas phosphites and thioesters are considered as secondary antioxidants that can provide protection to both polymer and primary antioxidant. Usually primary ("long-term antioxidants") and secondary antioxidants ("processing antioxidants") are used in combination. The first group scavenge the radical and brakes the chain reaction, whereas the latter group decomposes the peroxide.

The recent applications of antioxidants as stabilizers in plastic packaging are from the group of organophosphites, tris-nonylphenyl phosphite (TNPP) as most regularly used (Singh et al., 2012). The degradation products of antioxidants are regulated in the Risk Assessment of non-listed substances (NLS) and non-intentionally added substances under Article 19 of Commission Regulation (EU) N° 10/2011 of 14 January 2011 on plastic materials and articles intended to come into contact with food. They are considered as mostly nontoxic or non-carcinogenic except organophosphates, which might have neurotoxic properties and their exposure limit is 18 μg/person/day or a SML of 5 mg/kg (Europe Plastics, 2014). In this context, TNPP is approved by FDA with 0.1% residuals in food (Singh et al., 2012).

4.2.5 Light and ultra-violet (UV) stabilizers

Plastic material that is long term exposed to sunlight, UV radiation or different weather conditions can undergo photo-oxidative reactions and degrade. This leads to change of plastic material appearance, fading the color, loss of gloss, creation of defects or cracks, loss of tensile strength and elongation, etc. UV light can deteriorate color, flavor and nutritional value of the product. Photo-oxidation as an analogue of thermal oxidation has similar aspects.

The oxidative reactions are initiated by free radicals formed by photons, while hydro peroxide is generated as intermediary compound. The substances acting as light stabilizers scavenge the radical or absorb the UV light instead of the polymer (Singh et al., 2012). The degradation occurs at a wavelength between 290 and 400 nm, with different types of degradation reactions according to the different wavelengths that are absorbed. The products of oxidation reactions are mono- and dihydroxy-derivatives, carbon monoxide (CO) and carbon dioxide (CO_2). They are used in a concentration of 0.05 to 2%; still the added amount depends on the total quantity of other additives used (Begley et al., 2004).

The mechanism of action of light stabilizers involve scavenging the free radical or absorbing the energy and releasing the excessive energy converted to heat. Light stabilizers comprise 2-hydroxybenzophenones, 2-hydroxyphenyl benotriazoles, organic nickel compounds and most recently sterically hindered amines (HALS) (Drobny, 2014). New types of compounds are commercialized, which are effective in blocking the light up to 390 nm in comparison to the typical ones that have an effect up to 370–380 nm. Ultra-violet stabilizers have strong photostability and absorb the UV radiation themselves. The light absorption starts at 360 nm, intensifies at 320 nm and becomes noticeable below 300 nm.

Different polymers are stabilized with different compounds (Singh et al., 2012). Thus, polyolefins are stabilized with nickel compounds and hindered amine light stabilizers, benzotriazole-type absorbers stabilize PET, whereas PVC is preserved with UV absorbers. Polymeric hindered amines comprising Tinuvin 622, Tinuvin 234, Tinuvin 770, Tinuvin 1577 and Chimasorb 944 is another group of light stabilizers that has found and immense application in transparent plastic bottles and films. Hindered amines are approved by EU and FDA to be employed in plastic materials envisioned to come into contact with foodstuff. According to FDA, European Chemicals Agency and EU Legislation, benzotriazoles are approved to be used at 0.25% (w/w) in PVC and PS and 0.5% (w/w) in PC with SML of 1.5 mg/kg (EC, 2011; ECHA, n.d.; FDA, 2010).

4.2.6 Antistatic agents

Plastic packaging as insulating materials often build up electrostatic charge due to generated friction, which influence the processability (risks of electric shocks) and affinity for dust. To soothe the problem, antistatic chemicals are incorporated in the base polymer that are supposed to prevent the accumulation of static charge on the surface of the film. They act as facilitators of static charge dissipation by forming conductive transmission paths inside the polymer network. Additionally, they migrate to the surface of the polymer, where they bond with thin layer of water. Moreover, antistatic compounds assist in improving mold release (Singh et al., 2012). To form those conductive fields, they are frequently used in large quantities (above 5%) that negatively influence the transparency and mechanical properties of the film.

Most of the components used for this role are surfactants composed of hydrophilic and hydrophobic fraction. In these components, hydrophilic-lipophilic balance plays an important role in the antistatic action. They can be divided as cationic (quaternary ammonium, phosphonium or sulfonium salts), anionic (sodium salts of sulfonates, phosphates, and carboxylic acids), non-ionic (esters of fatty acids, and ethoxylated tertiary amines) and ampholytic antistatic agents. If not immobilized to the polymer matrix (UV crosslinking), they can be depleted easily over time owing to the low-molecular weight, a process that decreases the antistatic effect (Zheng et al., 2012).

The field of antigenic agents is constantly developing and many substances are added to this group including carbon nanotubes, ionic polymers and intrinsically conductive polymers (Singh et al., 2012). The migration depends mostly on the crystallinity of the polymer, the thickness of the packaging and time. Except traditional antistatic, non-migratory products have been recently introduced and exhibiting greater advantages including use at high temperatures (better sealing), long lasting antistatic effects but should be used in higher quantities and have higher relative price. Some glycol derivatives are regulated with Regulation (EU) N° 10/2011 on plastic materials and articles intended to come into contact with food and have individual migration limits (10 mg/kg·bodyweight TDI), while quaternary ammonium compounds are not regulated (Consumer health protection committee, 2009).

4.2.7 Blowing agents

In polymer production industry, blowing agents are employed to create a foam structure or to expand the plastic material matrix by generating air cells. These types of agents can be divided in physical and chemical according to the process involved in gas creation. Gas in physical processes is formed by evaporation or sublimation, while chemical decomposition reaction involves chemical processes. Chemical blowing agents (CBA) decompose at elevated temperatures by endothermic and exothermic reaction. The temperature of CBA decomposition shall be in the same range as polymer processing temperature intended to be expanded liberating a gas in a form of CO_2 or nitrogen (N_2).

Previously azodicarbonamide was vastly used as CBA due to its good foaming properties and price (Singh et al., 2012). However, it decomposes into semicarbazide thus it was banned from use by European Commission Directive 2004/1/EC from January 2004 amending Directive 2002/72/EC as regards the suspension of the use of azodicarbonamide as blowing agent (EC, 2004a). The most frequent compounds used in PS foams are fluorocarbon or aliphatic hydrocarbon, benzene disulphonyl hydrazide and CO_2 (Hahladakis et al., 2018; Raj & Matche, 2012). Other compounds could be regarded as potential physical or chemical blowing agents such as sodium bicarbonate, zinc carbonate, hydrogen peroxide, isocyanate and water, metal and strong acid, poly(hydro siloxane), etc. (Coste et al., 2020). Blowing agents pose more of environmental issue since they break down into gases and tend to escape in the atmosphere causing air pollution and global warming.

4.2.8 Flame-retardants

Additives that can retard or completely stop the burning are known as flame-retardants and have been extensively used in plastics, textiles, electronic equipment like computers and televisions, in furniture, upholstery, plastic heat insulation, building foams, flooring and curtains and toys. Since the 1970s, a large number of regulations have rapidly expanded the global use of flame retardants, but then the potential adverse health and environmental impacts of these chemicals were still not recognized or fully understood (Shaw et al., 2010; Weber & Kuch, 2003). Today, to preserve and improve the quality of the environment, protect human health and utilize natural resources prudently and rationally, the Directive 2002/96/EC of 27 January 2003 of the European Parliament and of the Council on Waste Electrical and Electronic Equipment (WEEE) is implemented to regulate their use (Kim, Osako, & Sakai, 2006).

Flame-retardants can reveal their characteristics to the polymers in either condensed or gas state. When in the condensed state, they remove thermal energy from the substrate and act like a heat sink or they can participate in the formation of char, creating a barrier for the heat and mass transfer.

Their flame-retarding capacity can also be achieved by mass dilution, evaporation, conduction or participation in endothermic reactions (Jansen, 1987). In turn, in the gas state, flame-retardants break the fire by disrupting its combustion chemistry. Usually throughout combustion, the polymer fragments connect with the oxygen and other reactive components in a series of chain reactions to produce radicals, namely oxygen, hydrogen and hydroxyl radicals. Flame-retardants in the gas state mostly based on halogens and phosphorus can chemically react with the radicals and create components with lower energy and, consequently, break the chain propagation needed for fire initiation and its continuation (Ranken, 2001).

Due to the release of the hazardous gases from the flame-retardants, recently efforts have been made in exploring some eco-friendly retardants such as the nanocomposites and bishydroxydeoxybenzoin (BHDB) monomer. Nanocomposites have inferior peak heat-release rates than pure polymers and in small loadings (3 to 6%) exhibit flame retardancy and, simultaneously, not affecting physical properties of the used resin. The BHDB monomer releases water vapor rather during its flame retardancy instead of the usual hazardous gases (Ranken, 2001).

4.2.9 Colorants

Color is an important characteristic of the final packaging. Its attractiveness and appeal for the consumers can be improved by the use of colorants. In industrial polymer production, the colorants are usually divided in dyes and pigments. European Union did not set yet any rules towards the risk assessment and use of colorants in plastics. Consequently, the use of colorants should remain subject to national law (EC, 2011).

Dyes are soluble in the polymer matrix and form a molecular solution that gives bright and intense color. They are transparent and easy to use. Most dyes have relatively reduced light-fastness and limited heat stability, but tend to keep their color. It is well known that the fading of the color happens due to the exposure of the surface layer to light. Dyes are also subject to fading on the surface but their transparency gives a real depth of color and they are not affected by the surface influences. Dyes can also be subject of color migration, which is the material of legislation for critical products, e.g., food contact applications and toys (Singh et al., 2012).

In turn, pigments are organic and inorganic compounds that appear as dry powders, color concentrates, liquids, and pre-colored resins. They are usually insoluble in the polymers and the color is achieved by the dispersion of the fine particles (0.01–0.1 μm) throughout the polymer and are responsible for the opacity and translucence in the final product. Some pigments can act as nucleating agents, altering the mechanical properties and improving the clarity of the resin (Drobny, 2014; Singh et al., 2012).

4.2.10 Nucleating agents

Nucleating agents are usually used in the industrial production of plastics to enhance the crystallization rate, decrease processing cycle time and improve the general physicochemical characteristics of the polymers, for instance optical, mechanical and heat resistance properties (Singh et al., 2012; Zhou, Wang, Lu, Zhang, & Zhang, 2007).

Polymers exhibit different crystal growth rate, which can be slow, medium or rapid. Those polymers that have slow crystallization rate like polystyrene and polycarbonate always end-up in an amorphous form. Therefore, to enable better crystalline form, nucleating agents are added to induce nucleation. During the crystallization process, the polymer crystals (lamellae) start from a primary nucleus and later on form complex spherical macrostructures called spherulites. The number and size of the spherulites formed varies depending on whether the nucleus formation is homogenous or heterogeneous. The medium crystallizing polymers such as polyamide-6, polypropylene and polyethylene terephthalate, respond well to heterogeneous and to thermal nucleation in the presence of the nucleating agents, while for rapidly crystallizing polymers such as polyethylene, nucleating agents are used to improve the physicochemical characteristics (Drobny, 2014; Seven et al., 2016; Singh et al., 2012). The spontaneous nucleation (nucleation without addition of a nucleating agent) is triggered by foreign substances such as residues of different catalysts or oxidatively degraded polymer. Spontaneous nucleation differs from auto-nucleation because the latter is induced by incompletely melted down crystallites of the polymer. Nucleating agents increase crystallization rate and temperature of semi-crystalline polymers by changing their morphology. They are added up to 0.5% as pre-dispersed powder or powder mixtures, suspensions or solutions or in the form of a master batch (Jansen, 1987). There are no specific rules at EU level for the risk assessment and use of nucleating agents.

4.2.11 Fillers

Fillers are commonly low cost substances that are used in polymer industry in large quantities to improve the production of the polymers and adjust the volume, mass, price, appearance, color, surface, rheology, shrinking, expansion, conductivity, permeability as well as mechanical characteristics. Fillers, just like the colorants and nucleating agents, are not specifically regulated at EU level. They can be divided into inactive and active fillers. Inactive or extender fillers are used to lower the expenses of the production, while the active or functional/reinforcing fillers are supposed to help a certain property of the polymer to fulfill the specific requirements (Hohenberger, 2001), which are improvement of mechanical properties, increased stiffness and strength, decreased deformability, higher glass transition temperature, better crystallinity, optical properties, hardness and abrasiveness, electrical and magnetic properties, acid solubility, loss on ignition, pH and moisture content.

5. Conclusions

Due to the intrinsic physicochemical characteristics of plastics, they rapidly become ubiquitous in our contemporary way of life. The most mentioned characteristics are the easiness of shaping, chemical inertness, transparency, barrier and heat sealing, specific weight and low cost of production. Low-molecular weight components from plastic packaging are susceptible to migrate and/or react with foodstuff. The most important additives are the stabilizers, antioxidants, light stabilizers, slip agents, antistatic agents as well as optical brighteners.

In this context, countries created regulations and directives for the use of plastic packaging materials for food systems. In USA and EU, the regulations and directives of the materials and articles intended to come into contact with foodstuffs are based on the scientific advises from Food and Drugs Administration (FDA) and European Food Safety Authority (EFSA), respectively.

Several interactions may exist between food and plastic packaging as a result of migrating phenomena of compounds with toxicological (carcinogenic, mutagenic, etc.) concerns on the human health. This is a major concern to our society and environment, and continuous scientific knowledge and a dynamic monitoring and regulation is mandatory. The type of migration (i.e., the mass-transfer of chemicals from plastic food packaging to the food) occurs in four steps and eight types of migration are considered, viz. according to number of migrants; related to food nature; based on diffusion coefficient; contact migration; gas-phase migration; condensation/distillation migration; penetration migration; and set-off migration. In addition, several factors significantly affect the migration of plastic materials to the foodstuffs, for example the contact type, time and temperature, the type and characteristics of the packaging material, the initial concentration and characteristics of the migrating substance, the nature of food, or the ratio packaging surface area versus food volume. Moreover, phenomena that affect the migration of plastic materials to the foodstuffs encompasses molecular energy, free volume theory, diffusion and partition coefficient, swelling, and crystallization and presence of nanomaterials in the polymer package. A full understanding of such factors and phenomena is of utmost important to improve the quality control. Furthermore, the use of nanoparticles, fibers and macro fillers in the polymer matrixes may be a promising nanotechnology when they are capable, for example, to reduce plastic migration.

The migrating components are linked either to the polymers (the monomers and oligomers) or to the additives used in the plastics. Monomers and oligomers related with the polymeric used in the plastic packaging as well as additives incorporated in the polymers during the manufacturing are susceptible to migrate to packed food products. Monomers and oligomers results from the absence of their incorporation (un-reaction) into the polymeric chains. Nevertheless, additives are diverse compounds included during the manufacturing of plastics with important functions to the final product. These compounds carry out different types and levels of toxicity and ecotoxicity and adverse health effects. For that reasons, their residual levels in packaging are strictly restricted and regulated.

The industry of plastics is very innovative, which brings more concerns and requires more attention regarding the resulting novel plastics (in type and composition) for food packaging and potential hazardous effects for health and environment. Plastics are definitely part of our contemporary society and have unique attractive properties that make them essential in our way of life. Application of plastic packaging are one the back of highest reached standards of food quality and safety, thus in this sense, contributing to promote health of the consumers. However, their impact on health and environment requires, above all, that they are used correctly and sparingly, in terms of application and the lifetime of use, and, naturally, not forgetting the demand to guarantee its safe use, including in food packaging.

Acknowledgments

This work is based upon the work from COST Action 18101 SOURDOMICS—*Sourdough biotechnology network towards novel, healthier and sustainable food and bioprocesses* (https://sourdomics.com/; https://www.cost.eu/actions/CA18101/, where the author João Miguel F. Rocha is the Chair and Grant Holder Scientific Representative, Elena Velickova Nikova is the Science Communication Manager, and Mishela Temkov is a member. Cost Action SOURDOMICS is supported by COST (European Cooperation in

Science and Technology) (https://www.cost.eu/). COST is a funding agency for research and innovation networks. Regarding the author J.M.R., this work was also financially supported by LA/P/0045/2020 (Associate Laboratory in the field of Chemical Engineering, ALiCE) and UIDB/00511/2020 - UIDP/00511/2020 (Laboratory for Process Engineering, Environment, Biotechnology and Energy, LEPABE) funded by national funds through FCT/MCTES (PIDDAC).

References

Abou Omar, T. F., Sukhn, C., Fares, S. A., Abiad, M. G., Habib, R. R., & Dhaini, H. R. (2017). Bisphenol A exposure assessment from olive oil consumption. *Environmental Monitoring and Assessment*, *189*(7), 341. https://doi.org/10.1007/s10661-017-6048-6.

Alamri, M. S., Qasem, A. A. A., Mohamed, A. A., Hussain, S., Ibraheem, M. A., Shamlan, G., et al. (2021). Food packaging's materials: A food safety perspective. *Saudi Journal of Biological Sciences*, *28*(8), 4490–4499. https://doi.org/10.1016/j.sjbs.2021.04.047.

Alberto Lopes, J., Tsochatzis, E. D., Karasek, L., Hoekstra, E. J., & Emons, H. (2021). Analysis of PBT and PET cyclic oligomers in extracts of coffee capsules and food simulants by a HPLC-UV/FLD method. *Food Chemistry*, *345*, 128739. https://doi.org/10.1016/j.foodchem.2020.128739.

Alin, J., & Hakkarainen, M. (2011). Microwave heating causes rapid degradation of antioxidants in polypropylene packaging, leading to greatly increased specific migration to food simulants as shown by ESI-MS and GC-MS. *Journal of Agricultural and Food Chemistry*, *59*(10), 5418–5427. https://doi.org/10.1021/jf1048639.

Anderson, W. A. C., & Castle, L. (2003). Benzophenone in cartonboard packaging materials and the factors that influence its migration into food. *Food Additives & Contaminants*, *20*, 607–618. https://doi.org/10.1080/0265203031000109486.

Araújo, H. P., Félix, J. S., Manzoli, J. E., Padula, M., & Monteiro, M. (2008). Effects of γ-irradiation on caprolactam level from multilayer PA-6 films for food packaging: Development and validation of a gas chromatographic method. *Radiation Physics and Chemistry*, 77(7), 913–917. https://doi.org/10.1016/j.radphyschem.2008.03.001.

Arvanitoyannis, I. S., & Bosnea, L. (2004). Migration of substances from food packaging materials to foods. *Critical Reviews in Food Science and Nutrition*, *44*(2), 63–76. https://doi.org/10.1080/10408690490424621.

ATSDR. (2006). *Toxicological profile for Vinyl Chloride*. Retrieved from https://www.atsdr.cdc.gov/toxprofiles/tp20.pdf.

Aznar, M., Vera, P., Canellas, E., Nerín, C., Mercea, P., & Störmer, A. (2011). Composition of the adhesives used in food packaging multilayer materials and migration studies from packaging to food. *Journal of Materials Chemistry*, *21*(12). https://doi.org/10.1039/c0jm04136j.

Barnes, K., Sinclair, R., & Watson, D. H. (2007). *Chemical migration and food contact materials* (1st ed.). Cambridge, England: Woodhead Publishing Limited.

Beardsley, M., & Adams, C. D. (2003). Modelling and control of vinyl chloride in drinking water distribution systems. *Journal of Environmental Engineering*, *129*, 844–851. https://doi.org/10.1061/(ASCE)0733-9372(2003)129:9(844).

Begley, T. H., Biles, J. E., Cunningham, C., & Piringer, O. (2004). Migration of a UV stabilizer from polyethylene terephthalate (PET) into food simulants. *Food Additives & Contaminants*, *21*(10), 1007–1014. https://doi.org/10.1080/02652030400010447.

Begley, T. H., Dennison, J. L., & Hollifield, H. C. (1990). Migration into food of polyethylene terephthalate (PET) cyclic oligomers from PET microwave susceptor packaging. *Food Additives & Contaminants*, 7(6), 797–803. https://doi.org/10.1080/02652039009373941.

Begley, T. H., Gay, M. L., & Hollifield, H. C. (1995). Determination of migrants in and migration from nylon food packaging. *Food Additives & Contaminants*, *12*(5), 671–676. https://doi.org/10.1080/02652039509374355.
Begley, T. H., & Hollifield, H. C. (1990). Evaluation of polyethylene terephthalate cyclic trimer migration from microwave food packaging using temperature-time profiles. *Food Additives & Contaminants*, 7(3), 339–346. https://doi.org/10.1080/02652039009373898.
Beldi, G., Pastorelli, S., Franchini, F., & Simoneau, C. (2012). Time- and temperature-dependent migration studies of Irganox 1076 from plastics into foods and food simulants. *Food Additives and Contaminants: Part A, Chemistry, Analysis, Control, Exposure and Risk Assessment*, *29*(5), 836–845. https://doi.org/10.1080/19440049.2011.649304.
Belz, R., van Gemert, L. J., & Nettenbreijer, A. H. (1977). In R. V. Drinkwatervoorziening (Ed.), *Compilation of Odour Threshold Values in Air and Water* National Institute for Water Supply.
Bomfim, M. V. J., Zamith, H. P. S., & Abrantes, S. M. P. (2011). Migration of ε-caprolactam residues in packaging intended for contact with fatty foods. *Food Control*, *22*(5), 681–684. https://doi.org/10.1016/j.foodcont.2010.09.017.
Brede, C., Fjeldal, P., Skjevrak, I., & Herikstad, H. (2003). Increased migration levels of Bisphenol A from polycarbonate baby bottles after dishwashing, boiling and brushing. *Food Additives & Contaminants*, *20*(7), 684–689. https://doi.org/10.1080/0265203031000119061.
Brede, C., & Skjevrak, I. (2004). Migration of aniline from polyamide cooking utensils into food simulants. *Food Additives & Contaminants*, *21*(11), 1115–1124. https://doi.org/10.1080/02652030400019349.
Brown, M. E., Breder, C. V., & Mcneal, T. P. (1978). Gas-solid chromatographic procedures for determining acrylonitrile monomer in acrylonitrile-containing polymers and food simulating solvents. *Journal - Association of Official Analytical Chemists*, *61*(6). https://doi.org/10.1093/jaoac/61.6.1383.
Canellas, E., Vera, P., & Nerín, C. (2017). Migration assessment and the "threshold of toxicological concern" applied to the safe design of an acrylic adhesive for food-contact laminates. *Food Additives and Contaminants. Part A, Chemistry, Analysis, Control, Exposure and Risk Assessment*, *34*(10), 1721–1729. https://doi.org/10.1080/19440049.2017.1308017.
Castle, L., Kelly, M., & Gilbert, J. (1991). Migration of mineral hydrocarbons into foods. 1. Polystyrene containers for hot and cold beverages. *Food Additives & Contaminants*, *8*(6), 693–699. https://doi.org/10.1080/02652039109374026.
Castle, L., Mayo, A., Crews, C., & Gilbert, J. (1989). Migration of poly(ethylene terephthalate) (PET) oligomers from pet plastics into foods during microwave and conventional cooking and into bottled beverages. *Journal of Food Protection*, *52*(5), 337–342. https://doi.org/10.4315/0362-028X-52.5.337.
Chan, R. K. S., Anselmo, K. J., Reynolds, C. E., & Worman, C. H. (1978). Diffusion of vinyl chloride from PVC packaging material into food simulating solvents. *Polymer Engineering & Science*, *18*(7), 601–606. https://doi.org/10.1002/pen.760180709.
Chaudhry, Q., Scotter, M., Blackburn, J., Ross, B., Boxall, A., Castle, L., et al. (2008). Applications and implications of nanotechnologies for the food sector. *Food Additives & Contaminants*, *25*(3), 241–258. https://doi.org/10.1080/02652030701744538.
Cheng, F. L. (2011). Research of mathematical model for plastic packaging materials migration. *Journal of Beijing Technology and Business University*, *29*, 61–63.
Choi, J. O., Jitsunari, F., Asakawa, F., & Sun Lee, D. (2005). Migration of styrene monomer, dimers and trimers from polystyrene to food simulants. *Food Additives & Contaminants*, *22*(7), 693–699. https://doi.org/10.1080/02652030500160050.

Consumer health protection committee. (2009). *Policy statement concerning coatings intended to come into contact with foodstuffs (Version 3)*. Retrieved from https://rm.coe.int/16805156e5.

Cooper, I., & Tice, P. A. (1995). Migration studies on fatty acid amide slip additives from plastics into food simulants. *Food Additives & Contaminants, 12*(2), 235–244. https://doi.org/10.1080/02652039509374298.

Coste, G., Negrell, C., & Caillol, S. (2020). From gas release to foam synthesis, the second breath of blowing agents. *European Polymer Journal, 140*, 110029. https://doi.org/10.1016/j.eurpolymj. 2020.110029.

Cramer, G. M., & Ford, R. A. (1978). Estimation of toxic hazard. A decision tree approach. *Food and Cosmetics Toxicology, 16*, 255–276. https://doi.org/10.1016/s0015-6264(76)80522-6.

Cruz, R. M. S., Rico, B. P. M., & Vieira, M. C. (2019). Food packaging and migration. In C. M. Galanakis (Ed.), *Food Quality and Shelf Life* (pp. 281–301). Academic Press: London, United Kingdom.

Ćwiek-Ludwicka, K. (2015). Bisphenol A in food contact materials - new scientific opinion from EFSA regarding public health risk. *Roczniki Państwowego Zakładu Higieny, 66*(4), 299–307. PMID: 26656411.

De Cort, S., Godts, F., & Moreau, A. (2017). *Packaging materials: 1. Polyethylene terephthalate (PET) for food packaging applications*. Retrieved from Belgium https://ilsi.eu/wp-content/uploads/sites/3/2018/06/PET-ILSI-Europe-Report-Update-2017_interactif_FIN.pdf.

Dopico-Garcia, M. S., Lopez-Vilarino, J. M., & Gonzalez-Rodriguez, M. V. (2003). Determination of antioxidant migration levels from low-density polyethylene films into food simulants. *Journal of Chromatography A, 1018*(1), 53–62. https://doi.org/10.1016/j.chroma.2003.08.025.

Drobny, J. G. (2014). Additives. In *Handbook of Thermoplastic Elastomers* (2nd ed., pp. 17–32). Norwich, NY: William Andrew Publisher.

EC. (1990). *Commission Directive 90/128/EEC of 23 February 1990 relating to plastic materials and articles intended to come into contact with foodstuffs, 33 C.F.R.*

EC. (2004a). *Commission Directive 2004/1/EC of 6 January 2004 amending Directive 2002/72/EC as regards the suspension of the use of azodicarbonamide as blowing agent.*

EC. (2004b). *Commission Regulation (EC) N° 1935/2004 of the European Parliament and of the Council of 27 October 2004 on materials and articles intended to come into contact with food and repealing Directives 80/590/EEC and 89/109/EEC, 47 C.F.R.*

EC. (2011). *Commission Regulation (EU) N° 10/2011 of 14 January 2011 on plastic materials and articles intended to come into contact with food, 54 C.F.R.*

EC. (2016). *Commission regulation (EU) 2016/1416 amending and correcting Regulation (EU) N° 10/2011 on plastic materials and articles intended to come into contact with food, 5 C.F.R.*

EC. (2020). *Commission Regulation (EU) 2020/1149 of 3 August 2020 amending Annex XVII to Regulation (EC) N° 1907/2006 of the European Parliament and of the Council concerning the registration, evaluation, authorisation and restriction of chemicals (REACH) as regards, 63 C.F.R.*

Zhou, Z., Wang, S. F., Lu, L., Zhang, Y., & Zhang, Y. X. (2007). Isothermal crystallisation kinetics of polypropylene with silane-functionalized multi walled carbon nanotubes. *Journal of Polymer Science Part B: Polymer Physics, 45*(13), 1616–1624. https://doi.org/10.1002/polb.21128.

ECHA. Retrieved from https://echa.europa.eu/about-us.

EFSA. (2010). Scientific opinion on the safety evaluation of the substance bis(2,6-diisopropylphenyl) carbodiimide for use in food contact materials. *EFSA Journal, 8*(12), 1–11. https://doi.org/10.2903/j.efsa.2010.1928.

EFSA. (2012). Scientific opinion on exploring options for providing advice about possible human health risks based on the concept of Threshold of Toxicological Concern (TTC). *EFSA Journal, 10*(7), 1–103. https://doi.org/10.2903/j.efsa.2012.2750.

EFSA. (2015). *Scientific opinion on Bisphenol A*. Retrieved from https://www.efsa.europa.eu/sites/default/files/corporate_publications/files/factsheetbpa150121.pdf.

EFSA. (2021). *Bisphenol A: EFSA draft opinion proposes lowering the tolerable daily intake*. Retrieved from https://www.efsa.europa.eu/en/news/bisphenol-efsa-draft-opinion-proposes-lowering-tolerable-daily-intake.

Ehret-Henry, J., Ducruet, V., Luciani, A., & Feigenbaum, A. (1994). Styrene and ethylbenzene migration from polystyrene into dairy products by dynamic purge-and-trap gas chromatography. *Journal of Food Science, 59*(5), 990–992. https://doi.org/10.1111/j.1365-2621.1994.tb08174.x.

Europe Plastics. (2014). *Risk Assessment of non-listed substances (NLS) and non-intentionally added substances (NIAS) under Article 19 of Commission Regulation (EU) N° 10/2011 of 14 January 2011 on plastic materials and articles intended to come into contact with food*. Retrieved from https://plasticseurope.org/wp-content/uploads/2021/11/risk-assesment-of-non-listed-substances-and-non-assesed-substances.pdf.

Fankhauser-Noti, A., & Grob, K. (2006). Migration of plasticizers from PVC gaskets of lids for glass jars into oily foods: Amount of gasket material in food contact, proportion of plasticizer migrating into food and compliance testing by simulation. *Trends in Food Science and Technology, 17*(3), 105–112. https://doi.org/10.1016/j.tifs.2005.10.013.

FDA. (2010). 21 Food and drugs, Chapter I-Food and drug administration, Part 178 - Indirect food additives: Adjuvants, production aids, and sanitizers, Subpart C - Antioxidants and stabilizers, Sec. 178.2010 Antioxidants and/or stabilizers for polymers. *Code of Federal Regulations, 3*. doi:21CFR178.

FDA. (2022). 21 Food and drugs, Chapter I-Food and drug administration, Part 177 - Indirect food additives: Polymers, Subpart B - Substances for Use as Basic Components of Single and Repeated Use Food Contact Surfaces, Sec. 177.1020 Acrylonitrile/butadiene/styrene co-polymer. *Code of Federal Regulations, 3*. doi:21CFR177.1020.

Félix, J. S., Manzoli, J. E., Padula, M., & Monteiro, M. (2014). Evaluation of different conditions of contact for caprolactam migration from multilayer polyamide films into food simulants. *Packaging Technology and Science, 27*(6), 457–466. https://doi.org/10.1002/pts.2046.

Ferrara, G., Bertoldo, M., Scoponi, M., & Ciardelli, F. (2001). Diffusion coefficient and activation energy of Irganox 1010 in poly(propylene-co-ethylene) copolymers. *Polymer Degradation and Stability, 73*(3), 411–416. https://doi.org/10.1016/s0141-3910(01)00121-5.

Funk, M., Schlettwein, D., & Leist, U. (2016). Migration characteristics under long-term storage and a combination of UV and heat exposure of poly(amide)/poly(ethylene) composite films for food packaging. *Packaging Technology and Science, 29*(6), 289–302. https://doi.org/10.1002/pts.2204.

Gao, Y., Gu, Y., & Wei, Y. (2011). Determination of polymer additives-antioxidants and ultraviolet (UV) absorbers by high-performance liquid chromatography coupled with UV photodiode array detection in food simulants. *Journal of Agricultural and Food Chemistry, 59*(24), 12982–12989. https://doi.org/10.1021/jf203257b.

Garde, J. A., Catala, R., Gavara, R., & Hernandez, R. J. (2001). Characterizing the migration of antioxidants from polypropylene into fatty food simulants. *Food Additives & Contaminants, 18*(8), 750–762. https://doi.org/10.1080/02652030116713.

Geens, T., Apelbaum, T. Z., Goeyens, L., Neels, H., & Covaci, A. (2010). Intake of Bisphenol A from canned beverages and foods on the Belgian market. *Food Additives and Contaminants: Part A, Chemistry, Analysis, Control, Exposure and Risk Assessment, 27*(11), 1627–1637. https://doi.org/10.1080/19440049.2010.508183.

Gilbert, J., & Startin, J. R. (1982). Determination of acrylonitrile monomer in food packaging materials and in foods. *Food Chemistry*, *9*, 243–252. https://doi.org/10.1016/0308-8146(82)90075-9.

Gilbert, S. G., Miltz, J., & Giacin, J. R. (1980). Transport considerations of potential migrants from food packaging materials. *Journal of Food Processing and Preservation*, *4*, 27–49. https://doi.org/10.1111/j.1745-4549.1980.tb00594.x.

Goodson, A., Robin, H., Summerfield, W., & Cooper, I. (2004). Migration of Bisphenol A from can coatings – effects of damage, storage conditions and heating. *Food Additives & Contaminants*, *21*(10), 1015–1026. https://doi.org/10.1080/02652030400011387.

Grassi, M., Colombo, I., & Lapasin, R. (2001). Experimental determination of the theophylline diffusion coefficient in swollen sodium-alginate membranes. *Journal of Controlled Release*, *76*, 93–105. https://doi.org/10.1016/S0168-3659(01)00424-2.

Guazzotti, V., Ebert, A., Gruner, A., & Welle, F. (2021). Migration from acrylonitrile butadiene styrene (ABS) polymer: Swelling effect of food simulants compared to real foods. *Journal of Consumer Protection and Food Safety*, *16*(1), 19–33. https://doi.org/10.1007/s00003-020-01308-8.

Hahladakis, J. N., Velis, C. A., Weber, R., Iacovidou, E., & Purnell, P. (2018). An overview of chemical additives present in plastics: Migration, release, fate and environmental impact during their use, disposal and recycling. *Journal of Hazardous Materials*, *344*, 179–199. https://doi.org/10.1016/j.jhazmat.2017.10.014.

Hansen, E., Nilsson, N. H., Lithner, D., & Lassen, C. (2013). *Hazardous substances in plastic materials*. Retrieved from Oslo, Danmark https://www.byggemiljo.no/wp-content/uploads/2014/10/72_ta3017.pdf.

Heimrich, M., Nickl, H., Bönsch, M., & Simat, T. J. (2015). Migration of cyclic monomer and oligomers from polyamide 6 and 66 food contact materials into food and food simulants: Direct food contact. *Packaging Technology and Science*, *28*(2), 123–139. https://doi.org/10.1002/pts.2094.

Henkel. (2020). *Legislation on GLYMO will shape packaging in 2020 [Press release]*.

Hohenberger. (2001). Fillers and Reinforcements/ Coupling Agents. In H. Zweifel, R. D. Maier, & M. Schiller (Eds.), *Plastics additive handbook* ((5th ed.), pp. 919–966). Munich: Hanser Publishers.

Hoppe, M., de Voogt, P., & Franz, R. (2016). Identification and quantification of oligomers as potential migrants in plastics food contact materials with a focus in polycondensates – A review. *Trends in Food Science and Technology*, *50*, 118–130. https://doi.org/10.1016/j.tifs.2016.01.018.

Jansen, J. (1987). Nucleating agents for partially crystalline polymers. In R. Gächter, & H. Müller (Eds.), *Plastics Additives Handbook - Stabilizers, Processing Aids, Plasticizers, Fillers, Reinforcements, Colorants for Thermoplastics* ((2nd ed.), pp. 863–874). Munich: Hanser Publishers.

Jeon, D. H., Park, G. Y., Kwak, I. S., Lee, K. H., & Park, H. J. (2007). Antioxidants and their migration into food simulants on irradiated LLDPE film. *LWT- Food Science and Technology*, *40*(1), 151–156. https://doi.org/10.1016/j.lwt.2005.05.017.

JRC. (2004). *Acrylonitrile: Summary Risk Assessment Report*. Retrieved from Dublin, Ireland: https://echa.europa.eu/documents/10162/b05f71cc-e1b9-4d15-8f82-e7b3a426eec5.

Kapp, R. W. (2014). Isocyanates. In P. Wexler (Ed.), *Encyclopaedia of Toxicology* (3rd ed., pp. 1112–1131). London, United Kingdom: Academic Press.

Kato, L. S., & Conte-Junior, C. A. (2021). Safety of plastic food packaging: The challenges about non-intentionally added substances (NIAS) discovery, identification and risk assessment. *Polymers*, *13*(13), 2077. https://doi.org/10.3390/polym13132077.

Kawamura, Y., Ogawa, Y., & Mutsuga, M. (2017). Migration of nonylphenol and plasticizers from polyvinyl chloride stretch film into food simulants, rapeseed oil, and foods. *Food Science & Nutrition*, *5*(3), 390–398. https://doi.org/10.1002/fsn3.404.

Khaksar, M. R., & Ghazi-Khansari, M. (2009). Determination of migration monomer styrene from GPPS (general purpose polystyrene) and HIPS (high impact polystyrene) cups to hot drinks. *Toxicology Mechanisms and Methods*, *19*(3), 257–261. https://doi.org/10.1080/15376510802510299.

Kim, Y. J., Osako, M., & Sakai, S. (2006). Leaching characteristics of polybrominated diphenyl ethers (PBDEs) from flame-retardant plastics. *Chemosphere*, *65*(3), 506–513. https://doi.org/10.1016/j.chemosphere.2006.01.019.

Kubwabo, C., Kosarac, I., Stewart, B., Gauthier, B. R., Lalonde, K., & Lalonde, P. J. (2009). Migration of Bisphenol A from plastic baby bottles, baby bottle liners and reusable polycarbonate drinking bottles. *Food Additives and Contaminants: Part A, Chemistry, Analysis, Control, Exposure and Risk Assessment*, *26*(6), 928–937. https://doi.org/10.1080/02652030802706725.

Lau, O.-W., & Wong, S.-K. (2000). Contamination in food from packaging material. *Journal of Chromatography A*, *882*, 255–270. https://doi.org/10.1016/S0021-9673(00)00356-3.

Li, B., Wang, Z. W., Lin, Q. B., & Hu, C. Y. (2016). Study of the migration of stabilizer and plasticizer from polyethylene terephthalate into food simulants. *Journal of Chromatography Science*, *54*(6), 939–951. https://doi.org/10.1093/chromsci/bmw025.

Lickly, T. D., Lehr, K. M., & Welsh, G. C. (1995). Migration of styrene from polystyrene foam food-contact articles. *Food and Chemical Toxicology*, *33*(6), 475–481. https://doi.org/10.1016/0278-6915(95)00009-Q.

Lickly, T. D., Markham, D. A., & Rainey, M. L. (1991). The migration of acrylonitrile from acrylonitrile-butadiene-styrene polymers into food-simulating liquids. *Food and Chemical Toxicology*, *29*(1), 25–29. https://doi.org/10.1016/0278-6915(91)90059-G.

Lopez-Cervantes, J., Sanchez-Machado, D. I., Simal-Lozano, J., & Paseiro-Losada, P. (2003). Migration of ethylene terephthalate oligomers from roasting bags into olive oil. *Chromatographia*, *58*(5/6), 321–326. https://doi.org/10.1365/s10337-003-0034-6.

Lorite, G. S., Rocha, J. M., Miilumäki, N., Saavalainen, P., Selkälä, T., Morales-Cid, G., et al. (2017). Evaluation of physicochemical/microbial properties and life cycle assessment (LCA) of PLA-based nanocomposite active packaging. *LWT- Food Science and Technology*, *75*, 305–315. https://doi.org/10.1016/j.lwt.2016.09.004.

Maia, J., Rodríguez-Bernaldo de Quirós, A., Sendón, R., Cruz, J. M., Seiler, A., Franz, R., et al. (2016). Determination of key diffusion and partition parameters and their use in migration modelling of benzophenone from low-density polyethylene (LDPE) into different foodstuffs. *Food Additives and Contaminants: Part A, Chemistry, Analysis, Control, Exposure and Risk Assessment*, *33*(4), 715–724. https://doi.org/10.1080/19440049.2016.1156165.

Marangoni, L., Fávaro Perez, M.Â., Torres, C. D., Cristianini, M., Massaharu Kiyataka, P. H., Albino, A. C., et al. (2020). Effect of high-pressure processing on the migration of ε-caprolactam from multilayer polyamide packaging in contact with food simulants. *Food Packaging and Shelf Life*, *26*. https://doi.org/10.1016/j.fpsl.2020.100576.

Monteiro, M., Nerin, C., & Reyes, F. G. R. (1999). Migration of Tinuvin P, a UV stabilizer, from PET bottles into fatty-food simulants. *Packaging Technology and Science*, *12*, 241–248. https://doi.org/10.1002/(SICI)1099-1522(199909/10)12:5<241::AID-PTS478>3.0.CO;2-V.

Muncke, J., Backhaus, T., Geueke, B., Maffini, M. V., Martin, O. V., Myers, J. P., et al. (2017). Scientific challenges in the risk assessment of food contact materials. *Environmental Health Perspectives*, *125*(9), 095001. https://doi.org/10.1289/EHP644.

Munguia-Lopez, E. M., & Soto-Valdez, H. (2001). Effect of heat processing and storage time on migration of bisphenol A (BPA) and bisphenol A-diglycidyl ether (BADGE) to aqueous food simulant from Mexican can coatings. *Journal of Agricultural and Food Chemistry*, *49*, 3666–3667. https://doi.org/10.1021/jf0009044.

Mutsuga, M., Yamaguchi, M., & Kawamura, Y. (2014). Quantification of isocyanates and amines in polyurethane foams and coated products by liquid chromatography-tandem mass spectrometry. *Food Science & Nutrition*, *2*(2), 156–163. https://doi.org/10.1002/fsn3.88.

Nam, S. H., Seo, Y. M., & Kim, M. G. (2010). Bisphenol A migration from polycarbonate baby bottle with repeated use. *Chemosphere*, *79*(9), 949–952. https://doi.org/10.1016/j.chemosphere.2010.02.049.

Nasser, A. L., Lopes, L. M., Eberlin, M. N., & Monteiro, M. (2005). Identification of oligomers in polyethyleneterephthalate bottles for mineral water and fruit juice. Development and validation of a high-performance liquid chromatographic method for the determination of first series cyclic trimer. *Journal of Chromatography A*, *1097*(1-2), 130–137. https://doi.org/10.1016/j.chroma.2005.08.023.

Nerin, C., Canellas, E., & Vera, P. (2016). Plasticizer migration into foods. In G. Smithers (Ed.), *Reference module in food science* (pp. 1–13). Amsterdam: Elsevier.

Ohno, H., & Kawamura, Y. (2010). Analysis of acrylonitrile, 1,3-butadiene, and related compounds in acrylonitrile-butadiene-styrene copolymers for kitchen utensils. *Journal of AOAC International*, *93*(6), 1965–1971. https://doi.org/10.1093/jaoac/93.6.1965.

Paraskevopoulou, D., Achilias, D. S., & Paraskevopoulou, A. (2012). Migration of styrene from plastic packaging based on polystyrene into food simulants. *Polymer International*, *61*(1), 141–148. https://doi.org/10.1002/pi.3161.

Pearson, R. B. (1982). PVC as a food packing material. *Food Chemistry*, *8*, 85–96. https://doi.org/10.1016/0308-8146(82)90004-8.

Poustková, I., Poustka, J., Bablčka, L., & Dobláš, J. (2007). Acrylonitrile in food contact materials – two different legislative approaches: Comparison of direct determination with indirect evaluation using migration into food simulants. *Czech Journal of Food Science*, *25*(5), 265–271. https://doi.org/10.17221/678-CJFS.

Raj, B., & Matche, R. S. (2012). Safety and regulatory aspects of plastics as food packaging materials. In K. L. Yam, & D. S. Lee (Eds.), *Emerging Food Packaging Technologies* (pp. 335–357). Cambridge: Woodhead Publishing Series in Food Science, Technology and Nutrition: N°230.

Ranken, P. (2001). Flame Retardants. In I. H. Zweifel, R. D. Maier, & M. Schiller (Eds.), *Plastics additive handbook* ((5th ed.), pp. 701–717). Munich: Hanser Publishers.

Ritter, M., Schlettwein, D., & Leist, U. (2020). Specific migration of caprolactam and infrared characteristics of a polyamide/polyethylene composite film for food packaging under conditions of long-term storage before use. *Packaging Technology and Science*, *33*(12), 501–514. https://doi.org/10.1002/pts.2531.

Schaefer, A., Kuchler, T., Simat, T. J., & Steinhart, H. (2003). Migration of lubricants from food packagings. Screening for lipid classes and quantitative estimation using normal-phase liquid chromatographic separation with evaporative light scattering detection. *Journal of Chromatography A*, *1017*(1-2), 107–116. https://doi.org/10.1016/j.chroma.2003.08.004.

Seven, K. M., Cogen, J. M., & Gilchrist, J. F. (2016). Nucleating agents for high-density polyethylene. A review. *Polymer Engineering & Science*, *541–554*. https://doi.org/10.1002/pen.24278.

Shaw, S. D., Blum, A., Weber, R., Kannan, K., Rich, D., Lucas, D., et al. (2010). Halogenated flame-retardants: Do the fire safety benefits justify the risks? *Reviews on Environmental Health*, *25*(4). https://doi.org/10.1515/reveh.2010.25.4.261.

Shin, C., Kim, D. G., Kim, J. H., Kim, J. H., Song, M. K., & Oh, K. S. (2021). Migration of substances from food contact plastic materials into foodstuff and their implications for human exposure. *Food and Chemical Toxicology*, *154*, 112373. https://doi.org/10.1016/j.fct.2021.112373.

Silva, A. S., Freire, J. M. C., García, R. S., Franz, R., & Losada, P. P. (2007). Time–temperature study of the kinetics of migration of DPBD from plastics into chocolate, chocolate spread and margarine. *Food Research International*, *40*(6), 679–686. https://doi.org/10.1016/j.foodres.2006.11.012.

Singh, P., Saengerlaub, S., Abas Wani, A., & Langowski, H. C. (2012). Role of plastics additives for food packaging. *Pigment & Resin Technology*, *41*(6), 368–379. https://doi.org/10.1108/03699421211274306.

Song, H. J., Chang, Y., Lyu, J. S., Yon, M. Y., Lee, H. S., Park, S. J., et al. (2018). Migration study of caprolactam from polyamide 6 sheets into food simulants. *Food Science and Biotechnology*, *27*(6), 1685–1689. https://doi.org/10.1007/s10068-018-0403-4.

Stachak, P., Lukaszewska, I., Hebda, E., & Pielichowski, K. (2021). Recent advances in fabrication of non-isocyanate polyurethane-based composite materials. *Materials (Basel)*, *14*(13). https://doi.org/10.3390/ma14133497.

Statista. (2021). *Annual production of plastics worldwide from 1950 to 2020*. Retrieved from https://www.statista.com/statistics/282732/global-production-of-plastics-since-1950/.

Tawfik, M. S., & Huyghebaert, A. (1998). Polystyrene cups and containers: Styrene migration. *Food Additives & Contaminants*, *15*(5), 592–599. https://doi.org/10.1080/02652039809374686.

Tehrany, E. A., & Desobry, S. (2004). Partition coefficients in food/packaging systems: A review. *Food Additives & Contaminants*, *21*(12), 1186–1202. https://doi.org/10.1080/02652030400019380.

Tsochatzis, E. D., Alberto Lopes, J., Kappenstein, O., Tietz, T., & Hoekstra, E. J. (2020). Quantification of PET cyclic and linear oligomers in teabags by a validated LC-MS method - In silico toxicity assessment and consumer's exposure. *Food Chemistry*, *317*, 126427. https://doi.org/10.1016/j.foodchem.2020.126427.

Tsumura, Y., Ishimitsu, S., Kaihara, A., Yoshii, K., & Tonogai, Y. (2002). Phthalates, adipates, citrate and some of the other plasticizers detected in Japanese retail foods. A survey. *Journal of Health Science*, *48*(6), 493–502. https://doi.org/10.1248/jhs.48.493.

Ubeda, S., Aznar, M., Vera, P., Nerin, C., Henriquez, L., Taborda, L., et al. (2017). Overall and specific migration from multilayer high barrier food contact materials - kinetic study of cyclic polyester oligomers migration. *Food Additives and Contaminants: Part A, Chemistry, Analysis, Control, Exposure and Risk Assessment*, *34*(10), 1784–1794. https://doi.org/10.1080/19440049.2017.1346390.

US.EPA. (1992, updated in January 2000). Caprolactam. Retrieved from https://www.epa.gov/sites/default/files/2016-09/documents/caprolactam.pdf.

Velickova, E., Kuzmanova, S., & Winkelhausen, E. (2011). A lag-time model for substrate and product diffusion through hydroxyethylcellulose gels used for immobilization of yeast cells. *Macedonian Journal of Chemistry and Chemical Engineering*, *30*(1). https://doi.org/10.20450/mjcce.2011.73.

Vidal-Abarca Garrido, C., Kaps, R., Oyeshola, K., Oliver, W., Riera Maria, R., Hidalgo, C., et al. (2018). *Revision of the European ecolabel criteria for lubricants*. Retrieved from https://ec.europa.eu/jrc.

Vinson, C. G., & Banzer, J. D. (1981). Effect of a variable diffusivity on the migration of vinyl chloride from poly(vinyl chloride) pipe. *Journal of Vinyl Technology*, *3*(2), 143–147. https://doi.org/10.1002/vnl.730030212.

Weber, R., & Kuch, B. (2003). Relevance of BFRs and thermal conditions on the formation pathways of brominated and brominated–chlorinated dibenzodioxins and dibenzofurans. *Environment International*, *29*(6), 699–710. https://doi.org/10.1016/s0160-4120(03)00118-1.

WHO. (2004). *Vinyl chloride in drinking – water*. Retrieved from https://www.who.int/water_sanitation_health/dwq/chemicals/vinylchloride.pdf.

Yan, J. W., Hu, C., Tong, L. H., Lei, Z. X., & Lin, Q.-B. (2020). Migration test and safety assessment of polyurethane adhesives used for food-contact laminated films. *Food Packaging and Shelf Life, 23*, 100449. https://doi.org/10.1016/j.fpsl.2019.100449.

Yao, Z., Hong-fei, L., Li-jia, A., & Shi-chun, J. (2013). Progress in studies on structure and relaxation behavior of the amorphous phases in crystalline polymers. *Acta Polymerica Sinica, 013*(4), 462–472. https://doi.org/10.3724/sp.J.1105.2013.12362.

Zabihzadeh Khajavi, M., Mohammadi, R., Ahmadi, S., Farhoodi, M., & Yousefi, M. (2019). Strategies for controlling release of plastic compounds into foodstuffs based on application of nanoparticles and its potential health issues. *Trends in Food Science and Technology, 90*, 1–12. https://doi.org/10.1016/j.tifs.2019.05.009.

Zheng, A., Xu, X., Xiao, H., Li, N., Guan, Y., & Li, S. (2012). Antistatic modification of polypropylene by incorporating Tween/modified Tween. *Applied Surface Science, 258*(22), 8861–8866. https://doi.org/10.1016/j.apsusc.2012.05.105.

Microplastics (MPs) in marine food chains: Is it a food safety issue?

B.K.K.K. Jinadasa[a,b,*], Saif Uddin[c], and Scott W. Fowler[d,e]

[a]Analytical Chemistry Laboratory, National Aquatic Resources Research & Development Agency (NARA), Colombo, Sri Lanka
[b]Department of Food Science and Technology (DFST), Faculty of Livestock, Fisheries & Nutrition (FLFN), Wayamba University of Sri Lanka, Makandura, Gonawila, Sri Lanka
[c]Environment and Life Sciences Research Center, Kuwait Institute for Scientific Research (KISR), Kuwait City, Kuwait
[d]School of Marine and Atmospheric Sciences, Stony Brook University, Stony Brook, NY, United States
[e]Institute Bobby, Cap d'Ail, France
*Corresponding author: e-mail address: jinadasa76@gmail.com

Contents

Abstract

The enormous usage of plastic over the last seven decades has resulted in a massive quantity of plastic waste, much of it eventually breaking down into microplastic (MP) and nano plastic (NP). The MPs and NPs are regarded as emerging pollutants of serious concern. Both MPs and NPs can have a primary or secondary origin. Their ubiquitous presence and ability to sorb, desorb, and leach chemicals have raised concern over their presence in the aquatic environment and, particularly, the marine food chain. MPs and NPs are also considered vectors for pollutant transfer along with the marine food chain, and people who consume seafood have began significant concerns about the toxicity of seafood. The exact consequences and risk of MP exposure to marine foods are largely unknown and should be a priority research area. Although several studies have documented an effective clearance mechanism by defecation, significant aspect has been

Advances in Food and Nutrition Research, Volume 103
ISSN 1043-4526
https://doi.org/10.1016/bs.afnr.2022.07.005

less emphasized for MPs and NPs and their capability to translocate in organs and clearance is not well established. The technological limitations to study these ultra-fine MPs are another challenge to be addressed. Therefore, this chapter discusses the recent findings of MPs in different marine food chains, their translocation and accumulations potential, MPs as a critical vector for pollutant transfer, toxicology impact, cycling in the marine environment and seafood safety. Besides, the concerns and challenges that are overshadowed by findings for the significance of MPs were covered.

1. Introduction

Anthropocene can be considered the plastic age due to the enormous quantities and varied types of plastic produced; to have a quantitative perspective, 367 million tons of plastics were produced in 2020. The omnipresence of microplastics (MPs) in the aquatic environment and its persistent nature have attracted significant scientific interest, with thousands of scientific articles on MPs for aquatic environments (Oliveira & Almeida, 2019). MPs have become an immense environmental and ecological problem in aquatic environments (Horton, Walton, Spurgeon, Lahive, & Svendsen, 2017) since they have been found in all environmental matrixes, including water, suspended particulate in the water column, sediments, biota at different trophic levels, aerosols, and snow.

The primary input of MP into the aquatic environment is associated with an enormous quantity of plastic making its way from spuriously managed plastic waste disposal (Alimi, Farner Budarz, Hernandez, & Tufenkji, 2018; Eerkes-Medrano, Thompson, & Aldridge, 2015; Habib, Thiemann, & Kendi, 2020; Naji, Azadkhah, Farahani, Uddin, & Khan, 2021; Uddin, Fowler, & Behbehani, 2020; Uddin, Fowler, Uddin, Behbehani, & Naji, 2021). The hydrophobic contaminants tend to adsorb on the MP due to their chemical characteristics, significantly higher surface-to-volume ratio, and long residence time in the water column (Ferreira, Venâncio, Lopes, & Oliveira, 2019; Pan, Liu, Sun, Sun, & Lin, 2019), which has attracted much attention of scientists (Alimba & Faggio, 2019; Allen et al., 2019; Andrady, 2011, 2017; Auta, Emenike, & Fauziah, 2017; Barboza et al., 2019; Brennecke, Duarte, Paiva, Caçador, & Canning-Clode, 2016; de Sa, Oliveira, Ribeiro, Rocha, & Futter, 2018; Dris, Gasperi, Saad, Mirande, & Tassin, 2016; Enyoh, Verla, Verla, Ibe, & Amaobi, 2019; Fahrenfeld, Arbuckle-Keil, Naderi Beni, & Bartelt-Hunt, 2019; Hidalgo-Ruz, Gutow, Thompson, & Thiel,

2012; Lusher, 2015; Maes, Perry, Alliji, Clarke, & Birchenough, 2019; Naji et al., 2021; Rodrigues, Duarte, Santos-Echeandía, & Rocha-Santos, 2019; Schymanski, Goldbeck, Humpf, & Furst, 2018; Uddin et al., 2021; Uddin, Fowler, & Behbehani, 2020; Uddin, Fowler, & Saeed, 2020; Uddin, Fowler, Saeed, Naji, & Al-Jandal, 2020; Vroom, Koelmans, Besseling, & Halsband, 2017; Wagner et al., 2018; Zhang et al., 2020). Many aquatic organisms misidentify the MPs for prey, and sustained ingestion of these MPs can result in malnutrition and the possible transfer of toxic chemicals (Prinz & Korez, 2020), often reported to result in inflammations, impaired reproductive capacities, and other physiological functions (Abbasi et al., 2018; Barboza et al., 2020; Besseling, Quik, Sun, & Koelmans, 2017; Browne, Dissanayake, Galloway, Lowe, & Thompson, 2008; Mohsen et al., 2019; Vroom et al., 2017; Zakeri, Naji, Akbarzadeh, & Uddin, 2020; Zhang et al., 2019). In addition, there are concerns due to the toxicity caused by the polymer used for manufacturing plastic products (Devriese et al., 2015; Li et al., 2019; Lusher, Welden, Sobral, & Cole, 2017; Rochman, Hoh, Kurobe, & Teh, 2013; Sussarellu et al., 2016).

MPs in the environment occur either as 'primary' or 'secondary' particles. The primary MPs are manufactured in the size range from 0.001 to 5 mm to be incorporated into other products. In contrast, the secondary MPs are degraded and break down from larger plastic debris to a size <5 mm. Studies have investigated the pathways of MP reaching the aquatic environments, and wastewater input is overwhelmingly large (Naji et al., 2021; Uddin, Fowler, & Behbehani, 2020). However, a recent study has highlighted the role of precipitation and Eolian transport in transporting both the primary and secondary MPs (Freeman et al., 2020; Habibi, Uddin, Fowler, & Behbehani, 2022; Uddin et al., 2022).

The MPs occur as fibers, films, fragments, and spheres, in sizes varying from 1 to 5000 μm. The smaller microplastics usually outnumber large ones. The particles 10–20 μm in size can be identified using micro-FTIR and micro-ATR-FTIR, and the smaller MPs up to 1 μm can be identified using micro-Raman spectroscopy. There is a significant variation in chemical and physical characteristics of MP in the environment, which is suggested by Rochman et al. (2019) that MP is a suite of contaminants and not a single contaminant (Rochman et al., 2019). Thus, this chapter focuses on the recent findings of MPs in different marine food chains, their translocation and accumulations potential, MPs as a critical vector for pollutant transfer, toxicology impact, cycling in the marine environment and seafood safety. Additionally, the concerns and challenges that are overshadowed by findings for the significance of MP were covered.

2. Current status of microplastic research

There is presently an excess of 6500 scientific papers published on MPs, including only a few on NPs (Uddin, Fowler, Habibi, & Behbehani, 2022), which are usually reporting the occurrence of MPs in different aquatic systems. A large volume of literature on MPs in aquatic sediments and soils, uptake by marine organisms, and transfer in the marine food chain are available. In fact, these studies have helped to establish the scale and severity of the problem. A significant issue in MP research is developing harmonized protocols for sampling, sample preparation, identification, and reporting units (Uddin, Fowler, Saeed, Naji, & Al-Jandal, 2020). Several attempts have been made to standardize methodologies (Bessa et al., 2019; Frias et al., 2018, 2019; Masura, Baker, Foster, & Arthur, 2015; OSPAR, 2010; Pfeiffer & Fischer, 2020; Shim, Hong, & Eo, 2017; Uddin, Fowler, Saeed, Naji, & Al-Jandal, 2020; Van Cauwenberghe, Devriese, Galgani, Robbens, & Janssen, 2015; Vermeiren, Muñoz, & Ikejima, 2020; Zhang et al., 2019), but still, there are several unresolved issues.

Despite a consensus on the 1 μm lower cutoff for MP (GESAMP, 2015; Gigault et al., 2018; L. Van Cauwenberghe et al., 2015), standard guidelines like the EU Marine Strategy Framework Directive (MSFD) recommend a 333 μm plankton net for sampling and a mouth aperture of 60 cm for the sampling gear (Gago, Galgani, Maes, & Thompson, 2016). Such recommendations and protocols raise serious concern since 35–90% of MPs in the marine environment are smaller than 333 μm (Eriksen et al., 2014; Uddin, Fowler, Saeed, Naji, & Al-Jandal, 2020; Van Cauwenberghe et al., 2015). Employing this gear will lead to considerable underestimation of MPs in seawater. Most of the published literature has reported using 150–333 μm plankton net for MP sampling in the aquatic environment possibly leaving behind a considerable portion of MPs (Green et al., 2018), raising questions about the accuracy of these assessments.

The MPs are likely to be more dominant in the food chain as they are identical to prey size and are often misidentified as prey. Researchers have started discussing nano plastics (NP); this raises questions about the available technology to characterize nano-sized polymers in field samples (Uddin et al., 2022) as the best available tool is micro-Raman spectrometry, which cannot go below 1 μm size. It is worth noting that not a single study reported NPs from environmental samples, thus the suitability of experimental concentrations used in many studies may be quite different from real environmental concentrations. In several studies, identification of MP is limited to microscopy. Microscopic identification is reported to have a variance between 33 and 70% (Ballent, Corcoran, Madden, Helm, & Longstaffe, 2016; Burns & Boxall, 2018; Dekiff, Remy, Klasmeier, & Fries, 2014; Hidalgo-Ruz et al., 2012; Imhof, Rusek, Thiel, Wolinska, & Laforsch, 2017; Lenz, Enders, Stedmon, Mackenzie, & Nielsen, 2015; Lusher, 2015; Song et al., 2015). Raising concerns, how justified are the risk assessments in the absence of realistic environmental concentrations for MPs and where the environmental concentration is not known for NPs.

3. Distribution and assimilation of MPs in marine food

Significant concerns about MP contamination may compromise food safety and human health (Barboza, Dick Vethaak, Lavorante, Lundebye, & Guilhermino, 2018). Their presence in the aquatic environment may also affect the ecosystem's functioning and environmental impacts. MPs enters the environment mainly from the wastewater treatment effluent, washing synthetic cloths (textile fibers), and personal care products (e.g., cosmetics and toothpaste) (Sharma & Chatterjee, 2017). In addition, some industrial applications, organic fertilizers, leachates from landfills, road dust, transport infrastructures, and some natural fibers (plant and animal origin) have also contributed to the MPs (João, Paço, Santos, Duarte, & Rocha-Santos, 2018). During the covid-19 pandemic, the extensive use of facemask and personal protective equipments (PPEs) has contributed significantly to MP pollution (Ray, Lee, Huyen, Chen, & Kwon, 2022; Shen et al., 2021).

Several secondary MPs result in marine contamination through physical and biological fragmentation. Textiles, single-use plastic bags, packaging wastes, fishing nets, and ropes contribute to secondary MP (Coyle, Hardiman, & Driscoll, 2020; Lehtiniemi et al., 2018). The main source of MPs in the aquatic environment is Wastewater Treatment Plant (WWTP) outfalls. This input from a point source is spread over large spatial domains by coastal currents and waves. The aeolian input and river outflow also disperses these land-based MPs in the aquatic environment. MP is persistent in the marine environment, and due to its small size, it is bioavailable to the coral, zooplankton, fish, and other marine animals. It can potentially bioaccumulate in organisms at higher trophic levels (Sharma & Chatterjee, 2017). The size of MPs was affected more than the shape of ingested MPs in small predators by confusing them with natural prey items, especially in nonselective feeders (Lehtiniemi et al., 2018). The MP is difficult to degrade in the animal gut and 90% of orally ingested MP is excreted through animal feces (Alprol, Gaballah, & Hassaan, 2021). However, Browne et al. (2008) observed that the ingested MP could accumulate and become translocated in different tissues of the mussel (*Mytilus edulis*). Moreover, many studies have provided strong evidence for the effects of MP ingestion in different organisms e.g., zooplankton (Cole, Lindeque, Fileman, Halsband, & Galloway, 2015; Van Colen, Vanhove, Diem, & Moens, 2020), echinoderms (Kaposi, Mos, Kelaher, & Dworjanyn, 2014; Richardson, Burritt, Allan, & Lamare, 2021), jellyfish (Rapp et al., 2021), mussels (Patterson, Jeyasanta, Laju, & Edward, 2021; von Moos, Burkhardt-Holm, & Köhler, 2012), oyster (Sussarellu et al., 2016), crab (Watts et al., 2014) and fish (Batel, Linti, Scherer, Erdinger, & Braunbeck, 2016).

The hydrophobic organic and inorganic contaminants in the aquatic environment get sorbed on MPs, sometimes up to 100 times more than in the sediments. This preferential sorption capacity is due to the non-polar surface MPs possesses (Rochman, Hentschel, & Teh, 2014) besides the high surface-to-volume ratio. The presence of toxic pollutants on MP are well known. Hence, MPs serves as vectors for the transfer of toxic metals (Hildebrandt, Nack, Zimmermann, & Pröfrock, 2021; Liu et al., 2022), polybrominated diphenyl ethers (PDEs) (Singla, Díaz, Broto-Puig, & Borrós, 2020; Sun, Liu, Zhang, Leungb, & Zeng, 2021), polyaromatic hydrocarbons (PAHs) (Akhbarizadeh et al., 2021; Peng et al., 2022), bisphenols, phthalates and pesticides (Li, Wang, Li, Deng, & Zhang, 2021; Montoto-Martínez et al., 2021), among others. These chemical compounds can be released in the gut due to the acidic condition while numerous models and studies have investigated this transport behavior (Carbery, O'Connor, & Palanisami, 2018).

Moreover, there was a wide range of sub-lethal effects i.e., gastrointestinal track injuries, malnutrition, growth, and reproduction issues, and those effects are more difficult to assess (Panti et al., 2019). A large number of studies are providing information on the presence of MP abundance in some marine food chain taxa, and a few recent publications are presented in Table 1. MPs detected in those taxa shown are regularly characterized by their shapes, sizes, and colors; however, polymer concentration and composition/concentrations are not fully instrumentally validated.

Table 1 Abundance of MPs in various marine cohorts around the world (a few selected literature examples).

Cohort	Species/type	Location	Dominant MP type	Common MP type (size)	MP concentration (particles/individual, if not mentioned)	References
Zooplankton	Seven zooplankton groups (Calanoida, cyclopoida, harpacticoida, mysids, decapoda, Cladocera & Chaetognatha)	Malaysia	PA, PE & PP	Fiber ($361.7 \pm 226.8\,\mu m$) and Fragments ($96.8 \pm 28.1\,\mu m$)	0.01 ± 0.002–0.2 ± 0.14	Taha, Md Amin, Anuar, Nasser, & Sohaimi (2021)
	Copepods, chaetognaths, decapods, and fish larvae	Eastern Arabian Sea	PE, PP, PS, & PET	_	0.03 ± 0.01–0.57 ± 0.18	Rashid et al. (2021)
	Copepods (*Acartia tonsa*, *Paracalanus crassirostris* and *Centropages typicus*)	Hudson-Raritan estuary (North Atlantic Ocean)	PE & PP	Fragments and beads	0.30 to 0.82	Sipps et al. (2022)
	Fish larvae, cyclopoid, shrimps, polychaete, calanoid, and chaetognath	Malaysia	PA	Fragments and fiber	0.003–0.14	Payton, Beckingham, & Dustan (2020)
	Copepods (*Calanus euxinus* and Acartia (*Acartiura clausi*)	Black sea	PET, PA, PP, PAN, & PE	Fiber, films, and fragments (0.100 ± 0.153 & $0.062 \pm 0.056\,mm$)	0.024 ± 0.020 & 0.008 ± 0.006	Aytan, Esensoy, & Senturk (2022)

Continued

Table 1 Abundance of MPs in various marine cohorts around the world (a few selected literature examples).—cont'd

Cohort	Species/type	Location	Dominant MP type	Common MP type (size)	MP concentration (particles/individual, if not mentioned)	References
	Copepoda & pteropoda	East China Sea	Polymerized oxidized organic material, PS, PP	Fibers ($295.2 \pm 348.6\,\mu m$), pellets ($20.3 \pm 11.0\,\mu m$), and fragments ($82.4 \pm 80.5\,\mu m$)	0.13–0.35	Sun et al. (2018)
Polychaetes	*Namalycastis* sp	Malaysia	PP & PA	Filaments, fragment	20–46.79	Hamzah et al. (2021)
	Marphysa sanguinea	North Sea, South Korea	PS, PE, PP & NY	0.2–3.8 mm	43–483	Jang, Shim, Han, Song, & Hong (2018)
Sea urchin	*Diadema africanum*	Canary Islands, Spain	CE (46%), PP (24.3%), PET (24.3%)	Fiber (97.5%), fragment, film	2.8 ± 2.3 to 10 ± 4.5	Sevillano-González et al. (2021)
	Paracentrotus lividus	Eastern Aegean Sea	–	Fiber (97.12%),	$1.95 \pm 1.70\,g-1$	Hennicke, Macrina, Malcolm-Mckay, & Miliou (2021)
	Strongy locentrotus intermedius, Temnopleurus hardwickii, Temnopleurus reevesii, Hemicentrotus pulcherrimus	North China	CP (36.65%), PET, PE, PP	Fiber (92.9%), fragment, sheet	2.20 ± 1.50 to 10.04 ± 8.46	Feng et al. (2020)
	Diadema setosum	Indonesia	–	–	23.70 ± 2.99	Sawalman et al. (2021)

Sea cucumber	Japanese spiky sea cucumber (*Apostichopus japonicus*)	China	RY (27%), PE (19%), cellulose	Fragment (16%), film, spheres	0–15	Mohsen, Lin, Liu, & Yang (2022)
	”	”	CP, PES, PET	Fiber (97%), fragment, film	0–30 (intestine), 0–19 (coelomic fluid)	Mohsen et al. (2019)
	Holothuria sp.	Indonesia	–	Fiber, fragment, film, pellet	0.5–7*	Sayogo, Patria, & Takarina (2020)
	Black long sea cucumber (*Holothuria leucospilota*)	,	–	Fiber	3.32–6.82*	Wicaksono, Patria, & Suryanda (2021)
	Paracaudina sp.	,	–	Fiber, film, fragment	289–1380	Amin, Rukim, & Nurrachmi (2021)
	Green/black sea cucumber (*Stichopus horrens*)	Malaysia	PE, PMM	Filement (91%), fragment, film	60–70	Muhammad Husin, Mazlan, Shalom, Saud, & Abdullah Sani (2021)
	Florida sea cucumber (*Holothuria floridana*)	USA	–	Fragment (93%), fiber (7%)	8.4 ± 4.5	Coc, Rogers, Barrientos, & Sanchez (2021)
Mussels	Green mussel (*Perna viridis*)	Hong Kong	PP, PE, PS & PET	Fragment, fiber (40–1000 μm)	1.60–14.7	Leung et al. (2021)
	Mediterranean mussels (*Mytilus galloprovincialis*)	Marmara Sea	PET (66.38%), PA, PS, PAN	Fiber (81.16%), fragment, film, pellets	0.30–7.53	Gedik, Eryaşar, & Gözler (2022)

Continued

Table 1 Abundance of MPs in various marine cohorts around the world (a few selected literature examples).—cont'd

Cohort	Species/type	Location	Dominant MP type	Common MP type (size)	MP concentration (particles/individual, if not mentioned)	References
	Black mussels (*Choromytilus meridionalis*), Blue mussels (*Mytilus meridionalis*)	South Africa	PET, Latex, PE	Filaments (70%), fragment	3.5–4.26	Sparks, Awe, & Maneveld (2021)
	Mediterranean mussel (*Mytilus galloprovincialis*)	Aegean Sea	–	Fiber (87%), fragment, film	–	Yozukmaz (2021)
	Brown mussel (*Perna perna*) and Green mussel (*Perna viridis*)	Tamil Nadu, India	PE, PP	Fiber (0.5–1 mm)	0.87 ± 0.55 to 10.02 ± 4.15	Patterson et al. (2021)
Oyster	Portuguese oysters *(Crassostrea angulate)* and Rock oyster (*Saccostrea*)	Taiwan	PET (69.5%), PP, PVC	Fragment (67%), fiber	3.24 ± 1.02*	Liao, Chiu, & Huang (2021)
	–	Maowei Sea, China	PET (60.3%), PE & PS	Fiber, foam, and film	0.59 ± 0.08–7.05 ± 1.21*	Zhu et al. (2021)
	Pacific oyster (*Magallana gigas*)	Baja California, Mexico	PA, PET, PAN, PE, PP & PS	Fiber (93.3%), fragment	0.04 ± 0.02 to 0.08 ± 0.02*	Lozano et al. (2021)
	Philippine cupped oyster (*Magallana bilineata*)	Tuticorin coast, Gulf of Mannar	PE (94%), PP	Fiber, films, and fragments	6.9 ± 3.84	Patterson, Jeyasanta, Sathish, Booth, & Edward (2019)

	Eastern oyster (*Crassostrea virginica*)	Florida estuary, USA	PP, NY	Fiber, beads, fragment	16.5	Waite, Donnelly, & Walters (2018)
	Pacific cupped oyster (*Crassostrea gigas*), Portuguese oyster (*Crassostrea angulate*), Hong Kong oyster (*Crassostrea Hongkongensis*), Kumamoto oyster (*Crassostrea sikamea*)	China	CP, PE, PET, PP, PA, PS	Fiber (61%), fragment (20%), film (10%), pellet (9%)	2.93	Teng et al. (2019)
	Natal rock oyster (*Saccostrea cucullate*)	China	PET (34%), PP (19%), PE (14%), PS, CP, PVC, PE	Fiber (69%), fragment (20%)	1.4–7	Li et al. (2018)
	”	”	PE (23%), PP, PS, PVC, PET	Fiber (52%), fragment (45%), film, granule	0.14–7.9*	Wang, Su, et al. (2021)
	–	South Korea	PP (43%), PET (25%), PE, PS, PA	Fragment (69%), Fiber (31%)	0.33 ± 0.23*	Cho, Shim, Jang, Han, & Hong (2021)
	Pacific oyster (*Crassostrea gigas*)	Netherland	–	Fiber	10–100*	Leslie, Brandsma, van Velzen, & Vethaak (2017)

Continued

Table 1 Abundance of MPs in various marine cohorts around the world (a few selected literature examples).—cont'd

Cohort	Species/type	Location	Dominant MP type	Common MP type (size)	MP concentration (particles/individual, if not mentioned)	References
Clam	Asian clam (*Corbicula Fluminea*)	China	–	–	0.4–5	Su et al. (2018)
	Duck clam (*Mactra veneriformis*)	,	CP, PET, PAA	Fiber (56%), film, granule, pellet	3.50 ± 1.35*	Zhang, Yan, & Wang (2020)
	Manila clams (*Venerupis philippinarum*)	Canada	–	Fiber (90%), film, fragment	0.07–5.47*	Davidson & Dudas (2016)
	Razor clams (*Siliqua patula*)	USA	PET, acrylic, aramid, CP	–	8.84 ± 0.45	Baechler, Granek, Hunter, & Conn (2020)
	Bean clams (*Donax cuneatus*)	India	PE, PP, PA, PET, PS, PVC	Fiber (53%), fragment (22%), film (21%), foam (4%)	0.6–1.3*	Narmatha Sathish, Immaculate Jeyasanta, & Patterson (2020)
	Razor clams	USA	PET, CA, CP, PS, NY	Fiber, fragment	6.75 ± 0.60	Baechler, Granek, Mazzone, et al. (2020)
	Hard clam (*Meretrix lusoria*)	Taiwan	PP, PET, PEA, PMAA, PA	Fragment (91%), fiber, pellete	0.12*	Chen, Lee, & Walther (2020)
Cockles	Common cockle (*Cerastoderma edule*)	France	PE (36.8%), ABS (32.5%), SBR (26.3%), PP, PS PET,	Fiber	2.46 ± 1.16	Hermabessiere et al. (2019)

	Blood cockle (*Anadara antiquata*)	Tanzania	–	Fiber (75%), fragment (25%)	2.1 ± 1.8	Mayoma, Sørensen, Shashoua, & Khan (2020)
	Blood cockle (*Anadara granosa*)	Indonesia	–	Fiber (58%), film (27%), fragment (15%), granule	618.8 ± 121.4	Ukhrowi, Wardhana, & Patria (2021)
Scallops	Zhikong scallop (Chlamys farreri)	China	–	–	8–13	Sui et al. (2020)
Cuttlefish	Common cuttlefish (*Sepia officinalis*)	Portugal	PP, LDPE, HDPE	Fiber (87%), fragment (8.4%), film (4.6%)	27–52	Oliveira et al. (2020)
	Pharaoh cuttlefish (*Sepia pharaonis*)	Indonesia	–	Fiber, fragment, film	208 (intestine), 226 (gills), 226 (surface)***	Prasetyo & Putri (2021)
	Humboldt squid (*Dosidicus gigas*)	Peru	CP, PA, PET, PP	Fiber (93–98%), fragment (2–7%)	7.42 ± 4.88 (stomach), 5.42 ± 5.96 (gills), 3.96 ± 3.62 (intestine)	Gong et al. (2021)
Crab	Gazami crab (*Portunus trituberculatus*), Asian paddle crab (*Charybdis japonica*), Samurai crab (*Dorippe japonica*), & Moon crab (*Matuta planipes*)	Haizhou Bay, Lvsi and Yangtze River Estuary, China	CP (34.87%), PET, PE, PP, PA	Fibers, fragments, sheets, and microbeads.	2.00 ± 2.00 to 9.81 ± 8.08	Zhang, Sun, Song, et al. (2021)
	–	South China Sea	PE, PET	Fiber, film, palettes, flake	0.39–2.83 (small MP), 0.74–4.96 (large MP)	Zhang, Sun, Liu, & Li (2021)

Continued

Table 1 Abundance of MPs in various marine cohorts around the world (a few selected literature examples).—cont'd

Cohort	Species/type	Location	Dominant MP type	Common MP type (size)	MP concentration (particles/individual, if not mentioned)	References
	Atlantic mud crab (*Panopeus herbstii*)	Florida estuary, USA	PP, Nylon	Fiber, beads, fragment	4.2	Waite et al. (2018)
	Atlantic ghost crab (*Ocypode quadrata*)	Brazil	–	Fiber, fragment	1–158	Costa, Arueira, da Costa, Di Beneditto, & Zalmon (2019)
	Chesapeake blue crab (*Callinectes sapidus*)	USA	CE/RY, PES, PS	Fiber, fragment, film	0.87	Waddell, Lascelles, & Conkle (2020)
Shrimp	Brown shrimp (*Crangon crangon*)	North Sea	–	Fiber (63%), fil, granule	1.23 ± 0.99	Devriese et al. (2015)
	Blue and red shrimp (*Aristeus antennatus*)	Mediterranean Sea	PET (57.1%), NY (28.6%), RY (14.3%))	Fiber	–	Carreras-Colom et al. (2018)
	Indian white shrimps (*Fenneropenaeus indicus*)	India	PS, PA, PE, PP	Fiber (83%), fragment (15%), sheet	0.39 ± 0.6	Daniel, Ashraf, & Thomas (2020)
	Green tiger prawn (*Penaeus semisulcatus*)	Persian Gulf	–	Fiber, fragment	0.360*	Akhbarizadeh, Moore, & Keshavarzi (2019)

	Speckled shrimp (*Metapenaeus monoceros*), Kiddi shrimp (*Parapeneopsis stylifera*), Indian prawan (*Penaeus indicus*)	Arabian sea	PE, PP, PA, NY, PS, PET	Fiber, fragment, pellet, beads, film	6.78 ± 2.80	Gurjar et al. (2021)
	Brown shrimp (*Metapenaeus mon ocerous*), tiger shrimp (*Penaeus monodon*)	Bay of Bengal	PA (59%), RY (27%)	Filement (57–58%), Fiber (32–57%),	3.40 ± 1.23* (*M. monocerous*), 3.87 ± 1.05* (*P. monodon*)	Hossain et al. (2020)
Fish	–	South China Sea	PE, PET	Fiber, film, palette, flake	0.35–3.22 (small MP), 0.72–5.39 (large MP)	Zhang, Sun, Liu, et al. (2021)
	European pilchard (*Sardina pilchardus*), European anchovy (*Engraulis encrasicolus*), *European hake* (*Merluccius merluccius*), Adriatic sole (*Pegusa impar*), *Striped red mullet* (*Mullus surmuletus*), Rock goby (*Gobius paganellus*)	Northern Adriatic Sea	PE, PP	Fragment (91.4%), fiber	4.11 ± 2.85 to 1.75 ± 0.71	Mistri et al. (2022)

Continued

Table 1 Abundance of MPs in various marine cohorts around the world (a few selected literature examples).—cont'd

Cohort	Species/type	Location	Dominant MP type	Common MP type (size)	MP concentration (particles/individual, if not mentioned)	References
	Australian herring (*Arripis georgianus*), Australian salmon (*Arripis trutta*), South American pilchard (*Sardinops sagax*), Australasian snapper (*Chrysophrys auratus*), Black flathead (*Platycephalus fuscus*), Southern garfish (*Hyporhamphus melanochir*), King George whiting (*Sillaginodes punctatus*), Flathead gray mullet (*Mugil cephalus*), Tiger flathead (*Platycephalus richardsoni*)	Australia	PE (40.8%), PP (46.9%),	–	0.94	Wootton, Reis-Santos, Dowsett, Turnbull, & Gillanders (2021)
	16 species	Malaysia	PA, PET, PP, PE, PS	Fiber (80.2%), fragment, filament	9.88	Jaafar et al. (2021))
	9 Species	Xiamen Bay, China	PA (26.97%), PET, PVA, PP, PAM, PE, EVA	Fiber (59%), fragment, film, particles	1.07–8	Wei et al. (2022)

	Hilsa shad (*Tenualosa ilisha*)	Bay of Bengal	–	Fiber (50%), fragment, foam	19.13 ± 10.77	Siddique et al. (2022)
	Bleak, three-spined stickle back, perch and roach	Baltic Sea	–	–	1.34 ± 0.71	Sainio, Lehtiniemi, & Setälä (2021)
	Common carp (*Cyprinus carpio*), Flathead gray mullet (*M. cephalus*), European flounder (*Platichthys flesus*)	NE Atlantic coast	RY (22%), PS (17%), PE (10%), PA (10%), PP (7%) & CA (7%)	Fiber (84%)	2 ± 2 to 10 ± 9	Guilhermino et al. (2021)
	Hhort mackerel (*Rastrelliger brachysoma*)	Thailand	–	–	2.70 ± 16.62	Hajisamae, Soe, Pradit, Chaiyvareesajja, & Fazrul (2022)
	18 species	North-western Mediterranean Sea	PES (45%), PA (36%), PP	Fiber (91%), fragment (9%)	0–5.15	Constant, Reynaud, Weiss, Ludwig, & Kerhervé (2022)
Mammals	10 species	British coast	NY (60%), PET, PS	Fiber (84%), fragment	3.8 ± 2.5 (Stomach) to 1.7 ± 1.4 (intestine)	Nelms et al. (2019)
	Beluga whale (*Delphinapterus leucas*)	Beaufort Sea	PS (54%)	Fiber (78%)	1.42 ± 0.44	Moore et al. (2022)
	Spotted seal (*Phoca largha*)	Liaodong Bay, China	PET, PP, PAN, PVC	Fiber (60%), Fragment (33%), pellets	1.33 ± 1.52 (stomach), 2.00 ± 1.26 (small intestine), 4.83 ± 6.21 (large intestine)**	Wang, Yu, et al. (2021)

ABS: acrylonitrile-butadiene-styrene, CA: cellulose acetate, CE: cellulosic, CP: cellophane, EVA: ethylene-vinyl acetate copolymer, LDPE: low density poly ethylene, HDPE: high density poly ethylene, NY: nylon, PA: polyamide/polyacrylate, PAM: polyacrylamide, PAN: poly-acrylonitrile, PE: polyethylene, PEA: poly(ethyl acrylate), PES: polyester, PET: polyethylene terephthalate, PMM: poly(methyl methacrylate), PMMA: polymethacrylate, PP: polypropylene, PS: polystyrene, PVA: polyvinyl alcohol, RY: Rayon, SBR: styrene butadiene rubber * items/g, ** item/10 g, *** item/mL.

Hence, the abundance was always reported as items (particle) for individuals or item/g on a wet weight basis. The significant difference in average abundance of MPs in different taxa differs; however, there is a uniformity that they are predominantly fiber across the edible marine taxa such as fish, mussels, oysters, crab and sea urchins.

4. Microplastic in critical marine vectors

A plethora of literature exists on MP presence in marine and aquatic biota (Abbasi et al., 2018; Akhbarizadeh et al., 2019; Akhbarizadeh, Moore, & Keshavarzi, 2018; Alimi et al., 2018; Al-Lihaibi et al., 2019; Alomar et al., 2017; Al-Salem, Uddin, & Lyons, 2020; Alvarez, Barros, & Velando, 2018; Bagheri et al., 2020; Barboza et al., 2020; Besseling, Wegner, Foekema, van den Heuvel-Greve, & Koelmans, 2013; Carbery et al., 2018; Catarino, Macchia, Sanderson, Thompson, & Henry, 2018; Collard et al., 2017; de Sa et al., 2018; de Sa, Luís, & Guilhermino, 2015; Devriese et al., 2015; Maass, Daphi, Lehmann, & Rillig, 2017; Zakeri et al., 2020). There are two primary routes by which MPs get into biota, i.e., active uptake (usually by ingestion or inhalation) (de Sa et al., 2018; Hermsen, Sims, & Crane, 1994; Kühn et al., 2015) or passive uptake by adhesion or adsorption. Several researchers have highlighted the potential hazard of plastic acting as a vector of pollutants that includes metals and, more often, hydrophobic organic chemicals (HOCs) (Koelmans, Bakir, Burton, & Janssen, 2016). The three major issues that emanate from this finding are how efficiently MPs transport metals and HOCs, the scale of this process on an oceanic scale, and whether MP ingestion contributes to metal and HOCs bioaccumulation in marine organisms. These questions have attracted significant attention. The passive dosing of organic compounds from polymers to organisms has been long known by ecotoxicologists. Several laboratory investigations have shown that MPs are a vector and source of HOC and metals to marine organisms (Abbasi et al., 2020; Brennecke et al., 2016; Browne, Niven, Galloway, Rowland, & Thompson, 2013; Chua, Shimeta, Nugegoda, Morrison, & Clarke, 2014; Gao et al., 2019; Gouin, Roche, Lohmann, & Hodges, 2011).

A very pertinent discussion is that how significant is MP as a pollution source and vector compared to other pathways (Uddin et al., 2022). The published evidence is in both support and the negation of the hypothesis. Since secondary MPs often results from physico-chemical degradation of larger plastics, the question remains how much of HOCs and metals are on these degraded secondary MPs. There are no field data; hence, no clarity on these aspects. Much evidence on MP as a vector has come from experimental studies, where primarily non-degraded MPs have been used. The other aspect that needs consideration, most ingested MPs are often defecated in hours (Uddin et al., 2022). In contrast, a few gastropods have retained MPs in their gastrointestinal tract for 12 h and even up to 7 days (Weber et al., 2021), and the longest reported time is 40 days (Duis & Coors, 2016).

The relative abundance of MP in the marine environment is very low compared to other likely vectors and receptors such as phytoplankton, zooplankton and suspended particulates (Annabi-Trabelsi et al., 2021; Uddin et al., 2018a, 2018b, 2022; Uddin, Bebhehani, Sajid, & Karam, 2019), which have more significant partition coefficients than most MPs (Mato et al., 2001). Several studies have suggested that MPs play a minor role in transporting organic contaminants (Gouin et al., 2011; Koelmans et al., 2016; Koelmans, Besseling, Wegner, & Foekema, 2013). So far, no field studies have been conducted evaluating MPs as a vector of contaminant transfer. Hence, conclusions are based on laboratory experiments that have used non-environmental concentrations. Moreover, no assessments are done on transfer factors and coefficients of transfer for metals or organics bound to MPs.

A significant information gap exists on the likely bioaccumulation of organic contaminants and metals taken up by phytoplankton and zooplankton and those sorbed to MPs. How does the concentration of these contaminants compares with the plankton species that can themselves bio-magnify these pollutants and thus have both the absorbed and adsorbed components, unlike MPs which contain only the sorbed component?

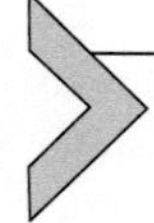

5. Biomagnification and biotransformation of microplastic through the trophic chain

Microplastics have been reported across the trophic level in marine organisms (Table 1). MP occupies the same size range as plankton, and they are found at different levels within the water column. Hence, they can be accessible organisms at different trophic levels and using different feeding strategies (Figueiredo & Vianna, 2018). Thus, it is likely that marine organisms might ingest unknown quantities of MP with their natural prey items. This will be obvious with non-selective filter feeders, which filter large quantities of water and sediment for organic nutrients (Alava, 2020).

In addition, MPs facilitate the initial uptake and transport of contaminants like polychlorinated bisphenols (PCBs), polyaromatic hydrocarbons (PAHs) as well as heavy metals.

A study demonstrates that some of these contaminants are up to six orders of magnitude higher than in the surrounding marine environment (Carbery et al., 2018). Supportive evidence for MP transfer across the trophic food chain comes from the quantification of MP in field-collected organisms, their natural predators, and controlled feeding studies endeavoring to represent the transfer of MP through artificial food chains. Some laboratory experimental studies and fieldwork have shown the transfer of MP between trophic levels through exposure and ingestion. Polyethylene microspheres (27–32 μm, 1.025 g/mL) were fed through the microalgae [two exposure routes using mysids (*Neomysis* spp.) and a benthic fish (*Myoxocephalus brandti*)]. It was observed that mysids ingested 2–3 times more PE beads from water containing the higher concentration, and fish ingested 3–11 times more polyethylene beads through trophic transfer than from the water column.

Moreover, it was demonstrated that prey such as mysids could fragment MP, and the smaller particles can translocate from the digestive tract into tissues (Hasegawa & Nakaoka, 2021). Virgin MP has been commonly used in trophic transfer experiments but it creates more items that result in lower ingestion rates than those using aged MP (Sucharitakul, Pitt, & Welsh, 2021). The experiment showed whether ephyrae (juvenile medusae of the jellyfish, *Aurelia* sp.) would ingest more MP by trophic transfer than direct ingestion and whether medusae would ingest more aged than virgin microbeads. It was shown that ephyrae ingested 35 times more microbeads through trophic transfer than by direct ingestion. In the second experiment, it was concluded that aged MP showed more potential for ingestion than virgin MP. Many studies show that trophic transfer is the main pathway for bioaccumulation of MP in the marine environment (da Costa Araújo, de Andrade Vieira, & Malafaia, 2020; Elizalde-Velázquez et al., 2020; Nelms, Galloway, Godley, Jarvis, & Lindeque, 2018; Stienbarger et al., 2021; Xu, Fang, Wong, & Cheung, 2022).

Even though biotransformation has occurred, it may not be a significant risk at higher trophic levels due to the rapid depuration of MP. Wang, Hu, et al. (2021) assessed the impact of water filtration and food consumption on MP accumulation in a predatory marine crab (*Charybdis japonica*) and evaluated the combined effects of MP. The authors suggested that marine organisms have an inherent capability to counter the acute effects of MP.

However, there is a limit beyond which defense mechanisms fail and thus physiological functions are compromised. However, da Costa Araújo and Malafaia (2021) studied the MP accumulation at higher trophic levels (tadpoles to mice via fish). The results confirmed that MPs in water could reach terrestrial trophic levels, which have a negative influence on the survival of these animals. There was evidence that MP ingestion impeded food intake, blocked the digestive tract, caused physiological stress in zooplankton such as copepods (Bai, Wang, & Wang, 2021), and decreased the photosynthesis of marine algae such as red lettuce (Dong, Song, Liu, & Gao, 2021). Microplastics have now been revealed to influence the immune system of marine organisms and lead to a significant reduction in gamete and oocyte quality, fecundity, sperm swimming speed as well as quality of offspring (Sharifinia, Bahmanbeigloo, Keshavarzifard, Khanjani, & Lyons, 2020).

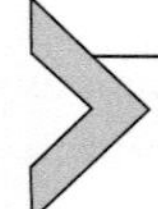

6. Toxicological impact and human health risk assessment

The uptake of MP particles by humans can occur through ingestion and inhalation (Danopoulos, Twiddy, West, & Rotchell, 2022). Seafood consumption represents one pathway for human MP exposure. Microplastic particles are often found concentrated in the digestive tract, and most humans consume bivalves, mussels, and small fish as whole organisms (e.g., sardines, anchovies, and sprat). Studying European pilchard (*Sardina pilchardus*) and European anchovy (*Engraulis encrasicolus*) caught along the Spanish Mediterranean coast Compa, Ventero, Iglesias, and Deudero (2018) found that 14–15% of the fish had MP or natural fibers in the gastrointestinal tract. The small fish, i.e., sprats and sardines, are commonly sold as canned products, and Karami et al. (2018) investigated the potential presence of micro and mesoplastic (size range 0.149–10 mm) in 20 different canned brands from 13 countries. They found that plastics were absent from the 16 canned brands, while the fish in four brands contained 1 to 3 particles per can. Piyawardhana et al. (2022) studied the occurrence of MP in 14 different dried marine fish species from 7 Asian countries. They reported that MP was observed in most dried fish and the highest count recorded was in dried fish from Japan (0.56 ± 0.03 per gram tissue). The MPs $<130\,\mu m$ in diameter can potentially translocate into human tissue, and particle sizes $<1.5\,\mu m$ can penetrate capillaries (Lundebye, Lusher, & Bank, 2022). However, the international scientific committee, such as the

Joint FAO/WHO Expert Committee on Food Additives (JECFA), has not yet evaluated the food safety concern based on MP. Nevertheless, some studies have assessed MP exposure, risk assessment, and the relative contribution of microplastic exposure to additives or chemicals (Smith, Love, Rochman, & Neff, 2018).

The amount of MPs ingestion by an individual human depends on highly variable factors, not only the characteristics of MPs but also age of the individual, size, demographics, cultural heritage, geographical area, diet, and surrounding environment (Senathirajah et al., 2021). On a global average basis, Senathirajah et al. (2021) estimated that humans ingest 0.1–5 g of MP weekly. Karami et al. (2018) concluded that 1–5 anthropogenic MP might enter human from the canned sardines and sprats. This is far less than the amount of MP ingestion through mollusk consumption by the top European fish consumers (11,000 MP particles/ year) (Van Cauwenberghe & Janssen, 2014). In another study, Cox et al. (2019) estimated the annual MP consumption range for Americans as 39,000–52,000 particles depending on age and sex. The human body's excretory system removes MP, likely disposing of >90% of ingested MP through feces. There were concerns whether MP was possible to remain in the human body, potentially accumulating in some organs. Studies in mice have found that MP around 5 μm in diameter could stay in the intestines or reach the liver. With minimal data available on how fast mice excrete MPs, Mohamed, Kooi, Diepens, and Koelmans (2021) proposed a probabilistic lifetime exposure model that children and adults might accumulate 553 particle/capita/day and 883 particle/capita/day, respectively. The first evidence of MP in the human placenta was reported by Ragusa et al. (2021), who found 12 MP fragments (5–10 μm) on the fetal side (5 nos) and the maternal side (4 nos) and the chorioamnionitis membranes (3 nos). However, Lim (2021) maintains that it is still not impossible that these particle numbers were the result of external contamination when the placentas were collected and analyzed. However, there is as yet no detailed human health risk assessment study associated with MP due to a lack of toxicity and toxicokinetic data available for both MP and NP (Chain, 2016).

MP exposure also can present exposure to correlated chemicals. Once MPs are in the gut, they can release absorbed toxins and additives and constitute monomers (Cox et al., 2019). Several studies have assessed most of the organic and inorganic toxic compounds associated with the MP. However, few studies have assessed the relative contribution of MP exposure to additives or chemicals found in organisms versus alternative exposure pathways (Smith et al., 2018). The laboratory studies have shown that MP increased toxicity from the combination of MP and associated chemicals. The uptake of pollutants, additives, and MP damaged the eco-physiological functions of certain organisms (Browne et al., 2013; Gallo et al., 2018). However, it is difficult to estimate whether toxicological effects transfer to humans. In animals, the quantity of chemicals from MP is minimal compared to that from other diet components.

Moreover, researchers do not fully understand how MP interacts with human biological tissue, even if some published findings on individual effects are available. However, it is challenging to observe such effects at the population level. As one example, there is a significant association between urine Bisphenol-A (BPA) levels and both cardiovascular disease and type II diabetes (Provvisiero et al., 2016). One of the human low dose exposure assessments for BPA is by MPs (Smith et al., 2018); thus, it is necessary to undertake further research to assess the risk and effects of such MP dietary exposure.

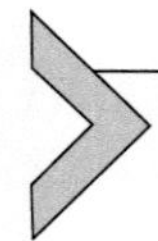

7. Prospects and challenges for seafood microplastic research

Importantly still, at present, there is a definite lack of harmonized and consistent methods for MP measurements in seafood. There are some recommendations presented for the isolation of MPs, and it is of great importance to organizing an effort to inter-compare the field and laboratory-based methods (sampling, isolation, and analysis) one to another to ascertain the level of comparability and overall efficiency and reliability (Dehaut, Hermabessiere, & Duflos, 2019; Lundebye et al., 2022). Not only for the analytical tools, but also there are limitations with the definition, especially particle morphology-based shapes and sizes (Filella, 2015). Currently, there is no standardized procedure for measurement units, production of certified reference materials, and initiating inter-laboratory testing for complex matrices such as seafood (Dehaut et al., 2019). Future directions are related to improving reproducible methodologies and are focused on accurately detecting s-MPs and image recognition.

Microplastic comprised of several raw materials and forms may cause damage to marine organisms to varying degrees. The absorption mechanism depends on several intrinsic and extrinsic factors of seafood varieties. A few studies show the trophic transport mechanisms in the natural environment.

Although there have been some risk assessments and associated studies, most have not reported quantitative toxicological risk assessments. Hence, it is a necessary and valuable strategy to evaluate precisely and express the risk associated with MPs, highlight the potential uncertainties and number of assumptions in future research, and focus primarily on known particle species-specific effect mechanisms. Moreover, it is necessary to address the research gap on human health hazards and exposure to MPs. This information will allow for simplistic communication in policy briefing and informing the general public.

This approach will lead researchers to organize innovative scientific research and boost solutions to find microorganisms that can break down MP and minimize MP pollution. It is necessary to formulate a global policy framework to reduce plastic waste at the source, promote degradable plastic products and plastic alternatives, and enhance the recycling capacity and reuse of plastic items.

8. Conclusions

Microplastics are inseparable items from our lives due to several direct and indirect applications. The ubiquitous presence of MPs in the aquatic environment and reports of their presence in seafood makes it a pressing issue to develop standardized protocols for analytical extraction methods, identification procedures and analytical techniques, and characterization as a priority research area. The fate and transport models, characterization of exposure pathways, and evaluation of toxicological exposures are necessary for a comprehensive understanding of MP pollution in the marine environment and seafood safety. One significant consideration is how effective MP is as a vector for pollutant transfer since this can only have adsorbed components. At the same time, other vectors like phytoplankton and zooplankton can also bioaccumulate and have both absorbed and adsorbed components. The other significant issue is an underestimation of the smaller MPs; this is essentially due to the sampling techniques used. The smaller MPs of $<1.5\,\mu m$ can cross the cell barrier, and there are hardly any sampling strategies to capture these ultrafine MPs, certainly drawing attention to the fact that harmonized methodologies need to be developed for capturing these MPs.

With the published evidence, it is obvious that the bulk of the MPs in the aquatic systems are sMP. It is postulated that lower trophic level species may be at the highest risk of MP contamination. The efficacy of the pollutant transfer from adsorbed biofilms to organisms and human receptors is yet to be established. In addition, most published literature has suggested an effective and quick clearing mechanism of ingested MPs. Another conspicuous observation is the significant attention given to MPs as the major contaminant of the Anthropocene, it is certainly in terms of quantity but their toxicity to human and non-human biota needs extensive assessments before the present focus of science and research funding is further diverted towards MPs. There are presently not many studies on MP in humans hence making any realistic health risk assessment is at best a theoretical estimate. Microplastic levels and their associated co-contaminants in seafood species and their resultant transfer to and consumption by humans are not well studied in the perspective and context of global seafood security and sustainability. The claims of translocation of ingested and inhaled MPs across organs and tissues need to be underpinned with more scientific evidence. It is wise to assess differences in the adsorption potential of different degraded MPs. Since most of the current experimental data come from laboratory experiments using fresh non-degraded MPs, future studies should consider these suggestions to provide a strong basis to assess the overall impact of MP on food security.

References

Abbasi, S., Moore, F., Keshavarzi, B., Hopke, P. K., Naidu, R., Rahman, M. M., et al. (2020). PET-microplastics as a vector for heavy metals in a simulated plant rhizosphere zone. *Science of the Total Environment*, *744*, 140984. https://doi.org/10.1016/j.scitotenv.2020.140984.

Abbasi, S., Soltani, N., Keshavarzi, B., Moore, F., Turner, A., & Hassanaghaei, M. (2018). Microplastics in different tissues of fish and prawn from the Musa estuary, Persian gulf. *Chemosphere*, *205*, 80–87. https://doi.org/10.1016/j.chemosphere.2018.04.076.

Akhbarizadeh, R., Dobaradaran, S., Amouei Torkmahalleh, M., Saeedi, R., Aibaghi, R., & Faraji Ghasemi, F. (2021). Suspended fine particulate matter (PM2.5), microplastics (MPs), and polycyclic aromatic hydrocarbons (PAHs) in air: Their possible relationships and health implications. *Environmental Research*, *192*, 110339. https://doi.org/10.1016/j.envres.2020.110339.

Akhbarizadeh, R., Moore, F., & Keshavarzi, B. (2018). Investigating a probable relationship between microplastics and potentially toxic elements in fish muscles from northeast of Persian gulf. *Environmental Pollution*, *232*, 154–163. https://doi.org/10.1016/j.envpol.2017.09.028.

Akhbarizadeh, R., Moore, F., & Keshavarzi, B. (2019). Investigating microplastics bioaccumulation and biomagnification in seafood from the Persian Gulf: a threat to human health? *Food Additives & Contaminants: Part A*, *36*(11), 1696–1708. https://doi.org/10.1080/19440049.2019.1649473.

Alava, J. J. (2020). Modeling the bioaccumulation and biomagnification potential of microplastics in a cetacean foodweb of the northeastern Pacific: A prospective tool to assess the risk exposure to plastic particles. *Frontiers in Marine Science*, 7. https://doi.org/10.3389/fmars.2020.566101.

Alimba, C. G., & Faggio, C. (2019). Microplastics in the marine environment: Current trends in environmental pollution and mechanisms of toxicological profile. *Environmental Toxicology and Pharmacology*, *68*, 61–74. https://doi.org/10.1016/j.etap.2019.03.001.

Alimi, O. S., Farner Budarz, J., Hernandez, L. M., & Tufenkji, N. (2018). Microplastics and nanoplastics in aquatic environments: Aggregation, deposition, and enhanced contaminant transport. *Environmental Science & Technology*, *52*, 1704.

Allen, S., Allen, D., Phoenix, V., Le Roux, G., Jiménez, P. D., & e. a. (2019). Atmospheric transport and deposition of microplastics in a remote mountain catchment. *Nature Geoscience*, *12*, 339–344.

Al-Lihaibi, S., Al-Mehmadi, A., Alarif, W. M., Bawakid, N. O., Kallenborn, R., & Ali, A. M. (2019). Microplastics in sediments and fish from the Red Sea coast at Jeddah (Saudi Arabia). *Environmental Chemistry*, *16*(8), 641–650. https://doi.org/10.1071/EN19113.

Alomar, C., Sureda, A., Capo, X., Guijarro, B., Tejada, S., & Deudero, S. (2017). Microplastic ingestion by *Mullus surmuletus* (Linnaeus, 1758) fish and its potential for causing oxidative stress. *Environmental Research*, *159*, 135–142.

Alprol, A. E., Gaballah, M. S., & Hassaan, M. A. (2021). Micro and Nanoplastics analysis: Focus on their classification, sources, and impacts in marine environment. *Regional Studies in Marine Science*, *42*, 101625. https://doi.org/10.1016/j.rsma.2021.101625.

Al-Salem, S. M., Uddin, S., & Lyons, B. (2020). Evidence of microplastics (MP) in gut content of major consumed marine fish species in the State of Kuwait (of the Arabian/Persian gulf). *Marine Pollution Bulletin*, *154*, 111052. https://doi.org/10.1016/j.marpolbul.2020.111052.

Alvarez, G., Barros, A., & Velando, A. (2018). The use of European shag pellets as indicators of microplastic fibers in the marine environment. *Marine Pollution Bulletin*, *137*, 444–448. https://doi.org/10.1016/j.marpolbul.2018.10.050.

Amin, B., Rukim, M., & Nurrachmi, I. (2021). Presence of microplastics in sea cucumber *Paracaudina sp* from Karimun Island, Kepulauan Riau, Indonesia. *IOP Conference Series: Earth and Environmental Science*, *934*(1), 012053. https://doi.org/10.1088/1755-1315/934/1/012053.

Andrady, A. L. (2011). Microplastics in the marine environment. *Marine Pollution Bulletin*, *62*(8), 1596–1605. https://doi.org/10.1016/j.marpolbul.2011.05.030.

Andrady, A. L. (2017). The plastic in microplastics: A review. *Marine Pollution Bulletin*, *119*(1), 12–22. https://doi.org/10.1016/j.marpolbul.2017.01.082.

Annabi-Trabelsi, N., Guermazi, W., Karam, Q., Ali, M., Uddin, S., Leignel, V., et al. (2021). Concentrations of trace metals in phytoplankton and zooplankton in the Gulf of Gabès, Tunisia. *Marine Pollution Bulletin*, *168*, 112392. https://doi.org/10.1016/j.marpolbul.2021.112392.

Auta, H. S., Emenike, C. U., & Fauziah, S. H. (2017). Distribution and importance of microplastics in the marine environment: A review of the sources, fate, effects, and potential solutions. *Environment International*, *102*, 165–176. https://doi.org/10.1016/j.envint.2017.02.013.

Aytan, U., Esensoy, F. B., & Senturk, Y. (2022). Microplastic ingestion and egestion by copepods in the Black Sea. *Science of the Total Environment*, *806*, 150921. https://doi.org/10.1016/j.scitotenv.2021.150921.

Baechler, B. R., Granek, E. F., Hunter, M. V., & Conn, K. E. (2020). Microplastic concentrations in two Oregon bivalve species: Spatial, temporal, and species variability. *Limnology and Oceanography Letters*, *5*(1), 54–65. https://doi.org/10.1002/lol2.10124.

Baechler, B. R., Granek, E. F., Mazzone, S. J., Nielsen-Pincus, M., & Brander, S. M. (2020). Microplastic exposure by razor clam recreational harvester-consumers along a sparsely populated coastline. *Frontiers in Marine Science*, 7. https://doi.org/10.3389/fmars.2020.588481.

Bagheri, T., Gholizadeh, M., Abarghouei, S., Zakeri, M., Hedayati, A., Rabaniha, M., et al. (2020). Microplastics distribution, abundance and composition in sediment, fishes and benthic organisms of the Gorgan Bay, Caspian Sea. *Chemosphere*, *257*, 127201. https://doi.org/10.1016/j.chemosphere.2020.127201.

Bai, Z., Wang, N., & Wang, M. (2021). Effects of microplastics on marine copepods. *Ecotoxicology and Environmental Safety*, *217*, 112243. https://doi.org/10.1016/j.ecoenv.2021.112243.

Ballent, A., Corcoran, P. L., Madden, O., Helm, P. A., & Longstaffe, F. J. (2016). Sources and sinks of microplastics in Canadian Lake Ontario nearshore, tributary and beach sediments. *Marine Pollution Bulletin*, *110*(1), 383–395. https://doi.org/10.1016/j.marpolbul.2016.06.037.

Barboza, L. G. A., Dick Vethaak, A., Lavorante, B. R. B. O., Lundebye, A.-K., & Guilhermino, L. (2018). Marine microplastic debris: An emerging issue for food security, food safety and human health. *Marine Pollution Bulletin*, *133*, 336–348. https://doi.org/10.1016/j.marpolbul.2018.05.047.

Barboza, L. G. A., Frias, J. P. G. L., Booth, A. M., Vieira, L. R., Masura, J., Baker, J., et al. (2019). Microplastics pollution in the marine environment. In C. Sheppard (Ed.), *Vol. 3. World seas: An environmental evaluation. Volume III: Ecological issues and environmental impacts* (pp. 329–351). London: Academic Press (Elsevier).

Barboza, L. G. A., Lopes, C., Oliveira, P., Bessa, F., Otero, V., Henriques, B., et al. (2020). Microplastic in wild fish from north East Atlantic Ocean and its potential for causing neurotoxic effects, lipid oxidative damage, and human health risks associated with ingestion exposure. *Science of the Total Environment*, *717*, 134625. https://doi.org/10.1016/j.scitotenv.2019.134625.

Batel, A., Linti, F., Scherer, M., Erdinger, L., & Braunbeck, T. (2016). Transfer of benzo[a]pyrene from microplastics to Artemia nauplii and further to zebrafish via a trophic food web experiment: CYP1A induction and visual tracking of persistent organic pollutants. *Environmental Toxicology and Chemistry*, *35*(7), 1656–1666. https://doi.org/10.1002/etc.3361.

Bessa, F., Frias, J., Kögel, T., Lusher, A., Andrade, J., Antunes, J., et al. (2019). *Harmonized protocol for monitoring microplastics in biota: JPI-oceans BASEMAN project.*

Besseling, E., Quik, J. T. K., Sun, M., & Koelmans, A. A. (2017). Fate of nano- and microplastic in freshwater systems: A modeling study. *Environmental Pollution*, *220*, 540–548. https://doi.org/10.1016/j.envpol.2016.10.001.

Besseling, E., Wegner, A., Foekema, E. M., van den Heuvel-Greve, M. J., & Koelmans, A. A. (2013). Effects of microplastic on fitness and PCB bioaccumulation by the lugworm Arenicola marina (L.). *Environmental Science & Technology*, *47*(1), 593.

Brennecke, D., Duarte, B., Paiva, F., Caçador, I., & Canning-Clode, J. (2016). Microplastics as vector for heavy metal contamination from the marine environment. *Estuarine, Coastal and Shelf Science*, *178*, 189–195. https://doi.org/10.1016/j.ecss.2015.12.003.

Browne, M. A., Dissanayake, A., Galloway, T. S., Lowe, D. M., & Thompson, R. C. (2008). Ingested microscopic plastic translocates to the circulatory system of the mussel, *Mytilus edulis* (L.). *Environmental Science & Technology*, *42*(13), 5026–5031. https://doi.org/10.1021/es800249a.

Browne, M. A., Niven, S. J., Galloway, T. S., Rowland, S. J., & Thompson, R. C. (2013). Microplastic moves pollutants and additives to Worms, reducing functions linked to health and biodiversity. *Current Biology*, *23*(23), 2388–2392. https://doi.org/10.1016/j.cub.2013.10.012.

Burns, E. E., & Boxall, A. B. A. (2018). Microplastics in the aquatic environment: Evidence for or against adverse impacts and major knowledge gaps. *Environmental Toxicology and Chemistry*, *37*(11), 2776–2796. https://doi.org/10.1002/etc.4268.

Carbery, M., O'Connor, W., & Palanisami, T. (2018). Trophic transfer of microplastics and mixed contaminants in the marine food web and implications for human health. *Environment International, 115*, 400–409. https://doi.org/10.1016/j.envint.2018.03.007.

Carreras-Colom, E., Constenla, M., Soler-Membrives, A., Cartes, J. E., Baeza, M., Padrós, F., et al. (2018). Spatial occurrence and effects of microplastic ingestion on the deep-water shrimp *Aristeus antennatus*. *Marine Pollution Bulletin, 133*, 44–52. https://doi.org/10.1016/j.marpolbul.2018.05.012.

Catarino, A. I., Macchia, V., Sanderson, W. G., Thompson, R. C., & Henry, T. B. (2018). Low levels of microplastics (MP) in wild mussels indicate that MP ingestion by humans is minimal compared to exposure via household fibres fallout during a meal. *Environmental Pollution, 237*, 675–684.

Chen, J. Y.-S., Lee, Y.-C., & Walther, B. A. (2020). Microplastic contamination of three commonly consumed seafood species from Taiwan: A pilot study. *Sustainability, 12*(22), 9543.

Cho, Y., Shim, W. J., Jang, M., Han, G. M., & Hong, S. H. (2021). Nationwide monitoring of microplastics in bivalves from the coastal environment of Korea. *Environmental Pollution, 270*, 116175. https://doi.org/10.1016/j.envpol.2020.116175.

Chua, E. M., Shimeta, J., Nugegoda, D., Morrison, P. D., & Clarke, B. O. (2014). Assimilation of polybrominated diphenyl ethers from microplastics by the marine amphipod, Allorchestes Compressa. *Environmental Science and Technology, 48*, 8127–8134.

Coc, C., Rogers, A., Barrientos, E., & Sanchez, H. (2021). Micro and macroplastics analysis in the digestive tract of a sea cucumber (Holothuriidae, *Holothuria floridana*) of the Placencia lagoon, Belize. *Caribbean Journal of Science, 51*(2), 166–174 (169).

Cole, M., Lindeque, P., Fileman, E., Halsband, C., & Galloway, T. S. (2015). The impact of polystyrene microplastics on feeding, function and fecundity in the marine copepod Calanus helgolandicus. *Environmental Science & Technology, 49*(2), 1130–1137. https://doi.org/10.1021/es504525u.

Collard, F., Gilbert, B., Compere, P., Eppe, G., Das, K., Jauniaux, T., et al. (2017). Microplastics in livers of European anchovies (*Engraulis encrasicolus, L.*). *Environmental Pollution, 229*, 1000–1005.

Compa, M., Ventero, A., Iglesias, M., & Deudero, S. (2018). Ingestion of microplastics and natural fibres in *Sardina pilchardus* (Walbaum, 1792) and Engraulis encrasicolus (Linnaeus, 1758) along the Spanish Mediterranean coast. *Marine Pollution Bulletin, 128*, 89–96. https://doi.org/10.1016/j.marpolbul.2018.01.009.

Constant, M., Reynaud, M., Weiss, L., Ludwig, W., & Kerhervé, P. (2022). Ingested microplastics in 18 local fish species from the northwestern Mediterranean Sea. *Microplastics, 1*(1), 186–197.

Costa, L. L., Arueira, V. F., da Costa, M. F., Di Beneditto, A. P. M., & Zalmon, I. R. (2019). Can the Atlantic ghost crab be a potential biomonitor of microplastic pollution of sandy beaches sediment? *Marine Pollution Bulletin, 145*, 5–13. https://doi.org/10.1016/j.marpolbul.2019.05.019.

Cox, K. D., Covernton, G. A., Davies, H. L., Dower, J. F., Juanes, F., & Dudas, S. E. (2019). Human consumption of microplastics. *Environmental Science & Technology, 53*(12), 7068–7074. https://doi.org/10.1021/acs.est.9b01517.

Coyle, R., Hardiman, G., & Driscoll, K. O. (2020). Microplastics in the marine environment: A review of their sources, distribution processes, uptake and exchange in ecosystems. *Case Studies in Chemical and Environmental Engineering, 2*, 100010. https://doi.org/10.1016/j.cscee.2020.100010.

da Costa Araújo, A. P., de Andrade Vieira, J. E., & Malafaia, G. (2020). Toxicity and trophic transfer of polyethylene microplastics from Poecilia reticulata to *Danio rerio*. *Science of the Total Environment, 742*, 140217. https://doi.org/10.1016/j.scitotenv.2020.140217.

da Costa Araújo, A. P., & Malafaia, G. (2021). Microplastic ingestion induces behavioral disorders in mice: A preliminary study on the trophic transfer effects via tadpoles and fish. *Journal of Hazardous Materials*, *401*, 123263. https://doi.org/10.1016/j.jhazmat.2020.123263.

Daniel, D. B., Ashraf, P. M., & Thomas, S. N. (2020). Abundance, characteristics and seasonal variation of microplastics in Indian white shrimps (*Fenneropenaeus indicus*) from coastal waters off Cochin, Kerala, India. *Science of the Total Environment*, *737*, 139839. https://doi.org/10.1016/j.scitotenv.2020.139839.

Danopoulos, E., Twiddy, M., West, R., & Rotchell, J. M. (2022). A rapid review and meta-regression analyses of the toxicological impacts of microplastic exposure in human cells. *Journal of Hazardous Materials*, *427*, 127861. https://doi.org/10.1016/j.jhazmat.2021.127861.

Davidson, K., & Dudas, S. E. (2016). Microplastic ingestion by wild and cultured Manila clams (*Venerupis philippinarum*) from Baynes sound, British Columbia. *Archives of Environmental Contamination and Toxicology*, *71*(2), 147–156. https://doi.org/10.1007/s00244-016-0286-4.

de Sa, L. C., Luís, L. G., & Guilhermino, L. (2015). Effects of microplastics on juveniles of the common goby (*Pomatoschistus microps*): Confusion with prey, reduction of the predatory performance and efficiency, and possible influence of developmental conditions. *Environmental Pollution*, *196*, 359–361.

de Sa, L. C., Oliveira, M., Ribeiro, F., Rocha, T. L., & Futter, M. N. (2018). Studies of the effects of microplastics on aquatic organisms: What do we know and where should we focus our efforts in the future? *Science of The Total Environonment*, *645*, 1029–1039. https://doi.org/10.1016/j.scitotenv.2018.07.207.

Dehaut, A., Hermabessiere, L., & Duflos, G. (2019). Current frontiers and recommendations for the study of microplastics in seafood. *TrAC Trends in Analytical Chemistry*, *116*, 346–359. https://doi.org/10.1016/j.trac.2018.11.011.

Dekiff, J. H., Remy, D., Klasmeier, J., & Fries, E. (2014). Occurrence and spatial distribution of microplastics in sediments from Norderney. *Environmental Pollution*, *186*, 248–256. https://doi.org/10.1016/j.envpol.2013.11.019.

Devriese, L. I., van der Meulen, M. D., Maes, T., Bekaert, K., Paul-Pont, I., Frère, L., et al. (2015). Microplastic contamination in brown shrimp (*Crangon crangon*, Linnaeus 1758) from coastal waters of the southern North Sea and channel area. *Marine Pollution Bulletin*, *98*(1), 179–187. https://doi.org/10.1016/j.marpolbul.2015.06.051.

Dong, Y., Song, Z., Liu, Y., & Gao, M. (2021). Polystyrene particles combined with di-butyl phthalate cause significant decrease in photosynthesis and red lettuce quality. *Environmental Pollution*, *278*, 116871. https://doi.org/10.1016/j.envpol.2021.116871.

Dris, R., Gasperi, J., Saad, M., Mirande, C., & Tassin, B. (2016). Synthetic fibers in atmospheric fallout: A source of microplastics in the environment? *Marine Pollution Bulletin*, *104*(1–2), 290–293. https://doi.org/10.1016/j.marpolbul.2016.01.006.

Duis, K., & Coors, A. (2016). Microplastics in the aquatic and terrestrial environment: Sources (with a specific focus on personal care products), fate and effects. *Environmental Sciences Europe*, *28*(1), 2. https://doi.org/10.1186/s12302-015-0069-y.

Eerkes-Medrano, D., Thompson, R. C., & Aldridge, D. C. (2015). Microplastics in freshwater systems: A review of the emerging threats, identification of knowledge gaps and prioritisation of research needs. *Water Research*, *75*, 63.

EFSA Panel on Contaminants in the Food Chain (CONTAM). (2016). Presence of microplastics and nanoplastics in food, with particular focus on seafood. *EFSA Journal*, *14*(6), e04501.

Elizalde-Velázquez, A., Carcano, A. M., Crago, J., Green, M. J., Shah, S. A., & Cañas-Carrell, J. E. (2020). Translocation, trophic transfer, accumulation and depuration of polystyrene microplastics in *Daphnia magna* and *Pimephales promelas*. *Environmental Pollution*, *259*, 113937. https://doi.org/10.1016/j.envpol.2020.113937.

Enyoh, C. E., Verla, A. W., Verla, E. N., Ibe, F. C., & Amaobi, C. E. (2019). Airborne microplastics: a review study on method for analysis, occurrence, movement and risks. *Environmental Monitoring and Assessment*, *191*(11), 668. https://doi.org/10.1007/s10661-019-7842-0.

Eriksen, M., Lebreton, L. C. M., Carson, H. S., Thiel, M., Moore, C. J., Borerro, J. C., et al. (2014). Plastic pollution in the World's oceans: More than 5 trillion plastic pieces weighing over 250,000 tons afloat at sea. *PLoS One*, *9*(12), e111913. https://doi.org/10.1371/journal.pone.0111913.

Fahrenfeld, N. L., Arbuckle-Keil, G., Naderi Beni, N., & Bartelt-Hunt, S. L. (2019). Source tracking microplastics in the freshwater environment. *TrAC Trends in Analytical Chemistry*, *112*, 248–254. https://doi.org/10.1016/j.trac.2018.11.030.

Feng, Z., Wang, R., Zhang, T., Wang, J., Huang, W., Li, J., et al. (2020). Microplastics in specific tissues of wild sea urchins along the coastal areas of northern China. *Science of the Total Environment*, *728*, 138660. https://doi.org/10.1016/j.scitotenv.2020.138660.

Ferreira, I., Venâncio, C., Lopes, I., & Oliveira, M. (2019). Nanoplastics and marine organisms: What has been studied? *Environmental Toxicology and Pharmacology*, *67*, 1–7. https://doi.org/10.1016/j.etap.2019.01.006.

Figueiredo, G. M., & Vianna, T. M. P. (2018). Suspended microplastics in a highly polluted bay: Abundance, size, and availability for mesozooplankton. *Marine Pollution Bulletin*, *135*, 256–265. https://doi.org/10.1016/j.marpolbul.2018.07.020.

Filella, M. (2015). Questions of size and numbers in environmental research on microplastics: Methodological and conceptual aspects. *Environmental Chemistry*, *12*(5), 527–538. https://doi.org/10.1071/EN15012.

Freeman, S., Booth, A., Sabbah, I., Tiller, R., Dierking, J., Klun, K., et al. (2020). Between source and sea: The role of wastewater treatment in reducing marine microplastics. *Journal of Environmental Management*, *266*. https://doi.org/10.1016/j.jenvman.2020.110642.

Frias, J., Filgueiras, A., Gago, J., Pedrotti, M. L., Suaria, G., Tirelli, V., et al. (2019). *Standardised protocol for monitoring microplastics in seawater*. JPI-Oceans BASEMAN project.

Frias, J., Pagter, E., Nash, R., O'Connor, I., Carretero, O., Filgueiras, A., et al. (2018). *Standardised protocol for monitoring microplastics in sediments*. JPI-Oceans BASEMAN project.

Gago, J., Galgani, F., Maes, T., & Thompson, R. C. (2016). Microplastics in seawater: Recommendations from the marine strategy framework directive implementation process. *Frontiers in Marine Science*, *3*(219). https://doi.org/10.3389/fmars.2016.00219.

Gallo, F., Fossi, C., Weber, R., Santillo, D., Sousa, J., Ingram, I., et al. (2018). Marine litter plastics and microplastics and their toxic chemicals components: The need for urgent preventive measures. *Environmental Sciences Europe*, *30*(1), 13. https://doi.org/10.1186/s12302-018-0139-z.

Gao, F., Li, J., Sun, C., Zhang, L., Jiang, F., Cao, W., et al. (2019). Study on the capability and characteristics of heavy metals enriched on microplastics in marine environment. *Marine Pollution Bulletin*, *144*, 61–67. https://doi.org/10.1016/j.marpolbul.2019.04.039.

Gedik, K., Eryaşar, A. R., & Gözler, A. M. (2022). The microplastic pattern of wild-caught Mediterranean mussels from the Marmara Sea. *Marine Pollution Bulletin*, *175*, 113331. https://doi.org/10.1016/j.marpolbul.2022.113331.

GESAMP. (2015). *Sources, fate and effects of microplastics in the marine environment: a global assessment (vol. 90)*. IMO/FAO/UNESCO-IOC/UNIDO/WMO/IAEA/UN/UNEP/UNDP Joint Group of Experts on the Scientific Aspects of Marine Environmental Protection.

Gigault, J., Halle, A. T., Baudrimont, M., Pascal, P. Y., Gauffre, F., Phi, T. L., et al. (2018). Current opinion: What is a nanoplastic? *Environmental Pollution*, *235*, 1030.

Gong, Y., Wang, Y., Chen, L., Li, Y., Chen, X., & Liu, B. (2021). Microplastics in different tissues of a pelagic squid (*Dosidicus gigas*) in the northern Humboldt current ecosystem. *Marine Pollution Bulletin*, *169*, 112509. https://doi.org/10.1016/j.marpolbul.2021.112509.

Gouin, T., Roche, N., Lohmann, R., & Hodges, G. A. (2011). Thermodynamic approach for assessing the environmental exposure of chemicals absorbed to microplastic. *Environmental Science and Technology*, *45*, 1466–1472.

Green, D. S., Kregting, L., Boots, B., Blockley, D. J., Brickle, P., da Costa, M., et al. (2018). A comparison of sampling methods for seawater microplastics and a first report of the microplastic litter in coastal waters of Ascension and Falkland Islands. *Marine Pollution Bulletin*, *137*, 695–701. https://doi.org/10.1016/j.marpolbul.2018.11.004.

Guilhermino, L., Martins, A., Lopes, C., Raimundo, J., Vieira, L. R., Barboza, L. G. A., et al. (2021). Microplastics in fishes from an estuary (Minho River) ending into the NE Atlantic Ocean. *Marine Pollution Bulletin*, *173*, 113008. https://doi.org/10.1016/j.marpolbul.2021.113008.

Gurjar, U. R., Xavier, M., Nayak, B. B., Ramteke, K., Deshmukhe, G., Jaiswar, A. K., et al. (2021). Microplastics in shrimps: a study from the trawling grounds of north eastern part of Arabian Sea. *Environmental Science and Pollution Research*, *28*(35), 48494–48504. https://doi.org/10.1007/s11356-021-14121-z.

Habib, R., Thiemann, T., & Kendi, R. (2020). Microplastics and wastewater treatment plants—A review. *Journal of Water Resource and Protection*, *12*, 1–35. https://doi.org/10.4236/jwarp.2020.121001.

Habibi, N., Uddin, S., Fowler, Scott W., & Behbehani, M. (2022). Microplastics in the atmosphere: A review. *Journal of Environmental Exposure Assessment*, *1*(6). https://doi.org/10.20517/jeea.2021.07. In this issue.

Hajisamae, S., Soe, K. K., Pradit, S., Chaiyvareesajja, J., & Fazrul, H. (2022). Feeding habits and microplastic ingestion of short mackerel, *Rastrelliger brachysoma*, in a tropical estuarine environment. *Environmental Biology of Fishes*, *105*(2), 289–302. https://doi.org/10.1007/s10641-022-01221-z.

Hamzah, S. R., Altrawneh, R. A. S., Anuar, S. T., Khalik, W. M. A. W. M., Kolandhasamy, P., & Ibrahim, Y. S. (2021). Ingestion of microplastics by the estuarine polychaete, Namalycastis sp. in the Setiu wetlands, Malaysia. *Marine Pollution Bulletin*, *170*, 112617. https://doi.org/10.1016/j.marpolbul.2021.112617.

Hasegawa, T., & Nakaoka, M. (2021). Trophic transfer of microplastics from mysids to fish greatly exceeds direct ingestion from the water column. *Environmental Pollution*, *273*, 116468. https://doi.org/10.1016/j.envpol.2021.116468.

Hennicke, A., Macrina, L., Malcolm-Mckay, A., & Miliou, A. (2021). Assessment of microplastic accumulation in wild Paracentrotus lividus, a commercially important sea urchin species, in the eastern Aegean Sea, Greece. *Regional Studies in Marine Science*, *45*, 101855. https://doi.org/10.1016/j.rsma.2021.101855.

Hermabessiere, L., Paul-Pont, I., Cassone, A.-L., Himber, C., Receveur, J., Jezequel, R., et al. (2019). Microplastic contamination and pollutant levels in mussels and cockles collected along the channel coasts. *Environmental Pollution*, *250*, 807–819.

Hermsen, W., Sims, I., & Crane, M. (1994). The bioavailability and toxicity to Mytilus edulis L. of two organochlorine pesticides adsorbed to suspended solids. *Marine Environment Research*, *38*, 61–69.

Hidalgo-Ruz, V., Gutow, L., Thompson, R. C., & Thiel, M. (2012). Microplastics in the marine environment: a review of the methods used for identification and quantification. *Environmental Science & Technology*, *46*(6), 3060–3075. https://doi.org/10.1021/es2031505.

Hildebrandt, L., Nack, F. L., Zimmermann, T., & Pröfrock, D. (2021). Microplastics as a trojan horse for trace metals. *Journal of Hazardous Materials Letters*, *2*, 100035. https://doi.org/10.1016/j.hazl.2021.100035.

Horton, A. A., Walton, A., Spurgeon, D. J., Lahive, E., & Svendsen, C. (2017). Microplastics in freshwater and terrestrial environments: Evaluating the current understanding to identify the knowledge gaps and future research priorities. *Science of the Total Environment*, *586*, 127–141. https://doi.org/10.1016/j.scitotenv.2017.01.190.

Hossain, M. S., Rahman, M. S., Uddin, M. N., Sharifuzzaman, S. M., Chowdhury, S. R., Sarker, S., et al. (2020). Microplastic contamination in penaeid shrimp from the Northern Bay of Bengal. *Chemosphere*, *238*, 124688. https://doi.org/10.1016/j.chemosphere.2019.124688.

Imhof, H. K., Rusek, J., Thiel, M., Wolinska, J., & Laforsch, C. (2017). Do microplastic particles affect Daphnia magna at the morphological, life history and molecular level? *PLoS One*, *12*(11), e0187590. https://doi.org/10.1371/journal.pone.0187590.

Jaafar, N., Azfaralariff, A., Musa, S. M., Mohamed, M., Yusoff, A. H., & Lazim, A. M. (2021). Occurrence, distribution and characteristics of microplastics in gastrointestinal tract and gills of commercial marine fish from Malaysia. *Science of the Total Environment*, *799*, 149457. https://doi.org/10.1016/j.scitotenv.2021.149457.

Jang, M., Shim, W. J., Han, G. M., Song, Y. K., & Hong, S. H. (2018). Formation of microplastics by polychaetes (Marphysa sanguinea) inhabiting expanded polystyrene marine debris. *Marine Pollution Bulletin*, *131*, 365–369. https://doi.org/10.1016/j.marpolbul.2018.04.017.

João, P.d. C., Paço, A., Santos, P. S., Duarte, A. C., & Rocha-Santos, T. (2018). Microplastics in soils: Assessment, analytics and risks. *Environmental Chemistry*, *16*(1), 18–30.

Kaposi, K. L., Mos, B., Kelaher, B. P., & Dworjanyn, S. A. (2014). Ingestion of microplastic has limited impact on a marine larva. *Environmental Science & Technology*, *48*(3), 1638–1645. https://doi.org/10.1021/es404295e.

Karami, A., Golieskardi, A., Choo, C. K., Larat, V., Karbalaei, S., & Salamatinia, B. (2018). Microplastic and mesoplastic contamination in canned sardines and sprats. *Science of the Total Environment*, *612*, 1380–1386. https://doi.org/10.1016/j.scitotenv.2017.09.005.

Koelmans, A. A., Bakir, A., Burton, G. A., & Janssen, C. R. (2016). Microplastic as a vector for Chemicals in the Aquatic Environment: Critical review and model-supported reinterpretation of empirical studies. *Environmental Science & Technology*, *50*(7), 3315–3326. https://doi.org/10.1021/acs.est.5b06069.

Koelmans, A. A., Besseling, E., Wegner, A., & Foekema, E. M. (2013). Correction to plastic as a carrier of POPs to aquatic organisms. In *Vol. 47. A model analysis* (pp. 8992–8993). Environment Science and Technology.

Kühn, S., Bravo Rebolledo, E. L., van Franeker, J. A., Bergmann, M., Gutow, L., & Klages, M. (2015). *Marine anthropogenic litter.*

Lehtiniemi, M., Hartikainen, S., Näkki, P., Engström-Öst, J., Koistinen, A., & Setälä, O. (2018). Size matters more than shape: Ingestion of primary and secondary microplastics by small predators. *Food Webs*, *17*, e00097. https://doi.org/10.1016/j.fooweb.2018.e00097.

Lenz, R., Enders, K., Stedmon, C. A., Mackenzie, D. M. A., & Nielsen, T. G. (2015). A critical assessment of visual identification of marine microplastic using Raman spectroscopy for analysis improvement. *Marine Pollution Bulletin*, *100*(1), 82–91. https://doi.org/10.1016/j.marpolbul.2015.09.026.

Leslie, H. A., Brandsma, S. H., van Velzen, M. J. M., & Vethaak, A. D. (2017). Microplastics en route: Field measurements in the Dutch river delta and Amsterdam canals, wastewater treatment plants, North Sea sediments and biota. *Environment International*, *101*, 133–142. https://doi.org/10.1016/j.envint.2017.01.018.

Leung, M. M.-L., Ho, Y.-W., Maboloc, E. A., Lee, C.-H., Wang, Y., Hu, M., et al. (2021). Determination of microplastics in the edible green-lipped mussel Perna viridis using an automated mapping technique of Raman microspectroscopy. *Journal of Hazardous Materials*, *420*, 126541. https://doi.org/10.1016/j.jhazmat.2021.126541.

Li, J. N., Lusher, A. L., Rotchell, J. M., Deudero, S., Turra, A., Brate, I. L. N., et al. (2019). Using mussel as a global bioindicator of coastal microplastic pollution. *Environmental Pollution*, *244*, 522.

Li, H.-X., Ma, L.-S., Lin, L., Ni, Z.-X., Xu, X.-R., Shi, H.-H., et al. (2018). Microplastics in oysters *Saccostrea cucullata* along the Pearl River estuary, China. *Environmental Pollution*, *236*, 619–625. https://doi.org/10.1016/j.envpol.2018.01.083.

Li, H., Wang, F., Li, J., Deng, S., & Zhang, S. (2021). Adsorption of three pesticides on polyethylene microplastics in aqueous solutions: Kinetics, isotherms, thermodynamics, and molecular dynamics simulation. *Chemosphere*, *264*, 128556. https://doi.org/10.1016/j.chemosphere.2020.128556.

Liao, C.-P., Chiu, C.-C., & Huang, H.-W. (2021). Assessment of microplastics in oysters in coastal areas of Taiwan. *Environmental Pollution*, *286*, 117437. https://doi.org/10.1016/j.envpol.2021.117437.

Lim, X. (2021). *Microplastics are everywhere-but are they harmful?*. Nature Publishing Group.

Liu, S., Huang, J., Zhang, W., Shi, L., Yi, K., Yu, H., et al. (2022). Microplastics as a vehicle of heavy metals in aquatic environments: A review of adsorption factors, mechanisms, and biological effects. *Journal of Environmental Management*, *302*, 113995. https://doi.org/10.1016/j.jenvman.2021.113995.

Lozano, H. E. A., Ramírez, Á. N., Rios Mendoza, L. M., Macías, Z. J. V., Sánchez, O. J. L., & Hernández, G. F. A. (2021). Microplastic concentrations in cultured oysters in two seasons from two bays of Baja California, Mexico. *Environmental Pollution*, *290*, 118031. https://doi.org/10.1016/j.envpol.2021.118031.

Lundebye, A.-K., Lusher, A. L., & Bank, M. S. (2022). Marine microplastics and seafood: Implications for food security. In M. S. Bank (Ed.), *Microplastic in the environment: Pattern and process* (pp. 131–153). Cham: Springer International Publishing.

Lusher, A. (2015). Microplastics in the marine environment: Distribution, interactions and effects. In M. Bergmann, L. Gutow, & M. Klages (Eds.), *Marine anthropogenic litter* (pp. 245–307). Cham: Springer International Publishing.

Lusher, A. L., Welden, N. A., Sobral, P., & Cole, M. (2017). Sampling, isolating and identifying microplastics ingested by fish and invertebrates. *Analytical Methods*, *9*(9), 1346.

Maass, S., Daphi, D., Lehmann, A., & Rillig, M. C. (2017). Transport of microplastics by two collembolan species. *Environmental Pollution*, *225*, 456.

Maes, T., Perry, J., Alliji, K., Clarke, C., & Birchenough, S. N. R. (2019). Shades of grey: Marine litter research developments in Europe. *Marine Pollution Bulletin*, *146*, 274–281. https://doi.org/10.1016/j.marpolbul.2019.06.019.

Masura, J., Baker, J., Foster, G., & Arthur, C. (2015). *Laboratory methods for the analysis of microplastics in the marine environment: Recommendations for quantifying synthetic particles in waters and sediments*. NOAA Technical Memorandum NOS-OR&R-48.

Mato, Y., Isobe, T., Takada, H., Kanehiro, H., Ohtake, C., & Kaminuma, T. (2001). Plastic resin pellets as a transport medium for toxic chemicals in the marine environment. *Environmental Science & Technology*, *35*(2), 318.

Mayoma, B. S., Sørensen, C., Shashoua, Y., & Khan, F. R. (2020). Microplastics in beach sediments and cockles (*Anadara antiquata*) along the Tanzanian coastline. *Bulletin of Environmental Contamination and Toxicology*, *105*(4), 513–521. https://doi.org/10.1007/s00128-020-02991-x.

Mistri, M., Sfriso, A. A., Casoni, E., Nicoli, M., Vaccaro, C., & Munari, C. (2022). Microplastic accumulation in commercial fish from the Adriatic Sea. *Marine Pollution Bulletin*, *174*, 113279. https://doi.org/10.1016/j.marpolbul.2021.113279.

Mohamed, N. N. H., Kooi, M., Diepens, N. J., & Koelmans, A. A. (2021). Lifetime accumulation of microplastic in children and adults. *Environmental Science & Technology*, *55*(8), 5084–5096. https://doi.org/10.1021/acs.est.0c07384.
Mohsen, M., Lin, C., Liu, S., & Yang, H. (2022). Existence of microplastics in the edible part of the sea cucumber Apostichopus japonicus. *Chemosphere*, *287*, 132062. https://doi.org/10.1016/j.chemosphere.2021.132062.
Mohsen, M., Wang, Q., Zhang, L., Sun, L., Lin, C., & Yang, H. (2019). Microplastic ingestion by the farmed sea cucumber *Apostichopus japonicus* in China. *Environmental Pollution*, *245*, 1071–1078. https://doi.org/10.1016/j.envpol.2018.11.083.
Montoto-Martínez, T., De la Fuente, J., Puig-Lozano, R., Marques, N., Arbelo, M., Hernández-Brito, J. J., et al. (2021). Microplastics, bisphenols, phthalates and pesticides in odontocete species in the Macaronesian region (eastern North Atlantic). *Marine Pollution Bulletin*, *173*, 113105. https://doi.org/10.1016/j.marpolbul.2021.113105.
Moore, R. C., Noel, M., Etemadifar, A., Loseto, L., Posacka, A. M., Bendell, L., et al. (2022). Microplastics in beluga whale (*Delphinapterus leucas*) prey: An exploratory assessment of trophic transfer in the Beaufort Sea. *Science of the Total Environment*, *806*, 150201. https://doi.org/10.1016/j.scitotenv.2021.150201.
Muhammad Husin, M. J., Mazlan, N., Shalom, J., Saud, S. N., & Abdullah Sani, M. S. (2021). Evaluation of microplastics ingested by sea cucumber *Stichopus horrens* in Pulau Pangkor, Perak, Malaysia. *Environmental Science and Pollution Research*, *28*(43), 61592–61600. https://doi.org/10.1007/s11356-021-15099-4.
Naji, A., Azadkhah, S., Farahani, H., Uddin, S., & Khan, F. R. (2021). Microplastics in wastewater outlets of Bandar Abbas city (Iran): A potential point source of microplastics into the Persian Gulf. *Chemosphere*, *262*, 128039. https://doi.org/10.1016/j.chemosphere.2020.128039.
Narmatha Sathish, M., Immaculate Jeyasanta, K., & Patterson, J. (2020). Monitoring of microplastics in the clam Donax cuneatus and its habitat in Tuticorin coast of gulf of Mannar (GoM), India. *Environmental Pollution*, *266*, 115219. https://doi.org/10.1016/j.envpol.2020.115219.
Nelms, S. E., Barnett, J., Brownlow, A., Davison, N. J., Deaville, R., Galloway, T. S., et al. (2019). Microplastics in marine mammals stranded around the British coast: Ubiquitous but transitory? *Scientific Reports*, *9*(1), 1075. https://doi.org/10.1038/s41598-018-37428-3.
Nelms, S. E., Galloway, T. S., Godley, B. J., Jarvis, D. S., & Lindeque, P. K. (2018). Investigating microplastic trophic transfer in marine top predators. *Environmental Pollution*, *238*, 999–1007. https://doi.org/10.1016/j.envpol.2018.02.016.
Oliveira, M., & Almeida, M. (2019). The why and how of micro(nano)plastic research. *TrAC Trends in Analytical Chemistry*, *114*, 196–201. https://doi.org/10.1016/j.trac.2019.02.023.
Oliveira, A. R., Sardinha-Silva, A., Andrews, P. L. R., Green, D., Cooke, G. M., Hall, S., et al. (2020). Microplastics presence in cultured and wild-caught cuttlefish, Sepia officinalis. *Marine Pollution Bulletin*, *160*, 111553. https://doi.org/10.1016/j.marpolbul.2020.111553.
OSPAR. (2010). *Guideline for monitoring marine litter on the beaches in the OSPAR maritime area*. OSPAR Commission. ISBN: 90 3631 973 9.
Pan, Z., Liu, Q., Sun, Y., Sun, X., & Lin, H. (2019). Environmental implications of microplastic pollution in the northwestern Pacific Ocean. *Marine Pollution Bulletin*, *146*, 215–224. https://doi.org/10.1016/j.marpolbul.2019.06.031.
Panti, C., Baini, M., Lusher, A., Hernandez-Milan, G., Bravo Rebolledo, E. L., Unger, B., et al. (2019). Marine litter: One of the major threats for marine mammals. Outcomes from the European cetacean society workshop. *Environmental Pollution*, *247*, 72–79. https://doi.org/10.1016/j.envpol.2019.01.029.

Patterson, J., Jeyasanta, K. I., Laju, R. L., & Edward, J. K. P. (2021). Microplastic contamination in Indian edible mussels (Perna perna and Perna viridis) and their environs. *Marine Pollution Bulletin, 171*, 112678. https://doi.org/10.1016/j.marpolbul.2021.112678.

Patterson, J., Jeyasanta, K. I., Sathish, N., Booth, A. M., & Edward, J. K. P. (2019). Profiling microplastics in the Indian edible oyster, *Magallana bilineata* collected from the Tuticorin coast, gulf of Mannar, southeastern India. *Science of the Total Environment, 691*, 727–735. https://doi.org/10.1016/j.scitotenv.2019.07.063.

Payton, T. G., Beckingham, B. A., & Dustan, P. (2020). Microplastic exposure to zooplankton at tidal fronts in Charleston Harbor, SC USA. *Estuarine, Coastal and Shelf Science, 232*, 106510. https://doi.org/10.1016/j.ecss.2019.106510.

Peng, B., Hossain, K. B., Lin, Y., Zhang, M., Zheng, H., Yu, J., et al. (2022). Assessment and sources identification of microplastics, PAHs and OCPs in the Luoyuan Bay, China: Based on multi-statistical analysis. *Marine Pollution Bulletin, 175*, 113351. https://doi.org/10.1016/j.marpolbul.2022.113351.

Pfeiffer, F., & Fischer, E. K. (2020). Various digestion protocols within microplastic sample processing—Evaluating the resistance of different synthetic polymers and the efficiency of biogenic organic matter destruction. *Frontiers in environmental. Science, 8*(263). https://doi.org/10.3389/fenvs.2020.572424.

Piyawardhana, N., Weerathunga, V., Chen, H.-S., Guo, L., Huang, P.-J., Ranatunga, R. R. M. K. P., et al. (2022). Occurrence of microplastics in commercial marine dried fish in Asian countries. *Journal of Hazardous Materials, 423*, 127093. https://doi.org/10.1016/j.jhazmat.2021.127093.

Prasetyo, D., & Putri, L. S. (2021). Cuttlefish (*Sepia pharaonis* Ehrenberg, 1831) as a bio-indicator of microplastic pollution. *Aquaculture, Aquarium, Conservation & Legislation, 14*(2), 918–930.

Prinz, N., & Korez, Š. (2020). Understanding how microplastics affect marine biota on the cellular level is important for assessing ecosystem function: A review. In S. Jungblut, V. Liebich, & M. Bode-Dalby (Eds.), *YOUMARES 9 - The Oceans: Our Research, Our Future: Proceedings of the 2018 conference for YOUng MArine RESearcher in Oldenburg, Germany* (pp. 101–120). Cham: Springer International Publishing.

Provvisiero, D. P., Pivonello, C., Muscogiuri, G., Negri, M., de Angelis, C., Simeoli, C., et al. (2016). Influence of bisphenol a on type 2 diabetes mellitus. *International Journal of Environmental Research and Public Health, 13*(10), 989. https://doi.org/10.3390/ijerph13100989.

Ragusa, A., Svelato, A., Santacroce, C., Catalano, P., Notarstefano, V., Carnevali, O., et al. (2021). Plasticenta: First evidence of microplastics in human placenta. *Environment International, 146*, 106274. https://doi.org/10.1016/j.envint.2020.106274.

Rapp, J., Herrera, A., Bondyale-Juez, D. R., González-Pleiter, M., Reinold, S., Asensio, M., et al. (2021). Microplastic ingestion in jellyfish Pelagia noctiluca (Forsskal, 1775) in the North Atlantic Ocean. *Marine Pollution Bulletin, 166*, 112266. https://doi.org/10.1016/j.marpolbul.2021.112266.

Rashid, C. P., Jyothibabu, R., Arunpandi, N., Abhijith, V. T., Josna, M. P., Vidhya, V., et al. (2021). Microplastics in zooplankton in the eastern Arabian Sea: The threats they pose to fish and corals favoured by coastal currents. *Marine Pollution Bulletin, 173*, 113042. https://doi.org/10.1016/j.marpolbul.2021.113042.

Ray, S. S., Lee, H. K., Huyen, D. T. T., Chen, S.-S., & Kwon, Y.-N. (2022). Microplastics waste in environment: A perspective on recycling issues from PPE kits and face masks during the COVID-19 pandemic. *Environmental Technology and Innovation, 26*, 102290. https://doi.org/10.1016/j.eti.2022.102290.

Richardson, C. R., Burritt, D. J., Allan, B. J. M., & Lamare, M. D. (2021). Microplastic ingestion induces asymmetry and oxidative stress in larvae of the sea urchin Pseudechinus huttoni. *Marine Pollution Bulletin, 168*, 112369. https://doi.org/10.1016/j.marpolbul.2021.112369.

Rochman, C. M., Brookson, C., Bikker, J., Djuric, N., Earn, A., Bucci, K., et al. (2019). Rethinking microplastics as a diverse contaminant suite. *Environmental Toxicology and Chemistry*, *38*(4), 703–711. https://doi.org/10.1002/etc.4371.

Rochman, C. M., Hentschel, B. T., & Teh, S. J. (2014). Long-term sorption of metals is similar among plastic types: Implications for plastic debris in aquatic environments. *PLoS One*, *9*(1), e85433. https://doi.org/10.1371/journal.pone.0085433.

Rochman, C. M., Hoh, E., Kurobe, T., & Teh, S. J. (2013). Ingested plastic transfers hazardous chemicals to fish and induces hepatic stress. *Scientific Reports*, *3*, 3263.

Rodrigues, J. P., Duarte, A. C., Santos-Echeandía, J., & Rocha-Santos, T. (2019). Significance of interactions between microplastics and POPs in the marine environment: A critical overview. *TrAC Trends in Analytical Chemistry*, *111*, 252–260. https://doi.org/10.1016/j.trac.2018.11.038.

Sainio, E., Lehtiniemi, M., & Setälä, O. (2021). Microplastic ingestion by small coastal fish in the northern Baltic Sea, Finland. *Marine Pollution Bulletin*, *172*, 112814. https://doi.org/10.1016/j.marpolbul.2021.112814.

Sawalman, R., Werorilangi, S., Ukkas, M., Mashoreng, S., Yasir, I., & Tahir, A. (2021). Microplastic abundance in sea urchins (*Diadema setosum*) from seagrass beds of Barranglompo Island, Makassar, Indonesia. *IOP Conference Series: Earth and Environmental Science*, *763*(1), 012057. https://doi.org/10.1088/1755-1315/763/1/012057.

Sayogo, B. H., Patria, M. P., & Takarina, N. D. (2020). The density of microplastic in sea cucumber (Holothuria sp.) and sediment at Tidung Besar and Bira Besar island, Jakarta. *Journal of Physics: Conference Series*, *1524*(1), 012064. https://doi.org/10.1088/1742-6596/1524/1/012064.

Schymanski, D., Goldbeck, C., Humpf, H. U., & Furst, P. (2018). Analysis of microplastics in water by micro-Raman spectroscopy: Release of plastic particles from different packaging into mineral water. *Water Research*, *129*, 154–162. https://doi.org/10.1016/j.watres.2017.11.011.

Senathirajah, K., Attwood, S., Bhagwat, G., Carbery, M., Wilson, S., & Palanisami, T. (2021). Estimation of the mass of microplastics ingested – A pivotal first step towards human health risk assessment. *Journal of Hazardous Materials*, *404*, 124004. https://doi.org/10.1016/j.jhazmat.2020.124004.

Sevillano-González, M., González-Sálamo, J., Díaz-Peña, F. J., Hernández-Sánchez, C., Catalán Torralbo, S., Ródenas Seguí, A., et al. (2021). Assessment of microplastic content in Diadema africanum sea urchin from Tenerife (Canary Islands, Spain). *Marine Pollution Bulletin*, *113174*. https://doi.org/10.1016/j.marpolbul.2021.113174.

Sharifinia, M., Bahmanbeigloo, Z. A., Keshavarzifard, M., Khanjani, M. H., & Lyons, B. P. (2020). Microplastic pollution as a grand challenge in marine research: A closer look at their adverse impacts on the immune and reproductive systems. *Ecotoxicology and Environmental Safety*, *204*, 111109. https://doi.org/10.1016/j.ecoenv.2020.111109.

Sharma, S., & Chatterjee, S. (2017). Microplastic pollution, a threat to marine ecosystem and human health: a short review. *Environmental Science and Pollution Research*, *24*(27), 21530–21547. https://doi.org/10.1007/s11356-017-9910-8.

Shen, M., Zeng, Z., Song, B., Yi, H., Hu, T., Zhang, Y., et al. (2021). Neglected microplastics pollution in global COVID-19: Disposable surgical masks. *Science of the Total Environment*, *790*, 148130. https://doi.org/10.1016/j.scitotenv.2021.148130.

Shim, W. J., Hong, S. H., & Eo, S. (2017). Identification methods in microplastics analysis: a review. *Analytical Methods*, *3*(1384).

Siddique, M. A. M., Uddin, A., Rahman, S. M. A., Rahman, M., Islam, M. S., & Kibria, G. (2022). Microplastics in an anadromous national fish, Hilsa shad Tenualosa ilisha from the bay of Bengal, Bangladesh. *Marine Pollution Bulletin*, *174*, 113236. https://doi.org/10.1016/j.marpolbul.2021.113236.

Singla, M., Díaz, J., Broto-Puig, F., & Borrós, S. (2020). Sorption and release process of polybrominated diphenyl ethers (PBDEs) from different composition microplastics in aqueous medium: Solubility parameter approach. *Environmental Pollution*, *262*, 114377. https://doi.org/10.1016/j.envpol.2020.114377.

Sipps, K., Arbuckle-Keil, G., Chant, R., Fahrenfeld, N., Garzio, L., Walsh, K., et al. (2022). Pervasive occurrence of microplastics in Hudson-Raritan estuary zooplankton. *Science of the Total Environment*, *817*, 152812. https://doi.org/10.1016/j.scitotenv.2021.152812.

Smith, M., Love, D. C., Rochman, C. M., & Neff, R. A. (2018). Microplastics in seafood and the implications for human health. *Current Environmental Health Reports*, *5*(3), 375–386. https://doi.org/10.1007/s40572-018-0206-z.

Song, Y. K., Hong, S. H., Jang, M., Han, G. M., Rani, M., Lee, J., et al. (2015). A comparison of microscopic and spectroscopic identification methods for analysis of microplastics in environmental samples. *Marine Pollution Bulletin*, *93*(1–2), 202–209. https://doi.org/10.1016/j.marpolbul.2015.01.015.

Sparks, C., Awe, A., & Maneveld, J. (2021). Abundance and characteristics of microplastics in retail mussels from Cape Town, South Africa. *Marine Pollution Bulletin*, *166*, 112186. https://doi.org/10.1016/j.marpolbul.2021.112186.

Stienbarger, C. D., Joseph, J., Athey, S. N., Monteleone, B., Andrady, A. L., Watanabe, W. O., et al. (2021). Direct ingestion, trophic transfer, and physiological effects of microplastics in the early life stages of *Centropristis striata*, a commercially and recreationally valuable fishery species. *Environmental Pollution*, *285*, 117653. https://doi.org/10.1016/j.envpol.2021.117653.

Su, L., Cai, H., Kolandhasamy, P., Wu, C., Rochman, C. M., & Shi, H. (2018). Using the Asian clam as an indicator of microplastic pollution in freshwater ecosystems. *Environmental Pollution*, *234*, 347–355. https://doi.org/10.1016/j.envpol.2017.11.075.

Sucharitakul, P., Pitt, K. A., & Welsh, D. T. (2021). Trophic transfer of microbeads to jellyfish and the importance of aging microbeads for microplastic experiments. *Marine Pollution Bulletin*, *172*, 112867. https://doi.org/10.1016/j.marpolbul.2021.112867.

Sui, M., Lu, Y., Wang, Q., Hu, L., Huang, X., & Liu, X. (2020). Distribution patterns of microplastics in various tissues of the Zhikong scallop (*Chlamys farreri*) and in the surrounding culture seawater. *Marine Pollution Bulletin*, *160*, 111595. https://doi.org/10.1016/j.marpolbul.2020.111595.

Sun, B., Liu, J., Zhang, Y.-Q., Leungb, K. M. Y., & Zeng, E. Y. (2021). Leaching of polybrominated diphenyl ethers from microplastics in fish oil: Kinetics and bioaccumulation. *Journal of Hazardous Materials*, *406*, 124726. https://doi.org/10.1016/j.jhazmat.2020.124726.

Sun, X., Liu, T., Zhu, M., Liang, J., Zhao, Y., & Zhang, B. (2018). Retention and characteristics of microplastics in natural zooplankton taxa from the East China Sea. *Science of the Total Environment*, *640-641*, 232–242. https://doi.org/10.1016/j.scitotenv.2018.05.308.

Sussarellu, R., Suquet, M., Thomas, Y., Lambert, C., Fabioux, C., Pernet, M. E. J., et al. (2016). Oyster reproduction is affected by exposure to polystyrene microplastics. *Proceedings of the National Academy of Sciences*, *113*(9), 2430–2435. https://doi.org/10.1073/pnas.1519019113.

Taha, Z. D., Md Amin, R., Anuar, S. T., Nasser, A. A. A., & Sohaimi, E. S. (2021). Microplastics in seawater and zooplankton: A case study from Terengganu estuary and offshore waters, Malaysia. *Science of the Total Environment*, *786*, 147466. https://doi.org/10.1016/j.scitotenv.2021.147466.

Teng, J., Wang, Q., Ran, W., Wu, D., Liu, Y., Sun, S., et al. (2019). Microplastic in cultured oysters from different coastal areas of China. *Science of the Total Environment*, *653*, 1282–1292. https://doi.org/10.1016/j.scitotenv.2018.11.057.

Uddin, S., Bebhehani, M., Sajid, S., & Karam, Q. (2019). Concentration of 210Po and 210Pb in macroalgae from the northern gulf. *Marine Pollution Bulletin, 145*, 474–479. https://doi.org/10.1016/j.marpolbul.2019.06.056.
Uddin, S., Behbehani, M., Al-Ghadban, A., Sajid, S., Al-Zekri, W., Ali, M., et al. (2018a). ^{210}Po concentration in selected calanoid copepods in the northern Arabian gulf. *Marine Pollution Bulletin, 133*, 861–864. https://doi.org/10.1016/j.marpolbul.2018.06.061.
Uddin, S., Behbehani, M., Al-Ghadban, A. N., Sajid, S., Vinod Kumar, V., Al-Musallam, L., et al. (2018b). ^{210}Po concentration in selected diatoms and dinoflagellates in the northern Arabian gulf. *Marine Pollution Bulletin, 129*(1), 343–346. https://doi.org/10.1016/j.marpolbul.2018.02.051.
Uddin, S., Fowler, S. W., & Behbehani, M. (2020). An assessment of microplastic inputs into the aquatic environment from wastewater streams. *Marine Pollution Bulletin, 160*, 111538. https://doi.org/10.1016/j.marpolbul.2020.111538.
Uddin, S., Fowler, S. W., Habibi, N., Sajid, S., Dupont, S., & Behbehani, M. (2022). A preliminary assessment of size fractionated microplastics in indoor aerosol - Kuwait's baseline. *Toxics, 10*(2(71)). https://doi.org/10.3390/tox-ics10020071.
Uddin, S., Fowler, S. W., Habibi, N., & Behbehani, M. (2022). Micro-Nano plastic in the aquatic environment: Methodological problems and challenges. *Animals, 12*(3), 297.
Uddin, S., Fowler, S. W., & Saeed, T. (2020). Microplastic particles in the Persian/Arabian gulf – A review on sampling and identification. *Marine Pollution Bulletin, 154*, 111100. https://doi.org/10.1016/j.marpolbul.2020.111100.
Uddin, S., Fowler, S. W., Saeed, T., Naji, A., & Al-Jandal, N. (2020). Standardized protocols for microplastics determinations in environmental samples from the Gulf and marginal seas. *Marine Pollution Bulletin, 158*, 111374. https://doi.org/10.1016/j.marpolbul.2020.111374.
Uddin, S., Fowler, S. W., Uddin, M. F., Behbehani, M., & Naji, A. (2021). A review of microplastic distribution in sediment profiles. *Marine Pollution Bulletin, 163*, 111973. https://doi.org/10.1016/j.marpolbul.2021.111973.
Ukhrowi, H., Wardhana, W., & Patria, M. (2021). Microplastic abundance in blood cockle Anadara granosa (linnaeus, 1758) at Lada Bay, Pandeglang, Banten. *Paper presented at the Journal of Physics: Conference Series.*
Van Cauwenberghe, L., Devriese, L., Galgani, F., Robbens, J., & Janssen, C. R. (2015). Microplastics in sediments: a review of techniques, occurrence and effects. *Marine Environmental Research, 111*, 5–17.
Van Cauwenberghe, L., & Janssen, C. R. (2014). Microplastics in bivalves cultured for human consumption. *Environmental Pollution, 193*, 65–70. https://doi.org/10.1016/j.envpol.2014.06.010.
Van Colen, C., Vanhove, B., Diem, A., & Moens, T. (2020). Does microplastic ingestion by zooplankton affect predator-prey interactions? An experimental study on larviphagy. *Environmental Pollution, 256*, 113479. https://doi.org/10.1016/j.envpol.2019.113479.
Vermeiren, P., Muñoz, C., & Ikejima, K. (2020). Microplastic identification and quantification from organic rich sediments: A validated laboratory protocol. *Environmental Pollution, 262*, 114298. https://doi.org/10.1016/j.envpol.2020.114298.
von Moos, N., Burkhardt-Holm, P., & Köhler, A. (2012). Uptake and effects of microplastics on cells and tissue of the blue mussel Mytilus edulis L. after an experimental exposure. *Environmental Science & Technology, 46*(20), 11327–11335. https://doi.org/10.1021/es302332w.
Vroom, R. J. E., Koelmans, A. A., Besseling, E., & Halsband, C. (2017). Aging of microplastics promotes their ingestion by marine zooplankton. *Environmental Pollution, 231*(Pt 1), 987–996. https://doi.org/10.1016/j.envpol.2017.08.088.
Waddell, E. N., Lascelles, N., & Conkle, J. L. (2020). Microplastic contamination in Corpus Christi Bay blue crabs, *Callinectes sapidus. Limnology and Oceanography Letters, 5*(1), 92–102. https://doi.org/10.1002/lol2.10142.

Wagner, S., Huffer, T., Klockner, P., Wehrhahn, M., Hofmann, T., & Reemtsma, T. (2018). Tire wear particles in the aquatic environment - a review on generation, analysis, occurrence, fate and effects. *Water Research, 139*, 83.

Waite, H. R., Donnelly, M. J., & Walters, L. J. (2018). Quantity and types of microplastics in the organic tissues of the eastern oyster Crassostrea virginica and Atlantic mud crab Panopeus herbstii from a Florida estuary. *Marine Pollution Bulletin, 129*(1), 179–185. https://doi.org/10.1016/j.marpolbul.2018.02.026.

Wang, T., Hu, M., Xu, G., Shi, H., Leung, J. Y. S., & Wang, Y. (2021). Microplastic accumulation via trophic transfer: Can a predatory crab counter the adverse effects of microplastics by body defence? *Science of the Total Environment, 754*, 142099. https://doi.org/10.1016/j.scitotenv.2020.142099.

Wang, D., Su, L., Ruan, H. D., Chen, J., Lu, J., Lee, C.-H., et al. (2021). Quantitative and qualitative determination of microplastics in oyster, seawater and sediment from the coastal areas in Zhuhai, China. *Marine Pollution Bulletin, 164*, 112000. https://doi.org/10.1016/j.marpolbul.2021.112000.

Wang, F., Yu, Y., Wu, H., Wu, W., Wang, L., An, L., et al. (2021). Microplastics in spotted seal cubs (Phoca largha): Digestion after ingestion? *Science of the Total Environment, 785*, 147426. https://doi.org/10.1016/j.scitotenv.2021.147426.

Watts, A. J. R., Lewis, C., Goodhead, R. M., Beckett, S. J., Moger, J., Tyler, C. R., et al. (2014). Uptake and retention of microplastics by the shore crab Carcinus maenas. *Environmental Science & Technology, 48*(15), 8823–8830. https://doi.org/10.1021/es501090e.

Weber, A., von Randow, M., Voigt, A.-L., von der Au, M., Fischer, E., Meermann, B., et al. (2021). Ingestion and toxicity of microplastics in the freshwater gastropod Lymnaea stagnalis: No microplastic-induced effects alone or in combination with copper. *Chemosphere, 263*, 128040. https://doi.org/10.1016/j.chemosphere.2020.128040.

Wei, L., Wang, D., Aierken, R., Wu, F., Dai, Y., Wang, X., et al. (2022). The prevalence and potential implications of microplastic contamination in marine fishes from Xiamen Bay, China. *Marine Pollution Bulletin, 174*, 113306. https://doi.org/10.1016/j.marpolbul.2021.113306.

Wicaksono, K. B., Patria, M. P., & Suryanda, A. (2021). Microplastic ingestion in the black sea cucumber *Holothuria leucospilota* (Brandt, 1835) collected from Rambut Island, Seribu Islands, Jakarta, Indonesia. *IOP Conference Series: Materials Science and Engineering, 1098*(5), 052049. https://doi.org/10.1088/1757-899x/1098/5/052049.

Wootton, N., Reis-Santos, P., Dowsett, N., Turnbull, A., & Gillanders, B. M. (2021). Low abundance of microplastics in commercially caught fish across southern Australia. *Environmental Pollution, 290*, 118030. https://doi.org/10.1016/j.envpol.2021.118030.

Xu, X., Fang, J. K.-H., Wong, C.-Y., & Cheung, S.-G. (2022). The significance of trophic transfer in the uptake of microplastics by carnivorous gastropod Reishia clavigera. *Environmental Pollution, 298*, 118862. https://doi.org/10.1016/j.envpol.2022.118862.

Yozukmaz, A. (2021). Investigation of microplastics in edible wild mussels from İzmir Bay (Aegean Sea, Western Turkey): A risk assessment for the consumers. *Marine Pollution Bulletin, 171*, 112733. https://doi.org/10.1016/j.marpolbul.2021.112733.

Zakeri, M., Naji, A., Akbarzadeh, A., & Uddin, S. (2020). Microplastic ingestion in important commercial fish in the southern Caspian Sea. *Marine Pollution Bulletin, 160*, 111598. https://doi.org/10.1016/j.marpolbul.2020.111598.

Zhang, Y., Kang, S., Allen, S., Allen, D., Gao, T., & Sillanpää, M. (2020). Atmospheric microplastics: A review on current status and perspectives. *Earth-Science Reviews, 203*, 103118. https://doi.org/10.1016/j.earscirev.2020.103118.

Zhang, S., Sun, Y., Liu, B., & Li, R. (2021). Full size microplastics in crab and fish collected from the mangrove wetland of Beibu gulf: Evidences from Raman tweezers (1–20 μm) and spectroscopy (20–5000 μm). *Science of the Total Environment, 759*, 143504. https://doi.org/10.1016/j.scitotenv.2020.143504.

Zhang, T., Sun, Y., Song, K., Du, W., Huang, W., Gu, Z., et al. (2021). Microplastics in different tissues of wild crabs at three important fishing grounds in China. *Chemosphere*, *271*, 129479. https://doi.org/10.1016/j.chemosphere.2020.129479.

Zhang, S., Wang, J., Liu, X., Qu, F., Wang, X., Wang, X., et al. (2019). Microplastics in the environment: A review of analytical methods, distribution, and biological effects. *TrAC Trends in Analytical Chemistry*, *111*, 62–72. https://doi.org/10.1016/j.trac.2018.12.002.

Zhang, X., Yan, B., & Wang, X. (2020). Selection and optimization of a protocol for extraction of microplastics from *Mactra veneriformis*. *Science of the Total Environment*, *746*, 141250. https://doi.org/10.1016/j.scitotenv.2020.141250.

Zhu, J., Zhang, Q., Huang, Y., Jiang, Y., Li, J., Michal, J. J., et al. (2021). Long-term trends of microplastics in seawater and farmed oysters in the Maowei Sea, China. *Environmental Pollution*, *273*, 116450. https://doi.org/10.1016/j.envpol.2021.116450.

CHAPTER FOUR

Nano/micro-plastics: Sources, trophic transfer, toxicity to the animals and humans, regulation, and assessment

Anandu Chandra Khanashyam[a], M. Anjaly Shanker[b], and Nilesh Prakash Nirmal[c,*]

[a]Department of Food Science and Technology, Kasetsart University, Ladyao, Chatuchak, Bangkok, Thailand
[b]Department of Agriculture and Environmental Sciences, National Institute of Food Technology Entrepreneurship and Management (NIFTEM), Sonepat, Haryana, India
[c]Institute of Nutrition, Mahidol University, Nakhon Pathom, Thailand
*Corresponding author: e-mail address: nilesh.nir@mahidol.ac.th

Contents

Abstract

Being in an era of revolutionized production, consumption, and poor management of plastic waste, the existence of these polymers has resulted in an accumulation of plastic

Advances in Food and Nutrition Research, Volume 103
ISSN 1043-4526
https://doi.org/10.1016/bs.afnr.2022.07.003

litter in nature. With macro plastics themselves being a major issue, the presence of their derivatives like microplastics which are confined to the size limitations of less than 5 mm has ascended as a recent type of emergent contaminant. Even though there is size confinement, their occurrence is not narrowed and is extensively seen in both aquatic and terrestrial extents. The vast incidence of these polymers causing harmful effects on various living organisms through diverse mechanisms such as entanglement and ingestion have been reported. The risk of entanglement is mainly limited to smaller animals, whereas the risk associated with ingestion concerns even humans. Laboratory findings indicate the alignment of these polymers toward detrimental physical and toxicological effects on all creatures including humans. Supplementary to the risk involved with their presence, plastics also proceed as carters of certain toxic contaminants complemented during their industrial production process, which is injurious. Nevertheless, the assessment regarding the severity of these components to all creatures is comparatively restricted. This chapter focuses on the sources, complications, and toxicity associated with the presence of micro and nano plastics in the environment along with evidence of trophic transfer, and quantification methods.

Abbreviations

PP polypropylene
PS polystyrene
PE polyethylene
PET polyethylene terephthalate
PVC polyvinyl chloride
MPs microplastics
NPs nanoplastics
NMPs nano and microplastics
TRWP tire and road wear particles
PMB polymer modified bitumen
CME1 clathrin-mediated endocytosis
CME2 caveolae-mediated endocytosis
SOP standard operating procedure
DCS differential centrifugal sedimentation
FFF field-flow fractionation
SEC size exclusion chromatography
HDC hydrodynamic chromatography
SEM scanning electron microscope
FTIR Fourier-transform infrared spectroscopy
DSC differential scanning calorimetry
TGA thermogravimetry analyser

1. Introduction

With the introduction of synthetic plastic in 1907, it was unbolted up to an era of revolutionized polymer science, which has begotten a series of polymers and plastic formulations in our daily life. Plastics are essentially synthetic polymers shaped from the polymerization of monomer units that are obtained from oil or gas (Cole, Lindeque, Halsband, & Galloway, 2011). This series of polymers brought easiness to our life and construction, with its stable chemical properties, good insulation, lightweight, durability attributes, and most significant cost-effectiveness (Frias & Nash, 2019). To a greater degree, the acceptability of these diverse forms of plastics is very much associated with their properties, which endows them with an irreplaceable status in today's market. With the optimization of different cost-effective methods in mass production, there was an expansion in areas of application and utilization of this polymer. A well-defined upsurge in production from 322 million tonnes within a year gap to 335 million tonnes in 2016 (Europe, 2017) shows the most decisive predicament allied with the polymers. Even though a minor part, which is not more than 10% goes for recycling, the majority ends up in landfills demanding centuries for the breakdown and decomposition of the polymer. Literally, a fraction of these indiscriminate disposals reaches marine and terrestrial environments, which affects the total microflora. Harmful effects on habitats by altering species distribution, trapping organisms that cause damage, and even increased mortality from ingestion are some of the problems associated with the result of inappropriate plastic disposal. Accordingly, this radical matter had gradually turned out to be an environmental threat with pervasive distribution in different ecosystems. It was observed that subsequent to the physico-chemical and other favorable activities, the plastics are fragmented down to hazardous forms of micro/nano plastics which stances a further threat to the system.

Microplastics (MPs) are broadly described as synthetic polymers existing in a size range of less than 5 mm and deprived of any definite lower limit (Huang et al., 2021). They are labelled into two different sections; primary MPs which are factory-made in a size limit less than 5 mm and is largely found in textiles, and medicine, while secondary form are derived from disintegration of large plastic remains. Secondary MPs being the abundant forms that exists, the amount or proportions surge with the continuing poor handling and disposal of the plastic in different forms. Among the apprehensions related to the indiscriminate disposals, particular concern has been given to the infiltration of the plastic debris to the food chain causing potential threats to all animal and marine species including humans. Polymers that widely recognized in plastic clutter are polypropylene (PP), polystyrene (PS), polyethylene (PE), polyethylene terephthalate (PET) as well as polyvinyl chloride (PVC). MPs constituting to one of this form can be swallowed by any aquatic organism ascribing to its small size and thus inflowing into the aquatic food web. As there is no predictable metabolic pathway for the breakdown of these polymers, there is an episode of polymer build-up inside the body, which in turn is hazardous for the organism. The potential accretion of MPs also affects humans as animal and seafood meats are the part of their diet. Lack of clarity on the amount of these accumulated materials instigating terminal effects, and exposure to even small quantities of polymers is bothersome. Added to subjects related to the release of the monomer units, there is also risk attributed with the additives added to the polymer structure during the production. During the plastic manufacturing process, a wide range of chemicals including plasticizers, stabilizers, fillers, etc., are introduced to the polymer structure with the view of altering the properties. The additives that are not completely bound to the polymer structure have a tendency to leach out to the surrounding environment including air, water, food, and body tissues. This leaching out process is reliant on many components like temperature and pH of the surrounding environment, the polymer structure, volatility of the added component, etc. Although there is an awareness about the dependent factors, there is no actual information about the rate of migration as well as how it materializes which adds up to the potential threats.

Besides the allied health risks, it is also important to incorporate the economic impacts attributed to the incidence of plastic clutter. The economic angle of the plastic incidence is sometimes easy to assess like in the cases of the cost associated with cleaning up the environment whereas it is not easily tangible in instances involving the health impacts due to ingestion of this litter or the impact of deterioration or reduction in the worth of life. This makes the whole thing even more complex. Therefore, this chapter provides information about the sources, complications, and toxicity associated with the presence of micro and nano plastics in environment along with evidence of trophic transfer and quantification methods.

2. Sources

Incidence of plastic use can be identified in various areas of our life; from household coverings, personal care products, coating and wiring to packaging, handling films, containers, and large-scale products. The liberation of these widely used polymers takes place in the form of primary or secondary MPs. Sources of the secondary form of MPs are disparate, massive, and very vast, which makes it tough to enumerate. The degradation of plastics into this form can be impelled by decomposition pathways including photo-degradation, weathering, breakage by abrasive mechanical forces, etc. (Andrady, 2011; Auta, Emenike, & Fauziah, 2017; Wagner et al., 2014). Over and above the characteristic's environment in which the hap of hazardous accumulation ensues also affects the prevalence of these MPs. It is vital to recognize the sources of plastics and degradation process for appropriate quantification of exact source and types of MPs in the environment. With diminutive number of studies focused on the copiousness of MPs on terrestrial ecosystems, soils are considered to be a longstanding basin for MPs (Rochman, 2018). There is only a limited number of researches done on the effect of these polymers on terrestrial organisms, the soil, and the related activities (Huerta Lwanga, Gertsen, et al., 2017; Rillig, 2012). Through area-based depositions, sediment forms from household or personal care products, sludge applications, landfills or direct dumping of waste, etc., there is direct contribution of plastics to terrestrial ecosystems, and soils. Wastewater treatment plants can act as a bases that aids in their ingress to the environment through personal care products like shower gels, cleansers, toothpastes, shampoos, etc. Improper handling and poor waste management methods lead to the upsurge in accumulation of plastic resin bits and plastic powder forms. The intensity of general forms of industrial abrasives in industry like polyester, polystyrene, polyethylene, melamine, etc., can increase in the environment due to improper handling and unauthorized disposals. For example, higher strengths of plastic monomers and polymers were observed on nearby areas of processing sites (Duis & Coors, 2016). There are traces of other polymers that need to be assess by proper field sampling to understand the extent of contamination. Archetypally, the studies constituting the effects of MPs reflect terrestrial ecosystems as trials and pathways for the introduction of MPs to marine ecosystems. This is primarily due to the augmented series of direct and indirect disposals making them plentiful and occupied with traces of most types of the polymers present. Most of the marine plastic originates from land-based sources including industrial plants, waste treatment plants, disposals from the household, etc. (Andrady, 2011). Even though the process involves the removal of almost 90% of the plastic debris, wastewater treatment plants are one of the important sources that discharge a large number of plastics into the marine environment (Mintenig, Int-Veen, Loder, Primpke, & Gerdts, 2017). Other sources that constitute to the MP concentration of the marine environment can be plastic debris reaching through industrial and domestic drainage systems including cloth fibres, personal care products, paints, etc. Also, sewer runoffs, and direct disposals by humans which includes different forms of plastic waste (Bergmann, Gutow, & Klages, 2015). Laterally human activities proceeding in the coast including activity of harbors, shipping and fishing activities, misplaced or rejected materials from ship, aquaculture facilities, etc., adds up to the marine concentration of MPs (Driedger, Dürr, Mitchell, & Van Cappellen, 2015). Ship paints comprising synthetic polymers such as polyacrylate, alkyds, etc., have direct contact with water involving in the input (Sundt, Schulze, & Syversen, 2014). Addition to these, direct source of MPs to both terrestrial and marine atmospheres is disintegration and destruction of plastics involved in agricultural lands. For example; to increases yields, quality and attaining efficient water usage, proper temperature and moisture control in agricultural and horticultural purposes there is a usage of plastic mulches, which contributes to the potential pollution of MPs (Steinmetz et al., 2016). Other intakes sources from agriculture are silage and fumigation films, greenhouse setups, irrigation lines involving plastic pipes, fertilizer sacks, containers, etc. wholly has a backing possibility of MPs distribution within the environment (Muise, Adams, Côté, & Price, 2016).

3. Occurrence and transport of microplastics

Due to its high resilience and versatility, plastics polymers are considered as a very desirable material across different industrial sectors including agriculture, food, textile, packaging, electronics, construction, etc., causing a surge in plastic production and consumption (Drzyzga & Prieto, 2019). Although some plastic debris are recycled, the majority of the plastic waste ends up in landfills or water bodies, where it disintegrates and degrades over time. Based on the size, plastic wastes can be classified into four types (Mofijur et al., 2021), which are macro plastics (>1 cm), *meso* plastics (1 cm–1 mm), micro plastics (1 mm–1 μm), and nano plastics (1 μm–1 nm). The increased use of single plastic along with inadequate waste management and recycling practices have increased the plastic waste. The effect of larger macro and meso plastics have long been identified as a critical threat to the environment. Plastic polymers mostly disintegrate in the environment owing to weathering and ageing over a long time duration, resulting in the formation of nano and micro plastics (NPs/MPs). They can be classified into primary and secondary MPs based on their origin however, there are no precise concept for NPs and need to be thoroughly investigated. Primary

MPs are those which are intentionally developed for specific uses, whereas secondary MPs are formed by the fragmentation and degradation of macro plastics. The formation and trophic transfer are discussed in detail below.

3.1 Terrestrial

The occurrence of nano/microplastics (NMPs) in the terrestrial environment can be grouped into three main categories, which are formed as a result of agriculture practices, fragmentation of larger plastic waste, as well as resulting from erosion, and deposition. Plastic mulching is a common practice that is widely used to improve the crop yield and control the pest and weed infestation (de Souza Machado, Kloas, Zarfl, Hempel, & Rillig, 2018). The left-behind plastic film may accumulate and further fragments to produce NMPs. Sewage sludge is another major source of NMPs in a terrestrial environment. They enter the solid sludge from various domestic and industrial sources such as microbeads from face scrub and other cosmetic products, laundry dust, industrial process like blasting of plastics, etc., these are eventually conveyed to wastewater treatment plants. The sludge formed during the wastewater treatment has a very high NMP trapping efficiency. Studies have reported that around 90% of NMPs can be retained in sewage sludge (Carr, Liu, & Tesoro, 2016). As it is a common practice to use sewage sludge as agriculture fertilizer, approximately 50% of sludge ends up in agriculture fields (Nizzetto, Futter, & Langaas, 2016). Runoff from urban areas and erosion by a wind from landfills can also act as a potential sources in NMPs contamination. Another common source of NMPs are the in-situ formation of NMP due to degradation and fragmentation of bigger plastic materials. In fact, the degradation process is thought to be the primary source of nano plastics in the environment (da Costa, Santos, Duarte, & Rocha-Santos, 2016). A study on NMPs from soil collected from a vegetable farm in China's Wuhan province reported about 320–12,560 NMPs particles/kg (Chen, Leng, Liu, & Wang, 2020), whereas soil samples collected from industrial area of Sydney reported about 300–67,500 mg/kg of NMP (Fuller & Gautam, 2016). High NMPs contamination of 22,000–690,000 particles/kg were also reported from a vegetable farmland in Heilongjiang, China (Zhou, Liu, & Wang, 2019). The negative effect of these accumulating NMPs on terrestrial environment is unquantifiable. They can interact with various other organic pollutants, and hazardous materials in the soil doubling their potential and thereby changing soil physiochemical characteristics, contaminate groundwater, and affects plant growth and overall productivity.

3.2 Aquatic

It is important to mention that the bulk of the plastic wastes that wind up in aquatic bodies were created, used, and disposed indiscriminately on the land (de Souza Machado et al., 2018). Inadequate waste management, illicit dumping, and unintentional discharges from agriculture, industrial processing, and housekeeping activities are the primary cause of NMP contamination in aquatic environment. Moreover, sludge from the wastewater treatment plants and industries eventually ends up in aquatic systems. The major NMPs contributor in aquatic system were identified to be either polypropylene (PP), polyethylene (PE), polystyrene (PS), or polyvinyl chloride (Mammo et al., 2020). Tire and road wear particles (TRWP) are another source of NMPs and it is estimated to generate approximately 1Kg of TRWP per individual in a year in European cities, of which 18% reaches to rivers (Unice et al., 2019). These NMPs possesses large surface area, which can interact with organic matter and adsorb various toxic chemicals, transferring them between different marine habitats. NMPs once entered the ocean system could transfer both horizontally and vertically either alone or in combination with other bulk MPs. The movement of NMPs in ocean can be affected by many variables including: (1) physicochemical properties of the plastic such as density, hydrophobicity, and size, (2) the ocean dynamic conditions such as wind, tides, waves, and currents, (3) coastline characteristics (gravel, bed rock, vegetation, etc.) and ocean topography, (4) biotic interactions (biofilm, animal consumption, human interactions, etc.) (Xu, Ma, Ji, Pan, & Miao, 2020). NMPs might eventually be ingested by a various marine animals, making their way into the marine food web and presenting serious risks to all marine and terrestrial life. The distribution and physicochemical characteristics of these NMPs are investigated by many researchers and found to be varied significantly across the aquatic environment. Anderson et al. (2017) reported that the abundance of NMP on Lake Winnipeg, Canada was about 0.19 particles/m^2 of which 90% was fibers. Whereas a study from coastline of Tamilnadu, India reported about 2–178 particles/m^2 consisting of 47–50% fragments, 24–27% fibres and 10–19% foam (Karthik et al., 2018). MPs has also been identified in drinking water treatment plants in Czech Republic and drinking bottled waters from Germany (Pivokonsky et al., 2018; Schymanski, Goldbeck, Humpf, & Fürst, 2018).

Studies have also confirmed the presence of NMPs in many freshwater ecosystems around the globe. The distribution, sedimentation, and concentration of NMPs in a water body depend on the stream flow rate, geographical positions, winds, and currents (Bellasi et al., 2020). The variation in the concentration of MP in river Tame, England was primarily influenced by the flow velocity, as the concentrations of MP increased in regions where the flow velocity was low (Tibbetts, Krause, Lynch, & Sambrook Smith, 2018). However, increased flow rates can flush the MPs from the river beds (Hurley, Woodward, & Rothwell, 2018). Therefore, regions with low flow velocity, in general, function as MP sinks, while rivers and streams with high flow velocity can act as MP delivery systems. Rivers and streams can also lead to the formation of secondary NMPs due to the fragmentation of plastic by currents.

In addition, similar efforts have been made to measure NMP concentrations in various marine ecosystems. Rivers play an important role in transporting NMPs from the land to the marine ecosystem (Unice et al., 2019). When NMPs are released into the oceans, they are quickly transported by waves, currents, and tides throughout the marine ecosystem. Due to its surface area, oceanic transportation enables NMP to spread across a large range of distances and in addition, vertical transport allows NMP movement from the surface to deeper oceanic sediments (Kane & Clare, 2019; Shahul Hamid et al., 2018). The accumulation of MPs in the marine ecosystem depends on the density of the plastics, as less dense plastics tend to accumulate on the shorelines whereas the denser plastics accumulate on the deeper waters or seabed (Petersen & Hubbart, 2021). The marine ecosystem research can be focused on either surface water study, including beaches and coastal lines, or offshore study which includes deep-sea investigations. Surface water studies are frequently the focus of investigations due to their ease of access whereas, the difficulty and expenses connected with working in deeper seas, offshore ecosystems have received significantly less attention. The major MPs found on ocean water surfaces are polyethylene and polypropylene (Lebreton et al., 2018).

3.3 Atmospheric

Along with the terrestrial and oceanic transport, atmospheric transport is another major transportation pathway for worldwide distribution of NMPs. The NMPs distributed in the atmosphere are either dispersed directly in the form of aerosols from industrial sectors or are re-suspended from ground by wind and air currents called street dust (Wang, Enyoh,

Chowdhury, & Chowdhury, 2020; Zhang et al., 2020). The primary source of NMP in street dust are tyre treads formed by wearing of vehicle tires, the wear particles of polymer modified bitumen (PMB) which are usually used to increase the strength and adhesive property of roads in cold countries, and air deposition. In addition, synthetic textiles, degraded larger wastes in landfills, etc., are major source of NMPs in air. In contrary to water transportation, NMPs' flexible aerial movement has less topographic constraints and can travel in any directions, posing a serious threat to key fragile locations. The major polymer components of the atmospheric fall out micro plastic are reported to be polyethylene, polypropylene, polyethylene terephthalate and polystyrene (Chen, Feng, & Wang, 2020). A study from Shanghai reported that an estimate of 120.7 kg of MP has been transported annually through air (Allen et al., 2019). A similar study by Rachid Dris et al. (2015) reported an average MP fall out of 118 particles m^{-2}/day in greater Paris and over 90% of the microplastics found were fibres, with 50% of the strands being longer than 1000 μm. Moreover, Dris et al. (2017) conducted a comparative study in the microplastic concentration in indoor and outdoor air in Paris and found that fibre concentrations (1.0 and 60.0 fibres/m^3) was higher in indoor than the outdoor (0.3 and 1.5 fibres/m^3). These airborne NMPs can later deposit in aquatic or terrestrial systems, acting as a potential source of contamination.

4. Biological assimilation

MPs contamination has been documented in a wide variety of habitats. As a result, bioaccumulation, especially bio-magnification of MPs and related chemical additives, is frequently expected in aquatic, and terrestrial food webs. To assess the trophic transmission of NMPs, several abiotic and biotic processes involved in their ingestion, bioaccumulation, and bio-magnification must be investigated. As NMPs are water insoluble non-biodegradable synthetic polymers with high tendency toward fragmentation, they can be easily absorbed or ingested by microorganisms. The bioaccumulation and/or bioremediation of NMPs are confirmed across different terrestrial and aquatic ecosystems at various trophic levels (Paul-Pont et al., 2018).

4.1 Aquatic environment

The presence of NMPs is ubiquitous in the environment, and they can be easily absorbed by the marine organism due to their smaller size and large surface to volume ratio and thereby pose a great threat to the marine biota (Fig. 1).

In the case of marine microorganisms, the major pathway by which the NMPs interact with the organisms includes cellular uptake, bio-adsorption, or biodegradation (Avio, Gorbi, & Regoli, 2017). The cellular uptake of NMPs by marine microbiome largely depends on the cell membrane properties of the organism and physicochemical characteristics of NMPs (hydrophobicity, shape, size, charge). As the cell membrane of bacteria and microalgae are negatively charged, they can easily be attached to positively charged NMPs by the influence of electrostatic forces. In addition, benthic microorganisms can be attached to these plastic particles with the

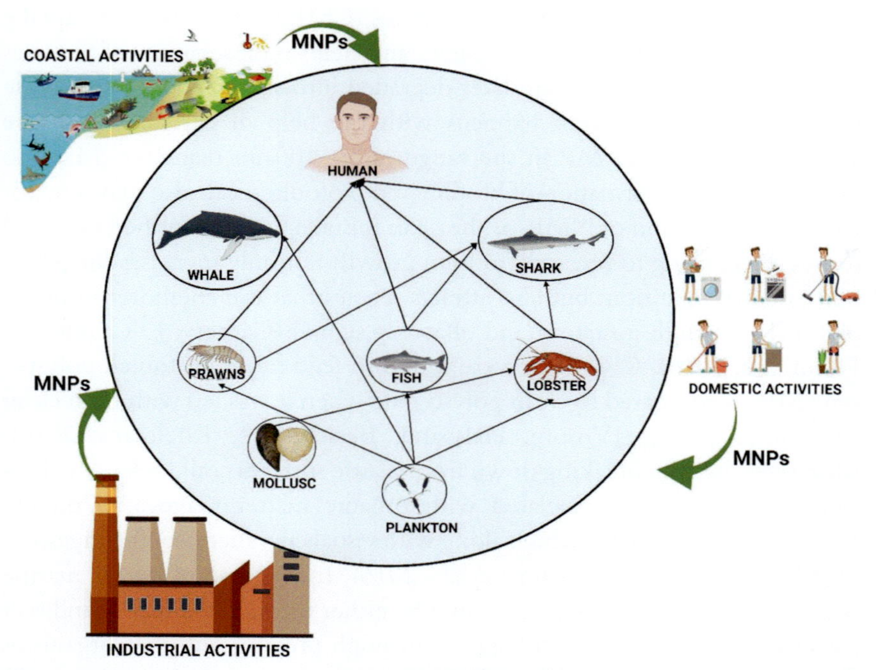

Fig. 1 Schematic diagram of source of MNPs and its trophic transfer in aquatic environment.

help of certain *exo*-polymers, and undergo bio-adsorption (Curren & Leong, 2019). As the size of NP and MP in the range of food particles generally consumed by planktons, bioaccumulation of NMPs on phytoplankton and zooplankton are also reported (Md Amin, Sohaimi, Anuar, & Bachok, 2020). Planktons can either be attached to the NMPs and form aggregates or they can ingest MP by suspended feeding, adsorption, or endocytosis (Mammo et al., 2020). The active uptake of MP by copepods zooplankton via suspension-feeding, and its ability for active uptake of polystyrene MP has been demonstrated by various research groups (Botterell et al., 2019; Cole et al., 2013). The bioaccumulation of NMPs ingestion in marine organism include endocytosis, pathways of phagocytosis, receptor mediated endocytosis, and pinocytosis. In eukaryotic bacteria, receptor-mediated endocytosis might possibly be a mechanism for active absorption of NMPs. Depending on the receptor type, uptake can be classified into two different types that are clathrin-mediated endocytosis (CME1) or caveolae-mediated endocytosis (CME2). In CME1, the uptake is mediated by clathrin coated vesicles and helps in absorption of particles larger than 100 nm which are later degraded into smaller NPs. In the case of CME2, the absorption happens with the help of certain membrane vesicles and trap tiny MP in the range of 50–100 nm diameter (Mammo et al., 2020). The formation of biofilm and biofouling can also play a major role in the ingestion of NMPs as they can mistakenly be identified as a food source. According to research, biofilm growth not only increases the possibility of MP ingestion, but also attracts species that use chemoreceptors to pick prey through gustatory and olfactory signals (Carbery, O'Connor, & Palanisami, 2018). Copepods were found to feed more on fouled polystyrene MPs as compared to clean polystyrene when it was fed with both clean and fouled plastics (Vroom, Halsband, Besseling, & Koelmans, 2016). Moreover, biofilm breaking down from plastic surfaces could release a characteristic odor often associated with organic matter (Savoca, Wohlfeil, Ebeler, & Nevitt, 2016), which along with visual cues increases the ingestion of NMPs in sea turtles (Nelms et al., 2016). In the case of bigger marine organisms, the NMPs ingestion could be either through a direct or indirect pathways. Indirect ingestion happens through preying on other organisms that have already ingested the NMPs or through means of respiration. In a marine environment, the transmission and accumulation of MPs at lower trophic levels may have a cascade effect on marine food webs. MPs are likely to bio-accumulate in bigger species due to predators' constant feeding of huge amounts of zooplankton, such as juvenile fish or even giant filter feeders like humpback whales (Desforges, Galbraith, & Ross, 2015). A recent meta-analysis study on bioaccumulation and bio-magnification of NMPs in marine organisms reported that 22,512 organisms (329 species) out of a total 23,049 individual organisms (411 species) were contaminated with NMPs (Miller, Hamann, & Kroon, 2020). The study also reported that there was no apparent evidence on bio-magnification across tropic levels, as there was no increase in the NMPs bioaccumulation with respect to increasing tropic levels. The highest bioaccumulation of MP was reported for secondary consumers (trophic level 2) while the minimum was for tertiary consumers. Among trophic levels a slight increase was observed only for tropic level 4 and besides, the accumulation was more influenced by feeding strategy, as the highest average individual MP contamination was reported for the filter feeding secondary consumers (Miller et al., 2020).

4.2 Terrestrial environment

In the terrestrial ecosystem, even though the NMP contamination is 4–23 times higher than aquatic ecosystems (Horton, Walton, Spurgeon, Lahive, & Svendsen, 2017), they have received far less attention from the scientific community. In the aquatic ecosystem, MPs can act as a transport medium for other contaminants and are particulate targets that can be ingested by the organisms which can potentially harm them. Emphasis on these criteria may have led to an underestimation of MPs risks to terrestrial animals. Fig. 2 shows the schematic presentation of sources of NMPs and their trophic transfer in the terrestrial environment.

The pioneer study in bioaccumulation of MPs in the terrestrial environment was reported by Huerta Lwanga, Mendoza Vega, et al. (2017) where field evidence of tropic transport of MP was reported in home gardens of Southwest Mexica. The study reported a gradual increase in the concentration of MPs as the concentration of MPs in soil was 0.87 ± 1.9 particles/g, which increased to 14.8 ± 28.8 particles/g in earthworm casts and finally to 128 ± 182.3 particles/g in chicken faeces. Another field study by D'Souza, Windsor, Santillo, and Ormerod (2020) on Eurasian dippers (*Cinclus cinclus*) reviled the trophic transfer of MP between aquatic and terrestrial food webs. The study estimated the MP concentration in faecal and regurgitate samples from dippers and reported the presence of MPs in 50% of regurgitate and 45% of faecal samples, respectively. These dippers are feds on aquatic

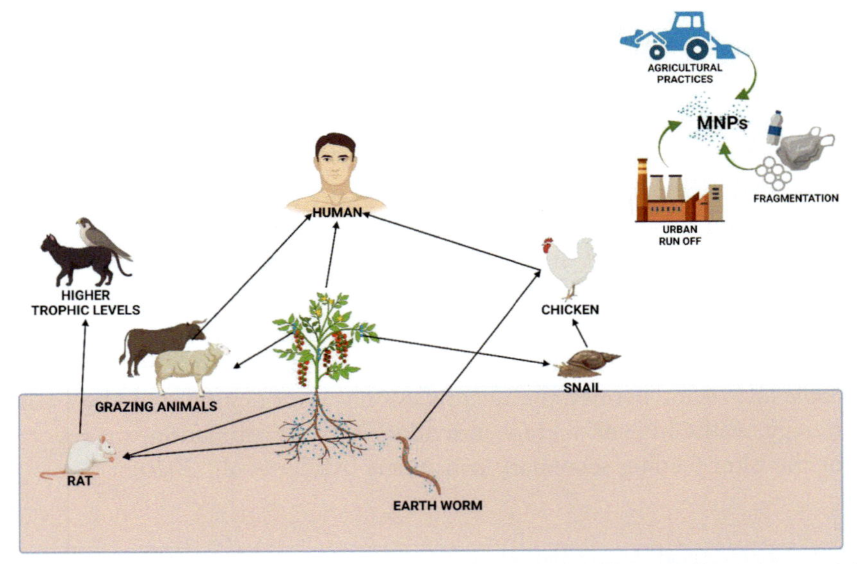

Fig. 2 Schematic diagram of source of MNPs and its trophic transfer in terrestrial environment.

insects including Ephemeroptera and Trichoptera order. Further analyses of these insects revealed the presence of an average of 16 ± 3 and 8 ± 2 particles/g of MPs, respectively. This indicates that aquatic insects are the source of MPs trophic transfer.

5. Toxicity studies

In conjunction with the cumulative disposition of these polymers in the environment, all the inhabitants are vulnerable to the threat postured by their existence. These forms affect the growth, development, enzyme activity, metabolic disorders, reproduction cycle, etc., of the living organisms in addition to the issues inclined with different degrees of toxicity. Commencing from smaller organisms extending all the way through the food chain, they are vigorously transferred into higher levels, spreading out to larger populations and finally affecting and destroying the whole environment. It is indeed of great magnitude to recognize the toxicity and the effect that these polymers have on every inhabitant including humans and overall effect on the ecosystem.

5.1 Aquatic animals and plants

Direct or indirect exposure to MPs brings about physical, chemical, and biological impacts on the aquatic inhabitants. The most visible and direct effect includes the entanglement and entrapment of aquatic creatures in plastic fibres, wires, rings, etc., which causes strangulation, suffocation, drowning, and even death of these creatures (Udyawer, Read, Hamann, Simpfendorfer, & Heupel, 2013). Entanglement often ensues in discarded forms of plastics, plastic sheets, bags, or lost fishing ropes, gears, etc., which limit the capacity of the organism to respire, move or feed. Ingestion is considered to be one of the indirect exposure trails of these polymers in marine organisms. When there is an increase NMPs content in water, the litter will get ingested into the body of marine species causing serious issues. Table 1 shows the effect of NMPs ingestion on marine species health. With many chronicle reports about the ingestion from 1960s, this intake of plastic litter is seen in marine organisms ranging from planktons, small crustaceans, shell fish, fishes, sea birds, carbs, lobsters, and whale. A study to measure the amount of NMPs uptake reported the presence of these polymers in 35% of fish sampled from the North Pacific central gyre region (Boerger, Lattin, Moore, & Moore, 2010). Regional reports on the presence of MPs in 13 out of 141 fishes caught from same region emphasis the level or degree of ingestion taking place (Davison & Asch, 2011). The NMP intake is known to cause different issues including augmented starvation, accumulation of plastics in tissues, effects on feeding and digestion potential, increased injury, etc. When this litter interacts with the body tissues causes hindrances in any metabolic pathways or activities, which could cause the death of an organism. The degree of the effect of these ingestions on marine life depends on a variety of factors and it is difficult to predict the trend followed. Another study reported that the algal feeding was hindered on a higher scale by the intake of MPs which in turn resulted in the negative effects in digestive system, eating patterns, and development (Cole et al., 2013). In addition to hindered feeding patterns, ingestion of polystyrene microbeads affected the reproductive system of different species of copepod (Cole, Lindeque, Fileman, Halsband, & Galloway, 2015; Lee, Shim, Kwon, & Kang, 2013). The study on the acquaintance of fishes to various levels of MPs revealed a variation in feeding patterns, amendments in the action of digestive enzymes, and low energy production, which adversely affected the species (Wen et al., 2018). The ingestion and accumulation of these microforms in tissues as well as other parts, varies within the species and it was also correlated with the properties of NMP size. These processes did not follow

Table 1 Effect of NMPs ingestion on health of marine animal.

Organism	Plastic type	Results	Reference
Zooplankton	Polystyrene	Causes immobilization in species. Ingestion decreased the feeding pattern	Cole et al. (2013)
Fresh water microalgae	Polypropylene & High density poly ethylene	Observed significant growth reduction	Lagarde et al. (2016)
Flat oysters	Poly lactic acid	Variations in respiration rates and patterns	Green (2016)
Green algae	Polystyrene	Reduced population growth and chlorophyll content	Besseling, Wang, Lurling, and Koelmans, (2014)
Marine fish	Polyethylene fragments	Sub-lethal effects on liver induced by the ingestion with accumulation showing signs of stress	Rochman, Hoh, Kurobe, and Teh, (2013)
Zebra fish	Polystyrene particles	Inhibition of acetylcholinesterase activity and also effected the nervous system	Chen, Yin, et al. (2017)
Zooplankton	Polystyrene	Reduced body sizes and severe alterations in reproductive patterns	Besseling et al. (2014)
Mussels	Polystyrene particles	Harmful alterations in neurotransmission and decrease in enzymatic activity	Brandts et al. (2018)
Pelagic and demersal fish	Polyethylene	The condition of prolonged starvation. Possible mechanical threats in the forms of abrasions	Rummel et al. (2016)
Whales	Poly ethylene/poly propylene particles	Increased toxicological related issues	Fossi et al. (2016)

any specific pattern or trend, hence it was critical to conclude the final remark on the toxicity and subsequent effects of MPs on specific species or whole. Moreover, the direct ingestion that is commonly seen among marine animals, a state of secondary ingestion also exist which contributes to the presence of MPs in organisms. This occurs when animals forage on prey that has previously swallowed plastic litter. Perry, Olsen, Richards, and Osenton (2013) stated an example of such a type of ingestion where a nylon fishing line was found inside a dovekie, a tiny water bird that was discovered inside the stomach of a goose fish. Owing to the size, possibility of the presence or exposure of NMPs to all types of marine species is unavoidable. Keeping this in mind, the probability of latter discussed type of ingestion will be high and needs to be explored on a larger scale.

5.2 Terrestrial animals and plants

Compared to the aquatic species, there is only a few number of research related to the effect of NMPs on growth and development of terrestrial organisms. With soils acting as long-time basins for the accumulation of NMPs, contamination caused here can be harsher than in aquatic ecosystems (Rillig, 2012). Ingestion of NMPs is contributing factor to toxicity that not only add risks to the consuming organism but also opens the threat of their introduction to the food chain even affecting humans. The reports on NMPs toxicity begin with their effects on soil organisms. Cao, Wang, Luo, Liu, and Zheng (2017) have reported the lethal effect of MPs (particularly polystyrene) on earthworms that damaged their self-defence system, growth, and increased mortality. MPs that cause changes in soil organisms also cause some changes in soil conditions (de Souza Machado et al., 2019). The vegetative and reproductive growth of wheat plant was affected on a larger scale with the presence of these polymers at different levels in the soil (Qi et al., 2018). The impact of NMPs on germination degree, root growth, and development in the terrestrial vascular plant (*Lepidium sativum*) was reported by Bosker, Bouwman, Brun, Behrens, and Vijver (2019) which inferred the adsorption was through root hairs. Similar effects on terrestrial plants were reported by Sun et al. (2020) which clears the fact that MPs detrimentally affects in every component of the ecosystem. A chance of intake or ingestion of MPs in higher mammals can happen through diet or water sources. The accumulation of NMPs was observed in gut, liver, and kidney of certain species of mice causing gut barrier malfunctioning, intestinal inflammation, and issues related with metabolic pathways (Sun et al., 2020; Yang, Chen, Lu, & Liao, 2019). Lacking research studies on the impact of NMPs in higher species, the generalization about the path and effect would be inappropriate.

5.3 In-vitro study

The in-vitro studies exercises NPs/MPs exposure events taking place at molecular and cellular levels without need of animal testing. The acceptability of these outcomes requests further validations concerning mechanisms, and in vivo responses. The presence of PVC, and PE MPs has shown a damaging influence on the immune activity of European sea bass and gilthead seabream during an in vitro study investigated by Espinosa, Garcia Beltran, Esteban, and Cuesta (2018). Weakening of immune peripheries was prophesied to be due to the oxidative stress developed at cellular levels. Cellular level evaluation of high-density polyethylene particles exposure in blue mussels affirmed the chance of inflammatory responses, and membrane de-stabilization with the upsurge of exposure times (von Moos, Burkhardt-Holm, & Kohler, 2012). The negative effect of MPs on aquatic organisms was confirmed by an in vitro study established by Pannetier et al. (2019) which showed DNA damage of that species. In contrast, the in vitro studies of the effect of polystyrene on mice showed that the concentration of 1, 4 and 10 μm does not pose a health risk in mice (Stock et al., 2019). Similar to these results, Alomar et al. (2017) reported that the cellular level activities of *Mullus surmuletus* Linnaeus and 1758 fish remain unaffected by the presence of MPs. These variations could be due to the variances in target organism, polymer material, exposure levels, time, and surrounding environment. For generalities and inferences, more information is obligatory which underlines the necessity of foremost research efforts on NMPs impacts at cellular levels.

5.4 Human risk assessment

Contiguous with introduction of NMPs in the food chain relating to terrestrial, and marine creatures, it is really hard to keep humans out of the loop. The exposure and ingestion pattern in different marine species were discussed in detail in above sections, which makes it significant to consider the incidence of NMPs in the form of food in human diet. Incidence of these polymers in certain plant species, and animals also justifies the presence of NMPs in different type of foods. Several studies confirmed the incidence of NMPs in different food products including seafood, honey, sea salt, sugar, beer as well as water sources (Kim, Lee, Kim, & Kim, 2018; Liebezeit & Liebezeit, 2015; Pivokonsky et al., 2018; Smith, Love, Rochman, & Neff, 2018). The main routes of exposure of NMPs in humans are occurred through ingestion and inhalation. The incidence of NMPs in different organism and plants and their prevalence in food chain justifies the ingestion trail in humans. With the lack of toxicity data about the effect of NMPs in in-vivo, there are certain studies which assessed their effects in human cells. These studies included degree of cellular uptake of NMPs which showed that lower concentrations does not affect the cellular functions, whereas a serious concern involved with higher concentrations (Magri et al., 2018; Prietl et al., 2014). These polymers are anticipated to enter the human gut through consumption. The pattern obtained from experiments on fish and mice, toxicity of these polymers can cause impairment in gut vascular barrier and thereby entering the other organs and tissues increasing the severity of the whole organ. Although there are studies stating that only improbable levels of absorptions are capable of such severity, the possibility of such incidence cannot be completely evaded with the presence of polymers everywhere. Main cause of the respiratory acquaintances is the air borne plastics that is present. Vianello, Jensen, Liu, and Vollertsen (2019) predicted the presence of MPs up to 272 particles per day from the indoor air itself, which unfastens the leeway of imagination about the incidence of these polymers from a highly polluted or mobbed arrangement. Dong, Chen, et al. (2020) found that polystyrene particles have increased the risks associated with chronic obstructive pulmonary diseases. The author predicted the possibility of lung disease risk even at low concentrations. Lim et al. (2019) also reported the possibility of lung issues related to the low concentrations of these polymers in the human body. The biological tenacity and exposure degrees are important factors that wheel the health risks associated with NMPs in the human body. There is no clear understanding and correlation about the level of concentration and the degree of risk associated with it. To have a specific conclusion and recommendations, detailed studies are necessary to evaluate the effect on the human body through animal model experiments or in-vivo studies.

6. Quantifications

Although there has been an increase in the number of studies on MNP contamination in various ecosystems in recent years, a standard operating procedure (SOP) has not yet been established for the analysis of these particles from the environmental media. Even though several papers have provided various analytical approaches for MNP plastic analysis, the differences in procedures might make it difficult to compare results across studies. Therefore, developing standard protocol for extracting, quantifying, and characterizing of MNP in terms of their size, chemistry, distribution, and surface properties remain as a key challenge. The following section discusses the extraction and characterization techniques for NMPs (Table 2).

6.1 Extraction protocols

Sampling of airborne NMPs are mainly done by using three main techniques, namely active collection, passive collection and dust sampling (Shim, Hong, & Eo, 2017). In active sampling, a known amount of air is pumped through a filter over a fixed period for the collection of suspended particles. As parameters such as volume, flow rate, mesh size, etc., can be controlled, this technique has a high level of reproducibility and ensures a complete validation of the surrounding environment. Whereas in passive collection, the sampling is done with the funnel of a known surface area which is used to collect the atmospheric fallouts of airborne plastics. This method is preferred due to its less maintenance and reduced power consumption. However, the settling of plastics depends on its physicochemical properties and care should be taken for using appropriate statistical tool for data analysis. In dust sampling, for collecting deposited street dust, a dustpan and brush is utilized, however for indoor dusts, a vacuum pump is advised for sample collection. After sample collection, the removal of other organic and inorganic particle and concentration of the NMPs have to be done before identification and counting. Organic matter elimination can be done by addition of strong oxidants such as NaClO or H_2O_2, but due to the low concentration of organic particle in atmospheric samples, organic digestion are usually not implemented. However, the elimination of inorganic particles is usually done using density separation.

Soil samples are usually collected through bulk sampling where, a known mass or volume of soil is directly collected. Even though the sampling looks straight forward, a clear justification of sampling site, depth of soil collected, proximity to water sources, etc., should be provided for accuracy. The separation of NMPs from the soil matrix can be carried out using techniques such as sieving, flotation, filtration, drying, as well as density based separation (Enfrin et al., 2021). An alternative extraction procedure for retrieving

Table 2 Extraction and characterization techniques of NMPs.

Application	Technique	Comments	References
Extraction methods			
	Density separation	Used for extraction of MPs from sediments	Claessens, Van Cauwenberghe, Vandegehuchte, and Janssen (2013)
	Pressurized fluid extraction (PFE)	An average recovery of 84–94% from municipal waste were observed for all types of plastics tested	Fuller and Gautam (2016)
	Field-Flow fractionation (FFF)	Particles ranging from 1 to 800 nm were successfully screened with fractionation resolution obtained between 10 and 400 nm	Gigault, El Hadri, Reynaud, Deniau, and Grassl (2017)
		Polystyrene nano-plastics were identified at concentrations as low as 20 μg/L	Mintenig, Bäuerlein, Koelmans, Dekker, and van Wezel (2018)
	Hydrodynamic chromatography (HDC)	A combination of HDC and size-exclusion chromatography were used to separate the particles based on their size and their hydrodynamic radius	Pirok et al. (2017)
	Size exclusion chromatography (SEC)	Both the size distribution and the NP/MP composition per product with distribution gaps of around 55,000 g/mol were analyzed. This approach was able to yield robust size distributions with high precision	Lei et al. (2017)
Characterization			
	Dynamic light scattering (DLS)	To detect the microplastic aggregation in seawater	Summers, Henry, and Gutierrez (2018)
	Nanoparticle tracking analysis (NTA)	The study was reproduced in 12 different laboratories in Europe and showed high reproducibility of the results	Hole et al. (2013)

Continued

Table 2 Extraction and characterization techniques of NMPs.—cont'd

Application	Technique	Comments	References
	Flow cytometry (FCM)	FCM was used to detect and separate polystyrene micro-plastic particles from biofilm samples	Sgier, Freimann, Zupanic, and Kroll (2016)
	Optical microscopy	The optical microscope revealed the structure and surface texture of MPs	Li, Duo, Wufuer, Wang, and Pan (2022)
	Scanning electron microscopy (SEM)	SEM was used to study the size and weathering state of synthetic micro and nano fibres during washing	Napper and Thompson (2016)
	Atomic force microscopy (AFM)	Morphological changes on plastic surfaces after bacterial colonization were observed by AFM	Kumari, Chaudhary, and Jha (2019)
	Fluorescence spectroscopy	The amounts of nano-plastics in zebrafish larvae were measured using fluorescence spectrophotometer	Chen, Gundlach, et al. (2017)
	Hyperspectral imaging (HSI)	HSI was used to monitor micro plastics in marine environments	Bonifazi, Palmieri, Serranti, Mazziotti, and Ferrari (2017)
	Fourier-transform infrared spectroscopy (FTIR)	FTIR was used to study and characterize micro-plastics collected from different aquatic sources in China	Li et al. (2022)
	Raman spectroscopy	Even though Raman spectroscopy is a time-consuming analysis as compared to FTIR, it was more accurate in classifying micro-plastics in the range of 50–1 μm	Kappler et al. (2016)
	X-ray photoelectron spectroscopy (XPS)	X-ray photoelectron spectroscopy was used to investigate the adsorption of As(III) onto polystyrene nano particles	Dong, Gao, Song, and Qiu (2020)
	Pyrolysis-Gas Chromatography-Mass Spectrometry (Pyr-GC/MS)	To identify and optionally quantify MP in environmental samples	Fischer and Scholz-Bottcher (2017)

spiked MPs by using a two-stage pressurized fluid extraction followed by methanol and dichloromethane extraction is also proposed by Boucher and Friot (2017).

In the case of aquatic environments, samples collection can be performed in the water body or shores. The samples can be collected either by selective sampling, bulk sampling or by volume reduced sampling (Enfrin et al., 2021). Selective sample collection is done directly by visual monitoring and have higher risk of misidentification and less accuracy. Whereas in bulk sampling, the entire matrix is collected into a container of finite size without separating any constituents. One of the drawbacks of this method is the poor representation of samples for a broad area of interest. Volume-reduced sampling is the most widely used sampling technique, where the help of fine mesh attached to trawls, which are towed over the water, sampling a wide area of interest. The extraction protocol from water samples depends on the characterization technique. One of the widely used separation technique to separate NMPs from inorganic salts is density separation. Even though there are no standard protocol for digestion of organic matter, chemical and enzymatic digestion (H_2O_2, NaClO, NaOH, HCl, lipase, cellulose, etc.) techniques are widely used. For large NMPs, methods based on the hydrophobic affinity of MPs reported to have impressive recovery yield (Enfrin et al., 2021).

6.2 Characterization techniques

6.2.1 Size fraction and morphology

An alternative characterization techniques for measuring and separating micro and nano plastic based on their size include differential centrifugal sedimentation (DCS), chromatography, and field-flow fractionation (FFF) (Enfrin et al., 2021). In DCS, a centrifugal force is applied to the particles with the help of a spinning disc, which are then separated according to their size and density. In this method, the particles are injected to the centre of the disc, which spins at very high speeds. This induce the migration of the plastics to the outer wall of the disc, with larger and heavier particles being pushed more to the outer edges, whereas lighter particles remain close to the centre. Another method that is used to separate NMPs based on size is FFF that uses a perpendicular force to the flow direction of liquid matrix that is passing through a narrow channel of 50–300 μm. This causes the particles to migrate toward the section of the channel called as accumulation wall and stratify particles in the liquid matrix. Chromatography techniques such as Size Exclusion Chromatography (SEC) (Enfrin et al., 2021) and

Hydrodynamic Chromatography (HDC) (Brewer & Striegel, 2010) can also be used for separation of NMPs based on their size. Morphological and size-based characterization of micro and nano plastics could also be done by microscopy techniques. Microscopy is an easy, fast, and simple technique that is used for identification of MPs and can provide details about its structure and surface texture (Li et al., 2022). A stereo microscope can be used to identify MPs in the size of hundreds of micron range. Even though stereo microscope can provide magnified images of MP, identification becomes difficult when particles size falls below 100 μm and when targeted particles show no color or typical shapes (Song et al., 2015). To overcome these problems, an electron microscope or polarized optical microscope are used. A scanning electron microscope (SEM) can be used for obtain high-magnification and high-resolution images of NMPs and can be used for differentiate NMPs from organic particles. The physical characterization of NMP can also be done with the help of atomic force microscopy, which uses a cantilever sharp tip that moves across the surface of the particle providing details about its surface topologies. In addition, another sophisticated microscopy method that is used to identify plastic materials is the polarized optical microscopy. The equipment can be used for predicting the crystallinity of the plastic material as the polarized light transmission of the plastic material changes with respect to its crystallinity.

6.2.2 Combined methods

It is quite challenging to accurately and reliably detect NMPs of varied size, shape and chemical properties using a single analytical technique. As a result, a combination of more than two analytical methodologies are commonly used. In general, the physical characterizations of the plastics are usually done with the help of microscopic technique whereas the chemical characters are usually analysed by spectroscopic techniques or thermal analysis (Shim et al., 2017). Due to the elastic nature of the chemical bonds, every molecule experience certain vibration when irradiated with photons in the wavelength of visible and infra-red regions. The commonly used vibrational spectroscopy technique for identification of NMPs are Fourier-Transform Infrared Spectroscopy (FTIR) and Raman Spectroscopy. In FTIR, the vibrational modes will induce changes in the dipole moment, which is analysed to get the molecules fingerprints, whereas in Raman spectroscopy, the spectra are obtained by measuring frequencies of the back-scattered light. FTIR can reveal information about the abundance and composition of the NMPs (Browne, Galloway, & Thompson, 2010). Moreover, the changes in the chemical bonds of the polymer due to environmental factors can be monitored with the help of FTIR, as the changes in initial crystallinity will lead to formation or disappearance of peaks. This can be used to find the degree of ageing or weathering of polymers (Shim et al., 2017). Raman spectroscopy is another non-destructive chemical analysis similar to FTIR, but as it uses low wavelength lasers, Raman spectroscopy gives better spatial resolution than FTIR. However, additive and pigment present in MPs are susceptible to Raman spectroscopy, and interferes with the identification of the polymer (Shim et al., 2017). For chemical identification of specific polymer types, thermal analysis can be used as an alternative to spectroscopy. Thermal analysis, in contrast, is a damaging approach that prevents further investigation of MP materials. The most commonly used thermal analysis techniques for NMPs analysis are Differential Scanning Calorimetry (DSC) and Thermogravimetry Analyser (TGA) (Castañeda, Avlijas, Simard, Ricciardi, & Smith, 2014; Dumichen et al., 2015).

7. Regulations

Prevalence of issues related to MP concentration is equally divided among different levels in the ecosystem emphasizing the need of control and regulatory measures regulating these polymers. It is vital to restrain the formation and cycling of both primary and secondary forms of these polymers to avoid their exposure to the ecosystem. Methods preventing the formation of waste is equally important as recycling, resource recovery as that minimizes the risks and costs associated with the waste management. Preventing incidence of these polymers is an intricate process as it is a global environmental issue irrespective of the boundaries. It advocates the participation and contributions of countries across the globe. These factors can be stipulated by the patronage obtained from people whose familiarity about the harm caused by the incidence of MPs in environment literally being shallow. This should be strengthened to improve the reduced usage or discarding of plastics and to boost the recycling of plastic products. Alongside the public awareness, the policies or regulations regarding the control of plastics needs to be incorporated, which contributes positively to the management. While looking into a global picture of plastic control its only national policies that comes into play, but also the international policies existing and the harmony between these policies plays a significant role on the road to control and elimination of these polymers. The institution of "Plastics Restriction Order" in 2008 by China has helped in increasing the awareness among the people about the need and importance of protection of ecosystem from microforms of these polymers. Another important regulation or policy which helped in better recycling rates of plastic and better recycled products is "European Strategy for Plastics in a Circular Economy" adopted by Europe in 2018. In addition to this, they espoused an obligation "Plastic 2030" that aims at hindering the whole process of leakage plastic particles into the environment and also to improve the circulation degree of these polymers. There are additions like "Zero Plastics to Landfill" and "Zero Pellet Loss" initiatives that ensures the escape of MPs, which makes the plastic circulation intact. Certain obligations like "Opinions on Further Strengthening the Control of Plastic Pollution" by Ministry of Ecology and Environment (China) put forward an initiative aimed at interdicting and constraining the manufacture, sale and usage of some plastic products like use of ultra-thin plastic shopping bags (thickness less than 0.025 mm), daily chemical products containing plastic beads, etc., as in initial stage (Gong & Xie, 2020). Alongside the preferment of use of non-plastic products such as paper bags, degradable shopping bags, etc., and standardization the proper recycling and discarding of plastic waste was the other ideas that were functional in diminution of the usage and replacement of plastics.

8. Conclusion

Although the socioeconomic benefits of plastics are noteworthy, the ever-increasing plastic manufacturing, along with improper disposal, has resulted in plastics being one of the primary sources of environmental pollution. The degradation of plastics can lead to the production of NPs and MPs that have a different physicochemical property than their bulk material. These NMPs have become ubiquitous in the environment and have been isolated from terrestrial and aquatic ecosystems. They can be easily bio-accumulated, thereby penetrating the food chain and food webs resulting in detrimental effects on living organisms. Moreover, the trophic transfer of NMPs in the food chain to humans can lead to various health complications like inflammation, oxidative stress and apoptosis, and metabolic disorders. However, there is limited information on the subcellular and molecular effects of NMPs on the human body. The fate and effects of NPs and MPs after ingestion by humans and marine species are unclear. Contamination of NMPs that are not handled properly can lead to an imbalance of the ecosystem. Therefore, the correct application of strict regulations and policies along with the development of a standardized methodology for the identification and evaluation of NMPs from natural matrices is of the utmost importance.

References

Allen, S., Allen, D., Phoenix, V. R., Le Roux, G., Durántez Jiménez, P., Simonneau, A., et al. (2019). Atmospheric transport and deposition of microplastics in a remote mountain catchment. *Nature Geoscience*, *12*(5), 339–344.

Alomar, C., Sureda, A., Capo, X., Guijarro, B., Tejada, S., & Deudero, S. (2017). Microplastic ingestion by *Mullus surmuletus* Linnaeus, 1758 fish and its potential for causing oxidative stress. *Environmental Research*, *159*, 135–142. https://doi.org/10.1016/j.envres.2017.07.043.

Anderson, P. J., Warrack, S., Langen, V., Challis, J. K., Hanson, M. L., & Rennie, M. D. (2017). Microplastic contamination in Lake Winnipeg, Canada. *Environmental Pollution*, *225*, 223–231. https://doi.org/10.1016/j.envpol.2017.02.072.

Andrady, A. L. (2011). Microplastics in the marine environment. *Marine Pollution Bulletin*, *62*(8), 1596–1605. https://doi.org/10.1016/j.marpolbul.2011.05.030.

Auta, H. S., Emenike, C. U., & Fauziah, S. H. (2017). Distribution and importance of microplastics in the marine environment: A review of the sources, fate, effects, and potential solutions. *Environment International*, *102*, 165–176. https://doi.org/10.1016/j.envint.2017.02.013.

Avio, C. G., Gorbi, S., & Regoli, F. (2017). Plastics and microplastics in the oceans: From emerging pollutants to emerged threat. *Marine Environmental Research*, *128*, 2–11. https://doi.org/10.1016/j.marenvres.2016.05.012.

Bellasi, A., Binda, G., Pozzi, A., Galafassi, S., Volta, P., & Bettinetti, R. (2020). Microplastic contamination in freshwater environments: A review, focusing on interactions with sediments and benthic organisms. *Environments*, 7(4), 30. https://doi.org/10.3390/environments7040030.

Bergmann, M., Gutow, L., & Klages, M. (2015). *Marine anthropogenic litter*. Springer Nature.

Besseling, E., Wang, B., Lurling, M., & Koelmans, A. A. (2014). Nanoplastic affects growth of *S. obliquus* and reproduction of *D. magna*. *Environmental Science & Technology*, *48*(20), 12336–12343. https://doi.org/10.1021/es503001d.

Boerger, C. M., Lattin, G. L., Moore, S. L., & Moore, C. J. (2010). Plastic ingestion by planktivorous fishes in the North Pacific Central Gyre. *Marine Pollution Bulletin*, *60*(12), 2275–2278. https://doi.org/10.1016/j.marpolbul.2010.08.007.

Bonifazi, G., Palmieri, R., Serranti, S., Mazziotti, C., & Ferrari, C. R. (2017). Hyperspectral imaging based approach for monitoring of microplastics from marine environment. In *OCM* (p. 193).

Bosker, T., Bouwman, L. J., Brun, N. R., Behrens, P., & Vijver, M. G. (2019). Microplastics accumulate on pores in seed capsule and delay germination and root growth of the terrestrial vascular plant *Lepidium sativum*. *Chemosphere*, *226*, 774–781. https://doi.org/10.1016/j.chemosphere.2019.03.163.

Botterell, Z. L. R., Beaumont, N., Dorrington, T., Steinke, M., Thompson, R. C., & Lindeque, P. K. (2019). Bioavailability and effects of microplastics on marine zooplankton: A review. *Environmental Pollution*, *245*, 98–110. https://doi.org/10.1016/j.envpol.2018.10.065.

Boucher, J., & Friot, D. (2017). *Primary microplastics in the oceans: A global evaluation of sources. Vol. 43*. Gland, Switzerland: IUCN.

Brandts, I., Teles, M., Goncalves, A. P., Barreto, A., Franco-Martinez, L., Tvarijonaviciute, A., et al. (2018). Effects of nanoplastics on *Mytilus galloprovincialis* after individual and combined exposure with carbamazepine. *Science of the Total Environment, 643*, 775–784. https://doi.org/10.1016/j.scitotenv.2018.06.257.

Brewer, A. K., & Striegel, A. M. (2010). Hydrodynamic chromatography of latex blends. *Journal of Separation Science, 33*(22), 3555–3563. https://doi.org/10.1002/jssc.201000565.

Browne, M. A., Galloway, T. S., & Thompson, R. C. (2010). Spatial patterns of plastic debris along estuarine shorelines. *Environmental Science & Technology, 44*(9), 3404–3409.

Cao, D., Wang, X., Luo, X., Liu, G., & Zheng, H. (2017). Effects of polystyrene microplastics on the fitness of earthworms in an agricultural soil. *IOP Conference Series: Earth and Environmental Science, 61*, 012148. https://doi.org/10.1088/1755-1315/61/1/012148.

Carbery, M., O'Connor, W., & Palanisami, T. (2018). Trophic transfer of microplastics and mixed contaminants in the marine food web and implications for human health. *Environment International, 115*, 400–409. https://doi.org/10.1016/j.envint.2018.03.007.

Carr, S. A., Liu, J., & Tesoro, A. G. (2016). Transport and fate of microplastic particles in wastewater treatment plants. *Water Research, 91*, 174–182. https://doi.org/10.1016/j.watres.2016.01.002.

Castañeda, R. A., Avlijas, S., Simard, M. A., Ricciardi, A., & Smith, R. (2014). Microplastic pollution in St. Lawrence River sediments. *Canadian Journal of Fisheries and Aquatic Sciences, 71*(12), 1767–1771. https://doi.org/10.1139/cjfas-2014-0281.

Chen, G., Feng, Q., & Wang, J. (2020). Mini-review of microplastics in the atmosphere and their risks to humans. *Science of the Total Environment, 703*, 135504. https://doi.org/10.1016/j.scitotenv.2019.135504.

Chen, Q., Gundlach, M., Yang, S., Jiang, J., Velki, M., Yin, D., et al. (2017). Quantitative investigation of the mechanisms of microplastics and nanoplastics toward zebrafish larvae locomotor activity. *Science of the Total Environment, 584–585*, 1022–1031. https://doi.org/10.1016/j.scitotenv.2017.01.156.

Chen, Y., Leng, Y., Liu, X., & Wang, J. (2020). Microplastic pollution in vegetable farmlands of suburb Wuhan, Central China. *Environmental Pollution, 257*, 113449. https://doi.org/10.1016/j.envpol.2019.113449.

Chen, Q., Yin, D., Jia, Y., Schiwy, S., Legradi, J., Yang, S., et al. (2017). Enhanced uptake of BPA in the presence of nanoplastics can lead to neurotoxic effects in adult zebrafish. *Science of the Total Environment, 609*, 1312–1321. https://doi.org/10.1016/j.scitotenv.2017.07.144.

Claessens, M., Van Cauwenberghe, L., Vandegehuchte, M. B., & Janssen, C. R. (2013). New techniques for the detection of microplastics in sediments and field collected organisms. *Marine Pollution Bulletin, 70*(1–2), 227–233. https://doi.org/10.1016/j.marpolbul.2013.03.009.

Cole, M., Lindeque, P., Fileman, E., Halsband, C., & Galloway, T. S. (2015). The impact of polystyrene microplastics on feeding, function and fecundity in the marine copepod *Calanus helgolandicus*. *Environmental Science & Technology, 49*(2), 1130–1137. https://doi.org/10.1021/es504525u.

Cole, M., Lindeque, P., Fileman, E., Halsband, C., Goodhead, R., Moger, J., et al. (2013). Microplastic ingestion by zooplankton. *Environmental Science & Technology, 47*(12), 6646–6655. https://doi.org/10.1021/es400663f.

Cole, M., Lindeque, P., Halsband, C., & Galloway, T. S. (2011). Microplastics as contaminants in the marine environment: A review. *Marine Pollution Bulletin, 62*(12), 2588–2597. https://doi.org/10.1016/j.marpolbul.2011.09.025.

Curren, E., & Leong, S. C. Y. (2019). Profiles of bacterial assemblages from microplastics of tropical coastal environments. *Science of the Total Environment, 655*, 313–320. https://doi.org/10.1016/j.scitotenv.2018.11.250.

da Costa, J. P., Santos, P. S. M., Duarte, A. C., & Rocha-Santos, T. (2016). (Nano)plastics in the environment—Sources, fates and effects. *Science of the Total Environment, 566–567*, 15–26. https://doi.org/10.1016/j.scitotenv.2016.05.041.

Davison, P., & Asch, R. G. (2011). Plastic ingestion by mesopelagic fishes in the North Pacific Subtropical Gyre. *Marine Ecology Progress Series, 432*, 173–180. https://doi.org/10.3354/meps09142.

de Souza Machado, A. A., Kloas, W., Zarfl, C., Hempel, S., & Rillig, M. C. (2018). Microplastics as an emerging threat to terrestrial ecosystems. *Global Change Biology, 24*(4), 1405–1416. https://doi.org/10.1111/gcb.14020.

de Souza Machado, A. A., Lau, C. W., Kloas, W., Bergmann, J., Bachelier, J. B., Faltin, E., et al. (2019). Microplastics can change soil properties and affect plant performance. *Environmental Science & Technology, 53*(10), 6044–6052. https://doi.org/10.1021/acs.est.9b01339.

Desforges, J. P., Galbraith, M., & Ross, P. S. (2015). Ingestion of microplastics by zooplankton in the Northeast Pacific Ocean. *Archives of Environmental Contamination and Toxicology, 69*(3), 320–330. https://doi.org/10.1007/s00244-015-0172-5.

Dong, C. D., Chen, C. W., Chen, Y. C., Chen, H. H., Lee, J. S., & Lin, C. H. (2020). Polystyrene microplastic particles: In vitro pulmonary toxicity assessment. *Journal of Hazardous Materials, 385*, 121575. https://doi.org/10.1016/j.jhazmat.2019.121575.

Dong, Y., Gao, M., Song, Z., & Qiu, W. (2020). As(III) adsorption onto different-sized polystyrene microplastic particles and its mechanism. *Chemosphere, 239*, 124792. https://doi.org/10.1016/j.chemosphere.2019.124792.

Driedger, A. G. J., Dürr, H. H., Mitchell, K., & Van Cappellen, P. (2015). Plastic debris in the Laurentian Great Lakes: A review. *Journal of Great Lakes Research, 41*(1), 9–19. https://doi.org/10.1016/j.jglr.2014.12.020.

Dris, R., Gasperi, J., Mirande, C., Mandin, C., Guerrouache, M., Langlois, V., et al. (2017). A first overview of textile fibers, including microplastics, in indoor and outdoor environments. *Environmental Pollution, 221*, 453–458. https://doi.org/10.1016/j.envpol.2016.12.013.

Dris, R., Gasperi, J., Rocher, V., Saad, M., Renault, N., & Tassin, B. (2015). Microplastic contamination in an urban area: A case study in Greater Paris. *Environmental Chemistry, 12*(5), 2015. https://doi.org/10.1071/en14167.

Drzyzga, O., & Prieto, A. (2019). Plastic waste management, a matter for the 'community'. *Microbial Biotechnology, 12*(1), 66–68. https://doi.org/10.1111/1751-7915.13328.

D'Souza, J. M., Windsor, F. M., Santillo, D., & Ormerod, S. J. (2020). Food web transfer of plastics to an apex riverine predator. *Global Change Biology, 26*(7), 3846–3857. https://doi.org/10.1111/gcb.15139.

Duis, K., & Coors, A. (2016). Microplastics in the aquatic and terrestrial environment: Sources (with a specific focus on personal care products), fate and effects. *Environmental Sciences Europe, 28*(1), 2. https://doi.org/10.1186/s12302-015-0069-y.

Dumichen, E., Barthel, A. K., Braun, U., Bannick, C. G., Brand, K., Jekel, M., et al. (2015). Analysis of polyethylene microplastics in environmental samples, using a thermal decomposition method. *Water Research, 85*, 451–457. https://doi.org/10.1016/j.watres.2015.09.002.

Enfrin, M., Hachemi, C., Hodgson, P. D., Jegatheesan, V., Vrouwenvelder, J., Callahan, D. L., et al. (2021). Nano/micro plastics—Challenges on quantification and remediation: A review. *Journal of Water Process Engineering, 42*, 102128. https://doi.org/10.1016/j.jwpe.2021.102128.

Espinosa, C., Garcia Beltran, J. M., Esteban, M. A., & Cuesta, A. (2018). In vitro effects of virgin microplastics on fish head-kidney leucocyte activities. *Environmental Pollution, 235*, 30–38. https://doi.org/10.1016/j.envpol.2017.12.054.

Europe, P. (2017). *Plastics—The Facts 2017*. Retrieved from https://plasticseurope.org/knowledge-hub/plastics-the-facts-2017/.

Fischer, M., & Scholz-Bottcher, B. M. (2017). Simultaneous trace identification and quantification of common types of microplastics in environmental samples by pyrolysis-gas chromatography-mass spectrometry. *Environmental Science & Technology*, *51*(9), 5052–5060. https://doi.org/10.1021/acs.est.6b06362.

Fossi, M. C., Marsili, L., Baini, M., Giannetti, M., Coppola, D., Guerranti, C., et al. (2016). Fin whales and microplastics: The Mediterranean Sea and the Sea of Cortez scenarios. *Environmental Pollution*, *209*, 68–78. https://doi.org/10.1016/j.envpol.2015.11.022.

Frias, J., & Nash, R. (2019). Microplastics: Finding a consensus on the definition. *Marine Pollution Bulletin*, *138*, 145–147. https://doi.org/10.1016/j.marpolbul.2018.11.022.

Fuller, S., & Gautam, A. (2016). A procedure for measuring microplastics using pressurized fluid extraction. *Environmental Science & Technology*, *50*(11), 5774–5780. https://doi.org/10.1021/acs.est.6b00816.

Gigault, J., El Hadri, H., Reynaud, S., Deniau, E., & Grassl, B. (2017). Asymmetrical flow field flow fractionation methods to characterize submicron particles: Application to carbon-based aggregates and nanoplastics. *Analytical and Bioanalytical Chemistry*, *409*(29), 6761–6769. https://doi.org/10.1007/s00216-017-0629-7.

Gong, J., & Xie, P. (2020). Research progress in sources, analytical methods, eco-environmental effects, and control measures of microplastics. *Chemosphere*, *254*, 126790.

Green, D. S. (2016). Effects of microplastics on European flat oysters, *Ostrea edulis* and their associated benthic communities. *Environmental Pollution*, *216*, 95–103. https://doi.org/10.1016/j.envpol.2016.05.043.

Hole, P., Sillence, K., Hannell, C., Maguire, C. M., Roesslein, M., Suarez, G., et al. (2013). Interlaboratory comparison of size measurements on nanoparticles using nanoparticle tracking analysis (NTA). *Journal of Nanoparticle Research*, *15*, 2101. https://doi.org/10.1007/s11051-013-2101-8.

Horton, A. A., Walton, A., Spurgeon, D. J., Lahive, E., & Svendsen, C. (2017). Microplastics in freshwater and terrestrial environments: Evaluating the current understanding to identify the knowledge gaps and future research priorities. *Science of the Total Environment*, *586*, 127–141.

Huang, D., Tao, J., Cheng, M., Deng, R., Chen, S., Yin, L., et al. (2021). Microplastics and nanoplastics in the environment: Macroscopic transport and effects on creatures. *Journal of Hazardous Materials*, *407*, 124399. https://doi.org/10.1016/j.jhazmat.2020.124399.

Huerta Lwanga, E., Gertsen, H., Gooren, H., Peters, P., Salanki, T., van der Ploeg, M., et al. (2017). Incorporation of microplastics from litter into burrows of *Lumbricus terrestris*. *Environmental Pollution*, *220*(Pt. A), 523–531. https://doi.org/10.1016/j.envpol.2016.09.096.

Huerta Lwanga, E., Mendoza Vega, J., Ku Quej, V., Chi, J. L. A., Sanchez Del Cid, L., Chi, C., et al. (2017). Field evidence for transfer of plastic debris along a terrestrial food chain. *Scientific Reports*, 7(1), 14071. https://doi.org/10.1038/s41598-017-14588-2.

Hurley, R., Woodward, J., & Rothwell, J. J. (2018). Microplastic contamination of river beds significantly reduced by catchment-wide flooding. *Nature Geoscience*, *11*(4), 251–257. https://doi.org/10.1038/s41561-018-0080-1.

Kane, I. A., & Clare, M. A. (2019). Dispersion, accumulation, and the ultimate fate of microplastics in deep-marine environments: A review and future directions. *Frontiers in Earth Science*, 7. https://doi.org/10.3389/feart.2019.00080.

Kappler, A., Fischer, D., Oberbeckmann, S., Schernewski, G., Labrenz, M., Eichhorn, K. J., et al. (2016). Analysis of environmental microplastics by vibrational microspectroscopy: FTIR, Raman or both? *Analytical and Bioanalytical Chemistry*, *408*(29), 8377–8391. https://doi.org/10.1007/s00216-016-9956-3.

Karthik, R., Robin, R. S., Purvaja, R., Ganguly, D., Anandavelu, I., Raghuraman, R., et al. (2018). Microplastics along the beaches of southeast coast of India. *Science of the Total Environment*, *645*, 1388–1399. https://doi.org/10.1016/j.scitotenv.2018.07.242.

Kim, J. S., Lee, H. J., Kim, S. K., & Kim, H. J. (2018). Global pattern of microplastics (MPs) in commercial food-grade salts: Sea salt as an indicator of seawater MP pollution. *Environmental Science & Technology*, *52*(21), 12819–12828. https://doi.org/10.1021/acs.est.8b04180.

Kumari, A., Chaudhary, D. R., & Jha, B. (2019). Destabilization of polyethylene and polyvinylchloride structure by marine bacterial strain. *Environmental Science and Pollution Research International*, *26*(2), 1507–1516. https://doi.org/10.1007/s11356-018-3465-1.

Lagarde, F., Olivier, O., Zanella, M., Daniel, P., Hiard, S., & Caruso, A. (2016). Microplastic interactions with freshwater microalgae: Hetero-aggregation and changes in plastic density appear strongly dependent on polymer type. *Environmental Pollution*, *215*, 331–339. https://doi.org/10.1016/j.envpol.2016.05.006.

Lebreton, L., Slat, B., Ferrari, F., Sainte-Rose, B., Aitken, J., Marthouse, R., et al. (2018). Evidence that the Great Pacific Garbage Patch is rapidly accumulating plastic. *Scientific Reports*, *8*(1), 4666. https://doi.org/10.1038/s41598-018-22939-w.

Lee, K. W., Shim, W. J., Kwon, O. Y., & Kang, J. H. (2013). Size-dependent effects of micro polystyrene particles in the marine copepod *Tigriopus japonicus*. *Environmental Science & Technology*, *47*(19), 11278–11283. https://doi.org/10.1021/es401932b.

Lei, K., Qiao, F., Liu, Q., Wei, Z., Qi, H., Cui, S., et al. (2017). Microplastics releasing from personal care and cosmetic products in China. *Marine Pollution Bulletin*, *123*(1–2), 122–126. https://doi.org/10.1016/j.marpolbul.2017.09.016.

Li, W., Duo, J., Wufuer, R., Wang, S., & Pan, X. (2022). Characteristics and distribution of microplastics in shoreline sediments of the Yangtze River, main tributaries and lakes in China-from upper reaches to the estuary. *Environmental Science and Pollution Research International*, *29*, 48453–48464. https://doi.org/10.1007/s11356-021-18284-7.

Liebezeit, G., & Liebezeit, E. (2015). Origin of synthetic particles in honeys. *Polish Journal of Food and Nutrition Sciences*, *65*(2), 143–147. https://doi.org/10.1515/pjfns-2015-0025.

Lim, S. L., Ng, C. T., Zou, L., Lu, Y., Chen, J., Bay, B. H., et al. (2019). Targeted metabolomics reveals differential biological effects of nanoplastics and nanoZnO in human lung cells. *Nanotoxicology*, *13*(8), 1117–1132. https://doi.org/10.1080/17435390.2019.1640913.

Magri, D., Sanchez-Moreno, P., Caputo, G., Gatto, F., Veronesi, M., Bardi, G., et al. (2018). Laser ablation as a versatile tool to mimic polyethylene terephthalate nanoplastic pollutants: Characterization and toxicology assessment. *ACS Nano*, *12*(8), 7690–7700. https://doi.org/10.1021/acsnano.8b01331.

Mammo, F. K., Amoah, I. D., Gani, K. M., Pillay, L., Ratha, S. K., Bux, F., et al. (2020). Microplastics in the environment: Interactions with microbes and chemical contaminants. *Science of the Total Environment*, *743*, 140518. https://doi.org/10.1016/j.scitotenv.2020.140518.

Md Amin, R., Sohaimi, E. S., Anuar, S. T., & Bachok, Z. (2020). Microplastic ingestion by zooplankton in Terengganu coastal waters, southern South China Sea. *Marine Pollution Bulletin*, *150*, 110616. https://doi.org/10.1016/j.marpolbul.2019.110616.

Miller, M. E., Hamann, M., & Kroon, F. J. (2020). Bioaccumulation and biomagnification of microplastics in marine organisms: A review and meta-analysis of current data. *PLoS One*, *15*(10), e0240792. https://doi.org/10.1371/journal.pone.0240792.

Mintenig, S. M., Bäuerlein, P. S., Koelmans, A. A., Dekker, S. C., & van Wezel, A. P. (2018). Closing the gap between small and smaller: Towards a framework to analyse nano- and microplastics in aqueous environmental samples. *Environmental Science: Nano*, *5*(7), 1640–1649. https://doi.org/10.1039/c8en00186c.

Mintenig, S. M., Int-Veen, I., Loder, M. G. J., Primpke, S., & Gerdts, G. (2017). Identification of microplastic in effluents of waste water treatment plants using focal plane array-based micro-Fourier-transform infrared imaging. *Water Research*, *108*, 365–372. https://doi.org/10.1016/j.watres.2016.11.015.

Mofijur, M., Ahmed, S. F., Rahman, S. M. A., Arafat Siddiki, S. Y., Islam, A., Shahabuddin, M., et al. (2021). Source, distribution and emerging threat of micro- and nanoplastics to marine organism and human health: Socio-economic impact and management strategies. *Environmental Research*, *195*, 110857. https://doi.org/10.1016/j.envres.2021.110857.

Muise, I., Adams, M., Côté, R., & Price, G. W. (2016). Attitudes to the recovery and recycling of agricultural plastics waste: A case study of Nova Scotia, Canada. *Resources, Conservation and Recycling*, *109*, 137–145. https://doi.org/10.1016/j.resconrec.2016.02.011.

Napper, I. E., & Thompson, R. C. (2016). Release of synthetic microplastic plastic fibres from domestic washing machines: Effects of fabric type and washing conditions. *Marine Pollution Bulletin*, *112*(1–2), 39–45. https://doi.org/10.1016/j.marpolbul.2016.09.025.

Nelms, S. E., Duncan, E. M., Broderick, A. C., Galloway, T. S., Godfrey, M. H., Hamann, M., et al. (2016). Plastic and marine turtles: A review and call for research. *ICES Journal of Marine Science: Journal du Conseil*, *73*(2), 165–181. https://doi.org/10.1093/icesjms/fsv165.

Nizzetto, L., Futter, M., & Langaas, S. (2016). Are agricultural soils dumps for microplastics of urban origin? *Environmental Science & Technology*, *50*(20), 10777–10779. https://doi.org/10.1021/acs.est.6b04140.

Pannetier, P., Morin, B., Clerandeau, C., Laurent, J., Chapelle, C., & Cachot, J. (2019). Toxicity assessment of pollutants sorbed on environmental microplastics collected on beaches: Part II-adverse effects on Japanese medaka early life stages. *Environmental Pollution*, *248*, 1098–1107. https://doi.org/10.1016/j.envpol.2018.10.129.

Paul-Pont, I., Tallec, K., Gonzalez-Fernandez, C., Lambert, C., Vincent, D., Mazurais, D., et al. (2018). Constraints and priorities for conducting experimental exposures of marine organisms to microplastics. *Frontiers in Marine Science*, *5*, 252. https://doi.org/10.3389/fmars.2018.00252.

Perry, M. C., Olsen, G. H., Richards, R. A., & Osenton, P. C. (2013). Predation on dovekies by goosefish over deep water in the Northwest Atlantic Ocean. *Northeastern Naturalist*, *20*(1), 148–154. https://doi.org/10.1656/045.020.0112.

Petersen, F., & Hubbart, J. A. (2021). The occurrence and transport of microplastics: The state of the science. *Science of the Total Environment*, *758*, 143936. https://doi.org/10.1016/j.scitotenv.2020.143936.

Pirok, B. W. J., Abdulhussain, N., Aalbers, T., Wouters, B., Peters, R. A. H., & Schoenmakers, P. J. (2017). Nanoparticle analysis by online comprehensive two-dimensional liquid chromatography combining hydrodynamic chromatography and size-exclusion chromatography with intermediate sample transformation. *Analytical Chemistry*, *89*(17), 9167–9174. https://doi.org/10.1021/acs.analchem.7b01906.

Pivokonsky, M., Cermakova, L., Novotna, K., Peer, P., Cajthaml, T., & Janda, V. (2018). Occurrence of microplastics in raw and treated drinking water. *Science of the Total Environment*, *643*, 1644–1651. https://doi.org/10.1016/j.scitotenv.2018.08.102.

Prietl, B., Meindl, C., Roblegg, E., Pieber, T. R., Lanzer, G., & Frohlich, E. (2014). Nano-sized and micro-sized polystyrene particles affect phagocyte function. *Cell Biology and Toxicology*, *30*(1), 1–16. https://doi.org/10.1007/s10565-013-9265-y.

Qi, Y., Yang, X., Pelaez, A. M., Huerta Lwanga, E., Beriot, N., Gertsen, H., et al. (2018). Macro- and micro-plastics in soil-plant system: Effects of plastic mulch film residues on wheat (*Triticum aestivum*) growth. *Science of the Total Environment*, *645*, 1048–1056. https://doi.org/10.1016/j.scitotenv.2018.07.229.

Rillig, M. C. (2012). Microplastic in terrestrial ecosystems and the soil? *Environmental Science & Technology*, *46*(12), 6453–6454. https://doi.org/10.1021/es302011r.

Rochman, C. (2018). Microplastics research—From sink to source. *Science*, *360*(6384), 28–29.

Rochman, C. M., Hoh, E., Kurobe, T., & Teh, S. J. (2013). Ingested plastic transfers hazardous chemicals to fish and induces hepatic stress. *Scientific Reports*, *3*, 3263. https://doi.org/10.1038/srep03263.

Rummel, C. D., Loder, M. G., Fricke, N. F., Lang, T., Griebeler, E. M., Janke, M., et al. (2016). Plastic ingestion by pelagic and demersal fish from the North Sea and Baltic Sea. *Marine Pollution Bulletin*, *102*(1), 134–141. https://doi.org/10.1016/j.marpolbul.2015.11.043.

Savoca, M. S., Wohlfeil, M. E., Ebeler, S. E., & Nevitt, G. A. (2016). Marine plastic debris emits a keystone infochemical for olfactory foraging seabirds. *Science Advances*, *2*(11), e1600395.

Schymanski, D., Goldbeck, C., Humpf, H.-U., & Fürst, P. (2018). Analysis of microplastics in water by micro-Raman spectroscopy: Release of plastic particles from different packaging into mineral water. *Water Research*, *129*, 154–162.

Sgier, L., Freimann, R., Zupanic, A., & Kroll, A. (2016). Flow cytometry combined with viSNE for the analysis of microbial biofilms and detection of microplastics. *Nature Communications*, 7, 11587. https://doi.org/10.1038/ncomms11587.

Shahul Hamid, F., Bhatti, M. S., Anuar, N., Anuar, N., Mohan, P., & Periathamby, A. (2018). Worldwide distribution and abundance of microplastic: How dire is the situation? *Waste Management & Research*, *36*(10), 873–897. https://doi.org/10.1177/0734242X18785730.

Shim, W. J., Hong, S. H., & Eo, S. E. (2017). Identification methods in microplastic analysis: A review. *Analytical Methods*, *9*(9), 1384–1391. https://doi.org/10.1039/c6ay02558g.

Smith, M., Love, D. C., Rochman, C. M., & Neff, R. A. (2018). Microplastics in seafood and the implications for human health. *Current Environmental Health Reports*, *5*(3), 375–386. https://doi.org/10.1007/s40572-018-0206-z.

Song, Y. K., Hong, S. H., Jang, M., Han, G. M., Rani, M., Lee, J., et al. (2015). A comparison of microscopic and spectroscopic identification methods for analysis of microplastics in environmental samples. *Marine Pollution Bulletin*, *93*(1–2), 202–209. https://doi.org/10.1016/j.marpolbul.2015.01.015.

Steinmetz, Z., Wollmann, C., Schaefer, M., Buchmann, C., David, J., Troger, J., et al. (2016). Plastic mulching in agriculture. Trading short-term agronomic benefits for long-term soil degradation? *Science of the Total Environment*, *550*, 690–705. https://doi.org/10.1016/j.scitotenv.2016.01.153.

Stock, V., Bohmert, L., Lisicki, E., Block, R., Cara-Carmona, J., Pack, L. K., et al. (2019). Uptake and effects of orally ingested polystyrene microplastic particles in vitro and in vivo. *Archives of Toxicology*, *93*(7), 1817–1833. https://doi.org/10.1007/s00204-019-02478-7.

Summers, S., Henry, T., & Gutierrez, T. (2018). Agglomeration of nano- and microplastic particles in seawater by autochthonous and de novo-produced sources of exopolymeric substances. *Marine Pollution Bulletin*, *130*, 258–267. https://doi.org/10.1016/j.marpolbul.2018.03.039.

Sun, X. D., Yuan, X. Z., Jia, Y., Feng, L. J., Zhu, F. P., Dong, S. S., et al. (2020). Differentially charged nanoplastics demonstrate distinct accumulation in *Arabidopsis thaliana*. *Nature Nanotechnology*, *15*(9), 755–760. https://doi.org/10.1038/s41565-020-0707-4.

Sundt, P., Schulze, P.-E., & Syversen, F. (2014). *Sources of microplastic-pollution to the marine environment. Vol. 86* (p. 20). Mepex for the Norwegian Environment Agency.

Tibbetts, J., Krause, S., Lynch, I., & Sambrook Smith, G. (2018). Abundance, distribution, and drivers of microplastic contamination in urban river environments. *Water, 10*(11), 1597. https://doi.org/10.3390/w10111597.

Udyawer, V., Read, M. A., Hamann, M., Simpfendorfer, C. A., & Heupel, M. R. (2013). First record of sea snake (*Hydrophis elegans, Hydrophiinae*) entrapped in marine debris. *Marine Pollution Bulletin, 73*(1), 336–338. https://doi.org/10.1016/j.marpolbul.2013.06.023.

Unice, K. M., Weeber, M. P., Abramson, M. M., Reid, R. C. D., van Gils, J. A. G., Markus, A. A., et al. (2019). Characterizing export of land-based microplastics to the estuary—Part I: Application of integrated geospatial microplastic transport models to assess tire and road wear particles in the Seine watershed. *Science of the Total Environment, 646*, 1639–1649. https://doi.org/10.1016/j.scitotenv.2018.07.368.

Vianello, A., Jensen, R. L., Liu, L., & Vollertsen, J. (2019). Simulating human exposure to indoor airborne microplastics using a Breathing Thermal Manikin. *Scientific Reports, 9*(1), 8670. https://doi.org/10.1038/s41598-019-45054-w.

von Moos, N., Burkhardt-Holm, P., & Kohler, A. (2012). Uptake and effects of microplastics on cells and tissue of the blue mussel *Mytilus edulis* L. after an experimental exposure. *Environmental Science & Technology, 46*(20), 11327–11335. https://doi.org/10.1021/es302332w.

Vroom, R., Halsband, C., Besseling, E., & Koelmans, A. (2016). Effects of microplastics on zooplankton. In *ICES/PICES 6th zooplankton production symposium: New challenges in a changing ocean—Scandic Bergen City, Bergen, Norway*.

Wagner, M., Scherer, C., Alvarez-Muñoz, D., Brennholt, N., Bourrain, X., Buchinger, S., et al. (2014). Microplastics in freshwater ecosystems: What we know and what we need to know. *Environmental Sciences Europe, 26*(1), 1–9.

Wang, Q., Enyoh, C. E., Chowdhury, T., & Chowdhury, A. H. (2020). Analytical techniques, occurrence and health effects of micro and nano plastics deposited in street dust. *International Journal of Environmental Analytical Chemistry*, 1–19. https://doi.org/10.1080/03067319.2020.1811262.

Wen, B., Zhang, N., Jin, S. R., Chen, Z. Z., Gao, J. Z., Liu, Y., et al. (2018). Microplastics have a more profound impact than elevated temperatures on the predatory performance, digestion and energy metabolism of an Amazonian cichlid. *Aquatic Toxicology, 195*, 67–76. https://doi.org/10.1016/j.aquatox.2017.12.010.

Xu, S., Ma, J., Ji, R., Pan, K., & Miao, A. J. (2020). Microplastics in aquatic environments: Occurrence, accumulation, and biological effects. *Science of the Total Environment, 703*, 134699.

Yang, Y. F., Chen, C. Y., Lu, T. H., & Liao, C. M. (2019). Toxicity-based toxicokinetic/toxicodynamic assessment for bioaccumulation of polystyrene microplastics in mice. *Journal of Hazardous Materials, 366*, 703–713. https://doi.org/10.1016/j.jhazmat.2018.12.048.

Zhang, Y., Kang, S., Allen, S., Allen, D., Gao, T., & Sillanpää, M. (2020). Atmospheric microplastics: A review on current status and perspectives. *Earth-Science Reviews, 203*, 103118. https://doi.org/10.1016/j.earscirev.2020.103118.

Zhou, Y., Liu, X., & Wang, J. (2019). Characterization of microplastics and the association of heavy metals with microplastics in suburban soil of Central China. *Science of the Total Environment, 694*, 133798. https://doi.org/10.1016/j.scitotenv.2019.133798.

CHAPTER FIVE

Detection methods of micro and nanoplastics

Abdo Hassoun[a,b,*], Luisa Pasti[c], Tatiana Chenet[c], Polina Rusanova[d,e], Slim Smaoui[f], Abderrahmane Aït-Kaddour[g], and Gioacchino Bono[d,h]
[a]Sustainable AgriFoodtech Innovation & Research (SAFIR), Arras, France
[b]Syrian Academic Expertise (SAE), Gaziantep, Turkey
[c]Department of Environmental and Prevention Sciences, University of Ferrara, Ferrara, Italy
[d]Institute for Biological Resources and Marine Biotechnologies, National Research Council (IRBIM-CNR), Mazara del Vallo, TP, Italy
[e]Department of Biological, Geological and Environmental Sciences (BiGeA) – Marine Biology and Fisheries Laboratory of Fano (PU), University of Bologna (BO), Bologna, Italy
[f]Laboratory of Microbial Biotechnology and Engineering Enzymes (LMBEE), Center of Biotechnology of Sfax (CBS), University of Sfax, Sfax, Tunisia
[g]Université Clermont Auvergne, INRAE, VetAgro Sup, UMRF, Lempdes, France
[h]Dipartimento di Scienze e Tecnologie Biologiche, Chimiche e Farmaceutiche (STEBICEF), Università Di Palermo, Palermo, Italy
*Corresponding author: e-mail address: a.hassoun@saf-ir.com

Contents

Abstract

Plastics and related contaminants (including microplastics; MPs and nanoplastics; NPs) have become a serious global safety issue due to their overuse in many products and applications and their inadequate management, leading to possible leakage into the environment and eventually to the food chain and humans. There is a growing literature reporting on the occurrence of plastics, (MPs and NPs) in both marine and terrestrial organisms, with many indications about the harmful impact of these contaminants on plants and animals, as well as potential human health risks. The presence of MPs and NPs in many foods and beverages including seafood (especially finfish, crustaceans,

Advances in Food and Nutrition Research, Volume 103
ISSN 1043-4526
https://doi.org/10.1016/bs.afnr.2022.08.002

bivalves, and cephalopods), fruits, vegetables, milk, wine and beer, meat, and table salts, has become popular research areas in recent years. Detection, identification, and quantification of MPs and NPs have been widely investigated using a wide range of traditional methods, such as visual and optical methods, scanning electron microscopy, and gas chromatography–mass spectrometry, but these methods are burdened with a number of limitations. In contrast, spectroscopic techniques, especially Fourier-transform infrared spectroscopy and Raman spectroscopy, and other emerging techniques, such as hyperspectral imaging are increasingly being applied due to their potential to enable rapid, non-destructive, and high-throughput analysis. Despite huge research efforts, there is still an overarching need to develop reliable analytical techniques with low cost and high efficiency. Mitigation of plastic pollution requires establishing standard and harmonized methods, adopting holistic approaches, and raising awareness and engaging the public and policymakers. Therefore, this chapter focuses mainly on identification and quantification techniques of MPs and NPs in different food matrices (mostly seafood).

1. Introduction

Nowadays, plastics, including microplastics (MPs) and nanoplastics (NPs) are everywhere. As plastic is not a biodegradable material, its wastes can only be broken down to smaller particles, which accumulate in the environment and could end up in the food chain. MPs are particles with a size smaller than 5 mm, whereas NPs are those within a size ranging from 1 to 1000 nm (Jadhav et al., 2021; van Raamsdonk et al., 2020). According to their sources, MPs can be divided into two classes, namely, primary and secondary MPs. Primary MPs are plastic particles produced for a specific function (e.g., specific personal care and cosmetics products) while secondary MPs result from the breakdown of larger plastic debris (Dehaut et al., 2016; van Raamsdonk et al., 2020). MPs can be classified into different categories according to their different shapes, including fibers, fragments, pellets, flakes, bead, filament, foam, and granules (López-Martínez et al., 2021; Mercogliano et al., 2020).

Plastic materials are used in numerous food-related applications (e.g., agriculture and food packaging) and many other fields, such as automotive industry, building and construction, and even cosmetic and health care products (Chatterjee & Sharma, 2019; Gündogdu et al., 2022). However, improper applications, overuse, and poor management of plastic wastes are transforming earth planet into a "plastic planet" (Chatterjee & Sharma, 2019). Therefore, environmental ubiquity of MPs and NPs has become a critical concern (Gündogdu et al., 2022; Mercogliano et al., 2020). For example, plastics materials, such as polypropylene (PP), polyethylene (PE), polystyrene (PS), polyethylene terephthalate (PET), and polyvinyl chloride (PVC), among others, continue to be among the most used materials in conventional food packaging, despite their potential effect on human health and the environment (Sid et al., 2021; ZabihzadehKhajavi et al., 2019). MPs can be released from packaging materials and interact with human cells, increasing acute inflammation and cell damage (Jadhav et al., 2021).

Over the last few years, a considerable research attention has been paid to the potential risk of plastics and MPs ingested through food for human health and environment, which can be reflected by the huge number of publications, making plastic/MPs/NPs one of the hottest research topics. In fact, according to the data collected using the Scopus database, the number of publications on MPs and NPs in food has recently increased significantly. Fig. 1 shows this trend in the number of articles published during the last decade and the corresponding number of citations.

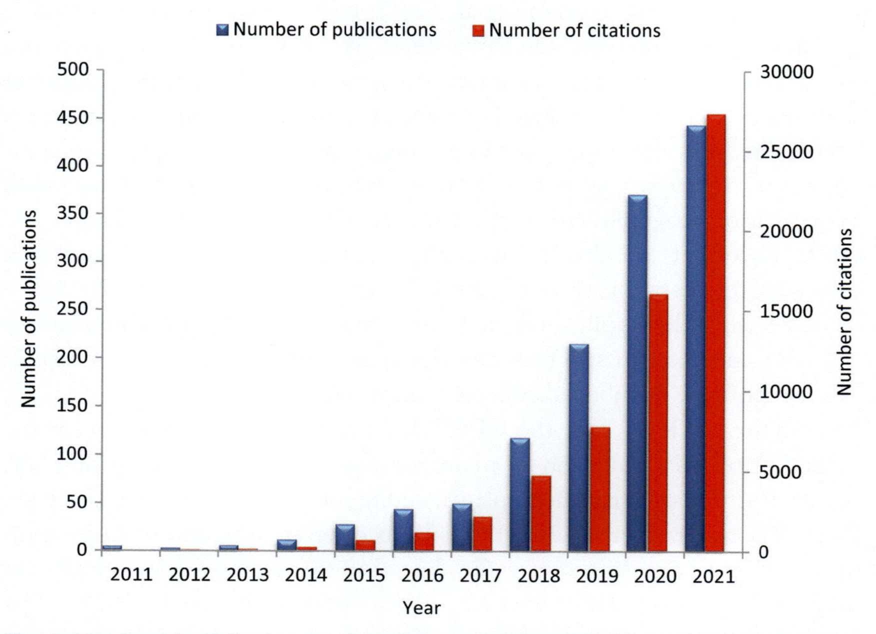

Fig. 1 Number of publications and citations per year on MPs/NPs in food/seafood over the last decade (search query was performed on 3rd May 2022). The following keywords search query was used in Scopus: TITLE-ABS-KEY (Microplastics) OR TITLE-ABS-KEY (Nanoplastics) AND TITLE-ABS-KEY (Food) OR TITLE-ABS-KEY (Seafood).

Many publications have been devoted to this topic, reporting the occurrence of MPs and NPs in different food categories such as fruit and vegetables (Oliveri Conti et al., 2020), packaged meat (Kedzierski et al., 2020), seafood (Dehaut, Hermabessiere, & Duflos, 2019; López-Martínez et al., 2021; Vázquez-Rowe, Ita-Nagy, & Kahhat, 2021), salt (Fadare, Okoffo, & Olasehinde, 2021; Renzi, Grazioli, et al., 2019), honey (Diaz-Basantes, Conesa, & Fullana, 2020), among many other foodstuffs.

Among these food groups, fish and other marine organisms have received a particular attention due to the fact that these foods are one of the most important routes of exposure for humans through the dietary intake (Dawson et al., 2021; Dehaut et al., 2019). For instance, the occurrences of MPs in canned fish (Akhbarizadeh et al., 2020), finfish, predominantly Atlantic salmon (*Salmo salar*), rainbow trout (*Oncorhynchus mykiss*), and channel catfish (*Ictalurus punctatus*) (Baechler et al., 2020), shellfish, such as mussel, oyster, clam, winkle, and scallop (Daniel et al., 2021; Li et al., 2021a, 2021b), and Mediterranean marine species including anchovy (*Sardina pilchardus*), sea bream (*Sparus aurata*), red mullet (*Mullus surmuletus*), and sole (*Solea solea*) (Ferrante et al., 2022) have been recently reported.

Most MPs studies include three steps: (i) extraction, (ii) detection and quantification, and (iii) identification and characterization. Different methods have been developed to determine and identify the MPs in recent years. The most commonly used techniques are visual and optical microscopy, spectroscopic methods as well as chromatographic techniques, such as gas chromatography/mass spectrometry (GC/MS) (Akoueson et al., 2021; Kwon et al., 2020; Mercogliano et al., 2020; van Raamsdonk et al., 2020; Veerasingam et al., 2021; Yuan, Nag, & Cummins, 2022).

Especially, the application of Fourier-transform infrared spectroscopy (FT-IR) and Raman spectroscopy has gained momentum since 2015, as can be noticed from data collected using the Scopus database (Fig. 2). During the last few years, the FT-IR has been extensively applied for the detection of MPs pollution in many various matrices, including dust/air, water, waste water treatment plants, sediment, biota, and salt (Bai et al., 2022; Veerasingam et al., 2021). FT-IR is often used in combination with optical microscopy to increase the detection performance, specifically for small MPs particles (Dehaut et al., 2019; Veerasingam et al., 2021). The Raman spectroscopy technique has been used in many applications for the analysis of very small microplastics ($<20\,\mu m$) (Anger et al., 2018; Araujo et al., 2018).

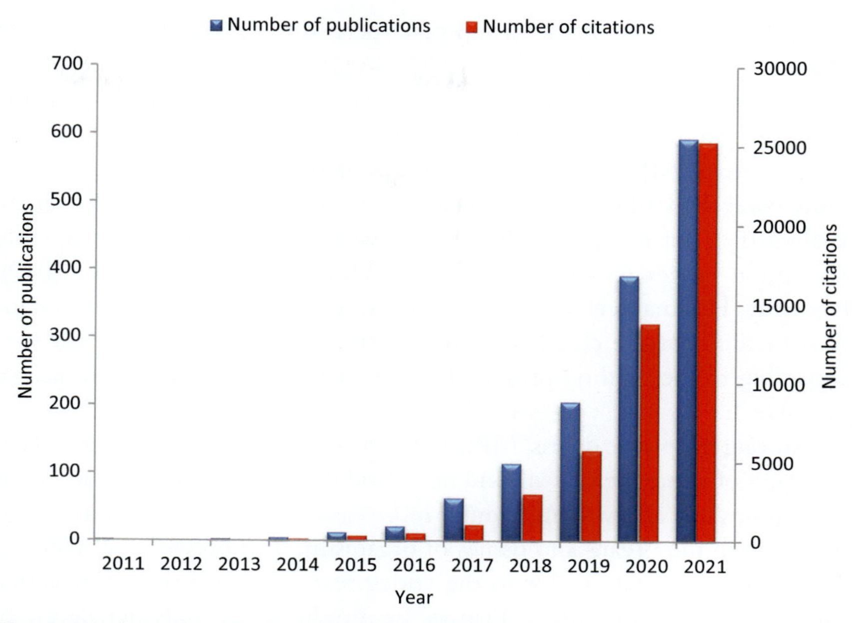

Fig. 2 Number of publications and citations per year on MPs/NPs using Fourier transform infrared spectroscopy or Raman spectroscopy over the last decade (search query was performed on 3rd May 2022). The following keywords search query was used in Scopus: TITLE-ABS-KEY (Fourier AND transform AND infrared) OR TITLE-ABS-KEY (FT-IR) OR TITLE-ABS-KEY (Raman) AND TITLE-ABS-KEY (Microplastics) OR TITLE-ABS-KEY (Nanoplastics).

2. Definition, sources, and types

Plastic debris is found in every environmental sphere, with a great variety of dimensions, shapes, composition, and color. The increased number of studies on the detection and effects of plastic debris in the last years, in particular on MPs, led to the need for a proper classification.

In general, plastic particles detected in the environment are classified according to their size, but there is a lack of a standard classification as many studies use different approaches. The term MPs was introduced in 2004 (Ivleva, 2021) referring to plastic fragments of small dimensions found in marine environment. In numerous studies published in the last 20 years, different upper limits have been considered to define the microplastic class, specifically from 0.5 to 5 mm, with most studies considering as MPs those particles with dimension lower than 5 mm (Hartmann et al., 2019).

In 2016, the UN advisory Group of Experts on the Scientific Aspects of Marine Environmental Protection (GESAMP) recommended the use of 5 mm as the cut-off value for MPs to preserve the information of the majority of the published works (Rosal, 2021).

Regarding NPs, many studies define as such those particles with dimensions lower than 0.1 μm (Ferreira et al., 2019; Nguyen et al., 2019; Venâncio et al., 2019), but in other studies, NPs have also been classified considering dimension ranges up to 1, 20 or even 335 μm (Hartmann et al., 2019). Recently, Hartmann et al. (2019) tackled this problem by suggesting a standard method for the classification of plastic fragments based on their size; according to the method proposed, plastic debris are classified as indicated in Table 1.

Among the plastic debris, MPs, and very recently NPs, have attracted the attention of researchers all around the world due to their ubiquity and their dimensions that allow them to enter more easily the food chain and even to penetrate in the organs and tissues of organisms. The presence of MPs and NPs in the ecosystem is due to the widespread use of plastic products in a vast variety of applications. Plastics are produced by polymerization of different monomers and additives, leading to a wide variety of lightweight and inexpensive polymeric materials characterized by their low density, low electrical conductivity, transparency, as well as toughness. The main polymers used, which account for 90% of the total world plastic production, are PET, high-density polyethylene (HDPE), PVC, low-density polyethylene (LDPE), PP, PS, and polyurethane (PU) (Boyle & Örmeci, 2020). Often, particular substances, such as phthalate acid esters, perfluoroalkyl substances, nonylphenol and bisphenol A, are added to the polymer in order to improve its properties; these additives, being of small molecular size, have the potential to leach from the plastic matrix during the degradation processes, and cause damage to the environment and biota (Boyle & Örmeci, 2020).

Table 1 Classification of plastic particles according to dimensions, as proposed by Hartmann et al. (2019).

Classification	Dimension range
Macroplastics	1 cm and larger
Mesoplastics	1 to <10 mm
Microplastics	1 to <1000 μm
Nanoplastics	1 to <1000 nm

Plastics are used in a variety of different applications, such as packaging, which represent the major sector for plastic consumption, building and construction, textile sector to produce clothes and other products, electrical/ electronic, industry and machinery, personal care products, marine coating, and others (Ryberg et al., 2019). Due to the wide variety of products containing polymeric materials, the main criterion for identifying plastic particles is based on their chemical composition. A distinction can be made by considering different classes: polymers include all synthetic and semi-synthetic polymers, copolymers are those materials produced from more than one type of monomers, and composites include those materials that contain synthetic polymer as an essential ingredient. All particles that fall into these categories should be considered as plastic debris (Hartmann et al., 2019).

Plastic debris has been detected in every environmental sphere all around the world. Plastic materials find their way to soils, sediments, freshwaters, oceans, and air through different transport pathways; the most significant inputs of plastics in the environment are losses from landfills, wastewater treatment plant (WWTP) effluents and sludge, and mishandled waste. Ryberg et al. (2019) estimated the global losses of plastics to the environment occur along the plastic value chains, identifying the municipal solid waste management as the major source of plastic loss. Rivers represent a major vector for the transportation of plastics to the oceans; it has been estimated that from 70% to 80% of the plastics that reach the oceans are transported through freshwater streams (Alimi et al., 2018).

Usually, MPs found in the environment can be divided according to their origin: primary or secondary MPs. Primary MPs are manufactured to be already of microscopic dimensions for different industrial and domestic applications, such as beads in scrubbers for paint or rust removal, in toothpaste formulations and in cosmetic products. Pollution of water bodies by primary MPs arises from the direct use of these products, while the contamination of agricultural land may occur through the use of fertilizer, or sludge generated by WWTPs, which can act as a sink for MPs during the water treatment processes (Sol et al., 2020). Conversely, secondary MPs are those particles originating from the degradation of bigger plastic debris through various natural processes, such as physical fragmentation, photodegradation, chemical, as well as biological degradation (Dehaut et al., 2016; van Raamsdonk et al., 2020). The same distinction can be applied to NPs: primary NPs are those produced with dimensions lower than 1 μm and used in different applications, whereas secondary NPs originate from the fragmentation of MPs (Gonçalves & Bebianno, 2021).

Once entered the environment, there are many pathways, abiotic or biotic, through which plastic debris can break down leading to the generation and dispersion of MPs and NPs, and different processes may often work synergistically (Alimi et al., 2018). The abiotic pathways by which plastic debris can be broken down into smaller fragments include mechanical breakdown and processes involving degradation through oxidation, hydrolytic, photo and thermal reactions. Mechanical degradation occurs due to weathering (e.g., water turbulence, freezing, thawing and changes in pressures) and it leads to morphological changes, but does not affect the polymeric bonds in the plastic matrix. This is particularly the case of disposable plastic bags, which represent one of the major plastic wastes often found in the environment (Ke et al., 2019).

Photodegradation is one of the main processes that damage the most plastic materials; in particular, the ultraviolet (UV; with wavelengths between 290 and 400 nm) and visible (between 400 and 800 nm) radiation is considered the primary factor in the degradation of polymers in the environment. Usually, the radiation is absorbed by chromophore groups or impurities present in the polymer matrix, leading to the formation of radicals that can react with oxygen generating peroxyl radicals, then hydroperoxides. Further reactions between radicals lead to polymer chain scission, cross-linking, chain branching and formation of groups containing oxygen (Masry et al., 2021). The oxidation reactions, thermal or photo-induced, bring about the introduction of carbonyl (C=O) and/or hydroxyl (OH) groups in the polymer chain, promoting biodegradation (Boyle & Örmeci, 2020).

Biotic degradation processes involve the action of enzymes secreted by microorganisms that, through hydrolytic reactions, break down the polymer chains, the progressive loss of molecular weight of the polymeric matrix triggers further microbial degradation. The continued loss of molecular structure results eventually in the formation of water-soluble oligomers, which in turn can be mineralized and assimilated by microorganisms as carbon and nitrogen (Boyle & Örmeci, 2020).

Another important input of MP/NP particles in the environment, especially in the atmosphere, is from the transport sector. In fact, the wearing of tires on the road over time due to abrasion is a common source of rubber particles, which enter the atmosphere and they can be transported to long distances by wind currents and pollute other ecosystems leading to adverse effects (Šourková, Adamcová, & Vaverková, 2021). Tire wear particles have been considered as pollutants for many years although they were recognized as MPs only recently (Knight et al., 2020). Järlskog et al. (2020) in their investigation on the occurrence of MPs on urban streets, sweep sand, and wash water found that tire tread wear and bitumen made up the largest proportion of anthropogenic particles detected.

MPs can be found in the environment in a wide variety of dimensions, shapes and colors depending on the initial material from which they originate, and the degradation processes they underwent. In general, when categorized according to their shape, MPs can be described as spheres, fibers, fragments, pellets, beads, films, flakes, and foams (Hartmann et al., 2019). MPs in the form of fibers are commonly found in the ecosystems due to the use of synthetic polymers for the production of textile fibers, and the presence of fibers of microscopic dimensions in the environment is caused by ordinary use of clothing and, especially in water, through the discharges of washing (Dalla Fontana, Mossotti, & Montarsolo, 2020). Fibers can have different compositional nature: they can be composed of exclusively synthetic polymers (e.g., polyesters or polyamides), of semi-synthetic materials (e.g., rayon), or of natural materials (e.g., cellulose). Fibers can pose a threat to organisms since they usually undergo further processes, such as dyeing; once in the environment, the pigments can leach from the fiber and lead to adverse effects toward the biota and the environment (Collard et al., 2018).

The term "fragments" refers to a broad category of plastic debris characterized by an irregular shape. They can originate from the degradation of bigger plastic debris, or they can be produced intentionally with irregular forms, for example, as abrasive particles in personal care products (Sun, Ren, & Ni, 2020), and enter the environment owing to the use of the products in which they are contained. Films and flakes are particles with irregular shapes and with a very thin thickness; they can be found in the environment due to fragmentation of paint, for instance, the coating of boats and ships can break down over time releasing small fragments into the water (Turner, 2021).

Another aspect taken into account when MPs and NPs are considered is their color; the coloration may help the identification of the sources of these particles, but it should be noticed that discoloration might occur due to degradation processes. While the classification of MPs and NPs by color is usually not crucial, it can be helpful in the biological field because colored plastic particles can be mixed with food and ingested by marine organisms. In fact, MPs have been found in wild aquatic species (Chenet et al., 2021; Mancia et al., 2020; Tsangaris et al., 2020) where the predominant colors of both filaments and fragments were blue, black, green, and red. Reinold et al. (2021) investigated the ingestion of MPs in *Dicentrarchus labrax* cultivated in aquaculture facilities, finding that 65% of the specimen analyzed ingested MPs mainly in the form of fibers with blue and yellow as predominant colors. In their study on the uptake routes of MPs by fishes, Roch, Friedrich, and Brinker (2020) considered, among other factors, the role presented by the color of the particles on the ingestion; they found that MPs with food-like coloration were ingested in higher amounts by the organisms considered.

Density represents another property of MPs and NPs that plays an important role on their distribution and possibility of ingestion by aquatic species (Roch et al., 2020). According to the composition of the matrix, particles, once released in water, can float or sink, becoming more prone to transportation through currents or to deposition in sediments. In the last case, these particles can pose a risk to marine species, which usually feed on organisms present on the seabed. The density of a given particle may change over time due to degradation processes that lead to a variation in the composition of the matrix, but also to the formation of biofilms onto the surface of the particles caused by microorganisms (Borges-Ramírez et al., 2020).

3. Occurrence of MPs and NPs in seafood products

3.1 MPs/NPs in finfish, crustaceans, molluscs, and other seafood products

Some 1300 scientific papers published between 2011 and 2021 (source: Scopus database) reveal how MPs are now ubiquitous in major seafood products (finfish, crustaceans, bivalves, cephalopods, gastropods, etc.) and other marine species (anemones, jellyfish, sea cucumbers, etc.) consumed around the world. Since 2014, 165 papers have also highlighted the threat of NPs (Fig. 3). Overall, the Scopus search revealed 1547 publications pertaining to MPs and NPs in the major seafood products listed above. Publications were grouped into 25 subject areas, 75% of which fall within the three basic areas of Environmental Science, Agricultural and Biological Sciences, and Earth and Planetary Sciences.

Another Scopus search using the combined terms "TITLE-ABS-KEY (microplastics) ORTITLE-ABS-KEY (nanoplastics) AND TITLE-ABS-KEY (crustaceans) OR TITLE-ABS-KEY (bivalves) OR TITLE-ABS-KEY

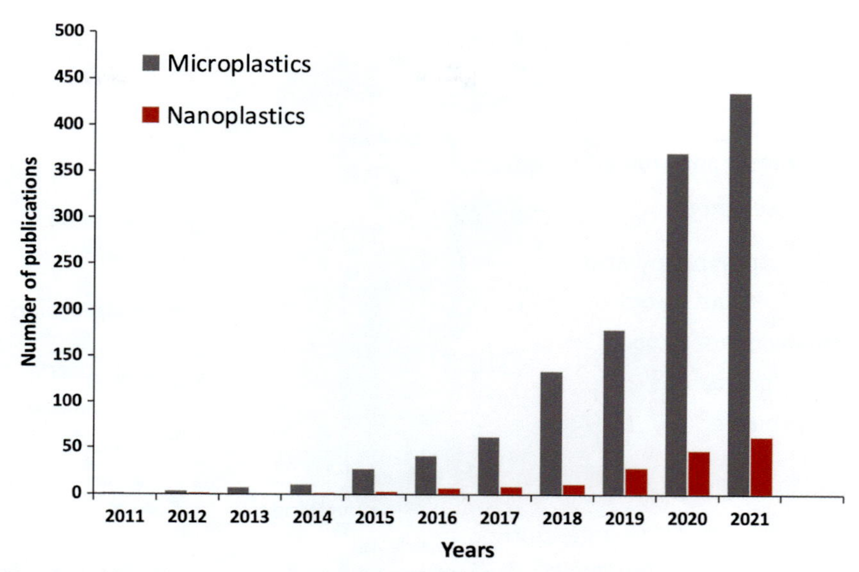

Fig. 3 Publications on microplastics and nanoplastics in fish, crustaceans, bivalves, cephalopods, gastropods, anemones, jellyfish, and sea cucumbers (source: Scopus database—search within "title, abstract and keywords") in the years 2011–2021.

(fish) OR TITLE-ABS-KEY (cephalopods) OR TITLE-ABS-KEY (sea AND cucumber) OR TITLE-ABS-KEY (gastropods) OR TITLE-ABS-KEY (anemones) OR TITLE-ABS-KEY (jellyfish)" was carried out in order to perform a cluster-based VOSviewer analysis.

According to the VOSviewer output, co-occurrence analysis identified 15,002 keywords, of which 557 occurred at least 20 times. As expected, "MPs" was the most frequent keyword identified (the size of the node indicates the frequency of the keyword), followed by "plastic," "animals," "fish," "polystyrene," "nanoplastic," "particle size," "polypropylene," "ingestion" and "oxidative stress" (Fig. 4). Three "clusters" were generated: these define thematic proximity among all considered keywords.

In this scenario, by comparing the huge amount of scientific data produced in just 10 years, it is clear that the millions of tons of plastic litter that has irresponsibly ended up in the oceans in the last half century is now returning to us (Ragusa et al., 2021) through the food web: zooplankton > shellfish > fish > humans.

Not all marine organisms respond in the same way to these emerging pollutants. MP and NP accumulation is strongly influenced by the environment in which organisms live (i.e., pelagic-neritic, demersal habitats), what,

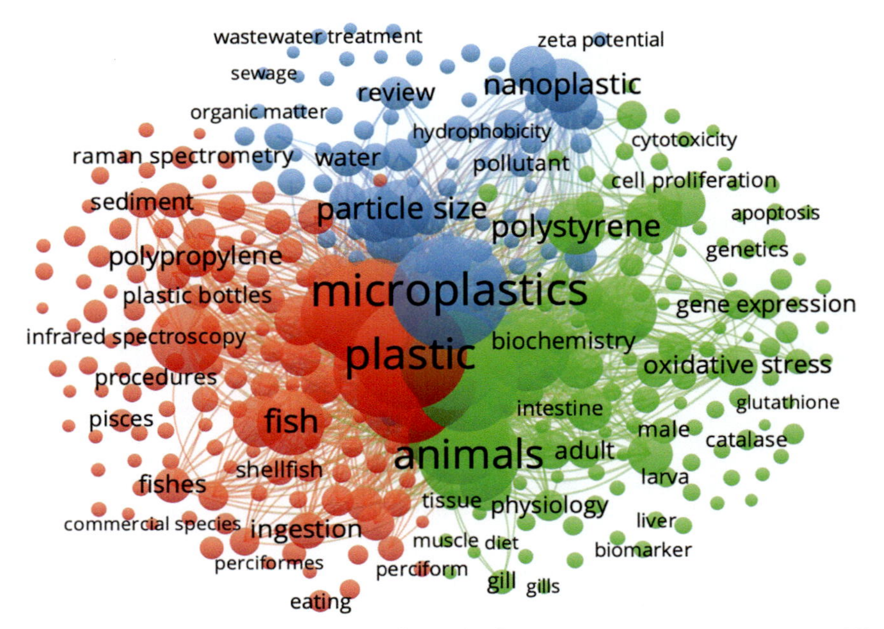

Fig. 4 Network map of 557 keywords (with the minimum occurrence set to 20) identified by VosViewer on the Scopus (CSV) database. The size of nodes indicates the frequency of occurrence, while the colors identify three clusters that define the thematic proximity and indicate research directions.

how and where they feed, as well as their trophic level and their anatomic features (Ribeiro et al., 2020). For example, mussels, clams, oysters and other minor bivalves are more subject to MP and NP contamination and accumulation because they feed by filtering significant quantities of coastal water, which is often more polluted than the open sea environment. As they are eaten whole, they directly transfer to humans a significant portion of MPs and NPs ingested during their life cycle. Other marine organisms, such as small fish, gastropods, cephalopods, and other small invertebrates, raise a similar concern because they are traditionally eaten whole (i.e., including the gastrointestinal tract), both fresh and/or dried, especially in low-income countries.

The risk of transferring these emerging pollutants to the food chain is obviously lower in the case of large marine species consumed after eliminating the gastrointestinal tract. It should be noted that there might be a microbiological risk associated with the consumption of contaminated seafood products. Recent studies have revealed that MPs and NPs are associated with a higher risk of transmitting pathogen microorganisms, such as viruses, bacteria, and fungi (Gündogdu et al., 2022). Considering that seafood products tend to be packed in a modified atmosphere and/or under vacuum skin packaging, two methods involving a high consumption of plastic [PP, ethylene vinyl alcohol (EVOH), polyamide (PA), PE], further studies are needed to understand how much these techniques, and the massive plastic polymers used, affect the final product in terms of contamination (Sobhani et al., 2020).

Given the growing interest in these new pollutants, Danopoulos et al. (2020) state that the issue of contamination by MPs and NPs should be included in food quality risk analysis and in the HACCP (Hazard Analysis and Critical Control Point) or HARPC (Hazard Analysis and Risk-based Preventive Controls for Human Food) programs for food safety.

From a physiological perspective, MP and NP ingestion could also alter endocrine system functions and the seafood microbiome, reduce the rate of fecundity, and cause DNA and neurological damage in various species (Chenet et al., 2021; Mancia et al., 2020). This is due to their intrinsic nature and to their ability to act as carriers of other pollutants, such as plastic additives (phthalates, triclosan, bisphenols, brominated and flame retardants), other persistent organic pollutants (POPs) and heavy metals dissolved in the aquatic marine environment (Guo & Wang, 2019; Kutralam-Muniasamy et al., 2021; Tourinho et al., 2019), with serious consequences for marine biota. Owing to their high toxicity, it is important to acquire greater insight into the effects of MPs and NPs on consumed marine organisms in terms of seafood safety and human health (Gündogdu et al., 2022; Piyawardhana et al., 2022). According to Gündogdu et al. (2022), another important aspect is the fate of MPs and NPs once they have been ingested by marine organisms. Specifically, how long they remain confined to the gastrointestinal tract before being excreted in feces, or whether they migrate through the circulatory system to other organs such as flesh muscle and gonads, as well as the liver and spleen. Therefore, although it seems that particles smaller than 150 μm in size can cross the intestinal epithelium, only 0.2–0.45% of the ingested MPs and NPs (<1.5 μm in size) can penetrate deeply into other organs (Food Safety Authority, 2016; Hazimah et al., 2021).

3.2 Connection between MPs/NPs and the most consumed wild seafood products

Based on FAO report (2020) "The state of world fisheries and aquaculture," Table 2 shows the most caught marine species in 2018. Starting from this

Table 2 Summary of reported microplastic ingestion by the most caught marine species in 2018 (FAO, 2020).

Common name	Scientific name	Average production in 2018 (million tons/percentage)	Habitat	Studied fishing ground	Presence of MPs (%)	Main recent references
Finfish						
Anchoveta	*Engraulis ringens*	7.045 (10%)	Pelagic-neritic	Pacific Ocean	0.8%	Ory et al. (2018)
Skipjack tuna	*Katsuwonus pelamis*	3.161 (4%)	Pelagic-oceanic	North Maluku Ocean	100%	Ridwan Lessy and Sabar (2021)
				Indian Ocean (Pantai Baron)	100%	Suwartiningsih, Setyowati, and Astuti (2020)
				South Pacific Ocean	23%	Markic et al. (2018)
				Southeast-south coast of Brazil	25.8%	Neto et al. (2020)
				Western equatorial Atlantic Ocean	0.75%	de Mesquita et al. (2021)
				North-East Atlantic	10%	Pereira et al. (2020)
Atlantic herring	*Clupea harengus*	1.820 (3%)	Benthopelagic	English Channel	50%	Collard et al. (2017)
				Baltic Sea	12.7%	Białowąs et al. (2022)
				Baltic Sea	20%	Beer et al. (2018)
Blue Whiting	*Micromesistius poutassou*	1.712 (2%)	Bathypelagic	English Channel	51.9%	Mercogliano et al. (2020)
				North-western Iberian Shelf Sea	0.02%	López-López et al. (2018)
				North Altlantic	0%	Murphy et al. (2017)
				Not reported	29.8%	Walkinshaw et al. (2020)
				Tyrrhenian coast	0%	Pittura et al. (2018)

European pilchard	*Sardina pilchardus*	1.608 (2%)	Pelagic-neritic	Adriatic Sea	96%	Renzi, Specchiulli, et al. (2019)
				Spanish Mediterranean coast	15.2%	Compa et al. (2018)
				Southern Tyrrhenian Sea	52.6%	Savoca et al. (2020)
				Northern Ionian Sea	47.2%	Digka et al. (2018)
				North Adriatic Sea	30%	Mistri et al. (2022)
				Mediterranean Sea	57%	Güven et al. (2017)
				Western Mediterranean Sea	17.39%	Rios-Fuster et al. (2019)
				Western and southern parts of Iberian coast	58%	Lopes et al. (2020)
				Northwestern Iberian continental shelf	87%	Filgueiras et al. (2020)
				South Adriatic Sea	50%	Anastasopoulou et al. (2018)
				Middle Adriatic Sea	37%	
				Ionian Sea (Greece)	47%	
Pacific chub mackerel	*Scomber japonicus*	1.557 (2%)	Pelagic-neritic	Southeast Pacific Ocean	3.3%	Ory et al. (2018)
				South-western Japan	62.5%	Yagi et al. (2022)
				Mediterranean Sea	57%	Güven et al. (2017)
				Mediterranean Sea	50%	Bray et al. (2019)
Yellowfin tuna	*Thunnus albacares*	1.458 (2%)	Pelagic-oceanic	South Pacific Ocean	2% (only mesoplastics >5 mm)	Chagnon et al. (2018)
Scads nei	*Decapterus* spp.	1.336 (2%)	Reef-associated, demersal, pelagic-oceanic benthopelagic	South Pacific	80%	Ory et al. (2017)
				Indonesia	29%	Rochman et al. (2015)

Continued

Table 2 Summary of reported microplastic ingestion by the most caught marine species in 2018 (FAO, 2020).—cont'd

Common name	Scientific name	Average production in 2018 (million tons/percentage)	Habitat	Studied fishing ground	Presence of MPs (%)	Main recent references
Atlantic cod	*Gadus morhua*	1.218 (2%)	Benthopelagic	Fogo Island, Newfoundland and Labrador	1.4%	Saturno et al. (2020)
				Baltic Sea	14.8%	Białowąs et al. (2022)
				Norwegian coast (Bergen City Harbor)	0–27%	Bråte et al. (2016)
				Newfoundland	1.7%	Liboiron et al. (2019)
				North Sea	0–13%	de Vries et al. (2020)
Largehead hairtail	*Trichiuruslepturus*	1.151 (2%)	Benthopelagic	Atlantic Ocean	1.4%	di Beneditto and da Silva Oliveira (2019)
				Oman Sea	14%	Ghattavi, Naji, and Kord (2019)
				South Atlantic Ocean	20%	Pegado et al. (2018)
Atlantic mackerel	*Scomber scombrus*	1.047 (1%)	Pelagic-neritic	Western Mediterranean Sea	49.2%	Palazzo et al. (2021)
				Portuguese coast	100%	Lopes et al. (2020)
				Coast of Portugal	31%	Neves et al. (2015)
Japanese anchovy	*Engraulis japonicus*	957 (1%)	Pelagic-neritic	Seto Inland Sea	90–100%	Ohkubo et al. (2022)
				Pacific Ocean (Tokyo Bay)	77%	Tanaka and Takada (2016)
				Yellow Sea	33%	Sun et al. (2019)

Sardinellas nei	*Sardinella* spp.	887 (1%)	Pelagic-neritic, reef-associated, pelagic	Southwest coast of India	38%	James et al. (2022)
				Pantai Indah Kapuk coast, Indonesia	100%	Hastuti, Lumbanbatu, and Wardiatno (2019)
				Northern Bay of Bengal	100%	Hossain et al. (2019)
				Thoothukudi region (Indian Ocean)	21%	Kalaiselvan et al. (2022)
				Indian Ocean	60–100%	Palermo et al. (2020)
				South eastern Arabian Sea, Indian coasts	5–9.1%	James et al. (2020)
				Central Atlantic Ocean, off the Coast of Ghana	26–41%	Adika et al. (2020)
Crustaceans						
Gazami crab	*Portunus trituberculatus*	493 (8%)	Benthopelagic	Yellow Sea	66.7–100%	Zhang et al. (2021)
Marine crabs nei	*Pachygrapsus transversus* *Carcinus maenas* *Eriocheir sinensis* *Callinectes sapidus*	314 (5%)	Benthic benthic benthic benthopelagic	Ponta Verde Beach, Brazil	47.4%	de Barros, dos Santos Calado, and de Sá Leitão Câmara de Araújo (2020)
				Thames Estuary	71.3%	McGoran et al. (2020)
				Thames Estuary	100%	McGoran et al. (2020)
				Adriatic Sea	83.3%	Renzi et al. (2020)
Blue swimming crab	*Portunuspelagicus*	298 (5%)	Reef-associated	Arabian Sea	13.3%	Daniel et al. (2021)

Continued

Table 2 Summary of reported microplastic ingestion by the most caught marine species in 2018 (FAO, 2020).—cont'd

Common name	Scientific name	Average production in 2018 (million tons/percentage)	Habitat	Studied fishing ground	Presence of MPs (%)	Main recent references
Argentine red shrimp	*Pleoticus muelleri*	256 (4%)	Benthic	Southwestern Atlantic Ocean	90%	Fernández Severini et al. (2020)
Molluscs						
Jumbo flying squid	*Dosidicus gigas*	892 (15%)	Pelagic	South Pacific Ocean	79.2%	Gong et al. (2021)
Marine Molluscs nei	*Mollusca*	664 (11%)	–	Liaohe Estuary (China)	67%	Wang et al. (2021a, 2021b)
Cuttlefish, bobtail squids nei	*Sepiidae, Sepiolidae*	348 (6%)	Benthic	Portugal	100% (cuttlefish)	Oliveira et al. (2020)

baseline, it was added the specific habitat (from pelagic to benthonic) of each taxa/species and what currently known about the presence of MPs and NPs (in the gastrointestinal tract, gills, liver, muscle tissue, gonads, etc.) within each taxa/species on a geographical basis.

In the literature of the last 5 years (2017–April 2022), only one study (Ory et al., 2018) focused on the most captured fish species in the world (*Engraulis ringens*) and yellowfin tuna (*Thunnus albacore*) (Chagnon et al., 2018); MP concentrations reported in these studies were 0.8% and 2%, respectively. No data are available for Alaska pollok (*Gadus chalcogrammus*), which is the second most captured species worldwide. Further studies are needed to discern whether and to what extent this species is affected by MPs and/or NPs. In studies (two to four) on Scad nei (*Decapterus* spp.), the largehead hairtail (*Trichiurus lepturus*), Atlantic mackerel (*Scomber scombrus*) and Japanese anchovy (*Engraulis japonicus*), MP concentrations ranged from 1.4% in largehead hairtail caught in the Atlantic Ocean (di Beneditto & da Silva Oliveira, 2019) to 100% in Japanese anchovy caught in the Seto Inland Sea (western Japan) (Ohkubo et al., 2022).

At least five published studies have reported MPs in the following marine species: In Skipjack tuna (*Katsuwonus pelamis*), MPs ranged from 0.75% in specimens from the western equatorial Atlantic Ocean (de Mesquita et al., 2021) to 100% in those from the Indian Ocean (Suwartiningsih et al., 2020). As for blue whiting (*Micromesistius poutassou*), MPs concentrations ranged from 0% in specimens from the Tyrrhenian Sea (central Mediterranean) (Pittura et al., 2018) to 51.9% in samples from the English Channel (Mercogliano et al., 2020). In the case of European pilchardus (*Sardina pilchardus*), MPs concentrations ranged from 15.24% in specimens from the Mediterranean coast of Spain (western Mediterranean) (Compa et al., 2018) to 96% in specimens from the Adriatic Sea (central Mediterranean) (Renzi, Specchiulli, et al., 2019). MP concentrations in Pacific chub mackerel (*Scomber japonicus*) ranged from 3.3% in specimens from the southeast Pacific Ocean (Ory et al., 2018) to 62.5% in samples from south western Japan (Yagi et al., 2022). The Atlantic cod (*Gadus morhua*) showed MP values ranging from 1.4% in specimens from Fogo Island, Newfoundland, and Labrador (Saturno et al., 2020) to 27% in specimens from Norwegian coast. MP concentrations in Sardinellas nei (*Sardinella* spp.) ranged from 5% in specimens from the South-eastern Arabian Sea, India (James et al., 2020), to 100% in specimens from the Pantai Indah Kapuk coast, Indonesia (Hastuti et al., 2019). When considering crustaceans, the only available study on the most harvested gazami crab (*Portunustri tuberculats*) reports MP concentrations of 66.7–100% (Zhang et al., 2021), whereas the blue swimming crab (*Portunus pelagicus*) caught in the Arabian Sea is impacted the least, with MP concentrations of up to 13% (Daniel et al., 2021). As for other invertebrates of particular commercial interest, such as cephalopods, jellyfish, and echinoderms, all three of these taxa have quite high MP contents ranging from 75% in the sea urchin nei (*Strongylocentrotus* spp.) of the northern coast of China (Feng et al., 2020) to 100% in cuttlefish (*Sepiidae*) caught from the coast of Portugal (Oliveira et al., 2020). To date, the literature on the incidence of NPs in marine organism, especially in the abovementioned major species/taxa, is still scarce. It is evident that more information is needed concerning the impact of these troubling contaminants on the food web.

3.3 MPs and NPs in marine aquaculture and processed seafood products

Considering the general decrease in global wild fishery captures and the increase in the total aquaculture production (from an average per year of 59.7 million tons in 2015 to 82.1 million tons in 2018) (FAO, 2020), as well as the forecast will reach 40 million tons in 2030, close attention should be paid to analyzing the impact of MPs and NPs on fed and non-fed marine aquaculture products. In 2018, the global production of the top five marine aquaculture species individually was 5.2 million tons for cupper oyster (*Crassostea* spp.), 5 million tons for whiteleg shrimp (*Litoenaeus vannamei*), 4.1 million tons for Japanese carpet shell (*Ruditapes philippinarum*), 2.4 million tons for Atlantic salmon (*Salmo salar*), and 1.9 million tons for scallops (Pectinidae) (FAO, 2020). For cupped oysters, the most aquacultured shellfish species worldwide, Walkinshaw et al. (2020) reported an average MPs value (50–100 μm size range) of 0.18–3.84 items/g (*w*/w) whereas the average value reported for the Japanese carpet shell ranged from 0.9 to 2.5 items/g (w/w).

For *Litopenaeus vannamei* farmed in Southern China, Li et al. (2021a, 2021b) reported a MPs abundance of 4–21 items/individual (mean value 10.87), whereas Valencia-Castañeda et al. (2022) found 7.6 items/individual in the gastrointestinal tract, 6.3 in the gills, and 4.3 in the exoskeleton, with an average of 18.5 MPs items per shrimp (1.06 items/g, wet weight [ww]). As for Atlantic salmon, the literature on the incidence of MPs is lacking in this fish species. Liboiron et al. (2019) found that the frequency of occurrence of plastic in the gastrointestinal tract of this highly consumed species that is 69 specimens collected in Newfoundland (Canada) was zero. Focusing on MPs abundance in Pectinidae, Ding et al. (2021) detected an average of 0.5–2.9 items/individual (0.4–3.4 items/g) in *Chlamys farreri* specimens collected around Qingdao, China.

Drying is one of the oldest techniques used for preserving foods at all latitudes to dehydrate different marine products. For example, it is traditionally used in the cold northern regions of Europe to produce stockfish (Inderhaug, 2020), and in the Mediterranean region to produce "bottarga," a delicious seafood-derived product based on salting and subsequent air drying of the raw roe of tuna (*Thunnus* spp.) and gray mullet (*Mugil* spp.). In some developing countries of South Asia and sub-Saharan Africa, this low-cost technique is often the only way to reduce water content (moisture) and prevent post-mortem degradation of seafood products. According to Karami, Golieskardi, Bin Ho, et al. (2017), seafood products prepared using these techniques tend to concentrate more MPs and NPs than other seafood products. This is due to post-harvest contamination during the long period spent in direct contact with common air impurities during the drying/salting process, which often occurs in environments that are not particularly healthy. This outcome has been recently confirmed by Piyawardhana et al. (2022), who studied the occurrence of MPs in 12 dried seafood consumed throughout the Asia-Pacific region, where such products are a staple in the diet. They observed that all species samples contained MPs both in the gastrointestinal tract and in the muscle. The authors also detected a positive correlation between the incidence of MPs and the weight of the analyzed specimens. By contrast, very little information is currently available on stockfish and salted fish (the Mediterranean salted sardine and anchovies), or on the abovementioned dried/salted fish roe (bottarga).

Canning is another traditional preservation technique used widely in fish and shellfish processing because it significantly extends the shelf life of highly perishable seafood products. In canned tuna and mackerel products, there is a high risk of contamination by MPs during the product cleaning and canning process (Akhbarizadeh et al., 2020).

4. Presence of MPs and NPs in other foods and beverages

In addition to seafood, the presence of MPs and NPs has been widely reported in other food products, such as fruit and vegetables, honey, salt, sugar, and beer (Table 3). A Scopus search using the combined terms

Table 3 Identification and quantification techniques of microplastics and nanoplastics in different food matrices.

Food	Origin	Type (species)	Type of MPs/NPs	Size	References
Fruits and vegetables	Italy	Apple (*M. domestica*))	MPs and NPs	2.17 μm	Oliveri Conti et al. (2020)
	Italy	Pear (*P. communis*)	MPs and NPs	1.99 μm	Oliveri Conti et al. (2020)
	Italy	Broccoli (*B. oleracea* italic)	MPs and NPs	2.10 μm	Oliveri Conti et al. (2020)
	Italy	Lettuce (*Lactuca sativa*)	MPs and NPs	2.52 μm	Oliveri Conti et al. (2020)
	Italy	Carrot (*Daucus carota*)	MPs and NPs	1.51 μm	Oliveri Conti et al. (2020)
	Chile	Maize (*Zeamays* L.)	PET microbeads	–	Urbina et al. (2020)
	The Netherlands	Garden cress (*Lepidium sativum* L.)	MPs and NPs	<100 nm <5 mm	Bosker et al. (2019)
	China	Seedling growth of wheat (*Triticum aestivum* L.)	PSNPs	88 nm	Lian et al. (2020)
	China	Lettuce (*Lactuca sativa*)	SPS and LPS	SPS: 100–1000 nm LPS: >10,000 nm	Gao et al. (2021)
	China	Lettuce (*Lactuca sativa* L.)	PVC	PVC- [18 μm −100 nm] PVC-b [18–150 μm]	Lian et al. (2020)
	China	Cucumber (*Cucumis sativus*)	PSNPs	100, 300, 500, and 700 nm	Lian et al. (2020)

Beverages	Ecuador	Skim milk	MPs	2.48–183.37 μm	Diaz-Basantes et al. (2020)
	Ecuador	Refreshing beverage	MPs	5.94–247.54 μm	Diaz-Basantes et al. (2020)
	Ecuador	Industrial Beer	MPs	3.50–202.29 μm	Diaz-Basantes et al. (2020)
	Ecuador	Craft Beer	MPs	6.15–160.345 μm	Diaz-Basantes et al. (2020)
	Mexico	Milk	PES andPSU	100–5000 μm	Kutralam-Muniasamy et al. (2020)
	Germany	Beer	PET	–	Liebezeit and Liebezeit (2014)
	Portugal	Wine	MPs	–	Prata et al. (2020)
	Mexico	Cold Tea	PA, PET, PEA and ABS	100–2000 μm	Shruti et al. (2020)
	Mexico	Soft drink	PA, PET, PEA and ABS	100–3000 μm	Shruti et al. (2020)
	Mexico	Energy drink	PA, PET, PEA and ABS	100–3000 μm	Shruti et al. (2020)
	Mexico	Beers	PA, PET, PEA and ABS	100–3000 μm	Shruti et al. (2020)
Honey	Ecuador	Industrial honey	MPs	5.63–182.96 μm	Diaz-Basantes et al. (2020)
	Ecuador	Craft honey	MPs	5.15–226.01 μm	Diaz-Basantes et al. (2020)
Salts	Eight African countries	Table salt	PVA, PP and PE	3.3–4660 μm	Fadare et al. (2021)
	Turkey	Lake salt	PE, PET, PU, PP, PMMA, PA-6 and PVC	20 μm–5 mm	Gündoğdu (2018)
	Turkey	Rock salt	PE, PET, PU, PP, PMMA, PA-6 and PVC	20 μm–5 mm	Gündoğdu (2018)
	Turkey	Sea salt	PE, PET, PU, PP, PMMA, PA-6 and PVC	20 μm–5 mm	Gündoğdu (2018)

Continued

Table 3 Identification and quantification techniques of microplastics and nanoplastics in different food matrices.—cont'd

Food	Origin	Type (species)	Type of MPs/NPs	Size	References
	Spain	Sea salt	polyethylene terephthalate (PET), polyethylene (PE) and polypropylene (PP)	30 μm–3.5 mm	Iñiguez, Conesa, and Fullana (2017)
	Spain	Well salt	PET, PE and PP	160–980 μm	Iñiguez et al. (2017)
	Republic of Korea	Lake salt	PP, PE, PS, PET	100–5000 μm	Kim et al. (2018)
	Republic of Korea	Rock salt	PP, PE, PS and PET	100–5000 μm	Kim et al. (2018)
	Republic of Korea	Sea salt	PP, PE, PS and PET	100–5000 μm	Kim et al. (2018)
	Italy	Salt	MPs	4–4628 μm	Renzi and Blašković (2018)
	Italy and Croatia	Salt	PET, PVC, PE, and PA,	10–150 μm	Renzi, Grazioli, et al. (2019)
Others	The Netherlands	Growth of sediment-rooted macrophytes	MPs and NPs	20–500 μm 50–190 nm	van Weert et al. (2019)
	China	Chicken meat	PS and PVC	3 μm and 100 μm	Huang et al. (2020)
	France	Packaged chicken breasts, packaged turkey escalopes (Extruded PS tray)	EPS and XPS	300–450 μm	Kedzierski et al. (2020)
	The Netherlands	Chicken gizzard and Chicken crop	MPs and NPs	1000–5000 μm	Lwanga et al. (2017)
	Italy	Terrestrail snail	MPs	100–2500 μm	Panebianco et al. (2019)

PE, Polyethylene; PSNPs, Polystyrene nanoplastics; SPS, Small polystyrene; LPS, Large polystyrene; PVC, Polyvinylchloride; PES, Polyethersulfone; PSU, Polysulfone; PET, Polyethylene-terephthalate; PA, Polyamide; PEA, Polyester-amide; ABS, Acrylonitrile-butadiene-styrene; PVA, Polyvinyl acetate; PP, Polypropylene; PU, Polyurethane; PMMA, Polymethyl-methacrylate; PA-6, Polyamide-6; EPS, Expanded polystyrene; XPS, Extruded polystyrene.

"TITLE-ABS-KEY (microplastics) OR TITLE-ABS-KEY (nanoplastics) AND TITLE-ABS-KEY (fruits) OR TITLE-ABS-KEY (vegetables) OR TITLE-ABS-KEY (salt) OR TITLE-ABS-KEY (sugar) OR TITLE-ABS-KEY (honey) OR TITLE-ABS-KEY (meat) OR TITLE-ABS-KEY (chicken) OR TITLE-ABS-KEY (wine) NOTTITLE-ABS-KEY(marine) AND NOT TITLE-ABS-KEY(sea) AND NOTTITLE-ABS-KEY(fish))" was carried out in order to perform a cluster-based VOSviewer analysis. Overall, the Scopus search revealed 156 publications pertaining to MPs and NPs in the food products listed above.

According to the VOSviewer output, co-occurrence analysis identified 2599 keywords, of which 173 occurred at least 5 times. As expected, "MPs" was the most frequent keyword identified, followed by "water pollutant," "soil," "environmental monitoring," "vegetable," "plant leaf," "fruits" and others (Fig. 5). Four "clusters" were generated: these define thematic proximity among all considered keywords.

Oliveri Conti et al. (2020) conducted a study to identify and quantify MPs in common edible fruit (apples and pears) and vegetables (carrot,

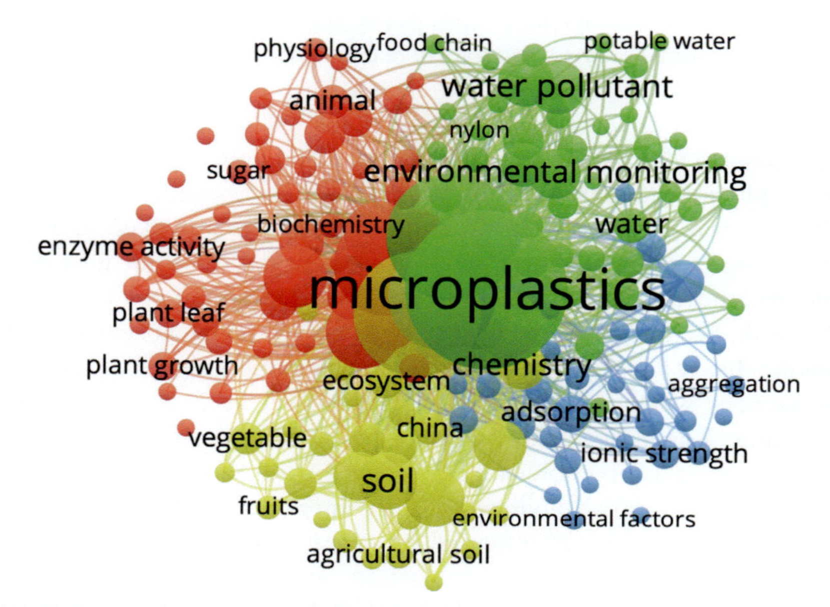

Fig. 5 Network map of 173 keywords (with the minimum occurrence set to 5) identified by VosViewer on the Scopus (CSV) database. The size of nodes indicates the frequency of occurrence, while the colors identify four clusters that define the thematic proximity and indicate research directions.

lettuce, broccoli, and potato). MPs particle sizes extended from 1.51 to 2.52 μm, with the smallest size (1.51 μm) of MPs being found in the carrot samples, while the biggest (2.52 μm) one was found in the lettuce. The authors reported that apple was the most contaminated fruit whereas carrot was the most contaminated vegetable. Depending on vegetable samples, MPs ranged from 52.050 to 233.000 particles/g. The authors supposed that MPs could be absorbed and translocated to the vegetal tissues in the same way as carbon nanomaterials.

Some studies have investigated the impact of plastics and related contaminants on plant growth and development. For example, Bosker et al. (2019) studied the impact of MPs with three different sizes (50, 500, and 4800 nm) on cress (*Lepidium sativum*). The results showed that the exposure to these contaminants could significantly affect germination and root growth, which was explained by a physical blockage of plant pores with plastic particles, inhibiting water uptake. In a similar study, Gao et al. (2021) investigated the harmful effects of small (100–1000 nm) and large (>10,000 nm) MPs on lettuce under dibutyl phthalate (DBP) stress. The findings suggested that the adhesion of plastic particles, especially the small MPs, on the root surface causes the physical clogging of root pores, aggravating the negative effects (decreased lettuce biomass and oxidative damage in lettuce leaves and roots) of DBP on lettuce growth.

In a study conducted by Urbina et al. (2020) in an experimental hydroponic culture of maize, the adsorption and potential uptake, as well as the physiological effects of PE microbeads were assessed. These authors observed that plastic accumulation in the rhizosphere provoked an important decrease in transpiration, nutrient uptake, and the plant growth. In another study, the effects of six different MPs, including polyester fibers, PA beads, PE, polyester terephthalate, PP, and PS, on the growth of spring onion (*Allium fistulosum*) and soil microbial activities, as well as soil–plant interactions, were investigated (De Souza Machado et al., 2019). The findings showed significant consequences of plastic contamination, in particular the contamination by polyester fibers and PA beads, for agro-ecosystems and general terrestrial biodiversity. The impact of biodegradable plastic and PS residues was also reported on wheat (*Triticum aestivum*) growth (Qi et al., 2018). The plastic contamination, notably the biodegradable plastic, showed an impact on the plant mass both above and below ground during the vegetative and reproductive growth.

The aforementioned examples, in addition to many others, reported in recently published review papers (Bai et al., 2022; Oliveri Conti et al., 2020; Silva et al., 2021; Vitali et al., 2022) demonstrate the occurrence and the negative impact of MPs and NPs on plant growth, biodiversity, and soil ecosystems. However, more research studies are still needed to confirm these findings as some investigations seem to indicate contradicting results. In fact, Lian et al. (2020) studied the influence of PS NPs at level ranges between 0.01 and 10 mg/mL on the germination of seed and growth seedling of wheat. Although the exposure to these NPs extremely altered the metabolic profiles in wheat leaves, the wheat seedling growth seemed to be significantly enhanced.

Some studies have reported the presence of MPs and NPs in sea salt, sugar, honey, milk, wine, beer, and other food and beverages, although little information is still available on MPs occurrence in these products. Kutralam-Muniasamy et al. (2021) tested the occurrence of a variety of MPs with different colors (blue, brown, red, and pink), shapes (fibers and fragments), and sizes (0.1–5 mm) in 23 milk samples in Mexico. The results showed that thermoplastic sulfone polymers (polyethersulfone and polysulfone) were common types of MPs due to their use in membrane materials in dairy processes. Blue fibers of size less than 0.5 mm were found to be prevalent in the studied samples.

The presence of MPs was first reported in Germany by Liebezeit and Liebezeit (2014) who examined 24 German beer brands for the contents of microplastic fibers, fragments, and granular. The results displayed a high variability among individual samples and samples from different production dates, and the counts ranged from 2 to 79/L for the fibers, 12 to 109/L for the fragments, and 2 to 66/L for the granules. Prata et al. (2020) detected MPs in 26 bottles of white wine capped with plastic stoppers. MPs were present in the examined samples with concentrations up to 5857 particles/L, and median dimensions of $26 \times 122\,\mu m$. Shruti et al. (2020) evaluated the MPs occurrence in Mexican soft drinks, cold tea, energy drinks, and they found that beer sample had the highest abundance. PA, poly (ester-amide) (PEA), acrylonitrile-butadiene-styrene (ABS), PET, and blue pigments were the most identified microplastic polymers, while synthetic textiles and packaging were reported as the contamination origins in these beverage products.

MPs contamination in commercial table salts has been monitored and reported in different countries around the world, such as Spain (Iñiguez et al., 2017), Turkey (Gündoğdu, 2018), India (Seth & Shriwastav, 2018), Italy and Croatia (Renzi & Blašković, 2018), and China (Yang et al., 2015). Fadare et al. (2021) analyzed 23 brands of table salts from 8 African countries: Nigeria (4), Cameroon (2), Ghana (3), Malawi (1), Zimbabwe (1), South Africa (6), Kenya (5) and Uganda (1). The highest concentration levels (0–1.33 particles/kg) were found in South Africa, followed by Nigeria, Cameroun, and Ghana with concentrations ranging between 0 and 0.33 particles/kg. Polyvinyl acetate (PVA), PP, and PE represented being the most prevalent MPs. Interestingly, the samples originated from the other countries (Malawi, Zimbabwe, Kenya and Uganda) had no detectable MPs at 0.3 μm filter pore size. In another study, Yang et al. (2015) reported that China salts are contaminated by MPs, with levels ranging from 550 to 681 particles/kg in sea salts, from 43 to 364 particles/kg in lake salts, and from 7 to 204 particles/kg in rock/well salts. The common MPs were PET, followed by PE and cellophane in sea salts, and 55% of particles were <200 μm. Seth and Shriwastav (2018) have reported several types of MPs, such PE, PET, PA, polyesters, and PS in Indian salt samples. The authors suggested a simple sand filtration to remove more than 90% of the contamination. MPs content of 16 brands of table salts obtained from the Turkish market was evaluated (Gündoğdu, 2018). MPs contents in sea salt, lake salt, and rock salt were in the range of 16–84, 8–102, and 9–16 item/kg, respectively, with PE and PP being the common plastic polymers. Karami, Golieskardi, Choo, et al. (2017) observed MPs in 88% of salt samples analyzed from eight countries (Australia, France, Iran, Japan, Malaysia, New Zeeland, Portugal, and South Africa). The MPs levels ranged from 0 particles (French sea salt) to 10 particles/kg (Portuguese sea salt).

It has been also reported that MPs can be present in chicken meat and meat products (Huang et al., 2020; Kedzierski et al., 2020). For instance, Huang et al. (2020) assessed the level of contamination by PS and PVC in homogenized chicken meat, while Kedzierski et al. (2020) identified and quantified MPs in meat packed in extruded PS trays, reporting a value range of 4.0–18.7 particles/kg. A recent study reported a severe MP contamination, represented by PP, PE, and polyester resin, in livestock and poultry farms in South China, highlighting the importance of understanding the contamination risk of MPs in livestock and poultry manure before its utilization as fertilizers during crop planting (Wu et al., 2021).

5. Identification and quantification of MPs

The presence of MPs in both terrestrial and aquatic ecosystems has become a widespread problem and an issue of great concern in recent years. MPs in marine habitats are those small plastic fragments that can be consumed by different aquatic species including fish, shellfish, marine invertebrates, bivalves, among others, and are ultimately transferred along the food chain to human beings (Chatterjee & Sharma, 2019; Dawson et al., 2021; López-Martínez et al., 2021; Savoca, McInturf, & Hazen, 2021; van Raamsdonk et al., 2020). To detect, identify and quantify MPs, several approaches have been developed, ranging from the simple visual inspection (by the naked eye or microscopically) to more advanced techniques such as chromatography coupled with mass spectrometry and spectroscopic techniques (Dawson et al., 2021; Lv et al., 2021; Razeghi et al., 2021; Silva et al., 2018; Wang & Wang, 2018). Many review papers have been published on this topic, comparing these different methods, their applications, advantages and limitations, etc. (Akoueson et al., 2021; Gong & Xie, 2020; la Nasa et al., 2020; Peñalver et al., 2020). The optical methods based on the visual inspection or the use of microscopes, scanning electron microscopy (SEM), and gas chromatography–mass spectrometry (GC–MS) have been widely used. The spectroscopic methods, especially the microscopic infrared and Raman spectroscopies remain the most commonly applied techniques (Cowger et al., 2020; Dehaut et al., 2019; Kwon et al., 2020; Li et al., 2019; Primpke et al., 2020; Schwaferts et al., 2019).

5.1 Traditional analysis techniques

5.1.1 Optical methods

A number of studies have demonstrated that MPs of size >100 μm can be easily detected and classified directly through a simple naked-eye observation, based on the examination of size, shape, and color of the microplastic particles. However, most often, this visual inspection is accompanied by optical microscopy (Lv et al., 2021; Silva et al., 2018; Zhu & Wang, 2020). For example, a stereomicroscope was used by Bono et al. (2020) to classify the microplastic particles ingested by Atlantic chub mackerel (*Scomber colias*) according to their size, texture, and shape into fragments, fibers, lines, paint and films.

Visual sorting is simple and cost effective but cannot be used to detect and identify particles of small size (<100 μm). In addition, the human subjectivity related to visual inspection is a challenging factor that can lead to variability in the detection results, and either under or overestimation of MPs can often occur (Cowger et al., 2020; Gong & Xie, 2020; Pinto da Costa et al., 2019; Strungaru et al., 2019; Zhu & Wang, 2020). It has been reported that the misidentifications can reach a high percentage rate (65–67%) for transparent polymers (Lakshmi Kavya, Sundarrajan, & Ramakrishna, 2020). Microscopic methods also proved to be a weak means to distinguish between the synthetic and natural fibers (e.g., PES vs dyed cotton). To overcome the limitations, several approaches have been developed in recent years. For example, Nile red staining and fluorescent microscopy has been applied in several studies in order to identify MPs in fresh and processed fish samples (Akhbarizadeh et al., 2020; TaghizadehRahmat Abadi et al., 2021).

5.1.2 Scanning electron microscopy (SEM)

SEM uses focused electron beams to investigate size, shape, and other physical parameters of MPs, providing high-resolution surface images under high-magnification overcoming all the limitation of optical microscopy. This technique is often coupled to energy dispersive X-ray spectroscopy (EDS or EDX) to obtain detailed information about the elemental composition of the microplastic particles (Primpke et al., 2020; Silva et al., 2018; Wang & Wang, 2018; Zhu & Wang, 2020) to ensure differentiation between plastics and gives the possibility to distinguish between organic and plastic particles.

Several studies have used SEM-EDS (or SEM-EDX) for the characterization of surface structure and elemental composition in MPs (Akhbarizadeh et al., 2020; Bermúdez-Guzmán et al., 2020; Jonathan et al., 2021; Karbalaei et al., 2019) (Table 4). For instance, in a recent one, the presence of MPs ingested by fishes of Magdalena bay was examined by means of SEM microscope in conjunction with EDS (Jonathan et al., 2021). SEM images showed angular and irregular edges of MPs, revealing the impact of mechanical disintegration of plastic fibers due to probably weathering processes, whereas the majority of the EDX spectra of the fibers presented carbon and oxygen peaks, suggesting them as plastics. Similarly, Martinez-Taveraand coauthors used the same techniques (i.e., SEM and EDX) to study the surface morphology and the elemental composition of MPs in freshwater tilapia (*Oreochromis niloticus*) that caught from a metropolitan reservoir in Central Mexico (Martinez-Tavera et al., 2021). However, these techniques are destructive and can be time-consuming and quite expensive and thus only few samples are usually examined in order to confirm the width and length of the fibers.

5.1.3 Gas chromatography-mass spectrometry (GC–MS)

Chromatographic techniques coupled with mass spectrometry have attracted increasing attention, and they have been widely used in the microplastic research to enable quantitative analysis. Two GC–MS-based approaches have been commonly applied: pyrolysis gas chromatography

Table 4 Examples of analytical methods used for the detection and identification of microplastics in fish/seafood and other food products.

Products category	Species/products	Analytical approach	Main conclusions	References
Fish and other seafood	Norway lobster	SEM, μ-Raman	83% of the examined samples contained plastic filaments	Murray and Cowie (2011)
	Laboratory fish (Medaka), Pelagic prey fish (Myctophid)	Optical microscopy, SEM/EDS, μ-FT-IR, μ-Raman	Optical microscopy is not enough to identify MPs, especially if the particles are of various morphologies	Wang et al. (2017)
	Pelagic and demersal fish	Py-GC–MS	The technique was able to identify the chemical composition of plastic polymers by analyzing its characteristic thermal degradation products and comparing with pyrolysis reference maps of known pure polymers	Fischer and Scholz-Böttcher (2017)
	Demersal and pelagic fish	Microscopic analyses	828 pieces of MPs were detected, most of them were in the form of fibers	Abbasi et al. (2018)
	Commercial marine fish from Malaysia	SEM-EDS Raman	MPs were detected in 9 out of 11 samples, and polyethylene was the most abundant plastic polymer	Karbalaei et al. (2019)
	Mussels and cockles	TED-GC–MS, μ-Raman	Up to 58% of the samples were contaminated by MPs. 13 plastic additives and 27 hydrophobic organic compounds were quantified in bivalves' flesh	Hermabessiere et al. (2019)
	Herring *Opisthonema* sp.	Optical microscopy, SEM-EDX, FT-IR-ATR	MPs were detected in all the sampled fishes; 20.5% were classified as particles and 79.5% as fibers	Bermúdez-Guzmán et al. (2020)

Continued

Table 4 Examples of analytical methods used for the detection and identification of microplastics in fish/seafood and other food products.—cont'd

Products category	Species/products	Analytical approach	Main conclusions	References
	Four fish species (*Clarias gariepinus*, *Cyprinus carpio*, *Carassius carassius*, and *Oreochromis niloticus*) from the Lake Ziway	ATR-FT-IR	Polypropylene, polyethylene, and alkyd-varnish were found to be the dominant polymers	Merga et al. (2020)
	Penaeid shrimp	μ-FT-IR	Polyamide-6 and rayon polymers were the most dominant polymers detected in the samples	Hossain et al. (2020)
	Sea bass	Fluorescence microscopy	Feeding juveniles of fish with a diet containing fluorescent microplastic particles showed that the entrance of MPs to lymphatic and/or vascular systems through the gastrointestinal tract was the most likely pathway for translocation of MPs	Zeytin et al. (2020)
	Commercial fish from Southeast coast of the Bay of Bengal	SEM FT-IR	20 plastic particulates (of polyethylene, polyamide and polyester types) were isolated from the gastrointestinal tract of 17 fish	Karuppasamy et al. (2020)
	Tuna and mackerel	SEM-EDX, fluorescence analysis, μ-Raman	At least one microplastic particle was found in 80% of the investigated samples, and polyethylene terephthalate, polystyrene, and polypropylene were the most abundant polymers	Akhbarizadeh et al. (2020)
	24 fish species from Beibu Gulf	μ-FT-IR	Polyester and nylon were the predominant polymer types and the fibers were the dominant form of particles	Koongolla et al. (2020)

Shrimp *Pleoticus muelleri*	SEM/EDS, μ-Raman	Complete studies of MPs require both physical and chemical analysis using microscopy (to measure color, size, shape) and spectroscopy (to identify polymer types), respectively	Fernández Severini et al. (2020)
Commercial fish from the Arabian Gulf	FT-IR	5.71%, of the examined fish was contaminated with MPs (especially polyethylene and polypropylene)	Baalkhuyur et al. (2020)
Commercial fish from Tunisian coasts	μ-Raman	Small MPs ($\leq 3\,\mu m$) were found in the in the gastrointestinal tracts and muscle of the examined fish	Zitouni et al. (2020)
Blue panchax fish (*Aplocheilus* sp.)	FT-IR	75% of the samples were contaminated by various types of microplastic shapes and sizes	Cordova, Riani, and Shiomoto (2020)
Oysters	Optical microscopy, μ-ATR-FT-IR	About 83% of MPs in samples were identified as plastic materials, including polyethylene, polypropylene, polystyrene, polyvinyl chloride, polyethylene terephthalate, rayon, and nylon	Wang et al. (2021a)
Kutum fish	SEM-EDS, fluorescent microscopy	80% of investigated fish contained MPs, and fibers were the dominant type followed by fragments and synthetic microbeads	TaghizadehRahmat Abadi et al. (2021)
Various seafood (fish, crustaceans, and molluscs)	Py-GC–MS, fluorescence microscopy, SEM, ATR-FT-IR, μ-Raman	Digestion efficiency was more than 98% in most of the analyzed seafood. Optimization of a protocol for the isolation of MPs present in the edible part of seafood	Süssmann et al. (2021)

Continued

Table 4 Examples of analytical methods used for the detection and identification of microplastics in fish/seafood and other food products.—cont'd

Products category	Species/products	Analytical approach	Main conclusions	References
Fruit and vegetables	Apples, pears, broccoli, lettuce and carrots	Scanning Electron Microscopy—SEM-EDX	Apples were the most contaminated fruit samples, whereas carrot was the most contaminated vegetable	Oliveri Conti et al. (2020)
	Maize	Isotope analysis (δ13C)	About 30% of the carbon in the rhizosphere of microplastic-exposed plants was derived from PE	Urbina et al. (2020)
	Carrots, lettuces, broccoli, potato apples and pears	SEM-EDX	Apple was the most contaminated samples, while carrots were the most contaminated vegetable. The smallest MPs size was found in carrots (1.51 μm) and the biggest ones in lettuce (2.52 μm)	Oliveri Conti et al. (2020)
Milk and dairy products	Milk	Visual identification and enumeration, SEM-EDS, μ-Raman	MPs exhibited variety of colors (blue, brown, red and pink), shapes (fibers and fragments) and sizes (0.1–5 mm) Blue colored fibers (<0.5 mm) were predominant. The most common microplastic particles were polyethersulfone and polysulfone	Kutralam-Muniasamy et al. (2020)
	Farm milk, reconstituted milk powder and skimmed milk	SEM–EDX, μ-Raman	PE, PES, PP, PTFE and PS were the most common MPs. Smaller quantities of PA, PU, PSU and PVA were detected. The concentration of MPs ranged varied between 204 and 1004 MPs per 100 mL of milk	da Costa Filho et al. (2021)
Meat and meat products	Homogenized chicken meat	ATR-MIR	PVC (particle size: 3 μm, 100 μm and 2–4 mm) and PS (particle size: 100 μm) can be detected and quantified by ATR-MIR at a concentration between 1% and 10%	Huang et al. (2020)

SEM, Scanning electron microscopy; EDS, Energy dispersive spectroscopy; EDX, Energy dispersive X-ray spectroscopy; GC–MS, Gas chromatography-mass spectrometry; Py-GC–MS, Pyrolysis-gas chromatography–mass spectrometry; FTIR, Fourier-transform infrared spectroscopy; ATR, Attenuated total reflection.

mass spectrometry (Py-GC–MS) and thermal extraction desorption gas chromatography mass spectrometry (TED-GC–MS). However, these techniques have been mainly used to analyze plastic and MPs in environmental samples (e.g., water or sediment) while few studies have been conducted on seafood (Dehaut et al., 2019; la Nasa et al., 2020; Peñalver et al., 2020; Yakovenko, Carvalho, & ter Halle, 2020). For instance, a Py-GC/MS approach was applied for polymer identification of microplastic ingested by benthivore fish from the Texas Gulf Coast (Peters et al., 2018). The results showed that PVC and PET were the most common microplastic polymers collected from the stomach content of the fish. In another study, Dehaut and coauthors investigated the potential of Py-GC–MS to identify polymers in mussel, crab, and fish tissues using different digestion protocols (Dehaut et al., 2016). The effect of the different approaches was tested on a set of 15 different plastic polymers, and the authors found that alkaline digestion procedures (KOH 10%) gave the best results, enabling a correct identification of all the polymer types, except in the case of cellulose acetate. The same protocol was later applied to characterize MPs in mussels and cockles (Hermabessiere et al., 2018). In a recent study, Ribeiro and coworkers succeeded in developing a Py-GC/MS methodology for the identification and quantification of MPs isolated from edible portions of oysters, prawns, squid, crabs, and sardines (Ribeiro et al., 2020). Five different plastics, including PS, PE, PVC, PP, and methyl methacrylate were appropriately identified and quantified. Nevertheless, it should be stressed that the technique is destructive and may be used only as a complementary technique to spectroscopic methods.

5.2 Spectroscopic methods

Spectroscopic techniques, especially vibrational spectroscopy (i.e., total reflection FT-IR and μFT-IR, Raman or stimulated Raman scattering; SRS) are tools that have been found to be useful for MPs identification and characterization.

5.2.1 Infrared (IR) spectroscopy

FT-IR spectroscopy is certainly the most popular technique for MPs quantification and characterization. This is probably due to the sensitivity of FT-IR to functional groups (C-H, N-H, O-H) providing a spectral fingerprint (distinct absorption bands) of the analyzed polymers (Cowger et al., 2020; Veerasingam et al., 2021). The identification of the polymer is also an easy task when using this technique due to a possible comparison to commercially available polymers library. However, the identification is generally complex because there is no standardized method defining different parameters, such as hit quality index (HQI) threshold or spectral pre-treatment. In addition, different forms (with irregular shapes), colors, chemical additives can be found in MPs, making their identification complex. Moreover, their affinity to chemical pollutants from seawater, denaturation by sun light, temperature variation, and biodegradation add complexity to this task (Kwon et al., 2020; Lv et al., 2021; Renner et al., 2019; Renner, Schmidt, & Schram, 2018; Veerasingam et al., 2021). Therefore, a consolidation of databases seems necessary.

FT-IR spectroscopy can be used in attenuated total reflection (ATR), transmission or reflection modes, noticing that each mode has some advantages and disadvantages (Gong & Xie, 2020; Renner et al., 2019). For instance, the ATR and reflectance modes are more suitable for thick and opaque samples compared to the transmission mode and needs no or little sample preparation. Furthermore, the use of ATR provides a stable and reliable spectral line data, even with surfaces presenting a rough texture (Lakshmi Kavya et al., 2020). An intimate contact must be established between the sample and the ATR crystal. This can be obtained by pressing the sample to the crystal; however, this can damage the fragile particles or alter polymers and cause tiny particles to stick to the crystal due to electrostatic forces (Shim, Hong, & Eo, 2017). A recent study conducted by Daniel et al. (2021) proved that this method is suitable for studying MP contaminants in various seafoods. They identified the presence of MPs in the size range of 100 μm–5 mm in four species of shellfishes (two species of shrimp, one species of crab, and one species of squid). Three common polymers (including PP, PE, and PS) were isolated from these shellfishes and identified.

However, it is well known that FT-IR measurements are limited to a size of MPs of about 20 μm. This drawback can be overcome by using FT-IR microscopy (μ-FT-IR). This technique combines FT-IR with microscopic imaging and provides both high spatial resolution and spectral information of the analyzed sample. Nevertheless, μ-FT-IR can be time-consuming (Pinto da Costa et al., 2019). There are essentially three FT-IR microscopy configurations: (1) point detector analysis in a confocal arrangement; (2) FT-IR mapping, performed by integrating the signal from successive locations of the specimen surface; and (3) FT-IR imaging, performed with bi-dimensional arrays, e.g., focal plane array (FPA) detectors. A wide range of studies using μ-FT-IR for MPs analysis has been reported and proved the effectiveness of this technique for polymer characterization and identification. For instance, μ-FT-IR was used to describe the distribution of MPs in gut, gills, and muscle of fish and the possibility of bioaccumulation in their tissues (Su et al., 2019). The μ-FT-IR was also used to analyze MPs in four species of finfish samples as well as water and sediment (Saha et al., 2021). This study proved the reliability of the μ-FT-IR to analyze a huge variety of MPs polymers, up to 37. ATR-FT-IR was also suitably used to differentiate between low density polyethylene (LDPE) and high density polyethylene (HDPE) in ingested plastic by sea turtles (Jung et al., 2018).

Despite the many advantages of the different techniques presented above, one of the biggest drawbacks is the need to apply robust methodologies to separate plastic products from organic-rich intestinal tract content. The separation and preparation of sample are often time-consuming. This is generally performed by density separation and visual inspection before positioning the sample on a filter. In recent years, Vis-NIR (Visible Near Infrared) hyperspectral imaging (HSI) has been investigated as a new approach for rapid and non-invasive, and online determination and identification of MPs (Huang et al., 2020; Zhang et al., 2019). Vis-NIR HSI allows to obtain 3D images(data cube) in the 400–1000 nm range. The data cube is characterized by two spatial axis and one wavelength axis. Different configurations exist; point scanning, line scanning and area scanning. Zhang and coauthors proposed this technique to separate, identify, and characterize MPs without prior separation from the intestinal tract content of crucian carps (*Carassius carassius*) (Zhang et al., 2019). The sample analysis was performed in a short time (sample preparation, image acquisition, and data analysis were performed in only 36 min). The analysis of data by a machine learning algorithm (support vector machine: SVM) allowed to obtain recall and precision factors higher than 96.22% for five MPs (PE, PS, PET, PP and PC). This method was capable of identifying only plastics with a size higher than 0.2 mm that is larger than the particle size sensitivity of the μ-FT-IR. This limit of detection was also reported for seawater study (Shan et al., 2019). However, it is admitted that the general MPs size found in fish is >0.1 mm (Garnier et al., 2019; Yuan et al., 2019; Zhu et al., 2019) meaning that this technique could lead to underestimation of MPs contamination in biomass.

5.2.2 Raman spectroscopy

Raman spectroscopy is another vibrational spectroscopic technique that has been frequently used to identify MPs in different environmental samples with high reliability. A Raman spectrum is obtained after irradiation with a monochromatic laser source, generally 455, 633, 532, 785, and 1064 nm (Anger et al., 2018; Araujo et al., 2018; Käppler et al., 2016). After molecule irradiation, a unique backscatter is recorded (Löder & Gerdts, 2015). Compared to FT-IR, Raman spectroscopy has several advantages, such as detection of smaller size of microplastic particles, higher spatial resolution, less interference of water, narrower spectral bands with wider spectral range. In addition, the technique provides detailed compositions of the polymers (Gong & Xie, 2020; Lv et al., 2021; Wang & Wang, 2018). Like FT-IR spectroscopy, Raman spectroscopy and its variants (surface-enhanced Raman, FT-Raman, tip-enhanced Raman, confocal Raman imaging, and others) have become a common technique to study MPs.

As with IR spectroscopy, Raman spectroscopy can be associated with microscopy (μ-Raman), making it possible to identify plastic particles of various and small sizes (below to 1 μm in size) (Anger et al., 2018; Lv et al., 2021). For instance, by using μ-Raman coupled with other techniques, Akhbarizadeh and coauthors showed that PET was the most abundant polymer in canned fish samples, followed by PS and PP (Akhbarizadeh et al., 2020). However, the long time required to acquire Raman spectra and the interference of fluorescence, as well as overlapping of signals due to the presence of additives or other contaminants and the high costs of instrumentation limit its applications for identification of microplastic polymers (Araujo et al., 2018; Peñalver et al., 2020; Zhu & Wang, 2020). To overcome these limitations, several studies have suggested the combined use of FT-IR and Raman for a reliable and complete chemical characterization of MPs, especially in the case of colored particles (Käppler et al., 2016; Xu et al., 2019). One example is a recent study conducted by Vinay Kumar and coauthors, who combined μ-Raman with μ-FT-IR to identify a broad size range (from 3 to 5000 μm) of MPs in commercial mussels (Vinay Kumar et al., 2021).

Recently, a new trend has been emerging in the field of analytical techniques: the development of portable devices and miniaturized systems that are suitable for in situ accurate detection and monitoring of MPs and NPs (Asamoah et al., 2021).

6. Conclusions

There is ample evidence that plastics, MPs, and NPs are everywhere. They can interact with vegetable and animal organisms, causing acute inflammation and cell damage, among other health concerns. Therefore, the environmental ubiquity of these emerging pollutants, in both aquatic and terrestrial environments, has become a critical concern, especially in the food sector. In addition, the COVID-19 pandemic has increased our utilization of plastics, specifically single-use plastics (Vanapalli et al., 2021), emphasizing the need for urgent solutions.

The majority of seafood, fruit, vegetables, meat, and beverages used today can be seriously impacted by MPs and other pollutants that use this microscopic material as a carrier. An intensive research interest has been recently devoted to this topic with a special focus being put on the occurrence of plastic materials in marine aquaculture and processed seafood products.

MPs and NPs ingestion by marine organisms could alter endocrine system functions and the seafood microbiome, reduce the rate of fecundity, and cause DNA and neurological damage in various species. Moreover, plastic materials could act as carriers of other pollutants, such as plastic additives and other persistent organic pollutants, including heavy metals, dissolved in the aquatic marine environment, with serious consequences for marine biota.

It has been reported that marine products of particular commercial interest, such as finfish, crustaceans, molluscs, jellyfish, and echinoderms, have quite high microplastic contents. Specifically, MPs accumulation in seafood products is strongly influenced by the environment in which organisms live, what, how and where they feed, as well as their trophic levels and anatomic features. In particular, close attention should be paid to some of them such as mussels, clams and oysters that seem more subject to MP and NP contamination and accumulation because they feed by filtering significant quantities of coastal water, which is often more polluted than the open sea environment. Moreover, drying, canning, salting and other traditional methods used to produce special seafood products seem to concentrate MPs and NPs. This is due to post-harvest contamination during their processing that often occurs in environments that are not particularly healthy.

The occurrence of MPs and NPs in terrestrial food (crops and livestock) has received less attention, although contaminations have been recently documented in fruit and vegetables, beer and wine, soft drinks, sugar, table salt, chicken and other meat products. Some publications reported that plastic particles could cause mechanical blocking of the pores in contaminated plants, thus reducing water and nutrient uptake and impairing growth.

To detect, identify, and quantify plastics, MPs, and NPs, several approaches, particularly optical methods and chromatography coupled with mass spectrometry have been developed. The optical detection, based on manual or microscopic counting, chromatographic-based methods and scanning electron microscopy have been widely used, but cannot be considered as standard methods due to their multiple disadvantages. Spectroscopic methods, specifically, Fourier-transform infrared spectroscopy and Raman spectroscopy, have been extensively suggested over the past few years as alternative techniques to achieve rapid and non-destructive measurements. However, spectroscopic techniques would not be a silver bullet as they still suffer from some limitations; e.g., the size of plastic particles must be higher than 10–20 μm and 1 μm to be detected and identified using Fourier-transform infrared spectroscopy and Raman spectroscopy, respectively (Silva et al., 2018). Therefore, more efforts are still needed to develop efficient, sensitive, and reliable detection, identification, and quantification methods. The ongoing development in hyperspectral imaging and the most recent trend of miniaturization and portability are promising strategies in this direction.

There is also a need for national and international authorities to provide stricter regulations so that plastic litter's impact on the environment and the food web could be better studied and managed. For instance, it can be recommended to include this new threat in food quality risk analysis and in the HACCP (Hazard Analysis and Critical Control Point) or HARPC (Hazard Analysis and Risk-based Preventive Controls for Human Food) programs for food safety. Promoting social awareness to reduce the consumption of single use plastics and increase the use of bio-plastics is one of the actions that should be encouraged. It is important to acquire greater insight into the MPs and NPs detection and their effects on food chain, both in terms of food safety and human health. Additionally, holistic approaches are needed and common standards and operating procedures for routine analysis should be implemented to address this global complex issue.

Finally, it is crucial to explore new technologies and innovative solutions that could contribute to solving the problem of plastics. Recently, a study published in Nature has opened up a new promising avenue, reporting on the possibility of breaking down plastics in days (instead of years), using an enzyme named FAST-PETase, acronymic for "functional, active, stable, and tolerant PETase" (Lu et al., 2022). This study is one example among others that show the significance of research and innovation to help solve the plastic dilemma.

References

Abbasi, S., et al. (2018). Microplastics in different tissues of fish and prawn from the Musa Estuary, Persian Gulf. *Chemosphere*, *205*, 80–87. https://doi.org/10.1016/j.chemosphere.2018.04.076.

Adika, S. A., et al. (2020). Microplastic ingestion by pelagic and demersal fish species from the Eastern Central Atlantic Ocean, off the Coast of Ghana. *Marine Pollution Bulletin*, *153*, 110998. https://doi.org/10.1016/J.MARPOLBUL.2020.110998.

Akhbarizadeh, R., et al. (2020). Abundance, composition, and potential intake of microplastics in canned fish. *Marine Pollution Bulletin, 160*. https://doi.org/10.1016/j.marpolbul.2020.111633.

Akoueson, F., et al. (2021). Identification and quantification of plastic additives using pyrolysis-GC/MS: A review. *Science of the Total Environment, 773*, 145073. https://doi.org/10.1016/j.scitotenv.2021.145073.

Alimi, O. S., et al. (2018). Microplastics and nanoplastics in aquatic environments: Aggregation, deposition, and enhanced contaminant transport. *Environmental Science & Technology, 52*, 1704–1724. https://doi.org/10.1021/acs.est.7b05559.

Anastasopoulou, A., et al. (2018). Assessment on marine litter ingested by fish in the Adriatic and NE Ionian Sea macro-region (Mediterranean). *Marine Pollution Bulletin, 133*, 841–851. https://doi.org/10.1016/J.MARPOLBUL.2018.06.050.

Anger, P. M., et al. (2018). Raman microspectroscopy as a tool for microplastic particle analysis. *TrAC Trends in Analytical Chemistry, 109*, 214–226. https://doi.org/10.1016/j.trac.2018.10.010.

Araujo, C. F., et al. (2018). Identification of microplastics using Raman spectroscopy: Latest developments and future prospects. *Water Research, 142*, 426–440. https://doi.org/10.1016/j.watres.2018.05.060.

Asamoah, B. O., et al. (2021). Towards the development of portable and in situ optical devices for detection of micro- and nanoplastics in water: A review on the current status. *Polymers, 13*(5), 730. https://doi.org/10.3390/POLYM13050730.

Baalkhuyur, F. M., et al. (2020). Microplastics in fishes of commercial and ecological importance from the Western Arabian Gulf. *Marine Pollution Bulletin, 152*, 110920. https://doi.org/10.1016/j.marpolbul.2020.110920.

Baechler, B. R., et al. (2020). Microplastic occurrence and effects in commercially harvested North American finfish and shellfish: Current knowledge and future directions. *Limnology and Oceanography Letters, 5*(1), 113–136. https://doi.org/10.1002/lol2.10122.

Bai, C. L., et al. (2022). Microplastics: A review of analytical methods, occurrence and characteristics in food, and potential toxicities to biota. *Science of the Total Environment, 806*, 150263. https://doi.org/10.1016/J.SCITOTENV.2021.150263.

Beer, S., et al. (2018). No increase in marine microplastic concentration over the last three decades—A case study from the Baltic Sea. *Science of the Total Environment, 621*, 1272–1279. https://doi.org/10.1016/J.SCITOTENV.2017.10.101.

Bermúdez-Guzmán, L., et al. (2020). Microplastic ingestion by a herring *Opisthonema* sp. in the Pacific coast of Costa Rica. *Regional Studies in Marine Science, 38*, 101367. https://doi.org/10.1016/j.rsma.2020.101367.

Białowąs, M., et al. (2022). Plastic in digestive tracts and gills of cod and herring from the Baltic Sea. *Science of the Total Environment, 822*, 153333. https://doi.org/10.1016/J.SCITOTENV.2022.153333.

Bono, G., et al. (2020). Microplastics and alien black particles as contaminants of deep-water rose shrimp (*Parapenaeus longistroris* Lucas, 1846) in the Central Mediterranean Sea. *Journal of Advanced Biotechnology and Bioengineering, 2*(8), 23–28.

Borges-Ramírez, M. M., et al. (2020). Plastic density as a key factor in the presence of microplastic in the gastrointestinal tract of commercial fishes from Campeche Bay, Mexico. *Environmental Pollution, 267*, 115659. https://doi.org/10.1016/J.ENVPOL.2020.115659.

Bosker, T., et al. (2019). Microplastics accumulate on pores in seed capsule and delay germination and root growth of the terrestrial vascular plant *Lepidium sativum*. *Chemosphere, 226*, 774–781. https://doi.org/10.1016/J.CHEMOSPHERE.2019.03.163.

Boyle, K., & Örmeci, B. (2020). Microplastics and nanoplastics in the freshwater and terrestrial environment: A review. *Water*. https://doi.org/10.3390/w12092633.

Bråte, I. L. N., et al. (2016). Plastic ingestion by Atlantic cod (*Gadus morhua*) from the Norwegian coast. *Marine Pollution Bulletin, 112*(1–2), 105–110. https://doi.org/10.1016/J.MARPOLBUL.2016.08.034.

Bray, L., et al. (2019). Determining suitable fish to monitor plastic ingestion trends in the Mediterranean Sea. *Environmental Pollution*, *247*, 1071–1077. https://doi.org/10.1016/J.ENVPOL.2019.01.100.
Chagnon, C., et al. (2018). Plastic ingestion and trophic transfer between Easter Island flying fish (*Cheilopogon rapanouiensis*) and yellowfin tuna (*Thunnus albacares*) from Rapa Nui (Easter Island). *Environmental Pollution*, *243*, 127–133. https://doi.org/10.1016/J.ENVPOL.2018.08.042.
Chatterjee, S., & Sharma, S. (2019). Microplastics in our oceans and marine health. *Field ACTions Science Reports*, 54–61.
Chenet, T., et al. (2021). Plastic ingestion by Atlantic horse mackerel (*Trachurus trachurus*) from Central Mediterranean Sea: A potential cause for endocrine disruption. *Environmental Pollution*, *284*, 117449. https://doi.org/10.1016/J.ENVPOL.2021.117449.
Collard, F., et al. (2017). Morphology of the filtration apparatus of three planktivorous fishes and relation with ingested anthropogenic particles. *Marine Pollution Bulletin*, *116*(1–2), 182–191. https://doi.org/10.1016/J.MARPOLBUL.2016.12.067.
Collard, F., et al. (2018). Anthropogenic particles in the stomach contents and liver of the freshwater fish *Squalius cephalus*. *Science of the Total Environment*, *643*, 1257–1264. https://doi.org/10.1016/J.SCITOTENV.2018.06.313.
Compa, M., et al. (2018). Ingestion of microplastics and natural fibres in *Sardina pilchardus* (Walbaum, 1792) and *Engraulis encrasicolus* (Linnaeus, 1758) along the Spanish Mediterranean coast. *Marine Pollution Bulletin*, *128*, 89–96. https://doi.org/10.1016/J.MARPOLBUL.2018.01.009.
Cordova, M. R., Riani, E., & Shiomoto, A. (2020). Microplastics ingestion by blue panchax fish (*Aplocheilus sp.*) from Ciliwung Estuary, Jakarta, Indonesia. *Marine Pollution Bulletin*, *161*, 111763. https://doi.org/10.1016/j.marpolbul.2020.111763.
Cowger, W., et al. (2020). Critical review of processing and classification techniques for images and spectra in microplastic research. *Applied Spectroscopy*, *74*(9), 989–1010. https://doi.org/10.1177/0003702820929064.
da Costa Filho, P. A., et al. (2021). Detection and characterization of small-sized microplastics ($\geq$5 μm) in milk products. *Scientific Reports*, *11*, 24046. https://doi.org/10.1038/s41598-021-03458-7.
Dalla Fontana, G., Mossotti, R., & Montarsolo, A. (2020). Assessment of microplastics release from polyester fabrics: The impact of different washing conditions. *Environmental Pollution*, *264*, 113960. https://doi.org/10.1016/J.ENVPOL.2020.113960.
Daniel, D. B., et al. (2021). Microplastics in the edible tissues of shellfishes sold for human consumption. *Chemosphere*, *264*, 128554. https://doi.org/10.1016/j.chemosphere.2020.128554.
Danopoulos, E., et al. (2020). Microplastic contamination of salt intended for human consumption: A systematic review and meta-analysis. *SN Applied Sciences*. https://doi.org/10.1007/s42452-020-03749-0. (Preprint).
Dawson, A. L., et al. (2021). Relevance and reliability of evidence for microplastic contamination in seafood: A critical review using Australian consumption patterns as a case study. *Environmental Pollution*, *276*, 116684. https://doi.org/10.1016/j.envpol.2021.116684.
de Barros, M. S. F., dos Santos Calado, T. C., & de Sá Leitão Câmara de Araújo, M. (2020). Plastic ingestion lead to reduced body condition and modified diet patterns in the rocky shore crab *Pachygrapsus transversus* (Gibbes, 1850) (Brachyura: Grapsidae). *Marine Pollution Bulletin*, *156*, 111249. https://doi.org/10.1016/J.MARPOLBUL.2020.111249.
de Mesquita, G. C., et al. (2021). Feeding strategy of pelagic fishes caught in aggregated schools and vulnerability to ingesting anthropogenic items in the western equatorial Atlantic Ocean. *Environmental Pollution*, *282*, 117021. https://doi.org/10.1016/J.ENVPOL.2021.117021.

De Souza Machado, A. A., et al. (2019). Microplastics can change soil properties and affect plant performance. *Environmental Science and Technology*, *53*(10), 6044–6052. https://doi.org/10.1021/ACS.EST.9B01339/SUPPL_FILE/ES9B01339_SI_003.PDF.

de Vries, A. N., et al. (2020). Microplastic ingestion by fish: Body size, condition factor and gut fullness are not related to the amount of plastics consumed. *Marine Pollution Bulletin*, *151*, 110827. https://doi.org/10.1016/J.MARPOLBUL.2019.110827.

Dehaut, A., Hermabessiere, L., & Duflos, G. (2019). Current frontiers and recommendations for the study of microplastics in seafood. *TrAC Trends in Analytical Chemistry*, *116*, 346–359. https://doi.org/10.1016/j.trac.2018.11.011.

Dehaut, A., et al. (2016). Microplastics in seafood: Benchmark protocol for their extraction and characterization. *Environmental Pollution*, *215*, 223–233. https://doi.org/10.1016/j.envpol.2016.05.018.

di Beneditto, A. P. M., & da Silva Oliveira, A. (2019). Debris ingestion by carnivorous consumers: Does the position in the water column truly matter? *Marine Pollution Bulletin*, *144*, 134–139. https://doi.org/10.1016/J.MARPOLBUL.2019.04.074.

Diaz-Basantes, M. F., Conesa, J. A., & Fullana, A. (2020). Microplastics in honey, beer, milk and refreshments in Ecuador as emerging contaminants. *Sustainability*, *12*(14), 5514. https://doi.org/10.3390/su12145514.

Digka, N., et al. (2018). Microplastics in mussels and fish from the Northern Ionian Sea. *Marine Pollution Bulletin*, *135*, 30–40. https://doi.org/10.1016/J.MARPOLBUL.2018.06.063.

Ding, J., et al. (2021). Microplastics in four bivalve species and basis for using bivalves as bioindicators of microplastic pollution. *Science of the Total Environment*, *782*, 146830. https://doi.org/10.1016/J.SCITOTENV.2021.146830.

EFSA Panel on Contaminants in the Food Chain (CONTAM). (2016). Presence of microplastics and nanoplastics in food, with particular focus on seafood. *EFSA Journal*. https://doi.org/10.2903/j.efsa.2016.4501.

Fadare, O. O., Okoffo, E. D., & Olasehinde, E. F. (2021). Microparticles and microplastics contamination in African table salts. *Marine Pollution Bulletin*, *164*, 112006. https://doi.org/10.1016/J.MARPOLBUL.2021.112006.

FAO. (2020). *The state of world fisheries and aquaculture 2020*. FAO. https://doi.org/10.4060/ca9229en.

Feng, Z., et al. (2020). Microplastics in specific tissues of wild sea urchins along the coastal areas of northern China. *Science of the Total Environment*, *728*, 138660. https://doi.org/10.1016/j.scitotenv.2020.138660.

Fernández Severini, M. D., et al. (2020). Chemical composition and abundance of microplastics in the muscle of commercial shrimp *Pleoticus muelleri* at an impacted coastal environment (Southwestern Atlantic). *Marine Pollution Bulletin*, *161*, 111700. https://doi.org/10.1016/j.marpolbul.2020.111700.

Ferrante, M., et al. (2022). Microplastics in fillets of Mediterranean seafood. A risk assessment study. *Environmental Research*, *204*, 112247. https://doi.org/10.1016/J.ENVRES.2021.112247.

Ferreira, I., et al. (2019). Nanoplastics and marine organisms: What has been studied? *Environmental Toxicology and Pharmacology*, *67*, 1–7. https://doi.org/10.1016/J.ETAP.2019.01.006.

Filgueiras, A. V., et al. (2020). Microplastic ingestion by pelagic and benthic fish and diet composition: A case study in the NW Iberian shelf. *Marine Pollution Bulletin*, *160*, 111623. https://doi.org/10.1016/J.MARPOLBUL.2020.111623.

Fischer, M., & Scholz-Böttcher, B. M. (2017). Simultaneous trace identification and quantification of common types of microplastics in environmental samples by pyrolysis-gas chromatography–mass spectrometry. *Environmental Science & Technology*, *51*(9), 5052–5060. https://doi.org/10.1021/acs.est.6b06362.

Gao, M., et al. (2021). Effect of polystyrene on di-butyl phthalate (DBP) bioavailability and DBP-induced phytotoxicity in lettuce. *Environmental Pollution, 268*, 115870. https://doi.org/10.1016/J.ENVPOL.2020.115870.

Garnier, Y., et al. (2019). Evaluation of microplastic ingestion by tropical fish from Moorea Island, French Polynesia. *Marine Pollution Bulletin, 140*, 165–170. https://doi.org/10.1016/j.marpolbul.2019.01.038.

Ghattavi, K., Naji, A., & Kord, S. (2019). Investigation of microplastic contamination in the gastrointestinal tract of some species of caught fish from Oman Sea. *Iranian Journal of Health and Environment, 12*(1), 141–150.

Gonçalves, J. M., & Bebianno, M. J. (2021). Nanoplastics impact on marine biota: A review. *Environmental Pollution, 273*, 116426. https://doi.org/10.1016/J.ENVPOL.2021.116426.

Gong, J., & Xie, P. (2020). Research progress in sources, analytical methods, eco-environmental effects, and control measures of microplastics. *Chemosphere, 254*, 126790. https://doi.org/10.1016/j.chemosphere.2020.126790.

Gong, Y., et al. (2021). Microplastics in different tissues of a pelagic squid (*Dosidicus gigas*) in the northern Humboldt current ecosystem. *Marine Pollution Bulletin, 169*, 112509. https://doi.org/10.1016/J.MARPOLBUL.2021.112509.

Gündoğdu, S. (2018). Contamination of table salts from Turkey with microplastics. *Food Additives & Contaminants: Part A, 35*(5), 1006–1014. https://doi.org/10.1080/19440049.2018.1447694.

Gündogdu, S., et al. (2022). The impact of nano/micro-plastics toxicity on seafood quality and human health: Facts and gaps. *Critical Reviews in Food Science and Nutrition*, 1–19. https://doi.org/10.1080/10408398.2022.2033684.

Guo, X., & Wang, J. (2019). Sorption of antibiotics onto aged microplastics in freshwater and seawater. *Marine Pollution Bulletin, 149*, 110511. https://doi.org/10.1016/J.MARPOLBUL.2019.110511.

Güven, O., et al. (2017). Microplastic litter composition of the Turkish territorial waters of the Mediterranean Sea, and its occurrence in the gastrointestinal tract of fish. *Environmental Pollution, 223*, 286–294. https://doi.org/10.1016/J.ENVPOL.2017.01.025.

Hartmann, N. B., et al. (2019). Are we speaking the same language? Recommendations for a definition and categorization framework for plastic debris. *Environmental Science & Technology, 53*(3), 1039–1047. https://doi.org/10.1021/acs.est.8b05297.

Hastuti, A. R., Lumbanbatu, D. T., & Wardiatno, Y. (2019). The presence of microplastics in the digestive tract of commercial fishes off Pantai Indah Kapuk coast, Jakarta, Indonesia. *Biodiversitas Journal of Biological Diversity, 20*(5). https://doi.org/10.13057/biodiv/d200513.

Hazimah, N., et al. (2021). Lifetime accumulation of microplastic in children and adults. *Environmental Science & Technology, 55*, 5096. https://doi.org/10.1021/acs.est.0c07384.

Hermabessiere, L., et al. (2018). Optimization, performance, and application of a pyrolysis-GC/MS method for the identification of microplastics. *Analytical and Bioanalytical Chemistry, 410*(25), 6663–6676. https://doi.org/10.1007/s00216-018-1279-0.

Hermabessiere, L., et al. (2019). Microplastic contamination and pollutant levels in mussels and cockles collected along the channel coasts. *Environmental Pollution, 250*, 807–819. https://doi.org/10.1016/j.envpol.2019.04.051.

Hossain, M. S., et al. (2019). Microplastics in fishes from the Northern Bay of Bengal. *Science of the Total Environment, 690*, 821–830. https://doi.org/10.1016/J.SCITOTENV.2019.07.065.

Hossain, M. S., et al. (2020). Microplastic contamination in Penaeid shrimp from the Northern Bay of Bengal. *Chemosphere, 238*, 124688. https://doi.org/10.1016/j.chemosphere.2019.124688.

Huang, Y., et al. (2020). Rapid measurement of microplastic contamination in chicken meat by mid infrared spectroscopy and chemometrics: A feasibility study. *Food Control, 113*, 107187. https://doi.org/10.1016/J.FOODCONT.2020.107187.

Inderhaug, T. (2020). Stockfish production, cultural and culinary values. *Food Ethics*, *5*(1–2), 6. https://doi.org/10.1007/s41055-019-00060-6.

Iñiguez, M. E., Conesa, J. A., & Fullana, A. (2017). Microplastics in Spanish table salt OPEN. *Scientific Reports*, 7(1), 1–7. https://doi.org/10.1038/s41598-017-09128-x.

Ivleva, N. P. (2021). Chemical analysis of microplastics and nanoplastics: Challenges, advanced methods, and perspectives. *Chemical Reviews*, *121*(19), 11886–11936. https://doi.org/10.1021/acs.chemrev.1c00178.

Jadhav, E. B., et al. (2021). Microplastics from food packaging: An overview of human consumption, health threats, and alternative solutions. *Environmental Nanotechnology, Monitoring & Management*, *16*, 100608. https://doi.org/10.1016/j.enmm.2021.100608.

James, K., et al. (2020). An assessment of microplastics in the ecosystem and selected commercially important fishes off Kochi, south eastern Arabian Sea, India. *Marine Pollution Bulletin*, *154*, 111027. https://doi.org/10.1016/J.MARPOLBUL.2020.111027.

James, K., et al. (2022). Microplastics in the environment and in commercially significant fishes of mud banks, an ephemeral ecosystem formed along the southwest coast of India. *Environmental Research*, *204*, 112351. https://doi.org/10.1016/J.ENVRES.2021.112351.

Järlskog, I., et al. (2020). Occurrence of tire and bitumen wear microplastics on urban streets and in sweeps and washwater. *Science of the Total Environment*, *729*, 138950. https://doi.org/10.1016/J.SCITOTENV.2020.138950.

Jonathan, M. P., et al. (2021). Evidences of microplastics in diverse fish species off the Western Coast of Pacific Ocean, Mexico. *Ocean and Coastal Management*, *204*, 105544. https://doi.org/10.1016/j.ocecoaman.2021.105544.

Jung, M. R., et al. (2018). Validation of ATR FT-IR to identify polymers of plastic marine debris, including those ingested by marine organisms. *Marine Pollution Bulletin*, *127*, 704–716. https://doi.org/10.1016/j.marpolbul.2017.12.061.

Kalaiselvan, K., et al. (2022). Environmental science and pollution research occurrence of microplastics in gastrointestinal tracts of planktivorous fish from the Thoothukudi region. *Environmental Science and Pollution Research*, *1*, 1–9. https://doi.org/10.1007/s11356-022-19033-0.

Käppler, A., et al. (2016). Analysis of environmental microplastics by vibrational microspectroscopy: FTIR, Raman or both? *Analytical and Bioanalytical Chemistry*, *408*(29), 8377–8391. https://doi.org/10.1007/s00216-016-9956-3.

Karami, A., Golieskardi, A., Bin Ho, Y., et al. (2017). Microplastics in eviscerated flesh and excised organs of dried fish. *Scientific Reports*, 7(1), 5473. https://doi.org/10.1038/s41598-017-05828-6.

Karami, A., Golieskardi, A., Choo, C. K., et al. (2017). The presence of microplastics in commercial salts from different countries. *Scientific Reports*, 7(1), 1–11. https://doi.org/10.1038/srep46173.

Karbalaei, S., et al. (2019). Abundance and characteristics of microplastics in commercial marine fish from Malaysia. *Marine Pollution Bulletin*, *148*, 5–15. https://doi.org/10.1016/j.marpolbul.2019.07.072.

Karuppasamy, P. K., et al. (2020). Baseline survey of micro and mesoplastics in the gastro-intestinal tract of commercial fish from Southeast coast of the Bay of Bengal. *Marine Pollution Bulletin*, *153*, 110974. https://doi.org/10.1016/j.marpolbul.2020.110974.

Ke, A. Y., et al. (2019). Impacts of leachates from single-use polyethylene plastic bags on the early development of clam *Meretrix meretrix* (Bivalvia: Veneridae). *Marine Pollution Bulletin*, *142*, 54–57. https://doi.org/10.1016/J.MARPOLBUL.2019.03.029.

Kedzierski, M., et al. (2020). Microplastic contamination of packaged meat: Occurrence and associated risks. *Food Packaging and Shelf Life*, *24*, 100489. https://doi.org/10.1016/J.FPSL.2020.100489.

Kim, J.-S., et al. (2018). Global pattern of microplastics (MPs) in commercial food-grade salts: Sea salt as an indicator of seawater MP pollution. *Environmental Science & Technology*, *52*(21), 12819–12828. https://doi.org/10.1021/acs.est.8b04180.

Knight, L. J., et al. (2020). Tyre wear particles: An abundant yet widely unreported microplastic? *Environmental Science and Pollution Research*, *27*, 18345–18354. https://doi.org/10.1007/s11356-020-08187-4.

Koongolla, J. B., et al. (2020). Occurrence of microplastics in gastrointestinal tracts and gills of fish from Beibu Gulf, South China Sea. *Environmental Pollution*, *258*, 113734. https://doi.org/10.1016/j.envpol.2019.113734.

Kutralam-Muniasamy, G., et al. (2020). Branded milks—Are they immune from microplastics contamination? *Science of the Total Environment*, *714*, 136823. https://doi.org/10.1016/J.SCITOTENV.2020.136823.

Kutralam-Muniasamy, G., et al. (2021). Overview of microplastics pollution with heavy metals: Analytical methods, occurrence, transfer risks and call for standardization. *Journal of Hazardous Materials*, *415*, 125755. https://doi.org/10.1016/J.JHAZMAT.2021.125755.

Kwon, J.-H., et al. (2020). Microplastics in food: A review on analytical methods and challenges. *International Journal of Environmental Research and Public Health*, *17*(18), 6710. https://doi.org/10.3390/ijerph17186710.

la Nasa, J., et al. (2020). A review on challenges and developments of analytical pyrolysis and other thermoanalytical techniques for the quali-quantitative determination of microplastics. *Journal of Analytical and Applied Pyrolysis*, *149*, 104841. https://doi.org/10.1016/j.jaap.2020.104841.

Lakshmi Kavya, A. N. V., Sundarrajan, S., & Ramakrishna, S. (2020). Identification and characterization of micro-plastics in the marine environment: A mini review. *Marine Pollution Bulletin*, *160*, 111704. https://doi.org/10.1016/j.marpolbul.2020.111704.

Li, J., et al. (2019). Using mussel as a global bioindicator of coastal microplastic pollution. *Environmental Pollution*, *244*, 522–533. https://doi.org/10.1016/j.envpol.2018.10.032.

Li, Q., et al. (2021a). Microplastics in shellfish and implications for food safety. *Current Opinion in Food Science*, *40*, 192–197. https://doi.org/10.1016/j.cofs.2021.04.017.

Li, Y., et al. (2021b). Microplastics environmental effect and risk assessment on the aquaculture systems from South China. *International Journal of Environmental Research and Public Health*, *18*(4), 1869. https://doi.org/10.3390/ijerph18041869.

Lian, J., et al. (2020). Impact of polystyrene nanoplastics (PSNPs) on seed germination and seedling growth of wheat (*Triticum aestivum L.*). *Journal of Hazardous Materials*, *385*, 121620. https://doi.org/10.1016/J.JHAZMAT.2019.121620.

Liboiron, M., et al. (2019). Low incidence of plastic ingestion among three fish species significant for human consumption on the island of Newfoundland, Canada. *Marine Pollution Bulletin*, *141*, 244–248. https://doi.org/10.1016/J.MARPOLBUL.2019.02.057.

Liebezeit, G., & Liebezeit, E. (2014). Synthetic particles as contaminants in German beers. *Food Additives & Contaminants: Part A*, *31*(9), 1574–1578. https://doi.org/10.1080/19440049.2014.945099.

Löder, M. G. J., & Gerdts, G. (2015). Methodology used for the detection and identification of microplastics—A critical appraisal. In *Marine anthropogenic litter* (pp. 201–227). Cham: Springer International Publishing. https://doi.org/10.1007/978-3-319-16510-3_8.

Lopes, C., et al. (2020). Microplastic ingestion and diet composition of planktivorous fish. *Limnology and Oceanography Letters*, *5*, 103–112. https://doi.org/10.1002/lol2.10144.

López-López, L., et al. (2018). Incidental ingestion of meso- and macro-plastic debris by benthic and demersal fish. *Food Webs*, *14*, 1–4. https://doi.org/10.1016/J.FOOWEB.2017.12.002.

López-Martínez, S., et al. (2021). Overview of global status of plastic presence in marine vertebrates. *Global Change Biology*, *27*(4), 728–737. https://doi.org/10.1111/gcb.15416.
Lu, H., et al. (2022). Machine learning-aided engineering of hydrolases for PET depolymerization. *Nature*, *604*(7907), 662–667. https://doi.org/10.1038/s41586-022-04599-z.
Lv, L., et al. (2021). Challenge for the detection of microplastics in the environment. *Water Environment Research*, *93*(1), 5–15. https://doi.org/10.1002/wer.1281.
Lwanga, E. H., et al. (2017). Field evidence for transfer of plastic debris along a terrestrial food chain. *Scientific Reports*, 7(1), 1–7. https://doi.org/10.1038/s41598-017-14588-2.
Mancia, A., et al. (2020). Adverse effects of plastic ingestion on the Mediterranean small-spotted catshark (*Scyliorhinus canicula*). *Marine Environmental Research*, *155*. https://doi.org/10.1016/j.marenvres.2020.104876.
Markic, A., et al. (2018). Double trouble in the South Pacific subtropical gyre: Increased plastic ingestion by fish in the oceanic accumulation zone. *Marine Pollution Bulletin*, *136*, 547–564. https://doi.org/10.1016/J.MARPOLBUL.2018.09.031.
Martinez-Tavera, E., et al. (2021). Microplastics and metal burdens in freshwater Tilapia (*Oreochromis niloticus*) of a metropolitan reservoir in Central Mexico: Potential threats for human health. *Chemosphere*, *266*, 128968. https://doi.org/10.1016/j.chemosphere.2020.128968.
Masry, M., et al. (2021). Characteristics, fate, and impact of marine plastic debris exposed to sunlight: A review. *Marine Pollution Bulletin*, *171*, 112701. https://doi.org/10.1016/J.MARPOLBUL.2021.112701.
McGoran, A. R., et al. (2020). High prevalence of plastic ingestion by *Eriocheir sinensis* and *Carcinus maenas* (Crustacea: Decapoda: Brachyura) in the Thames Estuary. *Environmental Pollution*, *265*, 114972. https://doi.org/10.1016/J.ENVPOL.2020.114972.
Mercogliano, R., et al. (2020). Occurrence of microplastics in commercial seafood under the perspective of the human food chain. A review. *Journal of Agricultural and Food Chemistry*, *68*(19), 5296–5301. https://doi.org/10.1021/acs.jafc.0c01209.
Merga, L. B., et al. (2020). Distribution of microplastic and small macroplastic particles across four fish species and sediment in an African lake. *Science of the Total Environment*, *741*, 140527. https://doi.org/10.1016/j.scitotenv.2020.140527.
Mistri, M., et al. (2022). Microplastic accumulation in commercial fish from the Adriatic Sea. *Marine Pollution Bulletin*, *174*, 113279. https://doi.org/10.1016/J.MARPOLBUL.2021.113279.
Murphy, F., et al. (2017). The uptake of macroplastic & microplastic by demersal & pelagic fish in the Northeast Atlantic around Scotland. *Marine Pollution Bulletin*, *122*(1–2), 353–359. https://doi.org/10.1016/J.MARPOLBUL.2017.06.073.
Murray, F., & Cowie, P. R. (2011). Plastic contamination in the decapod crustacean *Nephrops norvegicus* (Linnaeus, 1758). *Marine Pollution Bulletin*, *62*(6), 1207–1217. https://doi.org/10.1016/j.marpolbul.2011.03.032.
Neto, J. G. B., et al. (2020). Ingestion of plastic debris by commercially important marine fish in southeast-south Brazil. *Environmental Pollution*, *267*, 115508. https://doi.org/10.1016/J.ENVPOL.2020.115508.
Neves, D., et al. (2015). Ingestion of microplastics by commercial fish off the Portuguese coast. *Marine Pollution Bulletin*, *101*(1), 119–126. https://doi.org/10.1016/J.MARPOLBUL.2015.11.008.
Nguyen, B., et al. (2019). Separation and analysis of microplastics and nanoplastics in complex environmental samples. *Accounts of Chemical Research*, *52*(4), 858–866. https://doi.org/10.1021/acs.accounts.8b00602.
Ohkubo, N., et al. (2022). Microplastic uptake and gut retention time in Japanese anchovy (*Engraulis japonicus*) under laboratory conditions. *Marine Pollution Bulletin*, *176*, 113433. https://doi.org/10.1016/J.MARPOLBUL.2022.113433.

Oliveira, A. R., et al. (2020). Microplastics presence in cultured and wild-caught cuttlefish, *Sepia officinalis*. *Marine Pollution Bulletin*, *160*, 111553. https://doi.org/10.1016/J.MARPOLBUL.2020.111553.

Oliveri Conti, G., et al. (2020). Micro- and nano-plastics in edible fruit and vegetables. The first diet risks assessment for the general population. *Environmental Research*, *187*, 109677. https://doi.org/10.1016/j.envres.2020.109677.

Ory, N. C., et al. (2017). Amberstripe scad *Decapterus muroadsi* (Carangidae) fish ingest blue microplastics resembling their copepod prey along the coast of Rapa Nui (Easter Island) in the South Pacific subtropical gyre. *Science of the Total Environment*, *586*, 430–437. https://doi.org/10.1016/J.SCITOTENV.2017.01.175.

Ory, N., et al. (2018). Low prevalence of microplastic contamination in planktivorous fish species from the southeast Pacific Ocean. *Marine Pollution Bulletin*, *127*, 211–216. https://doi.org/10.1016/J.MARPOLBUL.2017.12.016.

Palazzo, L., et al. (2021). A novel approach based on multiple fish species and water column compartments in assessing vertical microlitter distribution and composition. *Environmental Pollution*, *272*, 116419. https://doi.org/10.1016/J.ENVPOL.2020.116419.

Palermo, J. D. H., et al. (2020). Susceptibility of Sardinella lemuru to emerging marine microplastic pollution. *Global Journal of Environmental Science and Management*, *6*(3), 373–384.

Panebianco, A., et al. (2019). First discoveries of microplastics in terrestrial snails. *Food Control*, *106*, 106722. https://doi.org/10.1016/J.FOODCONT.2019.106722.

Pegado, T., et al. (2018). First evidence of microplastic ingestion by fishes from the Amazon River estuary. *Marine Pollution Bulletin*, *133*, 814–821. https://doi.org/10.1016/J.MARPOLBUL.2018.06.035.

Peñalver, R., et al. (2020). An overview of microplastics characterization by thermal analysis. *Chemosphere*, *242*, 125170. https://doi.org/10.1016/j.chemosphere.2019.125170.

Pereira, J. M., et al. (2020). Microplastic in the stomachs of open-ocean and deep-sea fishes of the North-East Atlantic. *Environmental Pollution*, *265*, 115060. https://doi.org/10.1016/J.ENVPOL.2020.115060.

Peters, C. A., et al. (2018). Pyr-GC/MS analysis of microplastics extracted from the stomach content of benthivore fish from the Texas Gulf Coast. *Marine Pollution Bulletin*, *137*, 91–95. https://doi.org/10.1016/j.marpolbul.2018.09.049.

Pinto da Costa, J., et al. (2019). Micro (nano)plastics—Analytical challenges towards risk evaluation. *TrAC Trends in Analytical Chemistry*, *111*, 173–184. https://doi.org/10.1016/j.trac.2018.12.013.

Pittura, L. *et al.* (2018) Less Plastic More Mediterranean, 2017 Campaign on board the Rainbow Warrior Greenpeace Ship.

Piyawardhana, N., et al. (2022). Occurrence of microplastics in commercial marine dried fish in Asian countries. *Journal of Hazardous Materials*, *423*, 127093. https://doi.org/10.1016/J.JHAZMAT.2021.127093.

Prata, J. C., et al. (2020). Identification of microplastics in white wines capped with polyethylene stoppers using micro-Raman spectroscopy. *Food Chemistry*, *331*, 127323. https://doi.org/10.1016/J.FOODCHEM.2020.127323.

Primpke, S., et al. (2020). Critical assessment of analytical methods for the harmonized and cost-efficient analysis of microplastics. *Applied Spectroscopy*, *74*(9), 1012–1047. https://doi.org/10.1177/0003702820921465.

Qi, Y., et al. (2018). Macro- and microplastics in soil-plant system: Effects of plastic mulch film residues on wheat (*Triticum aestivum*) growth. *Science of the Total Environment*, *645*, 1048–1056. https://doi.org/10.1016/J.SCITOTENV.2018.07.229.

Ragusa, A., et al. (2021). Plasticenta: First evidence of microplastics in human placenta. *Environment International*, *146*, 106274. https://doi.org/10.1016/J.ENVINT.2020.106274.

Razeghi, N., et al. (2021). Scientific studies on microplastics pollution in Iran: An in-depth review of the published articles. *Marine Pollution Bulletin*, *162*, 111901. https://doi.org/10.1016/j.marpolbul.2020.111901.

Reinold, S., et al. (2021). Evidence of microplastic ingestion by cultured European sea bass (*Dicentrarchus labrax*). *Marine Pollution Bulletin*, *168*, 112450. https://doi.org/10.1016/J.MARPOLBUL.2021.112450.

Renner, G., Schmidt, T. C., & Schram, J. (2018). Analytical methodologies for monitoring micro (nano)plastics: Which are fit for purpose? *Current Opinion in Environmental Science & Health*, *1*, 55–61. https://doi.org/10.1016/J.COESH.2017.11.001.

Renner, G., et al. (2019). Data preprocessing & evaluation used in the microplastics identification process: A critical review & practical guide. *TrAC Trends in Analytical Chemistry*, *111*, 229–238. https://doi.org/10.1016/j.trac.2018.12.004.

Renzi, M., & Blašković, A. (2018). Litter & microplastics features in table salts from marine origin: Italian versus Croatian brands. *Marine Pollution Bulletin*, *135*, 62–68. https://doi.org/10.1016/J.MARPOLBUL.2018.06.065.

Renzi, M., Grazioli, E., et al. (2019). Microparticles in table salt: Levels and chemical composition of the smallest dimensional fraction. *Journal of Marine Science and Engineering*, 7(9), 310. https://doi.org/10.3390/jmse7090310.

Renzi, M., Specchiulli, A., et al. (2019). Marine litter in stomach content of small pelagic fishes from the Adriatic Sea: sardines (*Sardina pilchardus*) and anchovies (*Engraulis encrasicolus*). *Environmental Science and Pollution Research International*, *26*(3), 2771–2781. https://doi.org/10.1007/s11356-018-3762-8.

Renzi, M., et al. (2020). Litter in alien species of possible commercial interest: The blue crab (*Callinectes sapidus* Rathbun, 1896) as case study. *Marine Pollution Bulletin*, *157*, 111300. https://doi.org/10.1016/J.MARPOLBUL.2020.111300.

Ribeiro, F., et al. (2020). Quantitative analysis of selected plastics in high-commercial-value Australian seafood by pyrolysis gas chromatography mass spectrometry. *Environmental Science & Technology*, *54*, 9408–9417. https://doi.org/10.1021/acs.est.0c02337.

Ridwan Lessy, M., & Sabar, M. (2021). Microplastics ingestion by Skipjack tuna (*Katsuwonus pelamis*) in ternate, North Maluku-Indonesia. In *IOP Conference Series: Materials Science and Engineering*. https://doi.org/10.1088/1757-899X/1125/1/012085. (Preprint).

Rios-Fuster, B., et al. (2019). Anthropogenic particles ingestion in fish species from two areas of the western Mediterranean Sea. *Marine Pollution Bulletin*, *144*, 325–333. https://doi.org/10.1016/J.MARPOLBUL.2019.04.064.

Roch, S., Friedrich, C., & Brinker, A. (2020). Uptake routes of microplastics in fishes: Practical and theoretical approaches to test existing theories. *Scientific Reports*, *10*, 3896. https://doi.org/10.1038/s41598-020-60630-1.

Rochman, C. M., et al. (2015). Anthropogenic debris in seafood: Plastic debris and fibers from textiles in fish and bivalves sold for human consumption. *Scientific Reports*, *5*, 14340. https://doi.org/10.1038/srep14340.

Rosal, R. (2021). Morphological description of microplastic particles for environmental fate studies. *Marine Pollution Bulletin*, *171*, 112716. https://doi.org/10.1016/J.MARPOLBUL.2021.112716.

Ryberg, M. W., et al. (2019). Global environmental losses of plastics across their value chains. *Resources, Conservation and Recycling*, *151*, 104459. https://doi.org/10.1016/J.RESCONREC.2019.104459.

Saha, M., et al. (2021). Microplastics in seafood as an emerging threat to marine environment: A case study in Goa, west coast of India. *Chemosphere*, *270*, 129359. https://doi.org/10.1016/j.chemosphere.2020.129359.

Saturno, J., et al. (2020). Occurrence of plastics ingested by Atlantic cod (*Gadus morhua*) destined for human consumption (Fogo Island, Newfoundland and Labrador). *Marine Pollution Bulletin*, *153*, 110993. https://doi.org/10.1016/J.MARPOLBUL.2020.110993.

Savoca, M. S., McInturf, A. G., & Hazen, E. L. (2021). Plastic ingestion by marine fish is widespread and increasing. *Global Change Biology*, *27*(10), 2188–2199. https://doi.org/10.1111/gcb.15533.

Savoca, S., et al. (2020). Plastics occurrence in juveniles of *Engraulis encrasicolus* and *Sardina pilchardus* in the Southern Tyrrhenian Sea. *Science of the Total Environment*, *718*, 137457. https://doi.org/10.1016/J.SCITOTENV.2020.137457.

Schwaferts, C., et al. (2019). Methods for the analysis of submicrometer- and nanoplastic particles in the environment. *TrAC Trends in Analytical Chemistry*, *112*, 52–65. https://doi.org/10.1016/j.trac.2018.12.014.

Seth, C. K., & Shriwastav, A. (2018). Contamination of Indian sea salts with microplastics and a potential prevention strategy. *Environmental Science and Pollution Research*, *25*(30), 30122–30131. https://doi.org/10.1007/S11356-018-3028-5.

Shan, J., et al. (2019). Simple and rapid detection of microplastics in seawater using hyperspectral imaging technology. *Analytica Chimica Acta*, *1050*, 161–168. https://doi.org/10.1016/j.aca.2018.11.008.

Shim, W. J., Hong, S. H., & Eo, S. E. (2017). Identification methods in microplastic analysis: A review. *Analytical Methods*, *9*(9), 1384–1391. https://doi.org/10.1039/C6AY02558G.

Shruti, V. C., et al. (2020). First study of its kind on the microplastic contamination of soft drinks, cold tea and energy drinks—Future research and environmental considerations. *Science of the Total Environment*, *726*, 138580. https://doi.org/10.1016/J.SCITOTENV.2020.138580.

Sid, S., et al. (2021). Bio-sourced polymers as alternatives to conventional food packaging materials: A review. *Trends in Food Science & Technology*, *115*, 87–104. https://doi.org/10.1016/j.tifs.2021.06.026.

Silva, A. B., et al. (2018). Microplastics in the environment: Challenges in analytical chemistry—A review. *Analytica Chimica Acta*, *1017*, 1–19. https://doi.org/10.1016/j.aca.2018.02.043.

Silva, G. C., et al. (2021). Microplastics and their effect in horticultural crops: Food safety and plant stress. *Agronomy*, *11*(8), 1528. https://doi.org/10.3390/agronomy11081528.

Sobhani, Z., et al. (2020). Microplastics generated when opening plastic packaging. *Scientific Reports*, *10*(1), 1–7. https://doi.org/10.1038/s41598-020-61146-4.

Sol, D., et al. (2020). Approaching the environmental problem of microplastics: Importance of WWTP treatments. *Science of the Total Environment*, *740*, 140016. https://doi.org/10.1016/J.SCITOTENV.2020.140016.

Šourková, M., Adamcová, D., & Vaverková, M. D. (2021). The influence of microplastics from ground Tyres on the acute, subchronical toxicity and microbial respiration of soil. *Environments*, *8*(11), 128. https://doi.org/10.3390/environments8110128.

Strungaru, S.-A., et al. (2019). Micro- (nano) plastics in freshwater ecosystems: Abundance, toxicological impact and quantification methodology. *TrAC Trends in Analytical Chemistry*, *110*, 116–128. https://doi.org/10.1016/j.trac.2018.10.025.

Su, L., et al. (2019). The occurrence of microplastic in specific organs in commercially caught fishes from coast and estuary area of East China. *Journal of Hazardous Materials*, *365*, 716–724. https://doi.org/10.1016/j.jhazmat.2018.11.024.

Sun, Q., Ren, S. Y., & Ni, H. G. (2020). Incidence of microplastics in personal care products: An appreciable part of plastic pollution. *Science of the Total Environment*, *742*, 140218. https://doi.org/10.1016/J.SCITOTENV.2020.140218.

Sun, X., et al. (2019). Characteristics and retention of microplastics in the digestive tracts of fish from the Yellow Sea. *Environmental Pollution*, *249*, 878–885. https://doi.org/10.1016/J.ENVPOL.2019.01.110.

Süssmann, J., et al. (2021). Evaluation and optimisation of sample preparation protocols suitable for the analysis of plastic particles present in seafood. *Food Control*, *125*, 107969. https://doi.org/10.1016/j.foodcont.2021.107969.
Suwartiningsih, N., Setyowati, I., & Astuti, R. (2020). Microplastics in pelagic and demersal fishes of Pantai Baron, Yogyakarta, Indonesia. *JurnalBiodjati*, *5*(1), 33–49.
TaghizadehRahmat Abadi, Z., et al. (2021). Microplastic content of Kutum fish, *Rutilus frisii kutum* in the southern Caspian Sea. *Science of the Total Environment*, *752*, 141542. https://doi.org/10.1016/j.scitotenv.2020.141542.
Tanaka, K., & Takada, H. (2016). Microplastic fragments and microbeads in digestive tracts of planktivorous fish from urban coastal waters. *Scientific Reports*. https://doi.org/10.1038/srep34351.
Tourinho, P. S., et al. (2019). Partitioning of chemical contaminants to microplastics: Sorption mechanisms, environmental distribution and effects on toxicity and bioaccumulation. *Environmental Pollution*, *252*, 1246–1256. https://doi.org/10.1016/J.ENVPOL.2019.06.030.
Tsangaris, C., et al. (2020). Using *Boops boops* (osteichthyes) to assess microplastic ingestion in the Mediterranean Sea. *Marine Pollution Bulletin*, *158*, 111397. https://doi.org/10.1016/J.MARPOLBUL.2020.111397.
Turner, A. (2021). Paint particles in the marine environment: An overlooked component of microplastics. *Water Research X*, *12*, 100110. https://doi.org/10.1016/J.WROA.2021.100110.
Urbina, M. A., et al. (2020). Adsorption of polyethylene microbeads and physiological effects on hydroponic maize. *Science of the Total Environment*, *741*, 140216. https://doi.org/10.1016/J.SCITOTENV.2020.140216.
Valencia-Castañeda, G., et al. (2022). Microplastics in the tissues of commercial semi-intensive shrimp pond-farmed *Litopenaeus vannamei* from the Gulf of California ecoregion. *Chemosphere*, *297*, 134194. https://doi.org/10.1016/J.CHEMOSPHERE.2022.134194.
van Raamsdonk, L. W. D., et al. (2020). Current insights into monitoring, bioaccumulation, and potential health effects of microplastics present in the food chain. *Food*, *9*(1), 72. https://doi.org/10.3390/foods9010072.
van Weert, S., et al. (2019). Effects of nanoplastics and microplastics on the growth of sediment-rooted macrophytes. *Science of the Total Environment*, *654*, 1040–1047. https://doi.org/10.1016/J.SCITOTENV.2018.11.183.
Vanapalli, K. R., et al. (2021). Challenges and strategies for effective plastic waste management during and post COVID-19 pandemic. *Science of the Total Environment*, *750*, 141514. https://doi.org/10.1016/j.scitotenv.2020.141514.
Vázquez-Rowe, I., Ita-Nagy, D., & Kahhat, R. (2021). Microplastics in fisheries and aquaculture: Implications to food sustainability and safety. *Current Opinion in Green and Sustainable Chemistry*, *29*, 100464. https://doi.org/10.1016/j.cogsc.2021.100464.
Veerasingam, S., et al. (2021). Contributions of Fourier transform infrared spectroscopy in microplastic pollution research: A review. *Critical Reviews in Environmental Science and Technology*, *51*(22), 2681–2743. https://doi.org/10.1080/10643389.2020.1807450.
Venâncio, C., et al. (2019). The effects of nanoplastics on marine plankton: A case study with polymethylmethacrylate. *Ecotoxicology and Environmental Safety*, *184*, 109632. https://doi.org/10.1016/J.ECOENV.2019.109632.
Vinay Kumar, B. N., et al. (2021). Analysis of microplastics of a broad size range in commercially important mussels by combining FTIR and Raman spectroscopy approaches. *Environmental Pollution*, *269*, 116147. https://doi.org/10.1016/j.envpol.2020.116147.

Vitali, C., et al. (2022). Microplastics and nanoplastics in food, water, and beverages; part I. Occurrence. *TrAC Trends in Analytical Chemistry*, 116670. https://doi.org/10.1016/J.TRAC.2022.116670.

Walkinshaw, C., et al. (2020). Microplastics and seafood: Lower trophic organisms at highest risk of contamination. *Ecotoxicology and Environmental Safety*, *190*, 110066. https://doi.org/10.1016/J.ECOENV.2019.110066.

Wang, W., & Wang, J. (2018). Investigation of microplastics in aquatic environments: An overview of the methods used, from field sampling to laboratory analysis. *TrAC Trends in Analytical Chemistry*, *108*, 195–202. https://doi.org/10.1016/j.trac.2018.08.026.

Wang, Z.-M., et al. (2017). SEM/EDS and optical microscopy analyses of microplastics in ocean trawl and fish guts. *Science of the Total Environment*, *603–604*, 616–626. https://doi.org/10.1016/j.scitotenv.2017.06.047.

Wang, D., et al. (2021a). Quantitative and qualitative determination of microplastics in oyster, seawater and sediment from the coastal areas in Zhuhai, China. *Marine Pollution Bulletin*, *164*, 112000. https://doi.org/10.1016/j.marpolbul.2021.112000.

Wang, F., et al. (2021b). Microplastic characteristics in organisms of different trophic levels from Liaohe Estuary, China. *Science of the Total Environment*, *789*, 148027. https://doi.org/10.1016/J.SCITOTENV.2021.148027.

Wu, R. T., et al. (2021). Occurrence of microplastic in livestock and poultry manure in South China. *Environmental Pollution*, *277*, 116790. https://doi.org/10.1016/J.ENVPOL.2021.116790.

Xu, J.-L., et al. (2019). FTIR and Raman imaging for microplastics analysis: State of the art, challenges and prospects. *TrAC Trends in Analytical Chemistry*, *119*, 115629. https://doi.org/10.1016/j.trac.2019.115629.

Yagi, M., et al. (2022). Microplastic pollution of commercial fishes from coastal and offshore waters in southwestern Japan. *Marine Pollution Bulletin*, *174*, 113304. https://doi.org/10.1016/J.MARPOLBUL.2021.113304.

Yakovenko, N., Carvalho, A., & ter Halle, A. (2020). Emerging use thermo-analytical method coupled with mass spectrometry for the quantification of micro (nano)plastics in environmental samples. *TrAC Trends in Analytical Chemistry*, *131*, 115979. https://doi.org/10.1016/j.trac.2020.115979.

Yang, D., et al. (2015). Microplastic pollution in table salts from China. *Environmental Science and Technology*, *49*(22), 13622–13627. https://doi.org/10.1021/ACS.EST.5B03163/SUPPL_FILE/ES5B03163_SI_001.PDF.

Yuan, Z., Nag, R., & Cummins, E. (2022). Human health concerns regarding microplastics in the aquatic environment—From marine to food systems. *Science of the Total Environment*, *823*, 153730. https://doi.org/10.1016/j.scitotenv.2022.153730.

Yuan, W., et al. (2019). Microplastic abundance, distribution and composition in water, sediments, and wild fish from Poyang Lake, China. *Ecotoxicology and Environmental Safety*, *170*, 180–187. https://doi.org/10.1016/j.ecoenv.2018.11.126.

ZabihzadehKhajavi, M., et al. (2019). Strategies for controlling release of plastic compounds into foodstuffs based on application of nanoparticles and its potential health issues. *Trends in Food Science & Technology*, *90*, 1–12. https://doi.org/10.1016/j.tifs.2019.05.009.

Zeytin, S., et al. (2020). Quantifying microplastic translocation from feed to the fillet in European sea bass *Dicentrarchus labrax*. *Marine Pollution Bulletin*, *156*, 111210. https://doi.org/10.1016/j.marpolbul.2020.111210.

Zhang, Y., et al. (2019). Hyperspectral imaging based method for rapid detection of microplastics in the intestinal tracts of fish. *Environmental Science & Technology*, *53*(9), 5151–5158. https://doi.org/10.1021/acs.est.8b07321.

Zhang, T., et al. (2021). Microplastics in different tissues of wild crabs at three important fishing grounds in China. *Chemosphere*, *271*, 129479. https://doi.org/10.1016/J.CHEMOSPHERE.2020.129479.

Zhu, J., et al. (2019). Microplastic pollution in the Maowei Sea, a typical mariculture bay of China. *Science of the Total Environment*, *658*, 62–68. https://doi.org/10.1016/j.scitotenv.2018.12.192.

Zhu, J., & Wang, C. (2020). Recent advances in the analysis methodologies for microplastics in aquatic organisms: Current knowledge and research challenges. *Analytical Methods*, *12*(23), 2944–2957. https://doi.org/10.1039/D0AY00143K.

Zitouni, N., et al. (2020). First report on the presence of small microplastics ($\leq$3 μm) in tissue of the commercial fish *Serranus scriba* (Linnaeus. 1758) from Tunisian coasts and associated cellular alterations. *Environmental Pollution*, *263*, 114576. https://doi.org/10.1016/j.envpol.2020.114576.

The risks of marine micro/nano-plastics on seafood safety and human health

Nariman El Abed[a,*] and Fatih Özogul[b]
[a]Laboratory of Protein Engineering and Bioactive Molecules (LIP-MB), National Institute of Applied Sciences and Technology (INSAT), University of Carthage, Tunis, Tunisia
[b]Department of Seafood Processing Technology, Faculty of Fisheries, Cukurova University, Adana, Turkey
*Corresponding author: e-mail address: elabed_nariman@yahoo.fr

Contents

Abstract

A considerable mass of plastics has been released into the marine environment annually through different human activities, including industrial, agriculture, medical, pharmaceutical and daily care products. These materials are decomposed into smaller particles such as microplastic (MP) and nanoplastic (NP). Hence, these particles can be transported and distributed in coastal and aquatic areas and are ingested by the majority of marine biotas, including seafood products, thus causing the contamination of the different parts of aquatic ecosystems. In fact, seafood involves a wide diversity of edible marine organisms, such as fish, crustaceans, molluscs, and echinoderms, which can ingest the micro/nanoplastics particles, and then transmit them to humans through dietary consumption. Consequently, these pollutants can cause several toxic and adverse

Advances in Food and Nutrition Research, Volume 103
ISSN 1043-4526
https://doi.org/10.1016/bs.afnr.2022.08.004

impacts on human health and the marine ecosystem. Therefore, this chapter provides information on the potential risks of marine micro/nanoplastics on seafood safety and human health.

1. Introduction

In the last decades, the natural ecosystems and their sustainability have been deteriorated by the different anthropogenic activities. Thus, human behavior is largely responsible for environmental pollution. Plastic pollution is considered one of the principal environmental problems due to the increasing production of disposable plastic products. Generally, plastics products are considered synthetic organic polymers, which are produced by the polymerization of hydrocarbons derived from natural gas, crude oil, and coal (Cole, Lindeque, Halsband, Tamara, & Galloway, 2011; Derraik, 2002; Rios, Moore, & Jones, 2007). Since the 1940s, plastic production has increased rapidly worldwide, and consequently, its amount reached 230 million tons in 2009 (Cole et al., 2011). It is estimated that approximately 8 million tons of plastic wastes has annually discharged into the aquatic environment (Huang et al., 2021; Vivekanand, Mohapatra, & Tyagi, 2021). The large plastic wastes found in different environments (terrestrial and marine) can be fragmented into smaller particles or microscopic plastic particles through different degradation processes, such as biodegradation, mechanical abrasion, photo-degradation, etc. (Dawson et al., 2018; Liu et al., 2020). Thus, these plastic fragments are known as microplastics, which are characterized by a size diameter of $<5\,\text{mm}$ (Thompson et al., 2004). Microplastics (MPs) are classified into two categories according to their environmental sources: primary and secondary microplastics (Cole et al., 2011). In fact, these MPs can be degraded into finer particles, known as nanoplastics (NPs). These fine fragments are characterized by a size diameter of $1\,\mu\text{m}$ or $<100\,\text{nm}$ (Cole et al., 2011).

Recently, the issue related to MP/NP particles received increasing research interest due to their ecological consequences. Essentially, there is increasing concern regarding the potential impacts of plastic pollution on marine biotas and human health. Several research studies have focused the distribution of these particles on the marine and coastal environment. In addition, they have investigated the transport, destiny, and impacts of these pollutants in both marine and freshwater systems. The MPs and NPs can be transported from terrestrial sources to the aquatic environment and freshwater systems by different pathways, such as through the domestic and industrial wastewater effluents, and wastewater treatment plants (WWTP) as well as by the surface runoff (Kukkola, Krause, Lynch, Sambrook Smith, & Nel, 2021). Besides, these pollutants are ubiquitous and widespread in the aquatic environment and thus can cause several harms and adverse effects to marine organisms, including the seafood products, due to their small size. MPs and NPs are considered bioavailable to marine biota throughout the trophic chain (Cole et al., 2011). Moreover, their physico-chemical and geometric properties make them susceptible to adhere the other toxic compounds, including the organic contaminants.

However, the ingestion of MPs and NPs by the seafood products leads to toxicity and negative impacts in different groups of seafood, including fish, crustaceans, molluscs, and echinoderms. Thus, these contaminants can cause the reduction of immune response, alteration of nervous functions, and generation of oxidative stress and other adverse effects (Vázquez & Rahman, 2021).

Furthermore, seafood products contain an important amount of functional and bioactive compounds, which have an important role in human health (Inanli, Aksun Tümerkan, El Abed, Regenstein, & Özogul, 2020; Šimat et al., 2020). However, the ingestion of the MPs and NPs particles by seafood can introduce toxins to the base of the food chain, thus, causing their potential bioaccumulation. Therefore, these particles have negative consequences for other trophic levels, including humans (Vázquez & Rahman, 2021). Consequently, there is a raised concern related to the ingestion of these contaminants by humans through the consumption of seafood products contaminated with these pollutants and the potential adverse impacts on the human health (Barboza, Vethaakd, Lavorante, Lundebye, & Guilhermino, 2018). This chapter focuses on the potential risks of MPs and NPs on seafood safety and human health.

2. Marine micro/nanoplastics (M/NPs)

The marine plastics in the form of M/NPs can reach the oceans through different vectors, such as industrial raw materials applied to produce various plastic products; hygiene and cosmetic products; and adhesives (Singh et al., 2022). According to Jambeck et al. (2015), the amount of marine plastics generated in 2010 varied between 4.8 and 12.7 million metric tons (MT). These appreciations are anticipated to increase by 2025 (Jambeck et al., 2015). Eriksen et al. (2014) estimated in 2014 that around

250.000 tons plastics was floating on the surface of the ocean. While the existence of plastic debris has been observed on the surface of the oceans since the 1970s, in the early 2000s when Richard Thompson highlighted the appearance of MP/NP particles, which are invisible to the naked eye (Allen, Allen, Karbalaei, Maselli, & Walker, 2022; Rochman et al., 2019; Thompson et al., 2004).

2.1 Microplastics (MPs)

2.1.1 Origins and transport of MPs in the marine environment

Generally, MPs are known as tiny plastic debris (Andrady, 2011). Thus, they are considered as a diversified group of particles, with different shapes, color, and size ranges (<5 mm), as well as they are characterized by a considerable variety in their chemical compositions (Smith, Love, Rochman, & Neff, 2018). In fact, previous studies have demonstrated that the atmospheric conditions, such as the exposition to UV light, and the mechanical degeneration, can considerably change the information correlated to micro-plastics collection (Schmid, Cozzarini, & Zambello, 2021; Vivekanand et al., 2021). Wang et al. (2019) have revealed that the occurrence of typhoon can cause the augmentation of the amount of MPs both in sediments and water, as well as inside the seafood, such as oysters. Thus, their concentration can change according to their sites or environments. Generally, it has been reported that there is an increased accumulation of tiny plastic debris in both terrestrial ecosystems and seawater (Vázquez & Rahman, 2021). Consequently, they are found in various parts of the environment (Vivekanand et al., 2021). The principal sources of MP materials, e.g., polymeric plastics, can be from the household discharge of cleaning and cosmetic products, raw materials used in the production of these products, air-blasting media, the plastics results from refuses sites, and industrial drainage systems, as well as they can result from the tire fragments that distribute via the highways. Besides, the synthetic fibers of the clothes, which can be considered an important source of MP debris (Cole et al., 2011; Vivekanand et al., 2021). In addition, the microplastics can be generated from the plastics employed in consumer products. Therefore, this small plastic debris has a high potential to reach the marine environment through wastewater systems, and rivers (Moore, 2008; Thompson, 2006). In fact, plastic debris with a terrestrial origin provides almost 80% of the MP debris present in seawater (Andrady, 2011; Cole et al., 2011). The MPs can reach the marine environment, such as rainwater runoff from surface waters. In addition, due to the occurrence of accidental spills during transportation for both at sea and on land, unsuitable use of many materials, including the packaging materials, and direct flow from processing factories, these raw materials can reach and be transferred to the aquatic ecosystems. Moreover, the agricultural activities can be lead to the generation of various forms of MPs that can be transported and thrown into the aquatic environment (Alprol, Gaballah, & Hassaan, 2021). Fig. 1 shows the different source of MP particles and their transport until the aquatic environment.

Marine microplastics are present in the sediments, the water column, and the sea surface (Andrady, 2011; Smith et al., 2018). These marine little plastics are found generally in oceans worldwide, such as even in Antarctica (Barnes, Galgani, Thompson, & Barlaz, 2009; Vázquez & Rahman, 2021; Zarfl & Matthies, 2010). The changes in the climate and the global warming as a result of human actions induce the fusion of Arctic Sea ice and therefore the latter can liberate the trapped MPs into the marine and aquatic environment (Geilfus et al., 2019; Obbard et al., 2014).

However, marine MPs can be generated either from indirectly or directly throwing of wastes and their transfer to the seas and oceans (Ryan, Moore, van Franeker, & Moloney, 2009). Thus, the marine MPs can possess either indirect or direct sources by which they can reach the aquatic environment (Cole et al., 2011). Besides, the MPs can be also generated from the biological processing (Vivekanand et al., 2021). Moreover, several analysis studies,

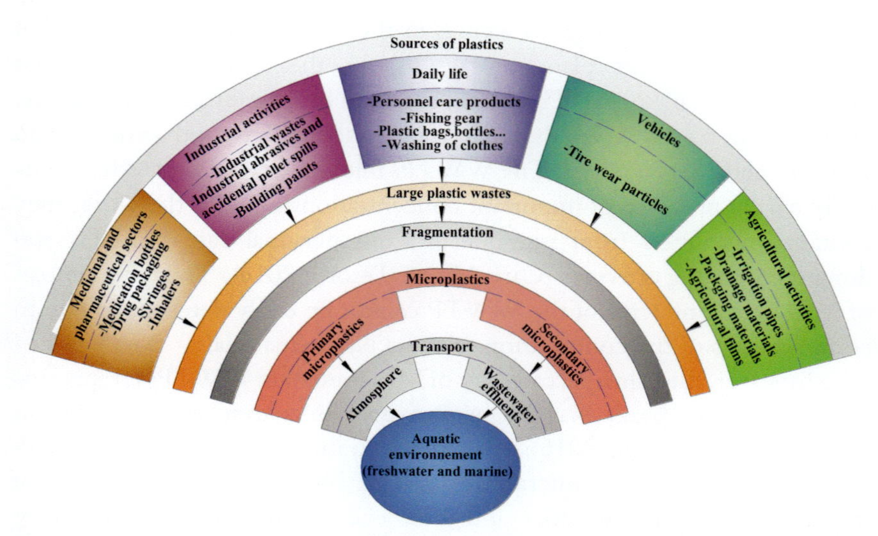

Fig. 1 Source of micro-plastic particles and their transport and disposal in aquatic environment.

realized in the period between 1950 and 2015, have revealed that the amount of overall types of MP materials, such as synthetic fibers, polymer resins, and additives, etc., has been estimated in the range of 8300 million metric tons (Mmt) (Vázquez & Rahman, 2021).

Furthermore, the principal part of micro-plastic amounts is disposed directly into the seawater and oceans in the form of effluent from industrial evacuation and wastewater processing factories (Vivekanand et al., 2021). Thus, the small plastic debris can be emerged on the surface of the water (Vivekanand et al., 2021). The micro-plastic debris arrived to the marine environment can be resulted from different sources, plastic pellets, fishing equipment, and the paints of the constructions. Commercial and recreational fishing, the coastal tourism, marine industries (such as oil-rigs and aquaculture, etc.), and marine vessels are considered as an important sources of the production and the direct transfer of MP debris to the marine environment (Cole et al., 2011). In fact, the marine vessels have been considered as an important contributor to the marine refuses and wastes (Cole et al., 2011). Thus, approximately 23,000 tons of plastic material was thrown away by the international commercial fishing fleet in the 1970s. (Cole et al., 2011). During 1990s, the amount of plastic materials dumped into the oceans was 6.5 million tons (Derraik, 2002).

Besides, the MPs are found in both ordinary water cycle and in anthropogenic water cycles owing to their ubiquitous properties. The MPs are present in domestic sewage as well as in the entrance and the way out of wastewater treatment plants (WWTPs) in the anthropogenic cycles (Vivekanand et al., 2021). The MPs debris can be transferred to the aquatic environment by the WWTPs due to their connection to both ordinary and anthropogenic water cycles. The WWTPs inlet has been characterized generally by the presence of important concentrations of MPs, therefore, they contain microfibers from microbeads from personal care products and domestic washing activities (Carr, Liu, & Tesoro, 2016). Thus, the sources of MPs in the influents of the WWTP cannot be easily determined due to complications of the composition of these plastic materials.

Moreover, the principal category of MPs present in WWTPs is represented by the microbeads that has been existed in marine water (Kärrman, Schönlau, & Engwall, 2016). The concentration of microbeads as MPs in the range of 94.500 microbeads can be liberated in the influents of the WWTPs from cosmetics and after every wash from personal care (Vivekanand et al., 2021). Thus, the principal type of micro-plastics present in the effluents is represented by PA fibers, the particles of polypropylene (PP), polyethylene (PE), and generic polyester (PES) as well as the polyester fibers, such as polyethylene terephthalate (PET) and PES (Schmid et al., 2021). Besides, the MPs can be represented by acrylonitrile butadiene rubber (NBR) from pipes and gaskets (automotive industry) (Schmid et al., 2021).

Nevertheless, the MP fragments can be transported and transferred by the river systems and arrived until the aquatic environments through either directly or indirectly ways. Thus, they can be resulted from wastewater effluent and transferred to the oceans and seawater (Cole et al., 2011). The unidirectional flux of freshwater systems leads to the displacement of plastic debris into the oceans (Browne, Galloway, & Thompson, 2010; Moore, Moore, Weisberg, Lattin, & Zellers, 2002). Moreover, the recreational activities and tourism are the sources of many MPs thrown away along the seaside resorts and the beaches (Derraik, 2002). The marine plastics detected on the beaches will also appear from different materials transported by ocean and coastal currents (Cole et al., 2011).

The principal sources of MPs generated in the marine environment are represented by fishing equipment (Andrady, 2011). The refuse and waste of fishing equipment, such as nylon netting, and plastic mono-filaments, are generally neutrally floating and therefore can drift to varying depths in the oceans (Cole et al., 2011). Thus, fishing gear or equipment can be degraded and generated various MPs by chemical, physical, and biological processes (Lundebye, Lusher, & Bank, 2022).

Moreover, the manufacture of plastic materials using the particles and small resin pellets, including the nibs, and their raw materials can be considered significant sources of various plastic debris that is transported and dumped in the marine environments (Ivar do Sul, Spengler, & Costa, 2009; Mato et al., 2001). For instance, approximately 2.9 million and 21.7 million pellets have been generated by the US alone, in 1960 and 1987, respectively (Cole et al., 2011).

Furthermore, the several accidental events, e.g., the typhoons and flooding can be considered considerable transporter of MPs and they participate in the distribution of such plastics in coastal marine ecosystems (Hurley, Woodward, & Rothwell, 2018; Wang et al., 2019).

2.2 Classification of microplastics

The microplastics are generally classified into two categories as primary and secondary types (Fig. 2). This classification of the MP particles is based on their origin (Smith et al., 2018; Vázquez & Rahman, 2021).

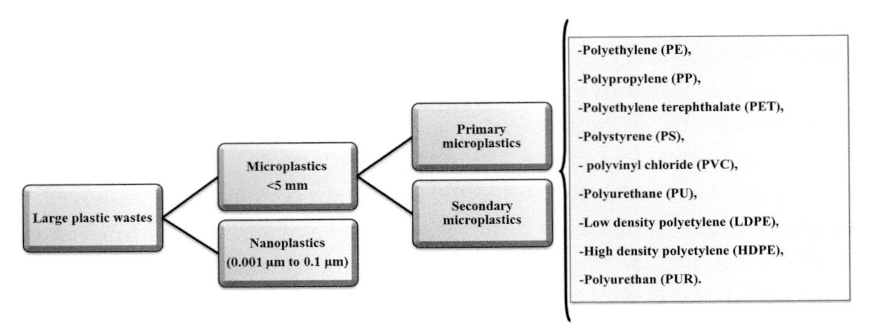

Fig. 2 Classification of microplastic and nanoplastics particles.

2.2.1 Primary microplastics

The primary MPs come from the sources like cosmetics, toothpastes, body washes, facial-cleansers, exfoliating cleansers, packed bed, post-industrial manufacturing plastic residues, industrial pre-production and abrasives of plastic packed bed of air-blasting scrubbers, as well as use medicine as vector for the drugs (Gregory, 1996; Patel, Goyal, Bhadada, Bhatt, & Amin, 2009; Vázquez & Rahman, 2021; Zitko & Hanlon, 1991). In this regard, the plastics that are originally manufactured to be <5 mm in size are identified as primary MPs (Smith et al., 2018; Vázquez & Rahman, 2021). Generally, according to broader significations of sizes related to the primary MPs, the pellets produced from the plastics manufacturing (typically 2–5 mm in diameter) can also be considered as primary MPs (Andrady, 2011; Costa et al., 2010). Besides, the microbeads used in personal care, particularly for scrubbing or exfoliating hand cleansers and facial scrubs, are considered among the primary micro-plastics (Fendall & Sewell, 2009; GESAMP, 2016; Smith et al., 2018). Rochman, Kross, et al. (2015) reported that the approximately 8 billion of microbeads from the USA quotidian were disgorged into the aquatic environment in 2015. Generally, the microbeads are characterized by their variation in size, shape, and composition according to the type of plastic products (Vázquez & Rahman, 2021). Earlier work has revealed that the cosmetic products include the granules of polypropylene and polyethylene (<5 mm) as well as polystyrene spheres (<2 mm) (Gregory, 1996). Besides, the cosmetic products can be characterized also by the presence of irregularly shaped microplastics with diameter of <5 mm and mode size <0.1 mm (Fendall & Sewell, 2009). Moreover, it was revealed by the work of Mason et al. (2016) that the microbeads could be degraded into smaller debris and pellets and they could be released in the US wastewater effluents with an amount in the range approximately of 23 billion MPs particles per day. Furthermore, the washing of synthetic clothing, the manufacture of carpets, and the utilization of ion exchange mediums in the healthcare industry and water purification/softening processes contribute to the dispersion of polystyrene resin beads and fibrous MPs (Browne et al., 2011; Vivekanand et al., 2021).

2.2.2 Secondary microplastics

The secondary MPs are known as the microscopic fragments of plastics derived from the breakdown of larger plastic items occurring on both terrestrial and sea environments (Ryan et al., 2009; Singh & Sharma, 2008; Smith et al., 2018; Thompson et al., 2004; Vázquez & Rahman, 2021). Due the chemical, physical and biological processes, the organizational integrity of plastic fragments can be reduced into smaller fragments by water, wind, sunlight, and other environmental factors (Vázquez & Rahman, 2021). In addition, the secondary MPs have several origins, such as tire dust, microfibers from textiles, and the big plastic particles that decomposed and therefore provide the microplastics, which considered smaller fragments (Duis & Coors, 2016).

Nevertheless, the amount of secondary MPs dumped into the marine environment is ranging from 68.500 to 275.000 tonnes per year (Alprol et al., 2021). The main commonly plastics used are polyvinyl chloride (PVC), polypropylene (PP), polyethylene terephthalate (PET), and polystyrene (PS), which represent approximately 90% of the total plastic generation, on the range of 4% as a charge of plastics additives (Bouwmeester, Hollman, Peters, & Bouwmeester, 2015; Costa & Duarte, 2017).

The lack and damage of the organizational integrity of such plastics can contribute extremely to their fragmentation and degradation resulting from wave-action and corrosion (Barnes et al., 2009; Browne, Galloway, & Thompson, 2007) (Fig. 3). The exposition of plastics to the sunlight for extended period can cause their photo-degradation and thus the obtaining of smaller debris. Besides, ultraviolet (UV) radiation present in sunlight can generate an oxidation of the plastic matrix and thus the obtaining of cleavage of the bonds between the composition of polymer matrix (Andrady, 2011; Browne et al., 2007; Moore, 2008; Rios et al., 2007; Vázquez & Rahman, 2021). Moreover, the plastic particles found on the beaches are characterized by the high oxygen availability and immediately their exposure to sunlight; consequently, they can be rapidly deteriorated (Andrady, 2011; Vázquez & Rahman, 2021). These operations have a highly impacts on the progressive deterioration of plastics and the latter become smaller during a considerable

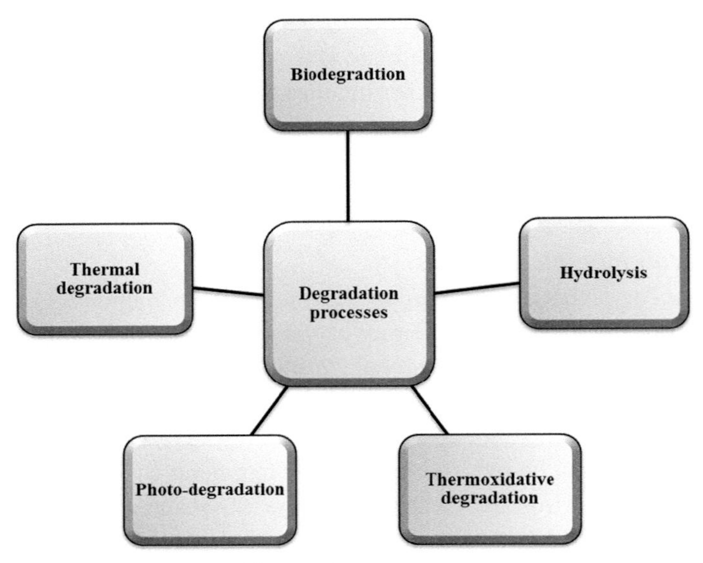

Fig. 3 Degradation processes of large plastic wastes.

period until they become micro-plastics with little size, which can be further degraded and become NPs in size (Vázquez & Rahman, 2021). In fact, the smallest MPs found and detected in the aquatic environment are characterized by diameter size of 1.6 μm (Galgani et al., 2010). Generally, biodegradable plastics, such as TDPA (Totally Degradable Plastic Additives), are consisted of vegetable oils, starch, synthetic polymers, and other chemical compounds; that is why the industrial composting plants rapidly degrade them under excellent conditions (humidity, temperature, aeration, etc.) (Derraik, 2002; Moore, 2008; Ryan et al., 2009; Thompson, 2006; Thompson et al., 2004). However, this decomposition is considered partial since the amount of synthetic polymers is hard to be deteriorated and they will be remain for long periods while only the starch and other biological compounds of the bio-plastics, which can be rapidly undergone deterioration (Andrady, 2011; Vázquez & Rahman, 2021). In addition, once the decomposition complete of the terrestrial MPs particles occurred, they will be dumped into the marine environment. Nonetheless, the decomposition period of bio-plastics differs according to the environmental conditions. The decomposition of the components of bio-plastics present in the marine environment will be prolonged due to the absence of terrestrial microorganisms and thus they will be accumulated and subsequently contaminated the aquatic ecosystems owing to the decrease of the penetration of UV on which the decomposition process depend on (Andrady, 2011; Vázquez & Rahman, 2021). Thus, even the plastic generation has been stopped and prevented the plastic residues dumping, the aquatic micro-plastics would continue to augment because of the macro-plastics decomposition into secondary MPs.

2.3 Nanoplastics (NPs)

The nanoplastic particles are characterized by size ranges from 1 to 100 nm (0.001–0.1 μm) (Alprol et al., 2021; Andrady, 2011; EFSA, 2016) (Fig. 2).

The NPs found in the marine ecosystems present miscellaneous sources. Therefore, the principal sources of such plastic samples are represented by the industrial applications and their products used in various processing (Alprol et al., 2021). Besides, several nanoparticle applications employed in the drug industry are characterized by the presence of polymeric nanoparticles, nanocapsules, and nanospheres (Guterres, Alves, & Pohlmann, 2007). Moreover, the nanoparticle may have also originated the use of the solid lipids, which known as biodegradable compounds and other products, such as the materials of medical diagnostics, biomedical products, adhesives, waterborne paints, re-dispersible lattices, coatings, the products of magnetic, electronic products, and optoelectronics. Thus, the NPs particles are emitted into the environment during the life cycle. Furthermore, the decomposition and degradation of the MPs particles with size diameter down to the <100 nm scale can be considered as another source of NP particles (Alprol et al., 2021).

However, earlier works proved that the thermal severing of polystyrene foam liberated the polymer particles with nanometer-size (Alprol et al., 2021; Zhang, Kuo, Gerecke, & Wang, 2012). The use of electro-pinning for designing plastics can provide the mats with nano-scale fibers, which can be decomposed further to provide the nano-scales. In fact, the biodegradable plastics can be used to replace traditional plastics (Thompson et al., 2004; Vázquez & Rahman, 2021). Hence, the exploitation of biodegradable plastics can be considered a solution to prevent the pollution of the environment.

The amount of the nano-plastics found in the marine ecosystems has known an elevation in the last years and thus they have negative impacts since they are considered pollutants on the base of the marine foods (Vázquez & Rahman, 2021).

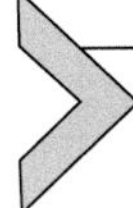

3. Physicochemical characteristics of marine micro/nanoplastics

Generally, various techniques have been used for the determination of the characteristics of both MPs and NPs particles (Singh et al., 2022). The different techniques used to evaluate the physicochemical properties of these material samples included the optical, chromatography, and spectroscopic analysis and therefore they can contribute to the determination of the quantitative and qualitative characteristics of such type of plastics (Mintenig, Bäuerlein, Koelmans, Dekker, & van Wezel, 2018).

In addition, the determinations of the size, density, and hydrophobicity properties of MPs and NPs are the principal and basic parameters to characterize them after their collection and pre-separation process (Singh et al., 2022). In fact, some methods identified can be applied for the separation of both MPs and NPs particles, such as gel electrophoresis, magnetic field flow fractionation (MFFF), and size-exclusion chromatography (SEC) (Nguyen et al., 2019; Robertson et al., 2016; Singh et al., 2022). Moreover, the characterization and quantification of M/NPs lead to the determination of the visual properties related to the morphology, including color, size, and shape, using the scanning and transmission electron microscopy and fluorescence microscopy (Nguyen et al., 2019; Singh et al., 2022).

Generally, the determination of the particle size, color, and shape characteristics are the primary step for the MPs and NPs particles screening in the detected environmental samples, which represent the physical characteristics of these particles (Nguyen et al., 2019; Zhao, Wei, Dong, Zhao, & Wang, 2022). The visualization is a significant method for the identification and distinction of the MPs in debris or sample materials, such as glass, sand, shell fragments, and algae (Fu, Min, Jiang, Li, & Zhang, 2020; Hidalgo-Ruz, Gutow, Thompson, & Thiel, 2012). The large MPs can be identified by the use of the naked eyes or also using an optical microscope (Fu et al., 2020). Moreover, the identification and distinction of micro-plastics from non-plastics can be done with the determination of the shape, color, and light transmission owing to the variability of these parameters between the non-plastic and plastic particles (Fu et al., 2020; Song et al., 2015).

MP particles can be divided into six classes based on their diameters, <50; 50–100; 100–200; 200–500; 500–1000; and $>1000\,\mu m$ (Zhao et al., 2022). The diameter size $>500\,\mu m$ can be detected and identified by naked eyes or optical microscopy (Song et al., 2015). The images of MPs particles by the use of optical microscopy can contribute to the analysis of their composition and structure (Zhao et al., 2022). The total number of particles per cubic meter for the MPs samples present in the wastewater can be determined by supporting different and all images for each of MPs particles and for each sieve fraction by means of polarized light optical microscopy (PLOM) (Fu et al., 2020; Rodríguez Chialanza, Sierra, Pérez Parada, & Fornaro, 2018; Sierra et al., 2020). In fact, the MPs particles found in the wastewater are characterized by the existence of an elevated number of suspended from both mineral and organic origins (Sierra et al., 2020). The PLOM has revealed that the majority of MPs particles present in the wastewater samples are range from 70 to 600 μm in Montevideo, Uruguay (Sierra et al., 2020).

Nevertheless, the micro-plastic particles derived from consumer products can be characterized by different color, such as white, blue, green, and purple, etc. The polypropylene (PP) fibers as a type of MPs can be characterized by their typical color: red or blue (Vianello et al., 2013). Therefore, the MPs characterized by milky white color, brightness, and slight shine can be easy to distinguish and identify them with naked eyes (Fu et al., 2020; Song et al., 2015). However, such sample materials can be sometimes confused with some minerals, for instance perlite (Fu et al., 2020). The majority of MPs samples have been characterized by the dominance of the white pellets, which followed by limpid, and colored pellets (Heo et al., 2013).

The overall characteristics of MPs particles, such as their gloss, color either unnatural or brightness; geometry properties; shape properties; hardness, elastic consistency, etc., can be investigated scanning electron microscopy (Shim, Hong, & Eo, 2017; Silva et al., 2018). Therefore, the combination of spectroscopic methods with visual examination can be valuable to determine the properties related to the MPs particles.

The combination of the scanning electron microscopy (SEM) with energy dispersive X-ray spectroscopy (SEM-EDS) are generally employed for the determination of the basic composition and the morphology of ultra-small materials (Goldstein et al., 2017). The differentiation and identification of the MPs from other sample materials can be determined by the use of SEM, which can provide high-resolution topography images of objects (Cooper & Corcoran, 2010). Besides, the energy dispersive X-ray spectroscopy (EDS) can contribute to the determination of elemental composition of the MP samples through the characteristic X-rays transmit from the elements present in the sample tested by the electron beam (Fu et al., 2020). Therefore, these techniques can contribute to the identification and characterization of MPs particles in sample matrixes (Fu et al., 2020).

The determination of the physicochemical characteristics of microplastics can also be investigated by the use of optical microscopes that develop an image contrast from light reflection on the surface of the tested samples (Fu et al., 2020). The fluorescence microscopy is used chemical reagents for the investigation of different characteristics of the MPs particles, and thus they can be useful to detect the presence of such plastic samples in the sediments from the aquatic environment (Qiu et al., 2015; Zhao et al., 2022). Qiu et al. (2015) showed that the MPs particles characterized by their brightly colored by the use of fluorescence microscope. In fact, the fluorescence microscopy collected the fluorescent signals from MPs samples, which excited by the stimulation wavelengths with appropriate selection of the filter lasers or cubes (Fu et al., 2020; Lichtman & Conchello, 2005). The surface of MP samples is absorbed by fluorescent dyes, giving them fluorescence (Zhao et al., 2022). Hence, the particles of the MPs samples can be identified and counted their number by image-analysis. Therefore, it is possible to detect and determine the diameter of plastic particles, which are known by their little size ranges to a few micrometers or smaller (Zhao et al., 2022). Besides, the use of the two methods of fluorescent dye combined with the density-based separation lead to the visualization of MPs particles with different size ranges.

The identification techniques used to characterize both the MPs and NPs are represented by Fourier-transform infrared spectroscopy (FTIR), optical microscopy, and Raman spectroscopy. In fact, the FTIR and Raman are considered spectroscopy processes that lead to determine the functional groups related to the different molecules in the analyst samples. Each technique possesses typical vibrational bands that make them useful for the identification. It is proven that the utilization of FTIR can contribute to the distinction and identification of the MPs/NPs particles from other debris. Besides, the staining of plastic particles with a lipophilic fluorescent dye, like Nile Red (NR), can facilitate their visualization under a microscope and hence, it quantifies and identifies their diverse polymer particles. However, NR can be incoherent and is not plastic-specific. Therefore, other studies need to be carried out to further understand the interactions between plastic and fluorescent dye and provide more information about the selected appropriated dye for the determination of plastic particles. (Erni-Cassola, Gibson, Thompson, & Christie-Oleza, 2017; Maes, Jessop, Wellner, Haupt, & Mayes, 2017; Nguyen et al., 2019).

However, the characterization of both marine MPs and NPs particles can be established by the use of dynamic light scattering (DLS), leading to the determination and investigation of the hydrodynamic size and zeta potential measurements in liquid environment (Fu et al., 2020). In fact, the DLS used a laser beam across a liquid suspension, including the analyst particles that disperse the incident laser at various scattering angles. The DLS has been extensively used in biological, physical and chemical fields (Xu, 2015), especially in the analysis of colloidal characterization of microparticles and nanoparticles. Therefore, this technique can contribute to the characterization both marine MPs and NPs. The DLS can be applied to determine and measure the size and the zeta potential of the MPs and NPs in biological matrices, for instance, in the planktonic crustaceans, Pacific oyster for the investigation of accumulation of such plastic particles (Gambardella et al., 2017; González-Fernández et al., 2018; Summers, Henry, & Gutierrez, 2018). Besides, an appropriate investigation and determination of the degradation of MPs can be provided by the application of DLS method. The DLS technique combined with photo-reactor can analyses the photo-degradation of marine MPs particles under various conditions without sampling (Fu et al., 2020; Gigault, Pedrono, Maxit, & Ter Halle, 2016). However, the application of a DLS technique for the study of food matrix cannot provide a valuable differentiation and distinction between the MP and NP and the non-plastic particles due to the complexity of these systems and the dispersion of laser beams and light (EFSA, 2016).

Overall, both MP and NP particles are characterized by similar chemical composition. However, the principal difference between these materials is related to their size ranges. In fact, this difference of the size diameters of these plastic samples is considered responsible for the transport of NPs particles, with smaller size, to far distant zones as compared to MP particles (Singh et al., 2022). Besides, under the impact of decomposition, the MPs are decomposed into smaller fragments and then they can have the shape of fibers, while the NPs cannot have a respective length of fibers, but they characterized by the shape of a sphere, granules, and fragments (Singh et al., 2022).

The motion in the water of the two types of plastics samples is not comparable. In fact, the MPs are characterized by vertical motion in the water owing to their density, however the NPs particles are relatively small and exhibit Brownian motion (Singh et al., 2022). In addition, the differences in genesis among various samples of plastics lead to their classification into primary and secondary MPs as well as NPs. Thus, MPs and NPs are produced with the appropriate size and diameter ranges for the purpose of their suitable usages (Singh et al., 2022).

4. Degradation of micro/nanoplastics in marine conditions

Generally, the degradation is considered a chemical reaction, which leads to the reduction of the size molecular of the samples. The polymers can undergo decomposition under the impact of the environmental conditions and therefore they exposed to reduction of their size and molecular weight (Andrady, 2011). Besides, the degradation processes of MPs and NPs can be chemical or biological processes by which the structure and properties of the polymers modified (Mofijur et al., 2021). The degradation mechanism of various polymers differs depending on their chemical structures (Mofijur et al., 2021). Therefore, the degradation of plastic or polymer samples contributed to the occurrence of various alterations, such as the generation of powder particles as well as the augmentation of brittleness of these materials. These plastic fragments can be visible to the naked eye (Mofijur et al., 2021). In addition, the secondary M/NPs are produced by progressive abiotic and biotic degradation of primary plastic samples during the time-frame (Singh et al., 2022).

The mechanisms of degradations can be classified into different categories according to the type of factors causing it. These mechanisms can be classified as (1) biodegradation, which can be owing to usually the action of microorganisms; (2) thermal degradation that can be due to the influence of high temperature; (3) the thermo-oxidative degradation, which considered slow oxidative deterioration under moderate temperature and could be a slow oxidative deterioration (Alprol et al., 2021; Arjula, Harsha, & Ghosh, 2008); (4) hydrolysis that can be owing to the action of water; (5) photo-degradation technique, which can be due to the impact of light (generally the exposition to sunlight is considered an effective mechanism for the decomposition of plastics both in ashore and air, although this technique of degradation in the marine environment is considerably retarded) (Andrady, 1994, 2011); and (6) finally these degradation mechanisms can be because of other factors that is not environmental. Therefore, these processes affect the chemical structure of the polymer fragments to break down into smaller molecular weight particles (Andrady, 2011). In fact, several degradation processes can appear for a specific polymer pending the total degradation process (Mofijur et al., 2021).

Moreover, the ultraviolet (UV) radiation mechanism has an appropriate energy to cut the bonds of carbon-carbon although the effectiveness of this type of degradation technique depends on the chemical structure of polymer as well as the wavelength of the light (Andrady, 2015).

In the marine ecosystem, the environmental conditions can affect the degradation process of plastic polymers (Mofijur et al., 2021). Several research studies have investigated the degradation processes of various plastic polymers, such as the polyethylene (PE) netting, nylon monofilament, and fishing gear, etc., in the marine ecosystems (Brinkhof et al., 2021; Deng, Dong, Zhang, Zhao, & He, 2021; Dong, You, & Hu, 2020; Figueroa-Pico, Tortosa, & Carpio, 2020; Ibrahim et al., 2018; Mofijur et al., 2021; Suzuki, Tachibana, Oba, Takizawa, & Kasuya, 2018; Yoshida et al., 2019; Zou et al., 2022). The plastics samples, including the nylons, high-density polyethylene (HDPE), polypropylene (PP), low-density polyethylene (LDPE) can be exposed first to photo-oxidative degradation, essentially by the action of UV-B radiation present in sunlight (Andrady, 2011). Additionally, the particles of these polymers can be exposed to the thermo-oxidative degradation. As long as the oxygen presents in the marine ecosystem, the polymers can expose to auto-catalytic degradation. Hence, the polymers can be decomposed and their molecular weight decreased as well as some oxygen functional groups are produced in these polymers. Moreover, hydrolysis as a degradation mechanism of the plastics samples is considered generally a usual process in the aquatic environment owing to the impact of water (Andrady, 2011; Mofijur et al., 2021). The degree of degradation based on the category of degradation of the polymer samples, such as the degradation induced by the light is considerably rapid compared to that of hydrolysis and the other degradation processes as well (Mofijur et al., 2021). Furthermore, the degradation of polymers with a high molecular weight is considerably slow, particularly in the aquatic environment (Mofijur et al., 2021).

All plastics begin to deteriorate from the moment they are produced. Thus, the degradation of these polymers in marine ecosystems is dependent on their chemical compositions and other factors. The plastic samples continue to deteriorate not only under the influence of solar radiation, but also activity of microorganisms. However, the degradation by solar UV radiation of plastic samples is fastest and most efficient when these materials are situated on the beach surface or exposed to the air compared to the plastic exposed to the sunlight and floating in the aquatic environments (Andrady, 2011; Resmeriță et al., 2018). The retardation of the degradation of plastic samples, which are floating in the marine environment, is generally due to the relatively lower oxygen concentration and temperature in water (Andrady, 2011). Besides, the floating plastics will immediately develop considerable surface contamination, quickly covering the surface firstly with a biofilm followed by an algal carpet and then a colony of invertebrates (Muthukumar et al., 2011). Moreover, the degradation mechanisms of plastics under the impact of microorganisms has differed from those induced by the oxidative degradation under the action of light (Andrady, 2011; Mofijur et al., 2021).

The degradation of miscellaneous polymer samples can generally be known as bio-degradation (Ahmed, Hall, & Ahmed, 2018a, 2018b). The bio-degradation of plastic samples is due to the microbial digestion mechanism under the impact of bacteria, algae and fungi (Schmid et al., 2021). All the bio-based plastics, which known as polymers of biological origin, cannot be biodegradable, while the fossil-based plastics, which made from petrochemicals, can be biodegradable (Schmid et al., 2021). The degradation of these polymer samples can contribute to the generation of CO_2 due to the microbial bio-decomposition. The CO_2 liberated are then consumed by marine biomass. In fact, the process of the whole consumption of carbon found in the plastic polymers is considered a total mineralization (Eubeler, Zak, Bernhard, & Knepper, 2009; Mofijur et al., 2021). Besides, the microbial cells can adapt and generate the CO_2, H_2O, and biomass compounds in aerobic conditions as well as the production of CO_2 and CH_4 in anaerobic environmental conditions (Tokiwa, Calabia, Ugwu, & Aiba, 2009).

In addition, the plastic polymers have heteroatoms (N, S, and O) as extended along their backbone, and in the biodegradable process. This chemical structure can perform the occurrence of the enzymatic and hydrolysis reactions, and thus these mechanisms can cause significantly in the reduction of the molecular structure of the polymer fragments in a short period (from days to several years) (Matsusaki, Kishida, Stainton, Ansell, & Akashi, 2001). In fact, the biodegradability based on both the chemical structures of the different plastic polymers and their physical properties (crystallinity, melting point, storage modulus, and glass transition temperature etc.) (Tokiwa et al., 2009).

Many strains of microorganisms are capable of producing the bio-degradation of PVC (Peng et al., 2020; Shah, Hasan, Hameed, & Ahmed, 2008) and the polyethylene (Sivan, 2011). It was reported that the *Actinomycetes Rhodococcus ruber*, which was strain C208 led to the decrease of 8 % of the dry weight of the polyolefin during 30 days of an incubation period in a liquid solution in a laboratory setting (Andrady, 2011; Mofijur et al., 2021; Orr, Hadar, & Sivan, 2004). Hence, this species secreted the compound of laccases, which could decrease the rate of molecular weight of plastic polymers and thus it proved their degradation through the separation of the main chains (Andrady, 2011). However, this process does not occur in the soil or aquatic environment, as the microbe strain is not present in sufficient concentration. (Andrady, 2011; Mofijur et al., 2021). In addition, the interaction established between MPs and the biota is distinct as the microorganisms have the ability to develop as biofilms on them; however, the NPs with smaller size compared to the microorganisms do not assist the development of later on them (Levin & Angert, 2015).

The Mater-Bi, which known as biodegradable plastic that constitute on vegetable polysaccharides, can be decomposed and degraded very slowly under the action of algae in the seabed (Balestri, Menicagli, Vallerini, & Lardicci, 2017). Besides, the poly butylene adipate co-terephthalate (PBAT), which recognized as biodegradable plastics, has a higher sorption and desorption ability of environmental pollutants, such as the phenanthrene, compared to the PS, and PE (de Oliveira et al., 2019). Generally, several plastics of biological origin can be degraded in particular composting places only, although in the marine environment their degradation is long and slow, and therefore can lead to the quantification of MPs (Cole et al., 2011). Moreover, the United Nations Environment Program report (UNEP, 2015) has assumed that the use of biodegradable plastics can decrease both the concentration of MPs in the seawater and the danger of chemical and physical effects on the marine ecosystems.

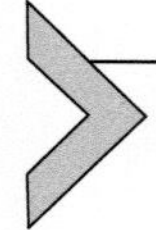

5. The risk of marine micro/nanoplastic on seafood safety

5.1 Toxicity effects of micro/nanoplastic ingestion and exposure

In between 1980 and 1990, the toxic impacts of several MPs and NPs particles of varying characteristics and sizes have been investigated and evaluated (Chae & An, 2017). In fact, these pollutants have become ubiquitous and widespread although their biological effects on organisms in the aquatic ecosystems have been studied in the last years (Barnes et al., 2009; Gregory, 1996; Ryan et al., 2009). Table 1 tabulates some research studies regarding the investigation of the impacts of micro/nanoplastic particles on seafood products. The particles of MPs and NPs suspended in the marine environment can be easily ingested by several marine biota due to their size ranges is similar to those of fish eggs (Boerger, Lattin, Moore, & Moore, 2010; Browne, Dissanayake, Galloway, Lowe, & Thompson, 2008).

Table 1 Overview of some research studies concerning the negative impacts of the MPs and NPs on some seafood products.

Seafood products	Seafood species	Micro/ nanoplastics	Concentrations	Major findings	References
Fish	*Pomatoschistus microps*	Micro-plastic: polyethylene (PE)	0.184 mg/L	– The micro-plastic fragments have a negative impact on the predatory performance of this fish species. They induce an alteration of the uptake and also caused toxic effects when they combined with nanoparticle of Au-NP	Ferreira, Fonte, Soares, Carvalho, and Guilhermino (2016)
	Danio rerio (zebrafish) "embryo"	Nano-plastic: polystyrene	0.1 mg/mL	– The nano-plastic of polystyrene caused the developmental abnormalities, marginal effects on the survival, the death of the Zebra fish cells, and hatching rate	Lee et al. (2019)
	Danio rerio (zebrafish)	Micro-plastic: polyethylene (PE)	10, 100, or 1000 beads/mL	– The PE as microplastic particles can induce adverse impacts which represent, by the reduction of the predatory performance via the decrease of the uptake of the feeds	Khan, Syberg, Shashoua, and Bury (2015)
	Danio rerio (zebrafish)	Micro-plastic: polyethylene (PE)	5–20 mg/L	– The PE can exhibit reproductive toxicity, gastrointestinal toxicity, and genotoxicity	LeMoine et al. (2018)
	Misgurnus anguillicaudatus	Microplastic: polyvinyl chloride (PVC)	50 mg/L	– The PVC induces the generation of the oxidative stress via the augmentation of the SOD and MDA, and the increase of the toxicity of venlafaxine	Qu et al. (2019)
	Oreochromis niloticus (tilapia)	Micro-plastic: polystyrene (PS)	1, 10, and 100 μg/L	– The micro-plastic can induce an oxidative stress. Thus, these pollutants can affect the metabolism of roxithromycin (ROX) in tilapia	Zhang, Ding, et al. (2019)

	Spaurus aurata	Nano-plastic: Poly (methyl methacrylate), Polystyrene	0–10 μg/mL	–	The poly (methyl methacrylate) can cause the alteration of the lipid metabolism pathways, increase the erythrocytic nuclear abnormalities, increase the cholesterol and triglycerides in plasma, and raise the genotoxicity in blood cells	Brandts et al. (2021)
	Spaurus aurata	Nano-plastic: polystyrene (PS)	0.001–10 μg/mL	–	The PS can induce the reduction of cell viability (25%) at the concentration of 0.001 mg/mL	Almeida, Martins, Soares, Cuesta, and Oliveira (2019)
Molluscs	*Mytilus galloprovincialis*	Nano-plastic: polystyrene (PS)	0.05 mg/L	–	The exposure of this species of molluscs to the nanoplastic of PS can induce a DNA damage, downregulation of the gene responsible for the biotransformation, and alteration of cell-tissue repair	Brandts, Teles, Gonçalves, et al. (2018)
	Mytilus edulis	Nano-plastic: polystyrene (PS)	100, 200 and 300 μg/mL	–	The nano-plastic can induce a production of pseudo-feces with the increase of the concentrations of these pollutants	Wegner, Besseling, Foekema, Kamermans, and Koelmans (2012)
	Perna viridis	Microplastic: polyvinyl chloride (PVC)	0, 21.6, 216, 2160 mg/L	–	The PVC can induce reproductive toxicity, gastrointestinal toxicity, and growth inhibition toxicity	Rist et al. (2016)
Crustaceans	*Amphibalanus amphitrite*	Nano-plastics: poly(methyl methacrylate)	25 μg/mL	–	The exposure of this species of crustaceans to the poly(methyl methacrylate) can generate a bioaccumulation of these particles in the organism's body	Bergami et al. (2017)
	Trigriopus japonicus	Nano-plastic: polystyrene (PS)	0.125–25 μg/mL	–	The nanoplastic of PS can induce embryo malformations, mortality, and decrease of the fecundity	Lee, Shim, Kwon, and Kang (2013)

The contamination of the aquatic species by these contaminant particles can be made by trophic transfer or by direct ingestion (Toussaint et al., 2019). Besides, the size characterized the fragments of both MPs and NPs ingested by marine biota is more significant compared to their shape (Lehtiniemi et al., 2018). Thus the sub-lethal and lethal effects of the MPs and NPs have been studied on the marine life, particularly on aquatic organisms, such as crustaceans, fish, shellfish, and other seafood (Chae & An, 2017). The ingestion of the particles of these materials by the edible marine animals can lead to their death (Alprol et al., 2021; Andrady, 2011). Hence these pollutants can provoke a hazard to the biota due to their little size, which make them accessible to a several category of marine organisms, and thus, cause a major scientific concern (Derraik, 2002; Thompson et al., 2004; Ng & Obbard, 2006; Betts, 2008; Lozano & Mouat, 2009; Barnes et al., 2009; Fendall & Sewell, 2009; Mofijur et al., 2021). In fact, the direct entrance of M/NP particles into the marine environment can cause their penetration and accumulation in the body of a wide variety of marine organisms through their ingestion (Gall & Thompson, 2015; Smith et al., 2018; Vivekanand et al., 2021). The analysis during the examination of the seafood products revealed the accumulation of the plastic fragments in their digestive tract (Smith et al., 2018). Thus, the little fragments, such as M/NP particles can persist in these marine biota and translocate from the intestinal tract to the surrounding tissue and the circulatory system (Duis & Coors, 2016; Lusher, Hollman, & Mendoza-Hill, 2017; Murray & Cowie, 2011; Van Cauwenberghe & Janssen, 2014). Therefore, these processes can cause a major risk and damage to the safety of seafood via the transition through the food chain (Singh et al., 2022). In addition, the trophic chain is considered a considerable reaction among the marine organisms, such as the seafood, crustaceans, shellfish, fish, shrimp, and bivalves and consequently, the different forms of both micro and nanoplastics can be ingested by these living organisms (Blight & Burger, 1997; Cole et al., 2011; Tourinho, Ivar do Sul, & Fillmann, 2010; Vivekanand et al., 2021). The polychlorinated biphenyls (PCBs) present on the surface of several types of plastic particles can be transferred to the marine biota through the trophic chains from various sources of plastics (Graham & Thompson, 2009). In fact, the ingestion of these material particles by microbiota can cause several problems (Andrady, 2011). In this regard, the ingestion of the micro/nano-sized plastics have a mechanical danger to marine biota due to their impact on the block of the transit of foods through the intestinal tract or provoke the pseudo-satiation, which can reduce the food intake (Barnes et al., 2009; Derraik, 2002; Fendall & Sewell, 2009; Thompson, 2006; Tourinho et al., 2010).

In addition to the potential unfavorable impacts and toxicity due to the ingestion of these contaminants, other negative effects can be resulted from the adherence of other external chemical pollutants to these fragments of M/NP (Mofijur et al., 2021). The investigation related to the impact of the fluorescent PS microspheres on cell viability of *Mytilus edulis* demonstrated that these little particles were accumulated easily in the tissues and cells of alive organisms (Browne et al., 2008). In fact, the *M. edulis* can ingest an amount of MPs ranging from 2 and 4 μm through the inhalant siphon and these pollutants can be transported until the labial palps in order to their digestion after their filtration by the gill of this marine organism (Cole et al., 2011). These contaminants are detected after exposure up to 480 days in the body of mussels, particularly in the circulatory system (Browne et al., 2008; Cole et al., 2011). Therefore, these pollutants cause some negative impacts on the living marine biota. Besides, the exposition to these pollutant particles can cause various damage to the blue mussels, especially in their immune response of this category of seafood. These particles can lead to the formation of granuloma in their digestive glands (Köhler, 2010).

Nevertheless, the lower-trophic level aquatic organisms are capable of ingesting both MPs and NPs since they do not have the ability to distinguish between the food and plastic particles (Moore, 2008). The toxicity of microplastics differs significantly from that of nanoplastics and thus can be due to differences in the numbers of the plastic particles as well as other factors, which can induce the toxicity (Chae & An, 2017). In addition, Oliviero et al. (2019) reported that the different colors characterizing these pollutants are due to the action of various additives, and therefore the latter may have different effects on biological systems.

Carpenter, Anderson, Harvey, Miklas, and Peck (1972) revealed that the analysis of the contents of the gastrointestinal tract of the species of larval fish has proven that this marine biota can ingest the plastic particles. In addition, Boerger et al. (2010) reported the presence of the micro/nanoplastics particles in the contents of the stomach of 35% of fish that had the plankton. Besides, another analysis study reported that the catfish caught from Brazilian estuaries had approximately 33% of the plastic fragments in their stomachs (Possatto, Barletta, Costa, Do Ivar, & Dantas, 2011).

The toxic impacts of the ingestion of the M/NPs are represented by the occurrence of physical deterioration, obstruction of the intestinal tract, which leads to the sensation of deceptive satiation, and thus, the starvation which can be potentially lethal (Oliveira & Almeida, 2019; Oliveira, Almeida, & Miguel, 2019). Besides, the ingestion of the M/NPs by the seafood products leads to the risk of their bio-magnification and bioaccumulation along with food chain with potentially several negative effects (Strafella et al., 2019). The ingestion of these pollutants can be occurred directly, when the edible marine organism ingested these samples (such as the particles of the balloons and plastic bags) due to similarity with their preys and the gelatinous animals, or when filtering their food (such as the particle beads), or indirectly when these marine biota feed on preys containing the fragments of plastics adhered to their surfaces (Rizzi et al., 2019). In this context, the fishes, can directly ingest the M/NP particles or indirectly by their alimentation on contaminated preys (Soares, Miguel, Venâncio, Lopes, & Oliveira, 2020). Besides, the exposure of the seafood products and other marine biota to the M/NP and their ingestion can contribute to the occurrence cell deterioration, malfunction of immune systems, and genotoxicity (Soares et al., 2020).

Moreover, amine-modified polystyrene nanoparticles as nanoplastic particles can lead to the occurrence of an oxidative damage (Mofijur et al., 2021). The exposure to the amino-modified polystyrene nanoparticles and their ingestion by the marine biotas cause a high level of toxicity in these marine organisms, and thus, can generate a genotoxicity, embryotoxicity, and the alteration of their gene expression (Brandts, Teles, Tvarijonaviciute, et al., 2018; Della Torre et al., 2014; Mofijur et al., 2021). The exposure of the marine organisms to the nanoplastic can cause the reproductive failure and the increase of the development of reactive oxygen species (Lenz, Enders, & Gissel Nielsen, 2016). Furthermore, the nanoplastic showed their negative impacts and toxicity when they exposed to several species of seafood, and thus, they revealed the generation of the kidney deterioration, inflammation, as well as alteration of the immune system (Alprol et al., 2021).

The exposure of seafood products to the M/NPs and their ingestion of these pollutants can generate the malregulation of the metabolic pathways related to immune functions and fatty acids metabolism, inducing an oxidative damage, alteration of nervous system and the neuro-transmission lead to the modification on histopathological and change the fish behavior both in their swimming and feeding (Brandts et al., 2021; Brandts, Teles, Tvarijonaviciute, et al., 2018; Luís, Ferreira, Fonte, Oliveira, & Guilhermino, 2015; Oliveira, Ribeiro, Hylland, & Guilhermino, 2013; Pedà et al., 2016; Yin, Chen, Xia, Shi, & Qu, 2018).

5.1.1 The impacts of marine micro/nanoplastic on crustaceans

The gut contents of the planktonic crustacean exposed to 62–1400 μm of the polyethylene terephthalate (PET) microplastics fibers for 48 h. It was found that these marine organisms could ingest an important amount of fibers (approximately 300 μm) and thus caused a deterioration and damage of their tissues (Jemec, Horvat, Kunej, Bele, & Kržan, 2016). Besides, the analysis of the stomach contents of the crabs demonstrated that this category of seafood could ingest a considerable amount of MPs particles (Rist, Carney Almroth, Hartmann, & Karlsson, 2018). The chronic exposition of some crustaceans to an important level of MP particles could contribute to the generation of the reproductive and gastrointestinal toxicities and to their mortality in some cases (Au, Bruce, Bridges, & Klaine, 2015; Jemec et al., 2016; Yuan, Nag, & Cummins, 2022). Moreover, the crustaceans, such as the crabs, were characterized by the presence of the edible part known by the hepatopancreas, which considered a vital organ for defence mechanisms (Wang et al., 2021; Yuan et al., 2022). In fact, this organ was considered highly susceptible to the contamination with microplastic (Barrento et al., 2008; Yuan et al., 2022).

Nephrops norvegicus obtained from the Clyde Sea area as an important commercially exploited crustacean contained an important amount of the microplastic fibers in their gut (Welden & Cowie, 2016). In addition, the *Maja squinado* (as spinous spider crabs), prawns, and shrimps considered among the commercially relevant seafood species ingested and accumulated the microplastic in their organisms (Cau et al., 2019; Devriese et al., 2015; Welden, Abylkhani, & Howarth, 2018; Zhang, Wang, et al., 2019). Therefore, several crustaceans obtained from the coastal environments revealed their contamination with the MPs and NPs (Lundebye et al., 2022; Soares et al., 2020). The shrimp "*Aristeus antennatus*" obtained from the Mediterranean and the urban areas contained 39.2 % and 100% microplastic, respectively (Carreras-Colom et al., 2018). Additionally, the ingested microplastic fragments by the two species of crustaceans "*A. antennatus*" and "*N. norvegicus*" were characterized by a significant difference in their composition and size (Cau et al., 2019). Therefore, the *N. norvegicus* could ingest and expose to a higher concentration of microplastic more than *A. antennatus* (Cau et al., 2019). Moreover, according to the laboratory research works on the crabs "*Carcinus maenas*" fed mussels exposed to the microplastics revealed that the polystyrene microspheres could be ingested by this crab species and accumulated in their foregut (Watts, Urbina, Corr, Lewis, & Galloway, 2015). The crustaceans could ingest and accumulate these pollutants in their stomachs and in their entire gastrointestinal tracts as well as in their tissues (Lundebye et al., 2022). The MP and NP were generally found in the digestive system, such as in the intestines, gills, and digestive tubules (Abbasi et al., 2018; Kolandhasamy et al., 2018). For example, the microspheres as microplastic particles could be ingested by the mysid shrimp and thus found in their intestine after 3 h of ingestion (Setälä, Fleming-Lehtinen, & Lehtiniemi, 2014). In fact, the fibers and the fragments of the plastics were considered the main particles of these contaminants accumulated in several crustacean species (Lundebye et al., 2022; McGoran, Clark, Smith, & Morritt, 2020; Welden & Cowie, 2016).

The exposition of the *Eriocheir sinensis* as species of crabs to different concentrations (0–40 mg/mL) of microplastics during 21 days could generate several toxic and negative effects, such as the reduction and inhibition of the activities of AChE and GPT (Ali et al., 2021). Besides, this research work revealed the increase of the activities of SOD, GOT, GPx, and GSH when this seafood product exposed to a high concentration (40 mg/mL). The expression of genes encoding the GPx, glutathione S-transferase, CAT, and antioxidants SOD was inhibited in the liver (Ali et al., 2021).

5.1.2 The impacts of marine micro/nanoplastic on fish

Several studies revealed the negative impacts of the marine MP and NP particles adsorbed by the surfaces of fish or ingested by them. The growth of larvae and juvenile fish was slow compared to the adult fish when exposed to the fluorescent polyethylene as microplastics with a size of diameter 1–5 μm (Ferreira et al., 2016; Steer, Cole, Thompson, & Lindeque, 2017). It was evaluated the contamination with the microplastic of three fish species (with a total of 150 samples), including *Scomber colias, Trachurus trachurus*, and *Dicentrarchus labrax* and they revealed the presence of these pollutants, distributed through the gills (with 36%), gastrointestinal tract (with 35%), and in the dorsal muscle (with 32%) (Barboza et al., 2020). In addition, the research related to the gut contents of *Hemiculter leucisculus,* as a sharp belly fish, obtained from the heavily industrialized area, which characterized by the production of plastics samples, demonstrated the presence of 1.9–6.1 particles/individual (Li et al., 2020). In fact, the ingestion of the MP and NP by the fish could generate a reduction of the cell viability. For example, the evaluation *in vitro* the toxic effect of the nanoplastics on the fish cell lines isolated from *Sparus aurata* (SAF-1) showed the decrease of their viability (Almeida et al., 2019).

The exposition of the oocytes of the rainbow trout (*Oncorhynchus mykiss*) to the bisphenol A (BPA) derived from the microplastics could contribute the disturbance of the energy into them and leads to genotoxic impacts (Aluru, Leatherland, & Vijayan, 2010; Hagger, Depledge, Oehlmann, Jobling, & Galloway, 2006). In addition, the exposition of the fish, particularly the innate immune system, to the nanoplastic could generate an oxidative damage and the liberation of the neutrophil granule contents by the process of exocytosis (Greven et al., 2016).

It was reported that several edible fish species obtained from worldwide showed that they accumulated a considerable amount of microplastic particles in their guts, such as Japanese anchovy (76.6%); yellowfin tuna (23.4%); Pacific chub mackerel (23.3%); horse mackerel and jack (24.5%); skipjack tuna (9.4%); Atlantic herring (8.8%); Atlantic cod (2.8%); and Peruvian anchovy (0.9%) (Lundebye et al., 2022).

5.1.3 The impacts of marine micro/nanoplastic on molluscs

The molluscs showed that their bioaccumulation of the M/NPs particles had many toxic effects (Thiagarajan, Alex, Seenivasan, Chandrasekaran, & Mukherjee, 2021). The exposure of the bivalve mollusc to the microplastic particle could generate immunotoxic effects (Thiagarajan et al., 2021). Besides, the micro-plastic fragments could cause the disruption of the reproduction in oysters as bivalve molluscs and thus the influencing on their offspring (Sussarellu et al., 2016). The exposure of the bivalve mytilus to the NPs and MPs particles could contribute to the reduction of the feeding activity, and phagocytic activity, as well as the generation of cytotoxicity (Mofijur et al., 2021). The bivalves had the ability to accumulate several nanoplastic particles presents in their environment (Gonçalves & Bebianno, 2021). Moreover, the exposure of the freshwater bivalve (*Corbicula fluninea*) to the mixture containing of microplastic particles (with concentrations varied from 0.2 to 0.7 mg/L) and florfenicol (with concentrations varied from 1.8 to 7.1 mg/L) induced several adverse impacts, such as the inhibition of feeding, isocitrate dehydrogenase activity, and cholinesterase activity, as well as the increase of MDA levels and the activity of the antioxidant enzymes (Thiagarajan et al., 2021). Furthermore, the exposure of the Mediterranean mussel (*Mytilus galloprovincialis*) particularly their gills to nanoplastic particle could induce genotoxic effects, which manifest by a DNA damage and considerable downregulation of the genes responsible for biotransformation, and cell-tissue repair (Brandts, Teles, Tvarijonaviciute, et al., 2018). The exposure of hemocytes of *M. galloprovincialis* to nanoplastic fragments during 30 min could induce a reduction of the stability of the lysosomal membrane and an increase of the generation of the toxic oxygen species, leading to rapid cellular deterioration, for instance, the misfortune of filopodia and membrane blebbing (Canesi et al., 2015). In another study, the exposure of bivalve molluscs to nanoplastic particles led to genotoxicity, cytotoxicity, harmful to the function of the reproductive system and immune system (Tallec et al., 2018).

6. Impact of marine micro/nanoplastic on human health

Despite the great interest and numerous research studies relating to the origin, occurrence, distribution, and transport of the MP and NP particles in the marine environment and their toxic and adverse impacts on marine biotas, researchers have only recently started to investigate their potential impacts on human health.

6.1 The human exposure pathways

Humans can be subjected to the harmful impacts of MP and NP particles by a wide range of the toxic compounds (such as water and seafood) present in their environment. The human bodies can be exposed to these particles through the drinking water, airborne dust, and seafood (Vivekanand et al., 2021). In fact, the consumption of seafood is considered one of the main pathways for the human exposure to these pollutants. The seafood is a main source of extremely nutritious foods and its consumption has been recently multiplying (Inanli et al., 2020). Generally, the seafood products are characterized by several functional health-promoting substances, such as n-3 polyunsaturated fatty acids (PUFA), digestible proteins, and vitamins and thus they provide several beneficial advantages for human health (Inanli et al., 2020). However, these products are exposed to contamination with several toxic compounds, like MP and NP particles, which represent a major concern due to their transfer through the trophic chain in the marine environment. These pollutants can be ingested by organisms at the lower trophic level and transferred to higher trophic levels such as fish and therefore eventually reach humans. (Oliveira & Almeida, 2019). The fish and shellfish (including bivalves and crustaceans), as common dietary foods for humans, are often contaminated with the MP and NP particles and can be the principal sources of the transfer of these pollutants to the human diet (Barboza et al., 2018; Oliveira & Almeida, 2019; Smith et al., 2018; Soares et al., 2020; Wakkaf et al., 2020). Hence, it was proven that the humans were exposed to these contaminants through the ingestion of seafood contaminated by these pollutants (Van Cauwenberghe & Janssen, 2014). For example, the commercial bivalves and fish collected from several marine environments was contaminated with higher amounts of synthetic particles, such as fiber (Soares et al., 2020). Besides, Devriese et al. (2015) reported that the consumption only of shrimp provided approximately 175 microplastic fragments per person per year and their sizes varied from 200 to 1000 μm.

In addition, the two commercial species of mussels, which harvested from five European countries (Spain, France, Italy, the Netherlands, and Denmark), both the *M. galloprovincialis* and *M. edulis* as common dietary foods by humans, showed a considerable accumulation of microplastic particles (Vandermeersch et al., 2015). Moreover, the mussels harvested from Belgium revealed an accumulation of microplastic fragments varied from 3 to 5 fibers/10 g mussels (de Witte et al., 2014). The other work showed the presence of the MP particles in marine molluscs obtained from several regions and consumed as dietary food by humans (Barboza et al., 2018). Thus, several recent research studies showed the multiple pathways of the exposure of human bodies to different categories of M/NPs particles through the consumption of foods, particularly seafood (Cox et al., 2019; Prata, da Costa, Lopes, Duarte, & Rocha-Santos, 2020; Rahman, Sarkar, Yadav, Achari, & Slobodnik, 2021). Consequently, these pollutant particles caused an important food safety concern (Lusher et al., 2017).

6.2 Toxicity influences on human cell lines

The consumption of seafood products can expose humans to various category of MP and NP particles. Hence, the accumulation of these contaminant particles in human bodies can contribute to potential adverse impacts on their health (Dong et al., 2020). The exposure to the MP and NP particles can induce inflammatory impacts in human lung epithelial BEAS-2B cells via the generation of ROS as well as these pollutants generate a pulmonary cytotoxicity (Ali et al., 2021). Besides, the exposure to MP particles at different concentration and size can lead to cytotoxicity in the human cells (such as and glioblastoma multiforme T98G cells and cervical HeLa cells) during 48 h (Schirinzi et al., 2017). In addition, Hwang, Choi, Han, Choi, and Hong (2019) reported that the exposure to different concentrations and sizes (varied from 25 to 200 and 20 μm) of MP fragments could cause the stimulation of immune system and increase the potential hypersensitivity through the augmentation of the concentration of histamines and cytokines in Raw 264.7, HMC-1, and PBM cells during 4 days at an elevate concentration as well as little amount of these pollutants. Moreover, the exposure to MP induced the deterioration of immune cells, whereas the exposure to NP was proven lethal as they induced disruption of cell membranes (Soares et al., 2020).

The little nanoplastic particles with size of 25 nm had the possibility to more quickly penetrate into the cytoplasm of the human alveolar epithelial

A549 cell compared to the big nanoplastic particles with size of 70 nm (Ali et al., 2021). Therefore, the MP and NP particles were highly influenced the cell viability, stimulated the gene responsible of transcription, triggered cell cycle S phrase arrest, and altered the expression of proteins connected with cell cycle and pro-apoptosis (Ali et al., 2021). The exposure of human cells to the MP and NP particles could cause the stimulation of the up-regulation of pro-inflammatory cytokines and pro-apoptotic proteins, and hence, these pollutants could generate tumor necrosis factor-α-(TNF-α-) associated with the apoptosis process (Ali et al., 2021; Xu et al., 2019).

6.3 Disease and health issues induced by micro/nanoplastics

The MP and NP particles can penetrate into the human bodies through the consumption of contaminated seafood products, and thus these pollutants pass through the gastrointestinal tract and then accumulated into the lymphatic system, tissues and blood, further into the organs (Vivekanand et al., 2021). The exposure of human to these particles can induce several health issues and create chronic diseases. Therefore, the accumulation of these contaminants in the body can cause inflammation and chronic obstructive pulmonary diseases (Dong et al., 2020; Mock, 2020; Rochman, Tahir, et al., 2015; Vivekanand et al., 2021). In fact, the exposure of human bodies to the MP fragments can reduce and deteriorate the pulmonary barrier by exhausting zonula occludens proteins (Ali et al., 2021). The penetration of MP into the human metabolism causes the alteration of the growth-factor transmission during and after implantation, immunity mechanisms, as well as the development of the embryo (Ilekis et al., 2016). Besides, Lee (2018) reported that the M/NP particles generated adverse impacts to human placentas and caused transgenerational impacts on reproduction and metabolisms. In addition, the ingestion of seafood products contaminated with MP particles caused the reduction in the in steroid hormone levels, damage fertility, and delay ovulation as well as inducing carcinogenicity effect (Vázquez & Rahman, 2021). Moreover, the consumption of fish contaminated with MP particles can threaten the human health by causing the cell necrosis (Rochman, Tahir, et al., 2015; Soares et al., 2020). Also, the smaller microplastics (<20 μm) can penetrate into few organs, and particles with a size of 10 μm can have access to the sensitive organs, such as liver, muscle, placenta, and brain through the blood-brain barrier and cell membranes (Vivekanand et al., 2021). However, the smallest particles (0.1 > 10 μm) are only able to access all organs (Barboza et al., 2018).

The microplastic particles can pass through living cells, for instance, the dendritic cells or M cells, to the lymphatic and circulatory system and then accumulate in secondary organs causing adverse effects on immune system and cell health (Soares et al., 2020). In addition, the exposure of human bodies to the NP particles causes several negatives impacts, such as alterations of endogenous metabolites, oxidative damage, inflammatory responses, alteration of reproductive system, cardiopulmonary responses, and adverse impacts on gut microflora and nutrient absorption (Soares et al., 2020). The penetration of the M/NP particles into organisms leads to immunotoxicity and hence provoke adverse effects, such as immune activation, immunosuppression, and abnormal inflammatory responses (Barboza et al., 2018).

Overall, the negative impacts of the exposure of human bodies to the M/NP particles can be varied based on their chemical and physical properties (size, length, and shape), their adhesion to other polymers and additives, as well as the individual susceptibility (Soares et al., 2020; Vivekanand et al., 2021).

7. Future trends and challenges

In the last decades, the pollution of marine environment by M/NP particles represents a major concern in the worldwide. The degradation of plastic products under the environmental factors can generate different types of MPs and NPs with different chemical and physical characteristics (size, shape, etc.) and therefore these particles can transport and reach the aquatic environment by several sources. However, there is a considerable difference in the abundance and distribution of M/NPs in the different sections of marine environment (such as seawater, aquatic biota, and corals, etc.) due to natural factors and anthropogenic activities. The M/NP particles have the ability to be transferred among the different levels trophic in the marine ecosystems. Therefore, they can enter the trophic chain and accumulate in different cells and tissues of the vital organisms. The MPs and NPs can induce significant eco-toxicological impacts on aquatic organisms and marine ecosystems. Several research works have revealed that these pollutants can induce many adverse effects on food safety, thus they can contaminate different species of seafood. Thus, the exposure of seafood products to these contaminants can generate the alteration of several vital systems (reproductive and nervous systems) and oxidative damage. In fact the presence of these particles in seafood products, such as fish, shellfish (e.g., bivalves and crustaceans), and echinoderms, consumed by humans is considered a global issue. Thus, the exposure of human bodies through the consumption of seafood products and other human food items can be vulnerable and generate several adverse impacts on human health, such as genotoxicity, immunotoxicity, cytotoxicity, inflammatory responses, oxidative damage, and alteration of reproductive system and other negative effects.

The exposure pathways of MPs and NPs particles to marine biotas, including seafood products and human bodies need to be more investigated. Therefore, more research works are necessary to provide a basis for more estimation of the risks of these pollutants. Further works are suggested to understand the potential cytotoxicity, genotoxicity, immunotoxicity, and other chronic and toxic impacts of the MPs and NPs particles on human health. Besides, more studies are required to understand better the adverse effects of other toxic compounds associated with the MPs and NPs fragments. Moreover, future studies are required to understand the potential toxic effects of nanoplastics due to their little size and limitation of analytical assays. Overall, it is a great challenge to understand the mechanisms and processes related to the penetration and assimilation of these contaminants in human cells and tissues and thus their potential impacts on human health.

8. Conclusions

Recently, scientific attention about plastics (particularly MP and NP) has gained an expanding information base due to their direct or indirect application in different domains worldwide, such as medicine, agriculture, industrial, and daily care. The plastics are released into the ecosystems after their utilization and they can be decomposed under several chemical and biological pathways into smaller particles (MPs and NPs) and therefore transported and reached the marine environment. These contaminants are widespread in the globe and have been detected by several analytical assays in diverse marine habitat. Besides, these pollutants have revealed different physicochemical characteristics, such as various sizes, shapes, colors, and geometric properties. MP and NPs particles can be ingested by seafood products, and consequently cause different adverse impacts on human health due to their consumption.

References

Abbasi, S., Soltani, N., Keshavarzi, B., Moore, F., Turner, A., & Hassanaghaei, M. (2018). Micro-plastics in different tissues of fish and prawn from the Musa Estuary, Persian Gulf. *Chemosphere*, *205*, 80–87.

Ahmed, S., Hall, A. M., & Ahmed, S. F. (2018a). Biodegradation of different types of paper in a compost environment. In *The 5th international conference on natural sciences and technology. iLab-Australia, Chittagong, Bangladesh* (pp. 26–30).

Ahmed, S., Hall, A. M., & Ahmed, S. F. (2018b). Comparative biodegradability assessment of different types of paper. *Journal of Natural Sciences Research*, *8*, 9–20.

Ali, I., Cheng, Q., Ding, T., Yiguang, Q., Yuechao, Z., Sun, H., et al. (2021). Micro/nano-plastics in the environment: Occurrence, detection, characterization and toxicity—A critical review. *Journal of Cleaner Production*, *313*, 127863.

Allen, S., Allen, D., Karbalaei, S., Maselli, V., & Walker, T. R. (2022). Micro (nano) plastics sources, fate, and effects: What we know after ten years of research. *Journal of Hazardous Materials Advances*, *6*, 100057.

Almeida, M., Martins, M. A., Soares, A. M. V. M., Cuesta, A., & Oliveira, M. (2019). Polystyrene nano-plastics alter the cytotoxicity of human pharmaceuticals on marine fish cell lines. *Environmental Toxicology and Pharmacology*, *69*, 57–65.

Alprol, A. E., Gaballah, M. S., & Hassaan, M. A. (2021). Micro and nano-plastics analysis: Focus on their classification, sources, and impacts in marine environment. *Regional Studies in Marine Science*, *42*, 101625.

Aluru, N., Leatherland, J. F., & Vijayan, M. M. (2010). Bisphenol A in oocytes leads to growth suppression and altered stress performance in juvenile rainbow trout. *PloS One*, *5*, e10741.

Andrady, A. L. (1994). Assessment of environmental biodegradation of synthetic polymers. *Journal of Macromolecule Science Part C*, *3*, 7–41.

Andrady, A. L. (2011). Micro-plastics in the marine environment. *Marine Pollution Bulletin*, *62*, 1596–1605.

Andrady, A. L. (2015). Persistence of plastic litter in the oceans. In M. Bergmann, L. Gutow, & M. Klages (Eds.), *Marine anthropogenic litter bergmann* (pp. 57–72). Berlin: Springer.

Arjula, S., Harsha, A. P., & Ghosh, M. K. (2008). Solid-particle erosion behavior of high performance thermoplastic polymers. *Journal of Materials Science*, *43*(6), 1757–1768.

Au, S. Y., Bruce, T. F., Bridges, W. C., & Klaine, S. J. (2015). Responses of Hyalella azteca to acute and chronic micro-plastic exposures. *Environmental Toxicology and Pharmacology*, *34*, 2564–2572.

Balestri, E., Menicagli, V., Vallerini, F., & Lardicci, C. (2017). Biodegradable plastic bags on the seafloor: A future threat for seagrass meadows? *Science of the Total Environment*, *605–606*, 755–763.

Barboza, L. G. A., Raimundo, J., Oliveira, P., Bessa, F., Henriques, B., Caetano, M., et al. (2020). Micro-plastics in wild fish from North East Atlantic Ocean and its potential for causing neurotoxic effects, lipid oxidative damage, and human health risks associated with ingestion exposure. *Science of the Total Environment*, *717*, 134625.

Barboza, L. G. A., Vethaakd, A. D., Lavorante, B. R. B. O., Lundebye, A.-K., & Guilhermino, L. (2018). Marine microplastic debris: An emerging issue for food security, food safety and human health. *Marine Pollution Bulletin*, *133*, 336–348.

Barnes, D. K. A., Galgani, F., Thompson, R. C., & Barlaz, M. (2009). Accumulation and fragmentation of plastic debris in global environments. *Philosophical Transactions of the Royal Society B*, *364*, 1985–1998.

Barrento, S., Marques, A., Teixeira, B., Vaz-Pires, P., Carvalho, M. L., & Nunes, M. L. (2008). Essential elements and contaminants in edible tissues of European and American lobsters. *Food Chemistry*, *111*, 862–867.

Bergami, B., Pugnalini, S., Vannuccini, M. L., Manfra, L., Faleri, C., Savorelli, F., et al. (2017). Long-term toxicity of surface-charged polystyrene nanoplastics to marine planktonic species *Dunaliella tertiolecta* and *Artemia franciscana*. *Aquatic Toxicology*, *189*, 159–169.

Betts, K. (2008). Why small plastic particles may pose a big problem in the oceans. *Environmental Science & Technology*, *42*, 8995.

Blight, L. K., & Burger, A. E. (1997). Occurrence of plastic particles in seabirds from the eastern North Pacific. *Marine Pollution Bulletin*, *34*, 323–325.

Boerger, C. M., Lattin, G. L., Moore, S. L., & Moore, C. J. (2010). Plastic ingestion by planktivorous fishes in the North Pacific Central Gyre. *Marine Pollution Bulletin*, *60*(12), 2275–2278.

Bouwmeester, H., Hollman, P. C. H., Peters, R. J. B., & Bouwmeester, H. (2015). Potential health impact of environmentally released micro/nano-plastics in the human food production chain: Experiences from nanotoxicology. *Environmental Science & Technology*, *49*(15), 8932–8947.

Brandts, I., Barría, C., Martins, M. A., Franco-Martínez, L., Barreto, A., Tvarijonaviciute, A., et al. (2021). Waterborne exposure of gilthead seabream (*Sparus aurata*) to polymethylmethacrylate nano-plastics causes effects at cellular and molecular levels. *Journal of Hazardous Materials*, *403*, 123590.

Brandts, I., Teles, M., Gonçalves, A. P., Barreto, A., Franco-Martinez, L., Tvarijonaviciute, A., et al. (2018). Effects of nano-plastics on Mytilus galloprovincialis after individual and combined exposure with carbamazepine. *Science of the Total Environment*, *643*, 775–784.

Brandts, I., Teles, M., Tvarijonaviciute, A., Pereira, M., Martins, M., Tort, L., et al. (2018). Effects of polymethylmethacrylate nano-plastics on *Dicentrarchus labrax*. *Genomics*, *110*, 435–441.

Brinkhof, J., Herrmann, B., Sistiaga, M., Larsen, R. B., Jacques, N., & Gjøsund, S. H. (2021). Effect of gear design on catch damage on cod (*Gadus morhua*) in the Barents Sea demersal trawl fishery. *Food Control*, *120*, 107562.

Browne, M. A., Crump, P., Niven, S. J., Teuton, E., Tonkin, A., Galloway, T., et al. (2011). Accumulation of micro-plastic on shorelines worldwide: Sources and sinks. *Environmental Science & Technology*, *45*, 9175–9179.

Browne, M. A., Dissanayake, A., Galloway, T. S., Lowe, D. M., & Thompson, R. C. (2008). Ingested microscopic plastic translocates to the circulatory system of the mussel, *Mytilus edulis* (L.). *Environmental Science & Technology*, *42*, 5026–5031.

Browne, M. A., Galloway, T., & Thompson, R. (2007). Micro-plastic—An emerging contaminant of potential concern? *Integrated Environmental Assessment and Management*, *3*, 559–561.

Browne, M. A., Galloway, T. S., & Thompson, R. C. (2010). Spatial patterns of plastic debris along estuarine shorelines. *Environmental Science & Technology*, *44*, 3404–3409.

Canesi, L., Ciacci, C., Bergami, E., Monopoli, M. P., Dawson, K. A., Papa, S., et al. (2015). Evidence for immunomodulation and apoptotic processes induced by cationic polystyrene nanoparticles in the haemocytes of the marine bivalve Mytilus. *Marine Environmental Research*, *111*, 34–40.

Carpenter, E. J., Anderson, S. J., Harvey, G. R., Miklas, H. P., & Peck, B. B. (1972). Polystyrene spherules in coastal waters. *Science*, *178*(4062), 749–750.

Carr, S. A., Liu, J., & Tesoro, A. G. (2016). Transport and fate of micro-plastic particles in wastewater treatment plants. *Water Research*, *91*, 174–182.

Carreras-Colom, E., Constenla, M., Soler-Membrives, A., Cartes, J. E., Baeza, M., Padrós, F., et al. (2018). Spatial occurrence and effects of micro-plastic ingestion on the deep-water shrimp *Aristeus antennatus*. *Marine Pollution Bulletin*, *133*, 44–52.

Cau, A., Avio, C. G., Dessì, C., Follesa, M. C., Moccia, D., Regoli, F., et al. (2019). Micro-plastics in the crustaceans *Nephrops norvegicus* and *Aristeus antennatus*: Flagship species for deep-sea environments? *Environmental Pollution*, *255*, 113107.

Chae, Y., & An, Y.-J. (2017). Effects of micro/nano-plastics on aquatic ecosystems: Current research trends and perspectives. *Marine Pollution Bulletin*, *124*, 624–632.

Cole, M., Lindeque, P., Halsband, C., Tamara, S., & Galloway, T. S. (2011). Micro-plastics as contaminants in the marine environment: A review. *Marine Pollution Bulletin*, *62*, 2588–2597.

Cooper, D. A., & Corcoran, P. L. (2010). Effects of mechanical and chemical processes on the degradation of plastic beach debris on the island of Kauai, Hawaii. *Marine Pollution Bulletin*, *60*, 650–654.

Costa, M. F., & Duarte, A. C. (2017). Micro-plastics sampling and sample handling. *Comprehensive Analytical Chemistry*, *75*, 25–47.

Costa, M., Ivar do Sul, J., Silva-Cavalcanti, J., Araújo, M., Spengler, Â., & Tourinho, P. (2010). On the importance of size of plastic fragments and pellets on the strandline: A snapshot of a Brazilian beach. *Environmental Monitoring and Assessment*, *168*, 299–304.

Cox, K. D., Covernton, G. A., Davies, H. L., Dower, J. F., Juanes, F., & Dudas, S. E. (2019). Human consumption of micro-plastics. *Environmental Science & Technology*, *53*, 7068–7074.

Dawson, A. L., Kawaguchi, S., King, C. K., Townsend, K. A., King, R., Huston, W. M., et al. (2018). Turning micro-plastics into nano-plastics through digestive fragmentation by Antarctic krill. *Nature Communications*, *9*(1), 1001.

de Oliveira, T. A., de Oliveira Mota, I., Mousinho, F. E. P., Barbosa, R., de Carvalho, L. H., & Alves, T. S. (2019). Biodegradation of mulch films from poly (butylene adipate co-terephthalate), carnauba wax, and sugarcane residue. *Journal of Applied Polymer Science*, *136*(47), 1–9.

de Witte, B., Devriese, L., Bekaert, K., Hoffman, S., Vandermeersch, G., Cooreman, K., et al. (2014). Quality assessment of the blue mussel (*Mytilus edulis*): Comparison between commercial and wild types. *Marine Pollution Bulletin*, *85*(1), 146–155.

Della Torre, C., Bergami, E., Salvati, A., Faleri, C., Cirino, P., Dawson, K. A., et al. (2014). Accumulation and embryotoxicity of polystyrene nanoparticles at early stage of development of sea urchin embryos *Paracentrotus lividus*. *Environmental Science & Technology*, *48*(20), 12302–12311.

Deng, Y.-H., Dong, L.-L., Zhang, Y.-J., Zhao, X.-M., & He, H.-Y. (2021). Enriched environment boosts the post-stroke recovery of neurological function by promoting autophagy. *Neural Regeneration Research*, *16*, 813–819.

Derraik, J. G. B. (2002). The pollution of the marine environment by plastic debris: A review. *Marine Pollution Bulletin*, *44*, 842–852.

Devriese, L. I., Van der Meulen, M. D., Maes, T., Bekaert, K., Paul-Pont, I., Frère, L., et al. (2015). Micro-plastic contamination in brown shrimp (*Crangon crangon*, Linnaeus 1758) from coastal waters of the southern North Sea and channel area. *Marine Pollution Bulletin*, *98*(1–2), 179–187.

Dong, S., You, X., & Hu, F. (2020). Effects of wave forces on knotless polyethylene and chain-link wire netting panels for marine aquaculture cages. *Ocean Engineering*, *207*, 107368.

Duis, K., & Coors, A. (2016). Micro-plastics in the aquatic and terrestrial environment: Sources (with a specific focus on personal care products), fate and effects. *Environmental Sciences Europe*, *28*(1), 2.

EFSA Contam Panel (EFSA Panel on Contaminants in the Food Chain). (2016). Statement on the presence of micro-plastics and nano-plastics in food, with particular focus on seafood. *EFSA Journal*, *14*(6), 4501. 30 pp.

Eriksen, M., Lebreton, L. C. M., Carson, H. S., Thiel, M., Moore, C. J., Borerro, J. C., et al. (2014). Plastic pollution in the world's oceans: More than 5 trillion plastic pieces weighing over 250,000 tons afloat at sea. *PLoS One*, *9*, 1–15.

Erni-Cassola, G., Gibson, M. I., Thompson, R. C., & Christie-Oleza, J. A. (2017). Lost, but found with nile red: A novel method for detecting and quantifying small micro-plastics (1 mm to 20 μm) in environmental samples. *Environmental Science & Technology*, *51*, 13641–13648.

Eubeler, J. P., Zak, S., Bernhard, M., & Knepper, T. P. (2009). Environmental biodegradation of synthetic polymers I. test methodologies and procedures. *TrAC Trends in Analytical Chemistry*, *28*(9), 1057–1072.

Fendall, L. S., & Sewell, M. A. (2009). Contributing to marine pollution by washing your face: Micro-plastics in facial cleansers. *Marine Pollution Bulletin*, *58*, 1225–1228.

Ferreira, P., Fonte, E., Soares, M. E., Carvalho, F., & Guilhermino, L. (2016). Effects of multi-stressors on juveniles of the marine fish *Pomatoschistus microps*: Gold nanoparticles, micro-plastics and temperature. *Aquatic Toxicology*, *170*, 89–103.

Figueroa-Pico, J., Tortosa, F. S., & Carpio, A. (2020). Coral fracture by derelict fishing gear affects the sustainability of the marginal reefs of Ecuador. *Coral Reefs*, *39*, 819–827.

Fu, W., Min, J., Jiang, W., Li, Y., & Zhang, W. (2020). Separation, characterization and identification of micro-plastics and nano-plastics in the environment. *Science of the Total Environment*, *721*, 137561.

Galgani, F., Fleet, D., Franeker, J. V., Katsanevakis, S., Maes, T., Mouat, J., et al. (2010). Task group 10 report: Marine litter. In N. Zampoukas (Ed.), *Marine strategy framework directive* JRC, IFREMER & ICES.

Gall, S. C., & Thompson, R. C. (2015). The impact of debris on marine life. *Marine Pollution Bulletin*, *92*(1–2), 170–179.

Gambardella, C., Morgana, S., Ferrando, S., Bramini, M., Piazza, V., Costa, E., et al. (2017). Effects of polystyrene microbeads in marine planktonic crustaceans. *Ecotoxicology and environmental safety*, *145*, 250–257.

Geilfus, N.-X., Munson, K. M., Sousa, J., Germanov, Y., Bhugaloo, S., Babb, D., et al. (2019). Distribution and impacts of micro-plastic incorporation within sea ice. *Marine Pollution Bulletin*, *145*, 463–473.

GESAMP. (2016). Sources, fate and effects of micro-plastics in the marine environment: Part two of a global assessment. In *IMO/FAO/UNESCO-IOC/UNIDO/WMO/IAEA/ UN/ UNEP/UNDP Joint group of experts on the scientific aspects of marine environmental protection. Report and studies, GESAMP No. 93* (p. 220).

Gigault, J., Pedrono, B., Maxit, B., & Ter Halle, A. (2016). Marine plastic litter: The unanalyzed nano-fraction. *Environmental Science: Nano*, *3*, 346–350.

Goldstein, J. I., Newbury, D. E., Michael, J. R., Ritchie, N. W., Scott, J. H. J., & Joy, D. C. (2017). *Scanning electron microscopy and X-ray microanalysis*. New York: Springer.

Gonçalves, J. M., & Bebianno, M. J. (2021). Nano-plastics impact on marine biota: A review. *Environmental Pollution*, *273*, 116426.

González-Fernández, C., Tallec, K., Le Goïc, N., Lambert, C., Soudant, P., Huvet, A., et al. (2018). Cellular responses of Pacific oyster (*Crassostrea gigas*) gametes exposed *in vitro* to polystyrene nanoparticles. *Chemosphere*, *208*, 764–772.

Graham, E., & Thompson, J. (2009). Deposit and suspension-feeding sea cucumbers (*Echinodermata*) ingest plastic fragments. *Journal of Experimental Marine Biology and Ecology*, *368*, 22–29.

Gregory, M. R. (1996). Plastic 'scrubbers' in hand cleansers: A further (and minor) source for marine pollution identified. *Marine Pollution Bulletin*, *32*, 867–871.

Greven, A. C., Merk, T., Karagöz, F., Mohr, K., Klapper, M., Jovanović, B., et al. (2016). Polycarbonate and polystyrene nano-plastic particles act as stressors to the innate immune system of fathead minnow (*Pimephales promelas*). *Environmental Toxicology and Chemistry*, *35*, 3093–3100.

Guterres, S. S., Alves, M. P., & Pohlmann, A. R. (2007). Polymeric nanoparticles, nanospheres and nanocapsules, for cutaneous applications. *Drug Target Insights*, *2*, 147–157.

Hagger, J. A., Depledge, M. H., Oehlmann, J., Jobling, S., & Galloway, T. S. (2006). Is there a causal association between genotoxicity and the imposex effect? *Environmental Health Perspectives*, *114*, 20–26.

Heo, N. W., Hong, S. H., Han, G. M., Hong, S., Lee, J., Song, Y. K., et al. (2013). Distribution of small plastic debris in cross-section and high strandline on Heungnam beach, South Korea. *Ocean Science Journal*, *48*, 225–233.

Hidalgo-Ruz, V., Gutow, L., Thompson, R. C., & Thiel, M. (2012). Micro-plastics in the marine environment: A review of the methods used for identification and quantification. *Environmental Science & Technology*, *46*, 3060–3075.

Huang, W., Chen, M., Song, B., Deng, J., Shen, M., Chen, Q., et al. (2021). Micro-plastics in the coral reefs and their potential impacts on corals: A mini-review. *Science of the Total Environment*, *762*, 143112.

Hurley, R., Woodward, J., & Rothwell, J. J. (2018). Micro-plastic contamination of river beds significantly reduced by catchment-wide flooding. *Nature Geoscience*, *11*, 251–257.

Hwang, J., Choi, D., Han, S., Choi, J., & Hong, J. (2019). An assessment of the toxicity of polypropylene micro-plastics in human derived cells. *Science of the Total Environment*, *684*, 657–669.

Ibrahim, A.-M., David, S., Bruno, G., Mark, C., Mohammed, A. M., & Lewis, L. V. (2018). Decline in oyster populations in traditional fishing grounds; is habitat damage by static fishing gear a contributory factor in ecosystem degradation? *Journal of Sea Research*, *140*, 40–51.

Ilekis, J. V., Tsilou, E., Fisher, S., Abrahams, V. M., Soares, M. J., Cross, J. C., et al. (2016). Placental origins of adverse pregnancy outcomes: potential molecular targets: An executive workshop summary of the eunice Kennedy shriver national institute of child health and human development. *American Journal of Obstetrics and Gynecology*, *215*(1), S1–S46.

Inanli, A. G., Aksun Tümerkan, E. T., El Abed, N., Regenstein, J. M., & Özogul, F. (2020). The Impact of chitosan on seafood quality and human health: A Review. *Trends in Food Science & Technology.*, *97*, 404–416.

Ivar do Sul, J. A., Spengler, A., & Costa, M. F. (2009). Here, there and everywhere, small plastic fragments and pellets on beaches of Fernando de Noronha (Equatorial Western Atlantic). *Marine Pollution Bulletin*, *58*, 1236–1238.

Jambeck, J. R., Geyer, R., Wilcox, C., Siegler, T. R., Perryman, M., Andrady, A., et al. (2015). Plastic waste inputs from land into the ocean. *Science*, *347*, 768–771.

Jemec, A., Horvat, P., Kunej, U., Bele, M., & Kržan, A. (2016). Uptake and effects of micro-plastic textile fibers on freshwater crustacean *Daphnia magna*. *Environmental Pollution*, *219*, 201–209.

Khan, F. R., Syberg, K., Shashoua, Y., & Bury, N. R. (2015). Influence of polyethylene micro-plastic beads on the uptake and localization of silver in zebrafish (*Danio rerio*). *Environnemental Pollution*, *206*, 73–79.

Köhler, A. (2010). Cellular fate of organic compounds in marine invertebrates. *Comparative Biochemistry and Physiology - Part A: Molecular & Integrative Physiology*, *157*, S8.

Kolandhasamy, P., Su, L., Li, J., Qu, X., Jabeen, K., & Shi, H. (2018). Adherence of micro-plastics to soft tissue of mussels: A novel way to uptake micro-plastics beyond ingestion. *Science of the Total Environment*, *610–611*, 635–640.

Kukkola, A., Krause, S., Lynch, I., Sambrook Smith, G. H., & Nel, H. (2021). Nano and micro-plastic interactions with freshwater biota—Current knowledge, challenges and future solutions. *Environment International*, *152*, 106504.

Kärrman, A., Schönlau, C., & Engwall, M. (2016). *Exposure and effects of micro-plastics on wildlife* (p. 39). Sweden: DiVA: Orebro University.

Lee, D. H. (2018). Evidence of the possible harm of endocrine-disrupting chemicals in humans: Ongoing debates and key issues. *Endocrinology and Metabolism*, *33*, 44–52.

Lee, W. S., Cho, H. J., Kim, E., Huh, Y. H., Kim, H. J., Kim, B., et al. (2019). Bioaccumulation of polystyrene nano-plastics and their effect on the toxicity of Au ions in zebrafish embryos. *Nanoscale*, *11*, 3200–3207.

Lee, K. W., Shim, W. J., Kwon, O. Y., & Kang, J. H. (2013). Size-dependent effects of micro polystyrene particles in the marine copepod *Tigriopus japonicus*. *Environmental Science & Technology*, *47*(19), 11278–11283.

Lehtiniemi, M., Hartikainen, S., Nakki, P., Engstrom-Ost, J., Koistinen, A., & Setala, O. (2018). Size matters more than shape: Ingestion of primary and secondary micro-plastics by small predators. *Food Webs*, *17*, e00097.

LeMoine, C. M., Kelleher, B. M., Lagarde, R., Northam, C., Elebute, O. O., & Cassone, B. J. (2018). Transcriptional effects of polyethylene micro-plastics ingestion in developing zebrafish (*Danio rerio*). *Environnemental Pollution*, *243*, 591–600.

Lenz, R., Enders, K., & Gissel Nielsen, T. (2016). Micro-plastic exposure studies should be environmentally realistic. *Proceedings of the National Academy of Sciences of the United States of America*, *113*, E4121–E4122.

Levin, P. A., & Angert, E. R. (2015). Small but mighty: Cell size and bacteria. *Cold Spring Harb. Perspectives in Biology*, *7*(7), 1–11.

Li, B., Su, L., Zhang, H., Deng, H., Chen, Q., & Shi, H. (2020). Micro-plastics in fishes and their living environments surrounding a plastic production area. *Science of the Total Environment*, *727*, 138662.

Lichtman, J. W., & Conchello, J.-A. (2005). Fluorescence microscopy. *Nature Methods*, *2*, 910.

Liu, P., Zhan, X., Wu, X., Li, J., Wang, H., & Gao, S. (2020). Effect of weathering on environmental behavior of micro-plastics: Properties, sorption and potential risks. *Chemosphere*, *242*, 125193.

Lozano, R. L., & Mouat, J. (2009). *Marine litter in the North-east Atlantic region: Assessment and priorities for response* (p. 127). London: United Kingdom.

Luís, L. G., Ferreira, P., Fonte, E., Oliveira, M., & Guilhermino, L. (2015). Does the presence of micro-plastics influence the acute toxicity of chromium(VI) to early juveniles of the common goby (*Pomatoschistus microps*)? A study with juveniles from two wild estuarine populations. *Aquatic Toxicology*, *164*, 163–174.

Lundebye, A. K., Lusher, A. L., Bank, M. S., & Bank, M. S. (2022). Marine micro-plastics and seafood: Implications for food security. In *Micro-plastic in the environment: Pattern and process, environmental contamination remediation and management* Springer.

Lusher, A., Hollman, P., & Mendoza-Hill, J. (2017). *Micro-plastics in fisheries and aquaculture: Status of knowledge on their occurrence and implications for aquatic organisms and food safety* (p. 615). FAO Fisheries and Aquaculture Technical Paper.

Maes, T., Jessop, R., Wellner, N., Haupt, K., & Mayes, A. G. (2017). A rapid-screening approach to detect and quantify micro-plastics based on fluorescent tagging with Nile Red. *Scientific Reports*, *7*, 1–10.

Mason, S. A., Garneau, D., Sutton, R., Chu, Y., Ehmann, K., Barnes, J., et al. (2016). Micro-plastic pollution is widely detected in US municipal wastewater treatment plant effluent. *Environmental Pollution*, *218*, 1045–1054.

Mato, Y., Isobe, T., Takada, H., Kanehiro, H., Ohtake, C., & Kaminuma, T. (2001). Plastic resin pellets as a transport medium for toxic chemicals in the marine environment. *Environmental Science & Technology*, *35*, 318–324.

Matsusaki, M., Kishida, A., Stainton, N., Ansell, C. W. G., & Akashi, M. (2001). Synthesis and characterisation of novel biodegradable polymers composed of hydroxycinnamic acid and D, L-lactic acid. *Journal of Applied Polymer Science*, *82*, 2357–2364.

McGoran, A. R., Clark, P. F., Smith, B. D., & Morritt, D. (2020). High prevalence of plastic ingestion by *Eriocheir sinensis* and *Carcinus maenas* (Crustacea: *Decapoda: Brachyura*) in the Thames estuary. *Environmental Pollution*, *265*, 114972.

Mintenig, S. M., Bäuerlein, P. S., Koelmans, A. A., Dekker, S. C., & van Wezel, A. P. (2018). Closing the gap between small and smaller: Towards a framework to analyse nano-and micro-plastics in aqueous environmental samples. *Environmental Science: Nano*, *5*(7), 1640–1649.

Mock, J. (2020). Micro-plastics are everywhere, but their health effects on humans are still unclear. *Discover Magazine*. https://www.discovermagazine.com/health/micro-plastics-are-everywhere-but-their-health-effects-on-humans-are-still.

Mofijur, M., Ahmed, S. F., Ashrafur Rahman, S. M., Yasir Arafat Siddiki, S. K., Saiful Islam, A. B. M., Shahabuddin, M., et al. (2021). Source, distribution and emerging threat of micro/nano-plastics to marine organism and human health: Socio-economic impact and management strategies. *Environmental Research*, *195*, 110857.

Moore, C. J. (2008). Synthetic polymers in the marine environment: A rapidly increasing, long-term threat. *Environmental Research*, *108*, 131–139.

Moore, C. J., Moore, S. L., Weisberg, S. B., Lattin, G. L., & Zellers, A. F. (2002). A comparison of neustonic plastic and zooplankton abundance in southern California's coastal waters. *Marine Pollution Bulletin*, *44*, 1035–1038.

Murray, F., & Cowie, P. R. (2011). Plastic contamination in the decapod crustacean *Nephrops norvegicus* (Linnaeus, 1758). *Marine Pollution Bulletin*, *62*(6), 1207–1217.

Muthukumar, T., Aravinthan, A., Lakshmi, K., Venkatesan, R., Vedaprakash, L., & Doble, M. (2011). Fouling and stability of polymers and composites in marine environment. *International Biodeterioration & Biodegradation*, *65*, 276–284.

Ng, K. L., & Obbard, J. P. (2006). Prevalence of micro-plastics in Singapore's coastal marine environment. *Marine Pollution Bulletin*, *52*, 761–767.

Nguyen, B., Claveau-Mallet, D., Hernandez, L. M., Xu, E. G., Farner, J. M., & Tufenkji, N. (2019). Separation and analysis of micro-plastics and nano-plastics in complex environmental samples. *Accounts of Chemical Research*, *52*, 858–866.

Obbard, R. W., Sadri, S., Wong, Y. Q., Khitun, A. A., Baker, I., & Thompson, R. C. (2014). Global warming releases micro-plastic legacy frozen in Arctic Sea ice. *Earths Future*, *2*(6), 315–320.

Oliveira, M., & Almeida, M. (2019). The why and how of micro- (nano)-plastic research. *TrAC Trends in Analytical Chemistry*, *114*, 196–201.

Oliveira, M., Almeida, M., & Miguel, I. (2019). Amicro (nano)-plastic boomerang tale: A never ending story? *TrAC Trends in Analytical Chemistry*, *112*, 196–200.

Oliveira, M. C. L., Ribeiro, A. L. P., Hylland, K., & Guilhermino, L. (2013). Single and combined effects of micro-plastics and pyrene on juveniles (0+ group) of the common goby *Pomatoschistus microps* (Teleostei, Gobiidae). *Ecological Indicators*, *34*, 641–647.

Oliviero, M., Tato, T., Schiavo, S., & Fern´andez, V., Manzo, S., & Beiras, R. (2019). Leachates of micronized plastic toys provoke embryo toxic effects upon sea urchin *Paracentrotus lividus*. *Environmental Pollution*, *247*, 706–715.

Orr, I. G., Hadar, Y., & Sivan, A. (2004). Colonization, biofilmformation and biodegradation of polyethylene by a strain of *Rhodococcus ruber*. *Applied Microbiology and Biotechnology*, *65*, 97–104.

Patel, M. M., Goyal, B. R., Bhadada, S. V., Bhatt, J. S., & Amin, A. F. (2009). Getting into the brain: Approaches to enhance brain drug delivery. *CNS Drugs*, *23*, 35–58.

Pedà, C., Caccamo, L., Fossi, M. C., Gai, F., Andaloro, F., Genovese, L., et al. (2016). Intestinal alterations in European sea bass *Dicentrarchus labrax* (Linnaeus, 1758) exposed to micro-plastics: Preliminary results. *Environmental Pollution*, *212*, 251–256.

Peng, B.-Y., Chen, Z., Chen, J., Yu, H., Zhou, X., Criddle, C. S., et al. (2020). Biodegradation of polyvinyl chloride (PVC) in Tenebrio molitor (*Coleoptera: tenebrionidae*) larvae. *Environment International*, *145*, 106106.

Possatto, F. E., Barletta, M., Costa, M. F., Ivar Do Sul, J. A., & Dantas, D. V. (2011). Plastic debris ingestion by marine catfish: An unexpected fisheries impact. *Marine Pollution Bulletin*, *62*(5), 1098–1102.

Prata, J. C., da Costa, J. P., Lopes, I., Duarte, A. C., & Rocha-Santos, T. (2020). Environmental exposure to micro-plastics: an overview on possible human health effects. *Science of the Total Environment*, *702*, 134455.

Qiu, Q., Peng, J., Yu, X., Chen, F., Wang, J., & Dong, F. (2015). Occurrence of micro-plastics in the coastal marine environment: First observation on sediment of China. *Marine Pollution Bulletin*, *98*, 274–280.

Qu, H., Ma, R., Wang, B., Yang, J., Duan, L., & Yu, G. (2019). Enantiospecific toxicity, distribution and bioaccumulation of chiral antidepressant venlafaxine and its metabolite in loach (*Misgurnus anguillicaudatus*) co-exposed to micro-plastic and the drugs. *Journal of Hazardous Materials*, *370*, 203–211.
Rahman, A., Sarkar, A., Yadav, O. P., Achari, G., & Slobodnik, J. (2021). Potential human health risks due to environmental exposure to nano- and micro-plastics and knowledge gaps: A scoping review. *Science of the Total Environment*, *757*(9), 143872.
Resmeriţă, A.-M., Coroaba, A., Darie, R., Doroftei, F., Spiridon, I., Simionescu, B. C., et al. (2018). Erosion as a possible mechanism for the decrease of size of plastic pieces floating in oceans. *Marine Pollution Bulletin*, *127*, 387–395.
Rios, L. M., Moore, C., & Jones, P. R. (2007). Persistent organic pollutants carried by synthetic polymers in the ocean environment. *Marine Pollution Bulletin*, *54*, 1230–1237.
Rist, S. E., Assidqi, K., Zamani, N. P., Appel, D., Perschke, M., Huhn, M., et al. (2016). Suspended micro-sized PVC particles impair the performance and decrease survival in the Asian green mussel Perna viridis. *Marine Pollution Bulletin*, *111*(1–2), 213–220.
Rist, S., Carney Almroth, B., Hartmann, N. B., & Karlsson, T. M. (2018). A critical perspective on early communications concerning human health aspects of micro-plastics. *Science of the Total Environment*, *626*, 720–726.
Rizzi, M., Rodrigues, F. L., Medeiros, L., Ortega, I., Rodrigues, L., Monteiro, D. S., et al. (2019). Ingestion of plastic marine litter by sea turtles in southern Brazil: Abundance, characteristics and potential selectivity. *Marine Pollution Bulletin*, *140*, 536–548.
Robertson, J. D., Rizzello, L., Avila-Olias, M., Gaitzsch, J., Contini, C., Magoń, M. S., et al. (2016). Purification of nanoparticles by size and shape. *Scientific Reports*, *6*, 1–9.
Rochman, C. M., Brookson, C., Bikker, J., Djuric, N., Earn, A., Bucci, K., et al. (2019). Rethinking micro-plastics as a diverse contaminant suite. *Environmental Toxicology and Chemistry*, *38*, 703–711.
Rochman, C. M., Kross, S. M., Armstrong, J. B., Bogan, M. T., Darling, E. S., Green, S. J., et al. (2015). Scientific evidence supports a ban on microbeads. *Environmental Science & Technology*, *49*(18), 10759–10761.
Rochman, C. M., Tahir, A., Williams, S. L., Baxa, D. V., Lam, R., Miller, J. T., et al. (2015). Anthropogenic debris in seafood: Plastic debris and fibers from textiles in fish and bivalves sold for human consumption. *Scientific Reports*, *5*, 14340.
Rodríguez Chialanza, M., Sierra, I., Pérez Parada, A., & Fornaro, L. (2018). Identification and quantitation of semi - crystalline micro-plastics using image analysis and differential scanning calorimetry. *Environmental Science and Pollution Research*, *25*, 16767–16775.
Ryan, P. G., Moore, C. J., van Franeker, J. A., & Moloney, C. L. (2009). Monitoring the abundance of plastic debris in the marine environment. *Philosophical Transactions of the Royal Society B: Biological Sciences*, *364*, 1999–2012.
Schirinzi, G. F., Perez-Pomeda, I., Sanchís, J., Rossini, C., Farre, M., & Barcelo, D. (2017). Cytotoxic effects of commonly used nanomaterials and micro-plastics on cerebral and epithelial human cells. *Environmental Research*, *159*, 579–587.
Schmid, C., Cozzarini, L., & Zambello, E. (2021). Micro-plastic's story. *Marine Pollution Bulletin*, *162*, 111820.
Setälä, O., Fleming-Lehtinen, V., & Lehtiniemi, M. (2014). Ingestion and transfer of micro-plastics in the planktonic food web. *Environmental Pollution*, *185*, 77–83.
Shah, A. A., Hasan, F., Hameed, A., & Ahmed, S. (2008). Biological degradation of plastics: A comprehensive review. *Biotechnology Advance*, *26*(3), 246–265.
Shim, W. J., Hong, S. H., & Eo, S. E. (2017). Identification methods in micro-plastic analysis: A review. *Analytical Methods*, *9*, 1384–1391.
Sierra, I., Chialanza, M. R., Faccio, R., Carrizo, D., Fornaro, L., & P´erez-Parada, A. (2020). Identification of micro-plastics in wastewater samples by means of polarized light optical microscopy. *Environmental Science and Pollution Research*, *27*, 7409–7419.

Silva, A. B., Bastos, A. S., Justino, C. I. L., da Costa, J. P., Duarte, A. C., & Rocha-Santos, T. A. P. (2018). Micro-plastics in the environment: Challenges in analytical chemistry—A review. *Analytica Chimica Acta, 1017*, 1–19.

Šimat, V., Elabed, N., Kulawik, P., Ceylan, Z., Jamroz, E., Yazgan, H., et al. (2020). Recent advances in marine-based nutraceuticals and their health benefits. *Marine Drugs, 18*(12), 627–667.

Singh, S., Kumar, V., Kapoor, D., Bhardwaj, S., Dhanjal, D. S., Pawar, A., et al. (2022). Fate and occurrence of micro/nano-plastic pollution in industrial wastewater. In I. Haq, A. Kalamdhad, & M. Shah (Eds.), *Biodegradation and detoxification of micropollutants in industrial wastewater* (pp. 27–38). Elsevier Inc.

Singh, B., & Sharma, N. (2008). Mechanistic implications of plastic degradation. *Polymer Degradation and Stability, 93*, 561–584.

Sivan, A. (2011). New perspectives in plastic biodegradation. *Current Opinion in Biotechnology, 22*, 422–426.

Smith, M., Love, D. C., Rochman, C. M., & Neff, R. A. (2018). Micro-plastics in seafood and the implications for human health. *Current Environmental Health Reports, 5*, 375–386.

Soares, J., Miguel, I., Venâncio, C., Lopes, I., & Oliveira, M. (2020). Perspectives on micro-(nano) plastics in the marine environment: Biological and societal considerations. *Water, 12*, 3208.

Song, Y. K., Hong, S. H., Jang, M., Han, G. M., Rani, M., Lee, J., et al. (2015). A comparison of microscopic and spectroscopic identification methods for analysis of micro-plastics in environmental samples. *Marine Pollution Bulletin, 93*, 202–209.

Steer, M., Cole, M., Thompson, R. C., & Lindeque, P. (2017). Microplastic ingestion in fish larvae in the western English Channel. *Environmental Pollution, 226*, 250–259.

Strafella, P., Fabi, G., Despalatovic, M., Cvitkovic, I., Fortibuoni, T., Gomiero, A., et al. (2019). Assessment of seabed litter in the Northern and Central Adriatic Sea (Mediterranean) over six years. *Marine Pollution Bulletin, 141*, 24–35.

Summers, S., Henry, T., & Gutierrez, T. (2018). Agglomeration of nano-and micro-plastic particles in seawater by autochthonous and de novo-produced sources of exopolymeric substances. *Marine pollution bulletin, 130*, 258–267.

Sussarellu, R., Suquet, M., Thomas, Y., Lambert, C., Fabioux, C., Pernet, M. E. J., et al. (2016). Oyster reproduction is affected by exposure to polystyrene microplastics. *Proceedings of the National Academy of Sciences of the United States of America, 113*(9), 2430–2435.

Suzuki, M., Tachibana, Y., Oba, K., Takizawa, R., & Kasuya, K.-I. (2018). Microbial degradation of poly (ε-caprolactone) in a coastal environment. *Polymer Degradation and Stability, 149*, 1–8.

Tallec, K., Huvet, A., Di Poi, C., Gonzalez-Fernandez, C., Lambert, C., Petton, B., et al. (2018). Nano-plastics impaired oyster free living stages, gametes and embryos. *Environmental Pollution, 242*, 1226–1235.

Thiagarajan, V., Alex, S. A., Seenivasan, R., Chandrasekaran, N., & Mukherjee, A. (2021). Interactive effects of micro/nano-plastics and nanomaterials/pharmaceuticals: Their ecotoxicological consequences in the aquatic systems. *Aquatic Toxicology, 232*, 105747.

Thompson, R. C. (2006). Plastic debris in the marine environment: Consequences and solutions. In J. C. Krause, H. Nordheim, & S. Bräger (Eds.), *Marine nature conservation in Europe. Federal agency for nature conservation* (pp. 107–115). Stralsund, Germany.

Thompson, R. C., Olsen, Y., Mitchell, R. P., Davis, A., Rowland, S. J., John, A. W. G., et al. (2004). Lost at sea: Where is all the plastic? *Science, 304*, 838.

Tokiwa, Y., Calabia, B. P., Ugwu, C. U., & Aiba, S. (2009). Biodegradability of plastics. *International Journal of Molecular Sciences, 10*(9), 3722–3742.

Tourinho, P. S., Ivar do Sul, J. A., & Fillmann, G. (2010). Is marine debris ingestion still a problem for the coastal marine biota of southern Brazil? *Marine Pollution Bulletin, 60*, 396–401.

Toussaint, B., Raffael, B., Angers-Loustau, A., Gilliland, D., Kestens, V., Petrillo, M., et al. (2019). Review of micro/nano-plastic contamination in the food chain. *Food Additives & Contaminants: Part A, 36*(5), 639–673.

UNEP. (2015). *Biodegradable plastics and marine litter: Misconceptions, concerns and impacts on marine environments.* Nairobi: *United Nations Environment Programme (UNEP).*

Van Cauwenberghe, L., & Janssen, C. R. (2014). Micro-plastics in bivalves cultured for human consumption. *Environmental Pollution, 193*, 65–70.

Vandermeersch, G., Van Cauwenberghe, L., Janssen, C. R., Marques, A., Granby, K., Fait, G., et al. (2015). A critical view on micro-plastic quantification in aquatic organisms. *Environmental Research, 143*, 46–53.

Vázquez, O. A., & Rahman, M. S. (2021). An ecotoxicological approach to micro-plastics on terrestrial and aquatic organisms: A systematic review in assessment, monitoring and biological impact. *Environmental Toxicology and Pharmacology, 84*, 103615.

Vianello, A., Boldrin, A., Guerriero, P., Moschino, V., Rella, R., Sturaro, A., et al. (2013). Micro-plastic particles in sediments of Lagoon of Venice, Italy: First observations on occurrence, spatial patterns and identification. *Estuarine, Coastal and Shelf Science, 130*, 54–61.

Vivekanand, A. C., Mohapatra, S., & Tyagi, V. K. (2021). Micro-plastics in aquatic environment: Challenges and perspectives. *Chemosphere, 282*, 131151.

Wakkaf, T., El Zrelli, R., Kedzierski, M., Balti, R., Shaiek, M., Mansour, L., et al. (2020). Micro-plastics in edible mussels from a southern Mediterranean lagoon: Preliminary results on seawater-mussel transfer and implications for environmental protection and seafood safety. *Marine Pollution Bulletin, 158*, 111355.

Wang, T., Hu, M., Xu, G., Shi, H., Leung, J. Y. S., & Wang, Y. (2021). Micro-plastic accumulation via trophic transfer: Can a predatory crab counter the adverse effects of micro-plastics by body defence? *Science of the Total Environment, 754*, 142099.

Wang, J., Lu, L., Wang, M., Jiang, T., Liu, X., & Ru, S. (2019). Typhoons increase the abundance of micro-plastics in the marine environment and cultured organisms: A case study in Sanggou Bay, China. *Science of the Total Environment, 667*, 1–8.

Watts, A. J. R., Urbina, M. A., Corr, S., Lewis, C., & Galloway, T. (2015). Ingestion of plastic microfibers by the crab Carcinus maenas and its effect on food consumption and energy balance. *Environmental Science &Technology, 49*(24), 14597–14604.

Wegner, A., Besseling, E., Foekema, E. M., Kamermans, P., & Koelmans, A. A. (2012). Effects of nanopolystyrene on the feeding behavior of the blue mussel (Mytilus edulis L.). *Environmental Toxicology and Chemistry, 31*(11), 2490–2497.

Welden, N. A., Abylkhani, B., & Howarth, L. M. (2018). The effects of trophic transfer and environmental factors on micro-plastic uptake by plaice, *Pleuronectes plastessa*, and spider crab, *Maja squinado*. *Environmental Pollution, 239*, 351–358.

Welden, N. A., & Cowie, P. R. (2016). Environment and gut morphology influence micro-plastic retention in langoustine, *Nephrops norvegicus*. *Environmental Pollution, 214*, 859–865.

Xu, R. (2015). Light scattering: A review of particle characterization applications. *Particuology, 18*, 11–21.

Xu, M., Halimu, G., Zhang, Q., Song, Y., Fu, X., Li, Y., et al. (2019). Internalization and toxicity: A preliminary study of effects of nano-plastic particles on human lung epithelial cell. *Science of the Total Environment, 694*, 133794.

Yin, L., Chen, B., Xia, B., Shi, X., & Qu, K. (2018). Polystyrene micro-plastics alter the behavior, energy reserve and nutritional composition of marine jacopever (*Sebastes schlegelii*). *Journal of Hazardous Materials, 360*, 97–105.

Yoshida, S., Koshima, I., Sasaki, A., Fujioka, Y., Nagamatsu, S., Yokota, K., et al. (2019). Mechanical dilation using nylon monofilament aids multisite lymphaticovenous anastomosis through improving the quality of anastomosis. *Annals of Plastic Surgery, 82*, 201–206.

Yuan, Z., Nag, R., & Cummins, E. (2022). Human health concerns regarding micro-plastics in the aquatic environment - from marine to food systems. *Science of the Total Environment, 823*, 153730.

Zarfl, C., & Matthies, M. (2010). Are marine plastic particles transport vectors for organic pollutants to the Arctic? *Marine Pollution Bulletin, 60*(10), 1810–1814.

Zhang, S., Ding, J., Razanajatovo, R. M., Jiang, H., Zou, H., & Zhu, W. (2019). Interactive effects of polystyrene micro-plastics and roxithromycin on bioaccumulation and biochemical status in the freshwater fish red tilapia (*Oreochromis niloticus*). *Science of the Total Environment*, 1431–1439.

Zhang, H., Kuo, Y., Gerecke, A. C., & Wang, J. (2012). Co-release of hexabromocyclododecane (HBCD) and nano- and microparticles from thermal cutting of polystyrene foams. *Environmental Science & Technology, 46*, 10990–10996.

Zhang, F., Wang, X., Xu, J., Zhu, L., Peng, G., Xu, P., et al. (2019). Food-web transfer of micro-plastics between wild caught fish and crustaceans in East China Sea. *Marine Pollution Bulletin, 146*, 173–182.

Zhao, K., Wei, Y., Dong, J., Zhao, P., & Wang, Y. (2022). Separation and characterization of micro-plastic and nano-plastic particles in marine environment. *Environmental Pollution, 297*, 118773.

Zitko, V., & Hanlon, M. (1991). Another source of pollution by plastics: Skin cleansers with plastic scrubbers. *Marine Pollution Bulletin, 22*, 41–42.

Zou, B., Zou, B., Nsangue, N., Thierry, B., Tang, H., Xu, L., et al. (2022). The Deformation characteristics and flow field around knotless polyethylene netting based on fluid structure interaction (FSI) one-way coupling. *Aquaculture and Fisheries, 7*, 89–102.

CHAPTER SEVEN

Occurrence of nano/microplastics from wild and farmed edible species. Potential effects of exposure on human health

Celia Rodríguez-Pérez[a,b,c,*], Miguel Sáenz de Rodrigáñez[d], and Héctor J. Pula[e,f]

[a]Department of Nutrition and Food Science, Faculty of Health Sciences, University of Granada (Melilla Campus), Melilla, Spain
[b]Biomedical Research Centre, Institute of Nutrition and Food Technology (INYTA) 'José Mataix', University of Granada, Granada, Spain
[c]Instituto de Investigación Biosanitaria ibs.GRANADA, Granada, Spain
[d]Department of Physiology, Faculty of Health Sciences, University of Granada (Melilla Campus), Melilla, Spain
[e]Fish Nutrition and Feeding Research Group, Faculty of Science, University of Granada, Granada, Spain
[f]Aula del Mar Cei-Mar of the University of Granada, Faculty of Sciences, Granada, Spain
*Corresponding author: e-mail address: celiarp@ugr.es

Contents

Abstract

The occurrence of nano/microplastics (N/MPs) has become a global concern due to their risk on the aquatic environment, food webs and ecosystems, thus, potentially affecting human health. This chapter focuses on the most recent evidence about the occurrence of N/MPs in the most consumed wild and farmed edible species, the occurrence of N/MPs in humans, the potential impact of N/MPs on human health as well as future research recommendations for assessing N/MPs in wild and farmed edible species. Additionally, the N/MP particles in human biological samples, which include the standardization of methods for collection, characterization, and analysis of N/MPs that might allow evaluating the potential risk of the intake of N/MPs in human health, are discussed. Thus, the chapter consequently includes relevant information about the content of N/MPs of more than 60 edible species such as algae, sea cucumber, mussels, squids, crayfish, crabs, clams, and fishes.

Advances in Food and Nutrition Research, Volume 103
ISSN 1043-4526
https://doi.org/10.1016/bs.afnr.2022.08.003

1. Introduction

We live in a society where plastic is an integral part of our daily life becoming indispensable. This incredibly versatile material is used in packaging, textiles, agricultural, automotive, construction or health sectors, among others (Rodrigues et al., 2019). As a result, its production has not stopped growing in recent years. In fact, the latest data reflect a global plastics production of over 367 million metric tons in 2020, which means an increase of 97 million metric tons during the last 10 years (Tiseo, 2022). The increased use of this material along with its inappropriate management and disposal is recognized as one of the main environmental problems today (Huang et al., 2021). This material has been found even in the Antarctic snow samples according to a recent study (Aves et al., 2022). In accordance with the "Plastic Garbage Project," the largest amount of plastic waste comes from the packaging industry.

Pollution with plastics is one of the major concerns in fresh and marine water systems that directly affect the quality of the edible species and, therefore, it might affect consumer's health. It is estimated that around 80% of plastics enter the ocean from land-based sources while some plastics come from maritime operations (Smith, Love, Rochman, & Neff, 2018), which represents up to 85% of marine litter (Auta, Emenike, & Fauziah, 2017). Most fishing gears such as fishing net and bucket, breeding, hatching, cultivating devices and others (Mohapatra, Sarkar, Barik, & Jayasankar, 2011; Mohapatra, Sarkar, & Singh, 2003) are plastic or contain plastic in their composition. For instance, the accumulation of the lost/discarded fishing gears (e.g., trawls, purse seines, Danish seines, gillnets, longline and traps/pots) increased up to over 4000 tons in 2016 in the oceans along the Norwegian commercial fishing alone (Deshpande, Philis, Brattebø, & Fet, 2020). Inputs of different kinds of external plastics also have become one of the important sources of pollution in the aquaculture environment (Ivleva, Wiesheu, & Niessner, 2017; Wang, Lee, Chiu, Lin, & Chiu, 2020). This is the case of the loss and scrap of the fishing plastic gears and the breeding facility equipment, which includes plastic equipment, bait machine, aerator, water pump or electric wire, among others but also the feeds natural or artificial (Van Cauwenberghe, Devriese, Galgani, Robbens, & Janssen, 2015; Wright, Thompson, & Galloway, 2013).

Plastic consists of polymers synthetically produced by materials such as mineral oil, coal, or natural gas, being the most common polyethylene

(PE), polypropylene (PP), polyvinyl chloride (PVC), polystyrene (PS), polyethylene terephthalate (PET), and polyurethane (PUR) (Farzi, Dehnad, & Fotouhi, 2019). Although many efforts have been made to define nano- and microplastics (NPs and MPs), the definition is still inconsistent. In fact, some authors have reflected the differences in the categorization of plastic according to size as applied in scientific literature and in institutional reports and highlighting the importance of a consensus to avoid ambiguous communication and the generation of incomparable data (Hartmann et al., 2019). However, to date, the most accepted definition of NPs usually refers to structures one dimension in a size range from 1 to 100 nm while MPs are particles with <5 mm with no defined lower size limit (Paul et al., 2020). At the same time, MPs are divided into primary (intentionally manufactured) or secondary (generated from larger plastics by photodegradation, mechanical abrasion, thermal degradation, or bacteria). Additionally, the definition includes a heterogeneous mixture of differently shaped materials referred to as fragments, fibers, spheroids, granules, pellets, flakes, or beads (EFSA, 2018). Other classes including type, shape, color, or erosion to describe MPs have been described in more detail by Yuan, Nag, and Cummins (2022).

The most frequent MPs polymers found in aquatic environment such as PP, PE, PVC, PET, PS, PUR, acrylonitrile-butadiene-styrene (ABS), PC, polyamide, and styrene acrylonitrile together with their characteristics, chemical structure, main use, and global production have been extensively reviewed recently (Yuan et al., 2022). Those microscopically small plastic particles like nano- and microplastics (N/MPs) are extremely persistent and they can be accumulated in all compartments of water ecosystems, contaminating the food web. To date, N/MPs or plastics have been found in the marine organisms over 233 species worldwide. Around 60% of the most farmed aquaculture species and 80% of most caught marine species have shown evidence of their capacity to ingest MP fragments (Walkinshaw, Lindeque, Thompson, Tolhurst, & Cole, 2020). Regarding the farmed species, most studies have been focused on marine shellfish and bivalve species although there are a very few studies focused on freshwater species (Carbery, O'Connor, & Palanisami, 2018; Cole et al., 2013; De Sá, Oliveira, Ribeiro, Rocha, & Futter, 2018; Granek, Brander, & Holland, 2020; Koelmans, 2015; Lavers & Bond, 2017; Lusher, Hollman, & Mendoza-Hill, 2017; Vandermeersch et al., 2015; Wilcox, Van Sebille, & Hardesty, 2015; Ziajahromi, Neale, & Leusch, 2016). Nano/microplastics can be later ingested as food by aquatic organisms and accumulated in the gastrointestinal tract of fish. In fact, they have been also found in zooplankton that belongs to the lower trophic levels (Amin, Sohaimi, Anuar, & Bachok, 2020; Botterell et al., 2019). However, they can also be transported to the liver and other organs (Kim, Yu, & Choi, 2021) as will be detailed in the following sections. They have been detected in fish, shellfish, crustaceans, mollusks, and even mammals (Danopoulos, Jenner, Twiddy, & Rotchell, 2020). Despite plastic polymers being considered biochemically inert, they can be accumulated in the tissues and act as vehicles for transport of pathogens, adsorb and accumulate toxic pollutants (Yuan et al., 2022).

It should be highlighted that forecasts are not encouraging, since it is expected that plastic waste generation and mismanagement of its disposal will triple by 2060 (Danopoulos et al., 2020). Additionally, it is estimated that microplastic contamination of aquatic environments will continue to increase in the near future.

Considering the objectives (O) O3, O6, O13, O14 and O15 from the Sustainable Development Goals (https://www.undp.org/sustainable-development-goals), the study of N/MPs is a hot topic, which has increased in recent years due to the growing awareness of their impact on the natural environment and their biological effects. Therefore, this section aims to briefly review the current status of N/MPs and to cover the latest evidence on the occurrence of N/MPs in marine and aquaculture edible species, humans as well as their potential impact on human health.

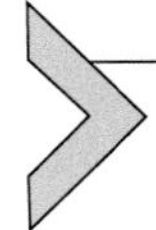

2. Occurrence of nano/microplastics in edible wild and farmed species

According to the National Oceanic and Atmospheric Administration (NOAA Fisheries, 2017), the European Commission (European Commission, 2018) and data from ourworldindata.org, the most consumed fish and seafood per capita worldwide are tuna (mostly canned), cod, salmon, Alaska pollock, shrimps, mussel, herring, hake, squid, mackerel, crabs, and clams (FAO Food and Agriculture Organization of the United Nations, 2020). Several studies have detected N/MPs in those species at different levels. In this regard, Tables 1 and 2 summarize the organism type, country, species, main polymer type, the quantity of N/MPs as well as the organ/tissue of wild and farmed edible species in which the polymers have been detected. Despite most research papers are from the last 3 years, for some species older bibliography has been included. Considering all data from

Table 1 Occurrence of N/MPs in wild edible species.

Organism type	Country	Species	Most common polymer type	Quantity MPs/g (mean ± SD)	Quantity MPs/Ind (mean ± SD)	Tissue	References
Wild algae	China	*Gracilaria lemaneiformis*	Fibers	1 ± 0.3	n.a.	All	Li, Su, Ma, Feng, and Shi (2022)
Wild algae	China	Laminaria (*Saccharina japónica*)	Fibers	4.42 ± 0.45	n.a.	All	Li et al. (2022)
Wild algae	China	*Ulva prolifera*	PE	4.66	n.a.	All	Feng et al. (2020)
Wild sea cucumber	Indonesia	Sandfish (*Holothuria scabra*)	PE	0.089 ± 0.058	2.01 ± 1.59	Viscera	Riani and Cordova (2022)
Wild sea cucumber	Croatia	Sandfish (*Holothuria tubulosa*)	Fibers	n.a.	0.74 ± 3.7	All	Renzi and Blašković (2020)
Wild mussels	Turkey	Mussel (*Mytilus galloprovincialis*)	PET fragment	0.23	0.69	Soft tissues	Gedik, Eryaşar, and Gözler (2022)
Wild mussels	Spain	Mussel (*Mytilus* sp.)	Fibers	2.07 ± 2.04	n.a.	Soft tissues	Reguera, Viñas, and Gago (2019)
Wild mussels	France	Mussel (*Mytilus edulis*)	Fibers	0.15 ± 0.06	0.76 ± 0.40	Soft tissues	Hermabessiere et al. (2019)
Wild mussels	France	*Cerastoderma edule*	Fibers	0.74 ± 0.35	2.46 ± 1.16	Soft tissues	Hermabessiere et al. (2019)
Wild mussels	Korea	Mussel (*M. edulis*)	PE	0.12 ± 0.11	n.a.	Soft tissues	Cho, Shim, Jang, Han, and Hong (2019)

Continued

Table 1 Occurrence of N/MPs in wild edible species.—cont'd

Organism type	Country	Species	Most common polymer type	Quantity MPs/g (mean ± SD)	Quantity MPs/Ind (mean ± SD)	Tissue	References
Wild mussels	Korea	Oyster (*Crassostrea gigas*)	PE	0.07 ± 0.06	n.a.	Soft tissues	Cho et al. (2019)
Wild oysters	EEUU	Oyster *(C. gigas)*	Fibers	0.35 ± 0.04	8.84 ± 0.45	Soft tissues	Baechler, Granek, Hunter, and Conn (2020)
Wild mussels	France	Mussel (*M. edulis*)	PP fragments	0.23 ± 0.20	0.60 ± 0.56	Soft tissues	Phuong, Poirier, Pham, Lagarde, and Zalouk-Vergnoux (2018)
Wild oysters	France	Oyster *(C. gigas)*	PP fragments	0.18 ± 0.16	2.10 ± 1.71	Soft tissues	Phuong et al. (2018)
Wild mussels	China	Mussel (*M. edulis*)	Fibers and fragments	2.75	4.55	Soft tissues	Li et al. (2016)
Wild mussels	Norway	Mussel (*Mytilus* spp.)	Fibers	0.97 ± 2.6	1.5 ± 2.3	Soft tissues	Bråte et al. (2018)
Wild mussels	Scotland	Mussel (*Mytilus* spp.)	Polyester or PE and Poly (ether-urethane)	3.0 ± 0.9	n.a.	Soft tissues	Catarino, Macchia, Sanderson, Thompson, and Henry (2018)
Wild mussels	Belgium	Commercial and wild Mussels (*Mytilus* spp.)	Orange Fibers	0.39	n.a.	Soft tissues	De Witte et al. (2014)

Wild mussels	Hong Kong	Mussel (*P. viridis*)	PP and PE fragments	n.a.	32.7 ± 29.3	Soft tissues	Phuong et al. (2018)
Wild mussels	French–Belgian–Dutch coastline	Mussel (*M. edulis*)	Low-density PE, high-density PE, and PS (from 20 to 90)	0.2 ± 0.3	n.a.	Soft tissues	Leung et al. (2021)
Wild/cultured mussels	Italy	Mussel (*M. galloprovincialis*)	n.a.	6.7	7.7	Soft tissues	Renzi, Guerranti, and Blašković (2018)
Wild squid	India	Indian squid (*Uroteuthis duvaucelii*)	PP	0.008 ± 0.02	0.18 ± 0.48	Edible tissue	Daniel, Ashraf, Thomas, and Thomson (2021)
Wild squid	Peru	Peruvian squid (*Dosidicus gigas*)	Fibers	0.45	5.7	Gill, intestine, and stomach	Gong et al. (2021)
Wild cuttlefish	Portugal	Cuttlefish (*Sepia officinalis*)	Fibers	0.14 ± 0.04	39	Digestive system	Oliveira et al. (2020)
Wild prawns	India	White shrimp (*Fenneropenaeus indicus*)	Red and blue Fibers	0.04 ± 0.07	n.a.	Soft tissue (with the GIT)	Daniel, Ashraf, and Thomas (2020)
Wild prawns	India	Kiddi shrimp (*Parapeneopsis stylifera*)	Fibers	5.36 ± 2.81	n.a.	GIT	Gurjar et al. (2021)

Continued

Table 1 Occurrence of N/MPs in wild edible species.—cont'd

Organism type	Country	Species	Most common polymer type	Quantity MPs/g (mean ± SD)	Quantity MPs/Ind (mean ± SD)	Tissue	References
Wild prawns	India	Pink shrimp (*Metapenaeus monoceros*)	Fibers	7.23 ± 2.63	n.a.	GIT	Gurjar et al. (2021)
Wild prawns	India	White shrimp (*Penaeus indicus*)	Fibers	7.40 ± 2.60	n.a.	GIT	Gurjar et al. (2021)
Wild prawns	Southwestern Atlantic	Shrimp (*Pleoticus muelleri*)	Fibers	1.31	n.a.	Abdominal muscle	Severini et al. (2020)
Wild prawns	Iran	Green tiger prawn (*Penaeus semisulcatus*)	Pink/red fibers (from 500 to 1000 μm) and fragments (<50 μm)	0.36	n.a.	Muscle	Akhbarizadeh, Moore, and Keshavarzi (2019)
Wild prawns	China	Spear shrimp (*Parapenaeopsis hardwickii*)	Fibers (<1mm)	n.a.	1.4 ± 0.37	Tissue	Wang, Lee, Chiu, Lin, and Chiu (2020)
Wild prawns	China	Pacific white shrimp (*Litopenaeus vannamei*)	Blue fibers in small size (<0.5 mm)	14.08 ± 5.70	n.a.	GIT	Yan et al. (2021)
Wild prawns	Iran	Green tiger prawn (*P. semisulcatus*)	Fibrous fragments (<100 μm to >1000 μm)	7.8/ 1.5	n.a.	Exoskeleton and muscle	Abbasi et al. (2018)

Wild prawns	India	Indian white shrimp (*F. indicus*)	n.f.	n.f.	n.f.	Tissue	Daniel et al. (2021)
Wild prawns	India	Indian white shrimp (*F. indicus*)	n.f.	n.f.	n.f.	Tissue	Daniel et al. (2021)
Wild prawns	India	Kadal shrimp (*Metapenaeus dobsoni*)	n.f.	n.f.	n.f.	Tissue	Daniel et al. (2021)
Wild prawns	India	Brown shrimp (*Metapenaeus monoceros*)	m-FTIR as polyamide	3.87 ± 1.05	n.a.	GIT	Hossain et al. (2020)
Wild prawns	India	Giant tiger prawn (*Penaeus monodon*)	m-FTIR as polyamide	3.40 ± 1.23	n.a.	GIT	Hossain et al. (2020)
Wild prawns	Iran	Jinga shrimp (*Metapenaeus affinis*)	Fibers (PET)	1.02	n.a.	Soft tissue	Keshavarzifard, Vazirzadeh, and Sharifinia (2021)
Wild crabs	Iran	Blue swimmer crab (*Portunus armatus*)	Fibers and fragments	0.256	n.a.	Muscle	Akhbarizadeh et al. (2019)
Wild crabs	Italy	Blue crab (*Callinectes sapidus*)	PE and PET	n.a.	2.5 ± 1.6	Stomach	Renzi et al. (2020)
Wild crabs	China	Gazami crab (*Portunus trituberculatus*)	Fibers (<1 mm)	0.16 ± 0.11	0.25 ± 0.14	Muscle	Wang et al. (2020)
Wild crabs	China	Gazami crab (*P. trituberculatus*)	Fibers	n.a.	n.a.	Gills, guts and muscle	Zhang et al. (2021)

Continued

Table 1 Occurrence of N/MPs in wild edible species.—cont'd

Organism type	Country	Species	Most common polymer type	Quantity MPs/g (mean ± SD)	Quantity MPs/Ind (mean ± SD)	Tissue	References
Wild crabs	China	Lady crab (*Charybdis japonica*)	Fibers	n.a.	n.a.	Gills, guts and muscle	Zhang et al. (2021)
Wild crabs	India	Blue swimmer crab (*Portunus pelagicus*)	n.a.	0.003 ± 0.01	0.14 ± 0.44	Muscle	Daniel et al. (2021)
Wild clams	India	Yellow clams (*Meretrix casta*)	Fibers and fragments	37.1 ± 35.1	n.a.	Soft tissue	Naidu, Xavier, Shukla, Jaiswar, and Nayak (2022)
Wild clams	India	Black clam (*Villorita cyprinoides*)	Fibers	Digestive gland: 22.8; gill: 29.6	n.a.	Digestive gland and gill	Joshy, Sharma, Mini, Gangadharan, and Pranav (2022)
Wild clams	China	Asian clam (*Corbicula fluminea*)	MicroFibers	2.6	2.7	Soft tissue	Su et al. (2018)
Wild clams	Brazil	*Anomalocardia flexuosa*	Fragments and Fibers	3.66 ± 2.59	5.15 ± 3.80	Soft tissue	Bruzaca et al. (2022)
Wild clams	China	*Ruditapes philippinarum*	Fibers and fragments	0.92	2.34	Soft tissue	Zhang et al. (2022)
Wild fishes	South Africa	Common carp (*Cyprinus carpio*)	HD-PE, fibers	0.042 ± 0.053	26.23 ± 12.57	GIT	Saad et al. (2022)
Wild fishes	Spain	Common carp (*C. carpio*)	Rayon	0.3 ± 0.3	8 ± 6	GIT	Guilhermino et al. (2021)

Wild fishes	Spain	Flathead grey mullet (*Mugil cephalus*)	Rayon	0.3 ± 0.2	10 ± 9	GIT	Guilhermino et al. (2021)
Wild fishes	Spain	Flounder (*Platichthys flesus*)	Rayon	0.2 ± 0.2	1 ± 1	GIT	Guilhermino et al. (2021)
Wild fishes	Greece	European sardine (*Sardina pilchardus*)	PE fragment	n.a.	1.8 ± 0.2	GIT	Digka, Tsangaris, Torre, Anastasopoulou, and Zeri (2018)
Wild fishes	EEUU	Northern anchovy (*Engraulis mordax*)	Fibers	n.a.	0.3 ± 0.5	Gut content	Rochman et al. (2015)
Wild fishes	EEUU	Striped bass (*Morone saxatilis*)	Fibers	n.a.	0.9 ± 1.2	Gut content	Rochman et al. (2015)
Wild fishes	EEUU	Chinook salmon (*Oncorhynchus tshawytscha*)	Fibers	n.a.	0.25 ± 0.5	Gut content	Rochman et al. (2015)
Wild fishes	Canada	Chinook salmon (*O. tshawytscha*)	Fibers	n.a.	1.2 ± 1.4	Gut content	Collicutt, Juanes, and Dudas (2019)
Wild fishes	Colombia	Nile tilapia (*Oreochromis niloticus*)	Lines or fibers	n.a.	13.5	GIT	Oliva-Hernández, Santos-Ruiz, Muñoz-Wug, and Pérez-Sabino (2021)

Continued

Table 1 Occurrence of N/MPs in wild edible species.—cont'd

Organism type	Country	Species	Most common polymer type	Quantity MPs/g (mean ± SD)	Quantity MPs/Ind (mean ± SD)	Tissue	References
Wild fishes	Ireland	Brown Trout (*Salmo trutta*)	Fibers	n.a.	1.88 ± 1.53	GIT	O'Connor et al. (2020)
Wild fishes	Northeast Greenland	Polar cod (*Boreogadus saida*)	Fibers (PS, PET, polyester, rayon)	n.a.	n.a.	GIT	Morgana et al. (2018)
Wild fishes	Norway	Cod (*Gadus morhua*)	PVC and PS	n.a.	n.a.	Liver, stomach, intestine and muscle	Haave, Lorenz, Primpke, and Gerdts (2018)
Wild fishes	Poland	*Clupea harengus* L.	Fragments and Fibers	n.a.	n.a.	GIT and gills	Białowąs et al. (2022)
Wild fishes	South Africa	West Coast round herring (*Etrumeus whiteheadi*)	Micro Fibers	n.a.	1.38	GIT	Bakir et al. (2020)
Wild fishes	Sweden	*Clupea harengus membras*	Fibers and fragments	n.a.	0.9	GIT	Ogonowski et al. (2019)
Wild fishes	Spain	Hake (*Merluccius merluccius*)	Fibers and fragments	3.9 ± 2.8	n.a.	GIT	Cabanilles et al. (2022)
Wild fishes	Italy	Hake (*M. merluccius*)	Fibers and fragments	n.a.	1.3 ± 0.4	GIT	Giani, Baini, Galli, Casini, and Fossi (2019)

Wild fishes	Italy	Hake (*M. merluccius*)	Fibers	n.a.	n.a.	Stomach	Mancuso, Savoca, and Bottari (2019)
Wild fishes	Spain	Atlantic chub mackerel (*Scomber colias*)	Fibers	n.a.	2.77	GIT	Herrera et al. (2019)
Wild fishes	Thailand	Mackerel (*Rastrelliger brachysoma*)	Blue Fibers and fragment (from 3 to 5.0-mm)	n.a.	2.70 ± 16.62	GIT	Hajisamae, Soe, Pradit, Chaiyvareesajja, and Fazrul (2022)
Wild fishes	Scotland	Atlantic mackerel (*Scomber scombrus*)	Fibers	0.29 ± 0.05	n.a.	Flesh tissue	Akoueson et al. (2020)
Wild fishes	Hong Kong	Japanese jack mackerel (*Trachurus japonicus*)	PP, PS, PE	n.a.	8.33 ± 7.09	Soft tissue	Leung et al. (2021)
Wild fishes	Portugal	Atlantic horse mackerel fish (*Trachurus trachurus*)	n.a.	n.a.	n.a.	Gill, heart, kidney, gut	Prata, da Costa, Duarte, and Rocha-Santos (2022)

Ind, individual; GIT, gastrointestinal tract; n.a., not available; n.f., not found; PC, polycarbonate; PE, Polyethylene; PET, polyethylene terephthalate; PP, polypropylene; PS, Polystyrene

Table 2 Occurrence of N/MPs in farmed edible species.

Organism type	Country	Species	Most common polymer type	Quantity MPs/g. (mean ± SD)	Quantity MPs/Ind. (mean ± SD)	Tissue	References
Farmed algae	China	*Nori* (*Pyropia yezoensis*)	PE, fibers	0.453 ± 0.213	n.a.	All	Feng et al. (2020)
Farmed algae	China	*Ulva prolifera*	PE, fibers	0.931 ± 0.343	n.a.	All	Feng et al. (2020)
Farmed algae	China	*Cladophora* spp.	PE, fibers	0.249 ± 0.018	n.a.	All	Feng et al. (2020)
Farmed algae	China	*Undaria pinnatifida*	PE, fibers	0.293 ± 0.021	n.a.	All	Feng et al. (2020)
Farmed algae	Thailand	*Gracilaria fisheri*	Fibers	1.817 ± 0.864	n.a.	All	Klomjit, Sutthacheep, and Yeemin (2021)
Farmed algae	China	*Nori* (*P. yezoensis*)	PE, fibers	1.8 ± 0.7	n.a.	All	Li, Feng, Zhang, Ma, and Shi (2020)
Farmed sea cucumber	China	Sandfish (*Apostichopus japonicus*)	Microfibers	n.a.	1.23 ± 0.75	All	Mohsen et al. (2019)
Farmed oysters	China	Oyster (*Crassostrea* sp.)	Cellophane, fibers	1.23 ± 0.75	n.a.	Soft tissues	Van Cauwenberghe et al. (2015)
Farmed mussels	China	Different species (*Perna viridis* & *Meretrix meretrix* & *Mytilus edulis*)	PET	0.29 ± 0.08	1.41 ± 0.36	Soft tissues	Fang et al. (2019)

Farmed cuttlefish	Portugal	Cuttlefish (*Sepia officinalis*)	Fibers	0.06 ± 0.02	28.5	Digestive system	Oliveira et al. (2020)
Farmed crayfish	China	Red swamp crayfish (*Procambarus clarkii*)	Fibers PP:PE	n.a.	0.71 ± 0.18	Digestive system	Zhang et al. (2021)
Farmed prawns	Mexico	Pacific whiteleg shrimp (*Litopenaeus vannamei*)	Fibers	261.7 ± 84.5	7.6 ± 0.6	GIT	Valencia-Castañeda et al. (2022)
Farmed prawns	Ecuador	Pacific whiteleg shrimp (*L. vannamei*)	Film	13 ± 1	n.a.	GIT	Curren, Leaw, Lim, and Leong (2020)
Farmed prawns	Malaysia	Pacific whiteleg shrimp (*L. vannamei*)	Film	21 ± 4	n.a.	GIT	Curren et al. (2020)
Farmed prawns	Indian Ocean	Indian white shrimp (*Fenneropenaeus indicus*)	Spheres	5570 ± 100	n.a.	GIT	Curren et al. (2020)
Farmed prawns	Thailand	Pacific whiteleg shrimp (*L. vannamei*)	PE, fibers	11.00 ± 4.60	4–23	GIT	Reunura and Prommi (2022)
Farmed prawns	Thailand	Giant Freshwater Prawn (*Macrobrachium rosenbergii*)	PE, fibers	33.37 ± 19.25	11–74	GIT	Reunura and Prommi (2022)
Farmed fish	Italy/ Croatia	Common carp (*Cyprinus carpio*)	Polyester	n.a.	0.11	GIT	Savoca et al. (2021)

Continued

Table 2 Occurrence of N/MPs in farmed edible species.—cont'd

Organism type	Country	Species	Most common polymer type	Quantity MPs/g. (mean $\pm$ SD)	Quantity MPs/Ind. (mean $\pm$ SD)	Tissue	References
Farmed fish	Italy/ Croatia	Gilthead Seabream (*Sparus aurata*)	Polyester	n.a.	0.48	GIT	Savoca et al. (2021)
Farmed fish	China	Grass carp (*Ctenopharyngodon idella*)	Fibers	0.044 ± 0.03	13 ± 9	GIT and gills	Aiguo et al. (2022)
Farmed fish	China	Silver carp (*Hypophthalmichthys molitrix*)	Fibers	0.035 ± 0.026	9.083 ± 4.699	GIT and gills	Aiguo et al. (2022)
Farmed fish	China	Nile tilapia (*Oreochromis niloticus*)	Fibers	n.a.	9.357 ± 7.143	GIT and gills	Aiguo et al. (2022)
Farmed fish	Spain	European sea bass (*Dicentrarchus labrax*)	Cellulose fiber	n.a.	1.65 ± 1.3	GIT	Reinold et al. (2021)

Ind, individual; GI, gastrointestinal tract; n.a., not available; n.f., not found; PC, polycarbonate; PE, Polyethylene; PET, polyethylene terephthalate; PP, polypropylene; PS, Polystyrene.

the literature analyzed, 98 out of 104 studies reported the detection of N/MPs while 64, 52 and 28 provided N/MPs abundance expressed by particles per gram sample, per animal or both, respectively. Additionally, Fig. 1 shows the places where the species were captured or collected.

2.1 Occurrence of nano/microplastics in edible wild marine species

Table 1 includes 54 different species from diverse organisms including wild algae, sea cucumber, mussels, squids, crayfish, crabs, clams, and fishes from 25 different countries. Although most studies found more than one polymer type in the same organ/tissue of the reviewed wild species, it should be mentioned that more than a half of the polymers found were fibers (~57%) followed by fragments (~11%). However, there was no available data of N/MPs for all species, which is the case of the blue swimmer crab (*Portunus pelagicus*) and two different species of shrimp (*Fenneropenaeus indicus, Metapenaeus dobsoni*) from India and the Atlantic horse mackerel (*Trachurus trachurus*) from Portugal. Fig. 2 shows the studied organs and/or tissues of the different species and the edible portion of the set of species included in Tables 1 and 2. It can be observed that around 41% of the reviewed studies used soft tissue for determining N/MPs in wild species followed by the analysis of the GIT (28%) and muscle (12%) (Fig. 2A). If it is considered that there are species consumed whole (e.g., mussels or clams) and others in which the GIT, gills and other parts of the fish are commonly discarded (e.g., cod, hake, or mackerel), it can be seen that 60% of the studies found N/MPs in the edible parts of the studied species (Fig. 2C).

Mussels are considered one of the species with higher levels of N/MPs. The most identified N/MPs were fibers and fragments. Interestingly, Li et al. (2016) found that wild mussels (*Mytilus edulis*) presented higher content of MPs than farmed one. In addition, they reported higher contamination in areas in which intensive humans' activities were present. Contrarily, Phuong et al., 2018 did not find differences between the content of MPs in mussels from different sampling sites, seasons, or mode of life effect. They reported higher frequency of MPs detection in cultivated mussels than wild. On their behalf, orange fibers were the most predominant in mussels from the Belgian coastline (De Witte et al., 2014) while grey fragments of PP and PE from 50 to 100 μm were found in higher concentration in mussels from the French Atlantic coast (Phuong et al., 2018). It was also found the MPs in mussels from Scotland (*M. edulis* and the protected mussel species *Modiolus modiolus*) where 48% of the observed 27 fibers were MPs

Fig. 1 Countries from which the reviewed species have been captured or farmed. The red spots represent the country. The corresponding bibliographical reference has been also included in the map.

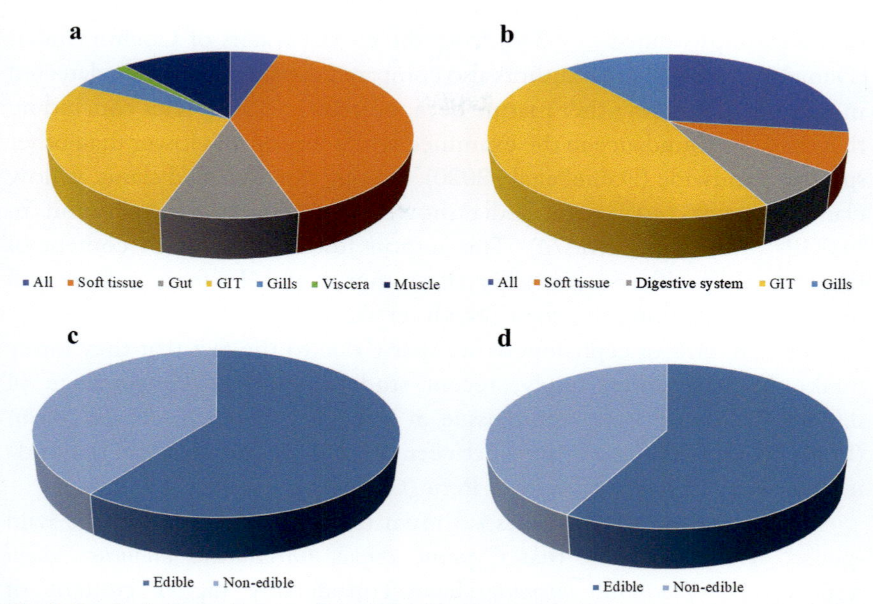

Fig. 2 Studied organs/tissues of the (A) wild species and (B) farmed species and edible portion of (C) wild species and (D) farmed species of N/MPs.

(Catarino et al., 2018). Several studies reviewed by Walkinshaw et al. (2020) showed quantities among 0.2–5.36 microplastics per gram w.w. in wild mussels. As bivalves, clams are also considered sources of N/MPs exposition and several authors have proposed them as indicators of water pollution (Su et al., 2018).

Since the capacity of mussels to filter large volumes of water, several studies have shown positive correlation between the N/MPs in water and corresponding mussel samples. Thus, mussels have been proposed to be considered as bioindicators of coastal microplastic pollution (Bråte et al., 2018; Catarino et al., 2018; Li et al., 2016). However, standardized methods are required for comparison between studies as the amount of N/MPs appears to be weight/size dependent (Catarino et al., 2018).

Despite fibers, fragments, pellets, beads, and films were found in different species of shrimps from the Arabian Sea, fibers occurrence was the most significant, accounting up to 47% in the GIT of *Parapeneopsis stylifera* (Gurjar et al., 2021). Additionally, fibers have been described as the predominant MPs in the muscle of the commercial shrimp (*Pleoticus muelleri*) (Severini et al., 2020). Those polymers were the majority MP found in nearly 31%

of the shrimps studied ($n=330$) from the coastal waters of Cochin (India) (Daniel et al., 2020). The authors also compared the average of MPs detected in their samples with other marine decapod crustacean's species concluding that the MPs abundance in the examined white shrimp was lower than other species worldwide (Daniel et al., 2020). Among the reviewed clams, yellow clams (*Meretrix casta*) from India showed the highest concentration in N/MPs (37.1 ± 35.1 MPs/g). The authors justified such high content of the lack of specific processes for the removal of MPs from sewage and domestic water along the coast (Naidu et al., 2022).

Data on MPs of cephalopods are sparse due to the fact that they rarely intake N/MPs. Interestingly, recent studies reported the presence of 0.008 ± 0.02 particles/g edible tissue in Indian squid (*Uroteuthis duvaucelii*) (Daniel et al., 2021) and a range between 0.2 and 0.7 particles/g of the studied tissues in *Dosidicus gigas* from Peru (Gong et al., 2021).

Average of N/MPs in fishes is shown in Fig. 3. Focusing on the wild species, it is remarkable that *Cyprinus carpio*, *Oreochromis niloticus*, *Mugil cephalus* and *Trachurus japonicus* have considerably higher content of N/MPs than the other reviewed species. It is noteworthy that the analysis of N/MPs have been carried out in the edible portion of only the last

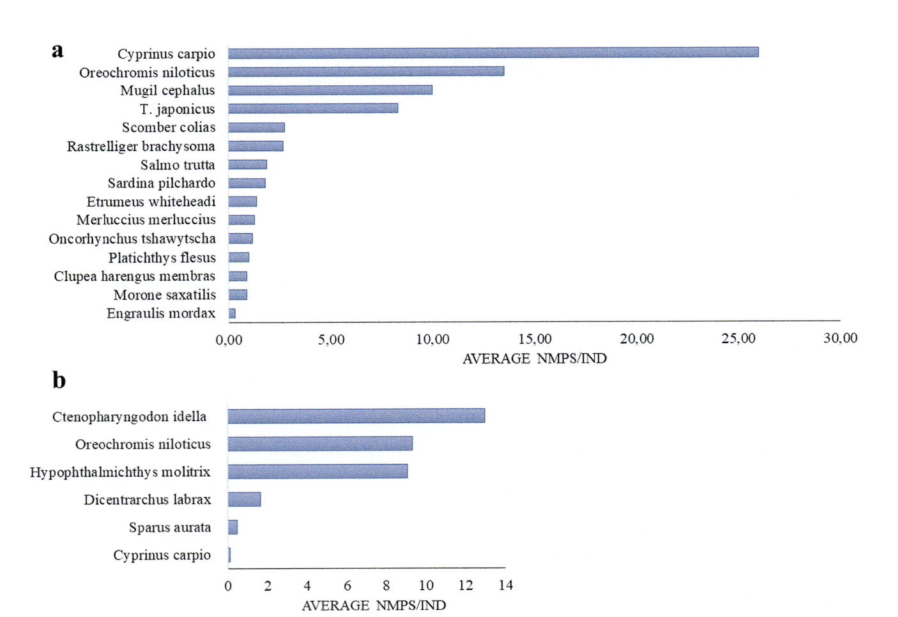

Fig. 3 Average of N/MPs in wild (A) and farmed (B) fishes expressed as N/MPs per individual.

one (*T. japonicus*) together with other mackerel species from Portugal (*Trachurus trachurus*). Mackerel have high commercial value both for fresh consumption and for canning. Some MPs have been found in mackerel species such as the Atlantic chub mackerel. In this regard, fibers, among other MPs, have been detected in nearly 78% of the studied samples from the Canary Island coast (Herrera et al., 2019). Additionally, MPs (mainly blue fibers and fragments) were found in 3.16% of the studied mackerel (*Rastrelliger brachysoma*) from the Pattani Bay (Thailand) (Hajisamae et al., 2022). Non-edible parts (i.e., gill, heart, kidney, and gut) of Atlantic horse mackerel captured from Portugal also showed variable levels of possible MPs from which 65.4% were particles <1–10 μm (Prata et al., 2022).

The content of N/MPs within the same species can vary considerably. This is the case of the GUT of *Oncorhynchus tshawytscha* from EEUU and Canada and the GIT of *C. carpio* from South Africa and Spain, thus, evidencing the influence of the geographical origin of the species in the content of N/MPs. It must be noted that the color of the materials is an important factor in the ingestion of these N/MPs by marine organisms, since they ingest elements of similar colors to those of their prey (Nelms, Galloway, Godley, Jarvis, & Lindeque, 2018).

2.2 Occurrence of nano/microplastics in edible aquaculture species

Aquaculture continues to grow at an average of 6% per year. Together with the growing world demand for food that is expected to increase by 60% by 2050 compared to 2013. It is estimated that the annual per capita consumption of fish and fisheries products can go from the current 20–25.5 kg in 2050, and that aquatic products will contribute to the food basket in greater proportion than at present to counteract the global food deficit (FAO Food and Agriculture Organization of the United Nations, 2020). In 2018, about 622 species were cultivated, including 387 fish, 111 mollusks, 64 crustaceans, 7 amphibians and reptiles, 10 other types of aquatic invertebrates, and 43 aquatic plants. In 2019, aquaculture placed 120.1 million tons in the market, surpassing extractive fishing for the seventh consecutive year by 26.5 million tons.

Like data from Table 1, Table 2 includes 18 different species from six diverse organisms such as farmed algae, sea cucumber, mussels, cuttlefish, crayfish, and fishes from 10 different countries. It is possible to observe the heterogeneity of the N/MPs described in the different species, being the main common polymer type fibers (60.87%) followed by PE (30.43%) and others.

It is remarkable the variability of data that could be explained not only for the different species included in each group but also by the origin of the studied species, the analyzed tissue and the methodology employed for the analysis (data not shown). Hence, around 46% of the studies analyzed the GIT of different species while nearly 27% analyzed all tissue while soft tissue was analyzed by 7.69% of the studies. As expected, algae were fully analyzed.

Fig. 3B shows the average of N/MPs per individual in farmed fishes, whenever data were available. *Oreochromis niloticus* is part of those with higher content of N/MPs, as the same that occurred with the wild one (Fig. 3A). Within the top positions is possible to find two types of carps (*Ctenopharyngodon idella* and *Hypophthalmichthys molitrix*). On its behalf, common farmed carp (*Cyprinus carpio*) from Italy and Croatia contained the lowest concentration of N/MPs in the studied species.

Prawns had higher concentrations of N/MPs compared to farmed mussels and algae. It can be explained because prawns are scavengers that eat everything they can catch and have tools to crush the particles until they fit in their mouths. Another explanation might be that shrimp have a good digestive system and the value increases when a few MP parts per gram of digestion are combined. However, in mussels and other bivalves, the whole parts are analyzed except for the shell. Alarming data were presented in the study of Curren et al., 2020 in which more than 5000 MPs/g were quantified in the GIT of Indian white shrimp (*Fenneropenaeus indicus*). They concluded that the high quantity could be related with the shape and size of the studied MPs, i.e., spheres which are smaller than, for instance, fibers (Curren et al., 2020).

Several studies compared microplastics in aquaculture animals and wild species. In this regard, some of these were not able to identify significant differences in the MPs content between farmed and wild mussels (Birnstiel, Soares-Gomes, & Gama, 2019), other studies showed similar results with other mytilid species (De Witte et al., 2014; Vandermeersch et al., 2015) and other bivalves (Davidson & Dudas, 2016). On the contrary, the average microplastic abundance in farmed mussels (*Mytilus edulis*) was reported to be higher than in wild mussels from Halifax Harbour, Nova Scotia (Mathalon & Hill, 2014). Prolonged depuration appears as an effective technique to reduce MPs concentration in mussels. However, improved reductions were observed in wild mussels compared to farmed ones even after this process (Renzi et al., 2018).

Significant differences were observed in N/MPs concentrations detected in farmed and wild clam (*Venerupis philippinarum*) in Baynes Sound

(Davidson & Dudas, 2016). Similar results were reported for demersal fish in Hong Kong (Chan, Dingle, & Not, 2019). Significantly higher numbers of microplastics were also observed in farmed blue mussels and Pacific oysters compared to their wild counterparts from Vancouver Island (Murphy, 2018). Reinold et al. (2021) found higher averages of ingested particles (from a minimum of 0.6 particles to a maximum of 2.7 items/fish) in cultivated *Dicentrarchus labrax*, compared to the wild sea bass (0.3 items/fish). This may lead to the suspicion that fish raised in aquaculture facilities may ingest more MPs than wild fish, even if two different environments are considered.

On the contrary, no significant difference was observed in microplastic concentrations detected in cultured and wild clam (*V. philippinarum*) in Baynes Sound (Davidson & Dudas, 2016) or (*Ruditapes philippinarum*) from Jiaozhou Bay (China) (Zhang et al., 2022). Different results were found in one study that wild mullets from the eastern coast of Hong Kong had a higher risk of MPs ingestion than their cultured mullets from fish farms (Cheung, Lui, & Fok, 2018).

In general, the concentration of MPs detected in aquaculture products are higher than those in their wild counterparts from surrounding environments, because of the special environments and frequent activities associated with aquaculture (Chen et al., 2018; Priscilla, Sedayu, & Patria, 2019). These results indicate that the farmed filter-feed aquatic animals are no less susceptible to MPs than wild aquatic animals and they are more easily affected by human activities under the same development conditions.

Some considerations relating the analysis of N/MPs in both wild and farmed edible species must be taken into account. The percentage of detection among species varied considerably among the reviewed samples ranging from 15% to 100% (data not shown). Data heterogeneity due to the different methodologies used to identify and quantify N/MPs makes it difficult to have a complete overview of their content in edible marine and aquaculture species. Additionally, the importance of controlling the possible contamination in the lab including the preparation of blank samples to correct the potential laboratory contamination must be considered. However, not all the studies indicate the precautions taken to avoid it, thus, it cannot be assumed that all the N/MPs detected in the selected species come from their contamination.

It should be noted that the methods used in most of the reviewed documents for the detection of MPs in fish are useful for particles larger than 55 μm, although it is necessary to increase studies on the presence of particles smaller than 1 μm in fishmeal so widely used in aquaculture

(Thiele, Hudson, Russell, Saluveer, & Sidaoui-Haddad, 2021). Hence, efficient methods are sought to digest the tissues and easily release these particles to be easily quantified (Murphy, 2018). Other studies attempt to standardize the extraction of N/MPs from tissues to obtain more homogeneous and therefore much more reliable values (Williams, 2019). For this same purpose, the use of a "Guide to Microplastic Identification" (Prior, 2020) is essential to be able to effectively compare the different studies.

Finally, given that aquaculture is a process in which variables affecting production are controlled to a certain degree, this industry needs to get a more specific idea of the amount and properties of N/MPs that can be found in their products. It can be a starting point to establish protocols that help reduce or control them. This point of view does not make sense in the case of extractive fishing, since the organisms depend entirely on the environmental conditions, which are more difficult to control than in farming systems.

When analyzing the published works from the last few years, it was found a rather contradictory. Logic would indicate that a species with higher productivity or economic importance should coincide with the one with a high number of published research, but this only happens in the case of crustacean and mollusk, especially in white leg shrimp (*Litopenaeus vannamei*) and Mediterranean mussel (*Mytilus galloprovincialis*). However, despite those species being well below the salmonids in production and economic importance, the simple characteristics for cultivation in humid laboratories make them more advantageous as experimental models.

3. Occurrence of N/MPs in humans

Although seafood is recognized, as a source of contaminants for the human diet, the occurrence of N/MPs in seafood is not regulated (Carbery et al., 2018). Previous sections of this chapter showed that the N/MPs could be accumulated in different organs and tissues from edible marine and aquaculture species. Zeytin et al. (2020) recently showed translocation of MPs into the fillets of European seabass juveniles fed with a diet containing MPs for the first time. Clark, Khan, Mitrano, Boyle, and Thompson (2022) also demonstrated the capacity of NPs to translocate out of the fish (concretely they used salmon) gut lumen (Clark et al., 2022). Considering the above mentioned, it is not surprising that those materials have been found in humans as well. Even though usually

N/MPs pass through the gastrointestinal tract (GI) and excrete through the biliary pathway, recent studies indicate that MPs $<10\,\mu m$ in size, and thus NPs can cross cell membranes and reach the circulatory system (Zhang et al., 2021). *In vivo* studies have already demonstrated that NPs can translocate to all organs (Lusher et al., 2017).

In this regard, Table 3 summarizes the main types of polymers (including size and quantity) found in human samples. Briefly, N/MPs have been recently found in stool samples of adults and infants (Schwabl et al., 2019; Zhang et al., 2021), human placentas (Ragusa et al., 2021), meconium (Zhang et al., 2021), human colectomy samples (Ibrahim et al., 2021) and human lung tissue (Amato-Lourenço et al., 2021). A recent study has also found MPs in human blood samples (Leslie et al., 2022). Considering that the human body's excretory system eliminates microplastics, likely disposing of >90% of ingested MNPs via feces (Smith et al., 2018), an exposition to N/MPs through foods or drinking water is a reality (Jiang et al., 2020). Among the mentioned studies, only Na Zhang and co-authors included a 3-days food intake dairy, thus, concluding that the presence of MPs in stool samples could be related with the intake of drinking water in their sample (Zhang et al., 2021). Even though it is not the objective of this chapter, it is noteworthy to mention that chemicals employed as additives from N/MPs such as bisphenol A (BPA) or phthalates, among other (Campanale, Massarelli, Savino, Locaputo, & Uricchio, 2020) have been also found in urine and fecal samples (Zhang et al., 2021). Polypropylene has been identified in placenta, lung tissue and feces while PET has been found in blood, feces, and meconium at variable concentration (Leslie et al., 2022; Schwabl et al., 2019; Zhang et al., 2021). It should be noted that the sample size of the available studies is small, and N/MPs are not found in all samples. In fact, those polymers were found in 98%, 77%, 67%, 67%, 65% and 100%, of the fecal, blood, meconium, placenta, lung tissue and colectomy samples, respectively (Table 3). Therefore, more studies are necessary to validate the previous ones.

Regarding the exposition through marine and aquaculture edible species, Gurjar et al. (2021) showed that MPs usually adhere to epithelial cells leading to human exposure to those polymers through gill or skin tissues of fish and shellfish ingestion. Danopoulos et al. (2020) recently conducted a systematic review and meta-analysis in which an estimation of about 55,000 MP particles from seafood (i.e., mollusks, crustaceans, fish, and Echinodermata) were annually consumed by humans. In the same line, Cox et al. (2020) determined an annual intake of MPs of 39,000–52,000

Table 3 N/MPs found in human samples.

Main type of polymer	Polymer size	Human sample	Sample size	Quantity	Probable source	Reference
Filaments or fibers	1100 ± 300 μm	Colectomy samples	11	28.1 ± 15.4 particles/g tissue	n.a.	Ibrahim et al. (2021)
PP and other non-defined	From 5 to 10 μm	Placenta	6	n.a.	n.a.	Ragusa et al. (2021)
PE and PP	From 1.60 to 5.56 μm	Lung tissue	20	n.a.	Air	Amato-Lourenço et al. (2021)
PET, PE, and PS	From 700 to 500.000 nm	Blood	22	Found in 17 samples: 1.6 μg/mL	n.a.	Leslie et al. (2022)
Mainly PP and PET as film of fragments	From 50 to 500 μm	Feces	8	2 MPs particles/g	n.a.	Schwabl et al. (2019)
PP	From 20 to 800 μm	Feces	26	From 1 particle/g–36 particles/g	Drinking water	Zhang et al. (2021)
PET, PC	n.a.	Feces	6 (Infants)	Median PET, 36,000 ng/g dw; Median PC, 78 ng/g dw	n.a.	Zhang et al. (2021)
PET, PC	n.a.	Faeces	10	Median PET, 2600 ng/g dw (8 samples); Median PC, 110 ng/g dw	n.a.	Zhang et al. (2021)
PET, PC	n.a.	Meconium	3	PET in 2 samples: 12,000 and 3200 ng/g dw; PC in 1 sample: 110 ng/g dw	n.a.	Zhang et al. (2021)

n.a., not available; dw, dry weight; PC, polycarbonate; PE, Polyethylene; PET, polyethylene terephthalate; PP, polypropylene; PS, Polystyrene.

particles (depending on age and sex, being male adults who consumed higher quantities) in the American diet. An annual intake of around 48 dietary MPs per person has been reported for the Peruvian population (De-la-Torre, Apaza-Vargas, & Santillán, 2020). In addition, Senathirajah et al. (2021) estimated the rate of MPs ingested from shellfish per person to be approximately 287 g per year. It was estimated an average intake of 3917.79 ± 144.71 N/MPs annually through mussel consumption in Indian people (Dowarah, Patchaiyappan, Thirunavukkarasu, Jayakumar, & Devipriya, 2020), while the annual average exposure for mussel consumers in Turkey was estimated as 1918 MPs item/per year (Gedik & Eryaşar, 2020). The estimated annual intake of MPs by shellfish consumption from the markets of several countries including Spain, France, UK, Belgium, South Korea, China, Canada, and Italy, ranged from 87 to 11,970 items person/year and Italy had the highest intake levels, followed by Canada and China (Cho et al., 2019). Karami et al. (2018) detected MPs (mainly PP and PET) in several brands of canned fish, which is estimated the intake from this source up to five anthropogenic particles annually, thus, concluding limited health risks of intake to the consumers (Karami et al., 2018).

Interestingly, Catarino et al. (2018) concluded that exposure to N/MPs via shellfish ingestion (concretely mussels) is minimal compared to household dust exposure. One of the studies of the Ocean Conservancy® project calculated an average intake of <1000 MPs/year based on a diet of eating near 220 g of fish twice per week (https://oceanconservancy.org/blog/2021/10/18/eating-microplastics/). In addition, they supported that although larger fish contained higher concentration of MPs, the fillets from the small fish could have more MPs/g of tissue. Clark et al. (2022) highlighted the higher biological impact of the NPs compared to MPs due to their smaller size and concluded that more *in vivo* studies in which the GI is exposed to N/MPs incorporated through diet are necessary to determine any effects on the gut itself or internal organs.

However, from the perspective of human consumption, most species (in which most species of bivalves, crustaceans and several species of small fish are not included) are consumed after removing the gastrointestinal tract thus, the presence of microplastics within fish intestines does not pose significant human health concerns. In this regard, Akoueson et al. found that the edible flesh samples from different fish species showed no significant differences from the number of particles isolated from the respective procedural blank samples while the non-edible parts, i.e., gills or GI tract showed higher N/MPs values (Akoueson et al., 2020).

4. Potential impact of N/MPs on human health

According to Vethaak and Legler (2021), after being internalized or being in contact with epithelial linings in the intestine, N/MPs may cause physical, chemical, and microbiological toxicity, which could also act cumulatively. In this regard, six different pathways have been reviewed by Prata, da Costa, Lopes, Duarte, and Rocha-Santos (2020) as the potential routes of N/MPs exposure, which include oxidative stress and cytotoxicity, energy and metabolism disruption, translocation, disruption of the immune system, neurotoxicity and neurodegenerative diseases and vector for organism and chemicals. The toxicity of N/MPs depend on dose, polymer type, size, surface chemistry, and hydrophobicity. However, individual susceptibility and hazard controls must be also considered (Smith et al., 2018). Despite the lack of detailed information, the European Food Safety Authority (EFSA) supports the idea that only a small percentage of 150 μm particles may be internalized in the mammalian gut. They can be absorbed by biota tissue, organs, and even cells (Yuan et al., 2022), while <1.5 μm particles may be more likely to be distributed to internal organs (EFSA Panel on Contaminants in the Food Chain CONTAM, 2016). Even though human risk assessments usually include epidemiological studies, in the case of N/MPs, only scientific toxicological data from *in vitro* (i.e., animal, or human cells) and *in vivo* (mainly focused on marine organisms or rodents) studies are available (Danopoulos et al., 2020).

The main described toxic effect of N/MPs is oxidative stress together with inflammation, which has been directly related to the chemical composition and size of N/MPs. In this regard, it seems to be that, on one hand, larger nano-sized particles produce more reactive oxygen species (ROS) and on the other hand, they are easily translocated (Yuan et al., 2022). The oxidative stress effect of PE and PS has been evaluated at cell level. Schirinzi et al. (2017) found that the presence of PS generated higher ROS in both human cells studied, i.e., cerebral (T98G) and epithelial (HeLa) human cells, while PE significantly increased ROS on the cerebral (T98G) human cells (Schirinzi et al., 2017). They also highlighted the well-established connection between oxidative stress and the proliferation of cancer cells. Those effects were found in mice after exposing them to different sizes of PS (50, 500 and 5000 nm) (Liang et al., 2021). Specifically, they concluded that the toxicity on intestinal barrier dysfunction was related to ROS-mediated epithelial cell apoptosis in effect in the mice when PS of 50 and 500 nm were combined (Liang et al., 2021). Additionally, Jin, Lu, Tu, Luo and Fu (2019) found damage in the intestinal barrier function of mice when they were exposed to 5 μm PS at dosage of 100 and 1000 μg/L. The same authors reported that PS could induce gut microbiota dysbiosis and metabolic disorders. Moreover, PE MPs exposition at 0.2 μg/g/day seemed to modulate the colon microflora composition in female mice, which could lead metabolic disorders and some other diseases (Sun, Chen, Yang, Xia, & Wu, 2021). Polystyrene MPs also induced oxidative stress negatively influencing the fertility by granulosa cells apoptosis of ovary in female Wistar rats (An et al., 2021). Interestingly, a recent meta-regression aimed to evaluate the impacts of microplastic exposure on human cells showed that irregular shape of MPs together with the duration of exposure and concentration could predict cell death, with Caco-2 cells exhibiting the highest association with MPs effects (Danopoulos, Twiddy, West, & Rotchell, 2021).

Hwang et al. (2020) measured the cytotoxicity of human-derived cell lines including human dermal fibroblasts (HDFs), human mast cell line-1 (HMC-1), primary peripheral blood mononuclear cells (PBMCs), and other cells to the PS particles. They found that PS particles were not toxic to human cells at the tested dosage (~500 μg/mL) and those particles with diameters of 460 nm and 1 μm, which have larger surface areas compared to the larger PS particles, affected red blood cells. Therefore, they concluded that PS particles act as immune stimulants that induce cytokine and chemokine production in a size and concentration-dependent manner (Hwang et al., 2020). However, earlier studies showed that PS particles lower than 240 nm can cross the placental barrier, thus, representing a potential risk for the fetus due to their toxicity through the induction of the enhancement of the immune response and, thus, reducing the response to pathogens, among other mechanisms (Ragusa et al., 2021; Wick et al., 2010; Wright & Kelly, 2017). In the same line, recent research found MPs fragments ranging from 5 to 10 μm in human placenta from which only PP was identified (Ragusa et al., 2021). Unfortunately, and despite they postulated that those MPs could have reached the placenta by uptake and translocation from the GI tract, they were unable to identify how the MPs reached the placenta or the origin of those microplastics. The cytotoxic effect of PS and PE have been also evaluated (Schirinzi et al., 2017). For that purpose, doses ranged from 50 μg/L to 10 mg/L were tested in cerebral and epithelial human cells that did not lead to a significant reduction of cell viability (Schirinzi et al., 2017). In the abovementioned review and meta-regression, characteristics such as MPs concentration and duration of exposure were related to cytotoxicity and the induction of immune responses (Danopoulos et al., 2021).

The complexity in the study of N/MPs toxicity lies not only in their direct exposition but also in the polymer matrix, employed additives, its degradation product and/or adsorbed contaminants (Rodrigues et al., 2019). One of the critical aspects in the study of N/MPs health effects is their quantification of N/MPs in organs and tissues to estimate thresholds of dose-response relationship. In this regard, several studies have tried to calculate the concentration of MPs using the mass of the MPs concentrations as recently reviewed by Danopoulos et al. (2020). Senathirajah et al. (2021) developed a method for estimating the average rate of microplastic ingestion in which the average number of MPs ingested (ANMP) × the average mass of individual particles was included.

If it is focused on the effect derived from fish and other marine and aquaculture species, only few authors risk giving vague recommendations in this regard. Although Akhbarizadeh et al., 2019 concluded the necessity of monitoring the consumption of high doses of some seafood in vulnerable groups such as pregnant/lactating women and their children due to the possible toxicity of MPs and their associated contaminant, some researchers have pointed out that organisms from higher trophic level such as humans accumulate less N/MPs than those from lower trophic level such as marine species (Akhbarizadeh et al., 2019). Moreover, it must be considered that the final N/MPs uptake by the human body would be less than the intake through ingestion (Danopoulos et al., 2020). The truth is that it is a hot topic, which needs more research. Nevertheless, it should not be forgotten that since the routes of N/MPs exposure include ingestion, inhalation, and skin contact, being ingestion the main one, the potential health effect cannot be only related to ingestion or an only food group. Furthermore, in accordance with the European Food Safety Agent (EFSA) (EFSA, 2016), the development of standardized analytical methods for assessing the presence and quantification of N/MPs in foods are necessary to further study their potential toxicity in humans. For that reason, the described potential toxic effects need to be placed into context, since there are many factors that must be considered prior to the ingestion of N/MPs from marine and aquaculture species responsible for the negative health impact.

5. Conclusions and future perspectives

Although most of the N/MPs from the studied edible species could be reduced if the intestine is completely removed before cooking and consumption, the exposure to those polymers through the intake of marine and aquaculture edible species is a reality. A first step to continue working with N/MPs should be focused on the establishment of a universal definition of those polymers. Additionally, the variable nature of N/MPs and the lack of standardized methods for collection, characterization, and analysis of those materials make the quantification of human ingestion difficult. In addition, since there is no legislation for N/MPs as contaminants in food, the levels of exposure that could affect human health are not well established. In addition, the determination of the exact exposure route in humans becomes difficult since N/MPs are widespread.

For that reason, the identified research challenges on this topic must be addressed in further research carrying out (1) the development of methods to accurately identification of N/MPs in environmental, food and biological samples; (2) the development of analytical methods to quantify the amount of N/MPs in the aforementioned samples; (3) the design and performance of epidemiological studies that allow the evaluation of the potential risk of the intake of N/MPs in human health which is necessary to establish a legislation about their presence in foods and their real risk of intake and (4) the standardization of all those strategies to be able to compare between different populations.

References

Abbasi, S., Soltani, N., Keshavarzi, B., Moore, F., Turner, A., & Hassanaghaei, M. (2018). Microplastics in different tissues of fish and prawn from the Musa Estuary, Persian Gulf. *Chemosphere, 205*, 80–87.

Aiguo, Z., Di, S., Chong, W., Yuliang, C., Shaolin, X., Peiqin, L., et al. (2022). Characteristics and differences of microplastics ingestion for farmed fish with different water depths, feeding habits and diets. *Journal of Environmental Chemical Engineering, 10*(2), 107189.

Akhbarizadeh, R., Moore, F., & Keshavarzi, B. (2019). Investigating microplastics bioaccumulation and biomagnification in seafood from the Persian Gulf: A threat to human health? *Food Additives & Contaminants: Part A, 36*(11), 1696–1708.

Akoueson, F., Sheldon, L. M., Danopoulos, E., Morris, S., Hotten, J., Chapman, E., et al. (2020). A preliminary analysis of microplastics in edible versus non-edible tissues from seafood samples. *Environmental Pollution, 263*, 114452.

Amato-Lourenço, L. F., Carvalho-Oliveira, R., Júnior, G. R., dos Santos Galvão, L., Ando, R. A., & Mauad, T. (2021). Presence of airborne microplastics in human lung tissue. *Journal of Hazardous Materials, 416*, 126124.

Amin, R. M., Sohaimi, E. S., Anuar, S. T., & Bachok, Z. (2020). Microplastic ingestion by zooplankton in Terengganu coastal waters, southern South China Sea. *Marine Pollution Bulletin, 150*, 110616.

An, R., Wang, X., Yang, L., Zhang, J., Wang, N., Xu, F., et al. (2021). Polystyrene microplastics cause granulosa cells apoptosis and fibrosis in ovary through oxidative stress in rats. *Toxicology, 449*, 152665.

Auta, H. S., Emenike, C. U., & Fauziah, S. H. (2017). Distribution and importance of microplastics in the marine environment: A review of the sources, fate, effects, and potential solutions. *Environment International*, *102*, 165–176.

Aves, A. R., Revell, L. E., Gaw, S., Ruffell, H., Schuddeboom, A., Wotherspoon, N. E., et al. (2022). First evidence of microplastics in Antarctic snow. *The Cryosphere*, *16*(6), 2127–2145.

Baechler, B. R., Granek, E. F., Hunter, M. V., & Conn, K. E. (2020). Microplastic concentrations in two Oregon bivalve species: Spatial, temporal, and species variability. *Limnology and Oceanography Letters*, *51*(1), 54–65.

Bakir, A., Van der Lingen, C. D., Preston-Whyte, F., Bali, A., Geja, Y., Barry, J., & Maes, T. (2020). Microplastics in commercially important small pelagic fish species from South Africa. *Frontiers in Marine Science*, *7*, 574663.

Białowąs, M., Jonko-Sobuś, K., Pawlak, J., Polak-Juszczak, L., Dąbrowska, A., & Urban-Malinga, B. (2022). Plastic in digestive tracts and gills of cod and herring from the Baltic Sea. *Science of the Total Environment*, *822*, 153333.

Birnstiel, S., Soares-Gomes, A., & Gama, B. A. (2019). Depuration reduces microplastic content in wild and farmed mussels. *Marine Pollution Bulletin*, *140*, 241–247.

Botterell, Z. L., Beaumont, N., Dorrington, T., Steinke, M., Thompson, R. C., & Lindeque, P. K. (2019). Bioavailability and effects of microplastics on marine zooplankton: A review. *Environmental Pollution*, *245*, 98–110.

Bråte, I. L. N., Hurley, R., Iversen, K., Beyer, J., Thomas, K. V., Steindal, C. C., et al. (2018). Mytilus spp. as sentinels for monitoring microplastic pollution in Norwegian coastal waters: A qualitative and quantitative study. *Environmental Pollution*, *243*, 383–393.

Bruzaca, D. N., Justino, A. K., Mota, G. C., Costa, G. A., Lucena-Frédou, F., & Gálvez, A. O. (2022). Occurrence of microplastics in bivalve molluscs *Anomalocardia flexuosa* captured in Pernambuco, Northeast Brazil. *Marine Pollution Bulletin*, *179*, 113659.

Cabanilles, P., Acle, S., Arias, A., Masiá, P., Ardura, A., & Garcia-Vazquez, E. (2022). Microplastics risk into a three-link food chain inside European hake. *Diversity*, *14*(5), 308.

Campanale, C., Massarelli, C., Savino, I., Locaputo, V., & Uricchio, V. F. (2020). A detailed review study on potential effects of microplastics and additives of concern on human health. *International Journal of Environmental Research and Public Health*, *17*(4), 1212.

Carbery, M., O'Connor, W., & Palanisami, T. (2018). Trophic transfer of microplastics and mixed contaminants in the marine food web and implications for human health. *Environment International*, *115*, 400–409.

Catarino, A. I., Macchia, V., Sanderson, W. G., Thompson, R. C., & Henry, T. B. (2018). Low levels of microplastics (MP) in wild mussels indicate that MP ingestion by humans is minimal compared to exposure via household fibres fallout during a meal. *Environmental Pollution*, *237*, 675–684.

Chan, H. S. H., Dingle, C., & Not, C. (2019). Evidence for non-selective ingestion of microplastic in demersal fish. *Marine Pollution Bulletin*, *149*, 110523.

Chen, M., Jin, M., Tao, P., Wang, Z., Xie, W., Yu, X., et al. (2018). Assessment of microplastics derived from mariculture in Xiangshan Bay, China. *Environmental Pollution*, *242*, 1146–1156.

Cheung, L. T., Lui, C. Y., & Fok, L. (2018). Microplastic contamination of wild and captive flathead grey mullet (*Mugil cephalus*). *International Journal of Environmental Research and Public Health*, *15*(4), 597.

Cho, Y., Shim, W. J., Jang, M., Han, G. M., & Hong, S. H. (2019). Abundance and characteristics of microplastics in market bivalves from South Korea. *Environmental Pollution*, *245*, 1107–1116.

Clark, N. J., Khan, F. R., Mitrano, D. M., Boyle, D., & Thompson, R. C. (2022). Demonstrating the translocation of nanoplastics across the fish intestine using palladium-doped polystyrene in a salmon gut-sac. *Environment International*, *159*, 106994.
Cole, M., Lindeque, P., Fileman, E., Halsband, C., Goodhead, R., Moger, J., et al. (2013). Microplastic ingestion by zooplankton. *Environmental Science & Technology*, *47*(12), 6646–6655.
Collicutt, B., Juanes, F., & Dudas, S. E. (2019). Microplastics in juvenile Chinook salmon and their nearshore environments on the east coast of Vancouver Island. *Environmental Pollution*, *244*, 135–142.
Cox, K. D., Covernton, G. A., Davies, H. L., Dower, J. F., Juanes, F., & Dudas, S. E. (2020). Correction to human consumption of microplastics. *Environmental Science & Technology*, *54*(17), 10974.
Curren, E., Leaw, C. P., Lim, P. T., & Leong, S. C. Y. (2020). Evidence of marine microplastics in commercially harvested seafood. *Frontiers in Bioengineering and Biotechnology*, *8*, 562760.
Daniel, D. B., Ashraf, P. M., & Thomas, S. N. (2020). Abundance, characteristics and seasonal variation of microplastics in Indian white shrimps (*Fenneropenaeus indicus*) from coastal waters off Cochin, Kerala, India. *Science of the Total Environment*, *737*, 139839.
Daniel, D. B., Ashraf, P. M., Thomas, S. N., & Thomson, K. T. (2021). Microplastics in the edible tissues of shellfishes sold for human consumption. *Chemosphere*, *264*, 128554.
Danopoulos, E., Jenner, L. C., Twiddy, M., & Rotchell, J. M. (2020). Microplastic contamination of seafood intended for human consumption: A systematic review and meta-analysis. *Environmental Health Perspectives*, *128*(12), 126002.
Danopoulos, E., Twiddy, M., West, R., & Rotchell, J. M. (2021). A rapid review and meta-regression analyses of the toxicological impacts of microplastic exposure in human cells. *Journal of Hazardous Materials*, *427*, 127861.
Davidson, K., & Dudas, S. E. (2016). Microplastic ingestion by wild and cultured Manila clams (*Venerupis philippinarum*) from Baynes Sound, British Columbia. *Archives of Environmental Contamination and Toxicology*, *71*(2), 147–156.
De Sá, L. C., Oliveira, M., Ribeiro, F., Rocha, T. L., & Futter, M. N. (2018). Studies of the effects of microplastics on aquatic organisms: What do we know and where should we focus our efforts in the future? *Science of the Total Environment*, *645*, 1029–1039.
De Witte, B., Devriese, L., Bekaert, K., Hoffman, S., Vandermeersch, G., Cooreman, K., et al. (2014). Quality assessment of the blue mussel (*Mytilus edulis*): Comparison between commercial and wild types. *Marine Pollution Bulletin*, *85*(1), 146–155.
De-la-Torre, G. E., Apaza-Vargas, D. M., & Santillán, L. L. (2020). Microplastic ingestion and feeding ecology in three intertidal mollusk species from Lima, Peru. *Revista de Biología Marina y Oceanografía*, *55*(2), 167–171.
Deshpande, P. C., Philis, G., Brattebø, H., & Fet, A. M. (2020). Using Material Flow Analysis (MFA) to generate the evidence on plastic waste management from commercial fishing gears in Norway. *Resources, Conservation & Recycling: X*, *5*, 100024.
Digka, N., Tsangaris, C., Torre, M., Anastasopoulou, A., & Zeri, C. (2018). Microplastics in mussels and fish from the Northern Ionian Sea. *Marine Pollution Bulletin*, *135*, 30–40.
Dowarah, K., Patchaiyappan, A., Thirunavukkarasu, C., Jayakumar, S., & Devipriya, S. P. (2020). Quantification of microplastics using Nile Red in two bivalve species *Perna viridis* and *Meretrix meretrix* from three estuaries in Pondicherry, India and microplastic uptake by local communities through bivalve diet. *Marine Pollution Bulletin*, *153*, 110982.
EFSA Panel on Contaminants in the Food Chain (CONTAM). (2016). Presence of microplastics and nanoplastics in food, with particular focus on seafood. *EFSA Journal*, *14*(6), e04501.
European Commission. (2018). Facts and figures on the common fisheries policy. Basic statistical data. *European Commission*.

European Food Safety Authority. (2018). The 2016 European Union report on pesticide residues in food. *EFSA Journal, 16*(7), e05348.

Fang, C., Zheng, R., Chen, H., Hong, F., Lin, L., Lin, H., et al. (2019). Comparison of microplastic contamination in fish and bivalves from two major cities in Fujian province, China and the implications for human health. *Aquaculture, 512*, 734322.

FAO (Food and Agriculture Organization of the United Nations). (2020). *The state of world fisheries and aquaculture 2020. Sustainability in action*. FAO. Rome.

Farzi, A., Dehnad, A., & Fotouhi, A. F. (2019). Biodegradation of polyethylene terephthalate waste using Streptomyces species and kinetic modelling of the process. *Biocatalysis and Agricultural Biotechnology, 17*, 25–31.

Feng, Z., Zhang, T., Shi, H., Gao, K., Huang, W., Xu, J., et al. (2020). Microplastics in bloom-forming macroalgae: Distribution, characteristics and impacts. *Journal of Hazardous Materials, 397*, 122752.

Gedik, K, & Eryaşar, A. R. (2020). Microplastic pollution profile of Mediterranean mussels (*Mytilus galloprovincialis*) collected along the Turkish coasts. *Chemosphere, 260*, 127570.

Gedik, K., Eryaşar, A. R., & Gözler, A. M. (2022). The microplastic pattern of wild-caught Mediterranean mussels from the Marmara Sea. *Marine Pollution Bulletin, 175*, 113331.

Giani, D., Baini, M., Galli, M., Casini, S., & Fossi, M. C. (2019). Microplastics occurrence in edible fish species (*Mullus barbatus* and *Merluccius merluccius*) collected in three different geographical sub-areas of the Mediterranean Sea. *Marine Pollution Bulletin, 140*, 129–137.

Gong, Y., Wang, Y., Chen, L., Li, Y., Chen, X., & Liu, B. (2021). Microplastics in different tissues of a pelagic squid (*Dosidicus gigas*) in the northern Humboldt Current ecosystem. *Marine Pollution Bulletin, 169*, 112509.

Granek, E. F., Brander, S. M., & Holland, E. B. (2020). Microplastics in aquatic organisms: Improving understanding and identifying research directions for the next decade. *Limnology and Oceanography Letters, 5*(1), 1–4.

Guilhermino, L., Martins, A., Lopes, C., Raimundo, J., Vieira, L. R., Barboza, L. G. A., et al. (2021). Microplastics in fishes from an estuary (Minho River) ending into the NE Atlantic Ocean. *Marine Pollution Bulletin, 173*, 113008.

Gurjar, U. R., Xavier, M., Nayak, B. B., Ramteke, K., Deshmukhe, G., Jaiswar, A. K., et al. (2021). Microplastics in shrimps: A study from the trawling grounds of northeastern part of Arabian Sea. *Environmental Science and Pollution Research, 28*(35), 48494–48504.

Haave, M., Lorenz, C., Primpke, S., & Gerdts, G. (2018). Microplastic occurrence and distribution from discharge points to deep basins in an urban model fjord. In *MICRO 2018. Fate and impact of microplastics: knowledge, actions and solutions* (p. 183). MSFS-RBLZ.

Hajisamae, S., Soe, K. K., Pradit, S., Chaiyvareesajja, J., & Fazrul, H. (2022). Feeding habits and microplastic ingestion of short mackerel, *Rastrelliger brachysoma*, in a tropical estuarine environment. *Environmental Biology of Fishes, 105*(2), 289–302.

Hartmann, N. B., Huffer, T., Thompson, R. C., Hassellöv, M., Verschoor, A., Daugaard, A. E., et al. (2019). Are we speaking the same language? Recommendations for a definition and categorization framework for plastic debris. *Environmental Science and Technology, 53*(3), 1039–1047.

Hermabessiere, L., Paul-Pont, I., Cassone, A. L., Himber, C., Receveur, J., Jezequel, R., et al. (2019). Microplastic contamination and pollutant levels in mussels and cockles collected along the channel coasts. *Environmental Pollution, 250*, 807–819.

Herrera, A., Ŝtindlová, A., Martínez, I., Rapp, J., Romero-Kutzner, V., Samper, M. D., et al. (2019). Microplastic ingestion by Atlantic chub mackerel (*Scomber colias*) in the Canary Islands coast. *Marine Pollution Bulletin, 139*, 127–135.

Hossain, M. S., Rahman, M. S., Uddin, M. N., Sharifuzzaman, S. M., Chowdhury, S. R., Sarker, S., et al. (2020). Microplastic contamination in Penaeid shrimp from the Northern Bay of Bengal. *Chemosphere, 238*, 124688.

Huang, D., Tao, J., Cheng, M., Deng, R., Chen, S., Yin, L., et al. (2021). Microplastics and nanoplastics in the environment: Macroscopic transport and effects on creatures. *Journal of Hazardous Materials*, *407*, 124399.
Hwang, J., Choi, D., Han, S., Jung, S. Y., Choi, J., & Hong, J. (2020). Potential toxicity of polystyrene microplastic particles. *Scientific Reports*, *10*(1), 1–12.
Ibrahim, Y. S., Tuan Anuar, S., Azmi, A. A., Wan Mohd Khalik, W. M. A., Lehata, S., Hamzah, S. R., et al. (2021). Detection of microplastics in human colectomy specimens. *Journal of Gastroenterology and Hepatology*, *5*(1), 116–121.
Ivleva, N. P., Wiesheu, A. C., & Niessner, R. (2017). Microplastic in aquatic ecosystems. *Angewandte Chemie International Edition*, *56*(7), 1720–1739.
Jiang, J., Wang, X., Ren, H., Cao, G., Xie, G., Xing, D., et al. (2020). Investigation and fate of microplastics in wastewater and sludge filter cake from a wastewater treatment plant in China. *Science of the Total Environment*, *746*, 141378.
Jin, Y., Lu, L., Tu, W., Luo, T., & Fu, Z. (2019). Impacts of polystyrene microplastic on the gut barrier, microbiota and metabolism of mice. *Science of the Total Environment*, *649*, 308–317.
Joshy, A., Sharma, S. K., Mini, K. G., Gangadharan, S., & Pranav, P. (2022). Histopathological evaluation of bivalves from the southwest coast of India as an indicator of environmental quality. *Aquatic Toxicology*, *243*, 106076.
Karami, A., Golieskardi, A., Choo, C. K., Larat, V., Karbalaei, S., & Salamatinia, B. (2018). Microplastic and mesoplastic contamination in canned sardines and sprats. *Science of the Total Environment*, *612*, 1380–1386.
Keshavarzifard, M., Vazirzadeh, A., & Sharifinia, M. (2021). Occurrence and characterization of microplastics in white shrimp, *Metapenaeus affinis*, living in a habitat highly affected by anthropogenic pressures, northwest Persian Gulf. *Marine Pollution Bulletin*, *169*, 112581.
Kim, J. H., Yu, Y. B., & Choi, J. H. (2021). Toxic effects on bioaccumulation, hematological parameters, oxidative stress, immune responses and neurotoxicity in fish exposed to microplastics: A review. *Journal of Hazardous Materials*, *413*, 125423.
Klomjit, A., Sutthacheep, M., & Yeemin, T. (2021). Occurrence of microplastics in edible seaweeds from aquaculture. *Ramkhamhaeng International Journal of Science and Technology*, *4*(2), 38–44.
Koelmans, A. A. (2015). Modeling the role of microplastics in bioaccumulation of organic chemicals to marine aquatic organisms. A critical review. *Marine Anthropogenic Litter*, 309–324.
Lavers, J. L., & Bond, A. L. (2017). Exceptional and rapid accumulation of anthropogenic debris on one of the world's most remote and pristine islands. *Proceedings of the National Academy of Sciences of the United States of America*, *114*(23), 6052–6055.
Leslie, H. A., Van Velzen, M. J., Brandsma, S. H., Vethaak, A. D., Garcia-Vallejo, J. J., & Lamoree, M. H. (2022). Discovery and quantification of plastic particle pollution in human blood. *Environment International*, *163*, 107199.
Leung, M. M. L., Ho, Y. W., Lee, C. H., Wang, Y., Hu, M., Kwok, K. W. H., et al. (2021). Improved Raman spectroscopy-based approach to assess microplastics in seafood. *Environmental Pollution*, *289*, 117648.
Li, J., Qu, X., Su, L., Zhang, W., Yang, D., Kolandhasamy, P., et al. (2016). Microplastics in mussels along the coastal waters of China. *Environmental Pollution*, *214*, 177–184.
Li, Q., Feng, Z., Zhang, T., Ma, C., & Shi, H. (2020). Microplastics in the commercial seaweed nori. *Journal of Hazardous Materials*, *388*, 122060.
Li, Q., Su, L., Ma, C., Feng, Z., & Shi, H. (2022). Plastic debris in coastal macroalgae. *Environmental Research*, *205*, 112464.

Liang, B., Zhong, Y., Huang, Y., Lin, X., Liu, J., Lin, L., et al. (2021). Underestimated health risks: Polystyrene micro-and nanoplastics jointly induce intestinal barrier dysfunction by ROS-mediated epithelial cell apoptosis. *Particle and Fibre Toxicology*, *18*(1), 1–19.

Lusher, A., Hollman, P., & Mendoza-Hill, J. (2017). Microplastics in fisheries and aquaculture: Status of knowledge on their occurrence and implications for aquatic organisms and food safety. *FAO*, *615*, 56–59.

Mancuso, M., Savoca, S., & Bottari, T. (2019). First record of microplastics ingestion by European hake *Merluccius merluccius* from the Tyrrhenian Sicilian coast (Central Mediterranean Sea). *Journal of Fish Biology*, *94*, 517–519.

Mathalon, A., & Hill, P. (2014). Microplastic fibres in the intertidal ecosystem surrounding Halifax Harbor, Nova Scotia. *Marine Pollution Bulletin*, *81*(1), 69–79.

Mohapatra, B. C., Sarkar, B., & Singh, S. K. (2003). Use of plastics in aquaculture. In K. K. Satapathy, & A. Kumar (Eds.), *Plasticulture intervention for agriculture development in North Eastern Region* (pp. 290–305). Umiam, Meghalaya, India: ICAR Research Complex for NEH Region.

Mohapatra, B. C., Sarkar, B., Barik, N. K., & Jayasankar, P. (2011). *Application of plastics in aquaculture* (1st ed.). Bhubaneswar, Odisha, India: The Director at Central Institute of Freshwater Aquaculture (Indian Council of Agricultural Research).

Mohsen, M., Wang, Q., Zhang, L., Sun, L., Lin, C., & Yang, H. (2019). Heavy metals in sediment, microplastic and sea cucumber *Apostichopus japonicus* from farms in China. *Marine Pollution Bulletin*, *143*, 42–49.

Morgana, S., Ghigliotti, L., Estévez-Calvar, N., Stifanese, R., Wieckzorek, A., Doyle, T., et al. (2018). Microplastics in the Arctic: A case study with sub-surface water and fish samples off Northeast Greenland. *Environmental Pollution*, *242*, 1078–1086.

Murphy, C. L. (2018). *A comparison of microplastics in farmed and wild shellfish near Vancouver Island and potential implications for contaminant transfer to humans* (Doctoral dissertation). Canada: Royal Roads University.

Naidu, B. C., Xavier, K. M., Shukla, S. P., Jaiswar, A. K., & Nayak, B. B. (2022). Comparative study on the microplastics abundance, characteristics, and possible sources in yellow clams of different demographic regions of the northwest coast of India. *Journal of Hazardous Materials Letters*, *3*, 100051.

Nelms, S. E., Galloway, T. S., Godley, B. J., Jarvis, D. S., & Lindeque, P. K. (2018). Investigating microplastic trophic transfer in marine top predators. *Environmental Pollution*, *238*, 999–1007.

NOAA Fisheries. (2017). *2021 North Atlantic right whale unusual mortality event*. NOAA Fisheries.

O'Connor, J. D., Murphy, S., Lally, H. T., O'Connor, I., Nash, R., O'Sullivan, J., et al. (2020). Microplastics in brown trout (*Salmo trutta Linnaeus*, 1758) from an Irish riverine system. *Environmental Pollution*, *267*, 115572.

Ogonowski, M., Wenman, V., Barth, A., Hamacher-Barth, E., Danielsson, S., & Gorokhova, E. (2019). Microplastic intake, its biotic drivers, and hydrophobic organic contaminant levels in the Baltic herring. *Frontiers in Environmental Science*, *7*, 134.

Oliva-Hernández, B. E., Santos-Ruiz, F. M., Muñoz-Wug, M. A., & Pérez-Sabino, J. F. (2021). Microplastics in Nile tilapia (*Oreochromis niloticus*) from Lake Amatitlán. *Revista Ambiente & Água*, *16*, 1–10.

Oliveira, A. R., Sardinha-Silva, A., Andrews, P. L., Green, D., Cooke, G. M., Hall, S., et al. (2020). Microplastics presence in cultured and wild-caught cuttlefish, *Sepia officinalis*. *Marine Pollution Bulletin*, *160*, 111553.

Paul, M. B., Stock, V., Cara-Carmona, J., Lisicki, E., Shopova, S., Fessard, V., et al. (2020). Micro-and nanoplastics–current state of knowledge with the focus on oral uptake and toxicity. *Nanoscale Advances*, *2*(10), 4350–4367.

Phuong, N. N., Poirier, L., Pham, Q. T., Lagarde, F., & Zalouk-Vergnoux, A. (2018). Factors influencing the microplastic contamination of bivalves from the French Atlantic coast: Location, season and/or mode of life? *Marine Pollution Bulletin*, *129*(2), 664–674.

Prata, J. C., da Costa, J. P., Duarte, A. C., & Rocha-Santos, T. (2022). Suspected microplastics in Atlantic horse mackerel fish (*Trachurus trachurus*) captured in Portugal. *Marine Pollution Bulletin*, *174*, 113249.

Prata, J. C., da Costa, J. P., Lopes, I., Duarte, A. C., & Rocha-Santos, T. (2020). Environmental exposure to microplastics: An overview on possible human health effects. *Science of the Total Environment*, *702*, 134455.

Prior, J. H. (2020). *Epigenetic effects of microplastics exposure on the common mysid shrimp, Americamysis bahia*. The University of West Florida.

Priscilla, V., Sedayu, A., & Patria, M. P. (2019, December). Microplastic abundance in the water, seagrass, and sea hare *Dolabella auricularia* in Pramuka Island, Seribu Islands, Jakarta Bay, Indonesia. *Journal of Physics: Conference Series*, *1402*(3), 033073. IOP Publishing.

Ragusa, A., Svelato, A., Santacroce, C., Catalano, P., Notarstefano, V., Carnevali, O., et al. (2021). Plasticenta: First evidence of microplastics in human placenta. *Environment International*, *146*, 106274.

Reguera, P., Viñas, L., & Gago, J. (2019). Microplastics in wild mussels (Mytilus spp.) from the north coast of Spain. *Scientia Marina*, *83*(4), 337–347.

Reinold, S., Herrera, A., Saliu, F., Hernández-González, C., Martinez, I., Lasagni, M., et al. (2021). Evidence of microplastic ingestion by cultured European sea bass (*Dicentrarchus labrax*). *Marine Pollution Bulletin*, *168*, 112450.

Renzi, M., & Blašković, A. (2020). Chemical fingerprint of plastic litter in sediments and holothurians from Croatia: Assessment & relation to different environmental factors. *Marine Pollution Bulletin*, *153*, 110994.

Renzi, M., Cilenti, L., Scirocco, T., Grazioli, E., Anselmi, S., Broccoli, A., & Specchiulli, A. (2020). Litter in alien species of possible commercial interest: The blue crab (Callinectes sapidus Rathbun, 1896) as case study. *Marine Pollution Bulletin*, *157*, 111300.

Renzi, M., Guerranti, C., & Blašković, A. (2018). Microplastic contents from maricultured and natural mussels. *Marine Pollution Bulletin*, *131*, 248–251.

Reunura, T., & Prommi, T. O. (2022). Detection of microplastics in *Litopenaeus vannamei* (Penaeidae) and *Macrobrachium rosenbergii* (Palaemonidae) in cultured pond. *PeerJ*, *10*, e12916.

Riani, E., & Cordova, M. R. (2022). Microplastic ingestion by the sandfish Holothuria scabra in Lampung and Sumbawa, Indonesia. *Marine Pollution Bulletin*, *175*, 113134.

Rochman, C. M., Tahir, A., Williams, S. L., Baxa, D. V., Lam, R., Miller, J. T., et al. (2015). Anthropogenic debris in seafood: Plastic debris and fibers from textiles in fish and bivalves sold for human consumption. *Scientific Reports*, *5*(1), 1–10.

Rodrigues, M. O., Abrantes, N., Gonçalves, F. J. M., Nogueira, H., Marques, J. C., & Gonçalves, A. M. M. (2019). Impacts of plastic products used in daily life on the environment and human health: What is known? *Environmental Toxicology and Pharmacology*, *72*, 103239.

Saad, D., Chauke, P., Cukrowska, E., Richards, H., Nikiema, J., Chimuka, L., et al. (2022). First biomonitoring of microplastic pollution in the Vaal river using Carp fish (*Cyprinus carpio*) "as a bio-indicator". *Science of the Total Environment*, *836*, 155623.

Savoca, S., Matanović, K., D'Angelo, G., Vetri, V., Anselmo, S., Bottari, T., et al. (2021). Ingestion of plastic and non-plastic microfibers by farmed gilthead sea bream (*Sparus aurata*) and common carp (*Cyprinus carpio*) at different life stages. *Science of the Total Environment*, *782*, 146851.

Schirinzi, G. F., Pérez-Pomeda, I., Sanchís, J., Rossini, C., Farré, M., & Barceló, D. (2017). Cytotoxic effects of commonly used nanomaterials and microplastics on cerebral and epithelial human cells. *Environmental Research, 159*, 579–587.

Schwabl, P., Köppel, S., Königshofer, P., Bucsics, T., Trauner, M., Reiberger, T., et al. (2019). Detection of various microplastics in human stool: A prospective case series. *Annals of Internal Medicine, 171*(7), 453–457.

Senathirajah, K., Attwood, S., Bhagwat, G., Carbery, M., Wilson, S., & Palanisami, T. (2021). Estimation of the mass of microplastics ingested–A pivotal first step towards human health risk assessment. *Journal of Hazardous Materials, 404*, 124004.

Severini, M. F., Buzzi, N. S., López, A. F., Colombo, C. V., Sartor, G. C., Rimondino, G. N., et al. (2020). Chemical composition and abundance of microplastics in the muscle of commercial shrimp *Pleoticus muelleri* at an impacted coastal environment (Southwestern Atlantic). *Marine Pollution Bulletin, 161*, 111700.

Smith, M., Love, D. C., Rochman, C. M., & Neff, R. A. (2018). Microplastics in seafood and the implications for human health. *Current Environmental Health Reports, 5*(3), 375–386.

Su, L., Cai, H., Kolandhasamy, P., Wu, C., Rochman, C. M., & Shi, H. (2018). Using the Asian clam as an indicator of microplastic pollution in freshwater ecosystems. *Environmental Pollution, 234*, 347–355.

Sun, H., Chen, N., Yang, X., Xia, Y., & Wu, D. (2021). Effects induced by polyethylene microplastics oral exposure on colon mucin release, inflammation, gut microflora composition and metabolism in mice. *Ecotoxicology and Environmental Safety, 220*, 112340.

Thiele, C. J., Hudson, M. D., Russell, A. E., Saluveer, M., & Sidaoui-Haddad, G. (2021). Microplastics in fish and fishmeal: An emerging environmental challenge? *Scientific Reports, 11*(1), 1–12.

Tiseo, I. (2022). *Global plastic production 1950-2020*. Available at: https://www.statista.com/statistics/282732/global-production-of-plastics-since-1950/.

Valencia-Castañeda, G., Ruiz-Fernández, A. C., Frías-Espericueta, M. G., Rivera-Hernández, J. R., Green-Ruiz, C. R., & Páez-Osuna, F. (2022). Microplastics in the tissues of commercial semi-intensive shrimp pond-farmed *Litopenaeus vannamei* from the Gulf of California ecoregion. *Chemosphere, 297*, 134194.

Van Cauwenberghe, L., Devriese, L., Galgani, F., Robbens, J., & Janssen, C. R. (2015). Microplastics in sediments: A review of techniques, occurrence and effects. *Marine Environmental Research, 111*, 5–17.

Vandermeersch, G., Lourenço, H. M., Alvarez-Muñoz, D., Cunha, S., Diogène, J., Cano-Sancho, G., et al. (2015). Environmental contaminants of emerging concern in seafood–European database on contaminant levels. *Environmental Research, 143*, 29–45.

Vethaak, A. D., & Legler, J. (2021). Microplastics and human health. *Science, 371*(6530), 672–674.

Walkinshaw, C., Lindeque, P. K., Thompson, R., Tolhurst, T., & Cole, M. (2020). Microplastics and seafood: Lower trophic organisms at highest risk of contamination. *Ecotoxicology and Environmental Safety, 190*, 110066.

Wang, Y. L., Lee, Y. H., Chiu, I. J., Lin, Y. F., & Chiu, H. W. (2020). Potent impact of plastic nanomaterials and micromaterials on the food chain and human health. *International Journal of Molecular Sciences, 21*(5), 1727.

Wick, P., Malek, A., Manser, P., Meili, D., Maeder-Althaus, X., Diener, L., et al. (2010). Barrier capacity of human placenta for nanosized materials. *Environmental Health Perspectives, 118*(3), 432–436.

Wilcox, C., Van Sebille, E., & Hardesty, B. D. (2015). Threat of plastic pollution to seabirds is global, pervasive, and increasing. *Proceedings of the National Academy of Sciences of the United States of America, 112*(38), 11899–11904.

Williams, S. J. (2019). *Extraction and identification of microplastics in bivalves harvested from aquaculture sites surrounding the lower.* Chesapeake Bay: Christopher Newport University.
Wright, S. L., & Kelly, F. J. (2017). Plastic and human health: A micro issue? *Environmental Science & Technology, 51*(12), 6634–6647.
Wright, S. L., Thompson, R. C., & Galloway, T. S. (2013). The physical impacts of microplastics on marine organisms: A review. *Environmental Pollution, 178*, 483–492.
Yan, M., Li, W., Chen, X., He, Y., Zhang, X., & Gong, H. (2021). A preliminary study of the association between colonization of microorganism on microplastics and intestinal microbiota in shrimp under natural conditions. *Journal of Hazardous Materials, 408*, 124882.
Yuan, Z., Nag, R., & Cummins, E. (2022). Human health concerns regarding microplastics in the aquatic environment-From marine to food systems. *Science of the Total Environment*, 153730.
Zeytin, S., Wagner, G., Mackay-Roberts, N., Gerdts, G., Schuirmann, E., Klockmann, S., et al. (2020). Quantifying microplastic translocation from feed to the fillet in European sea bass *Dicentrarchus labrax*. *Marine Pollution Bulletin, 156*, 111210.
Zhang, D., Fraser, M. A., Huang, W., Ge, C., Wang, Y., Zhang, C., et al. (2021). Microplastic pollution in water, sediment, and specific tissues of crayfish (*Procambarus clarkii*) within two different breeding modes in Jianli, Hubei province, China. *Environmental Pollution, 272*, 115939.
Zhang, K., Liang, J., Liu, T., Li, Q., Zhu, M., Zheng, S., et al. (2022). Abundance and characteristics of microplastics in shellfish from Jiaozhou Bay, China. *Journal of Oceanology and Limnology, 40*(1), 163–172.
Ziajahromi, S., Neale, P. A., & Leusch, F. D. (2016). Wastewater treatment plant effluent as a source of microplastics: Review of the fate, chemical interactions and potential risks to aquatic organisms. *Water Science and Technology, 74*(10), 2253–2269.

CHAPTER EIGHT

Migration of microplastics from plastic packaging into foods and its potential threats on human health

Shahida Anusha Siddiqui[a,b,*], Nur Alim Bahmid[c], Sayed Hashim Mahmood Salman[d], Asad Nawaz[e,f,g], Noman Walayat[h], Garima Kanwar Shekhawat[i], Alexey Alekseevich Gvozdenko[j], Andrey Vladimirovich Blinov[j], and Andrey Ashotovich Nagdalian[j,k]

[a]Technical University of Munich, Campus Straubing for Biotechnology and Sustainability, Straubing, Germany
[b]German Institute of Food Technologies (DIL e.V.), Quakenbrück, Germany
[c]Research Center for Food Technology and Processing, National Research and Innovation Agency (BRIN), Yogyakarta, Indonesia
[d]Department of Science, Arabian Pearl Gulf Private School, Bilad Al Qadeem, Bahrain
[e]College of Civil and Transportation Engineering, Shenzhen University, Shenzhen, China
[f]Shenzhen Key Laboratory of Marine Microbiome Engineering, Institute for Advanced Study, Shenzhen University, Shenzhen, China
[g]Institute for Innovative Development of Food Industry, Shenzhen University, Shenzhen, China
[h]College of Food Science and Technology, Zhejiang University of Technology, Hangzhou, China
[i]Department of Microbiology, School of Life Sciences, Central University of Rajasthan, Jaipur, India
[j]Food Technology and Engineering Department, North Caucasus Federal University, Stavropol, Russia
[k]Saint Petersburg State Agrarian University, St Petersburg, Russia
*Corresponding author: e-mail address: s.siddiqui@dil-ev.de

Contents

Advances in Food and Nutrition Research, Volume 103
ISSN 1043-4526
https://doi.org/10.1016/bs.afnr.2022.07.002

Abstract

Microplastics from food packaging material have risen in number and dispersion in the aquatic system, the terrestrial environment, and the atmosphere in recent decades. Microplastics are of particular concern due to their long-term durability in the environment, their great potential for releasing plastic monomers and additives/chemicals, and their vector-capacity for adsorbing or collecting other pollutants. Consumption of foods containing migrating monomers can lead to accumulation in the body and the build-up of monomers in the body can trigger cancer. The book chapter focuses the commercial plastic food packaging materials and describes their release mechanisms of microplastics from packaging into foods. To prevent the potential risk of microplastics migrated into food products, the factors influencing microplastic to the food products, e.g., high temperatures, ultraviolet and bacteria, have been discussed. Additionally, as many evidences shows that the microplastic components are toxic and carcinogenic, the potential threats and negative effects on human health have also been highlighted. Moreover, future trends is summarized to reduce the microplastic migration by enhancing public awareness as well as improving waste management.

1. Introduction

Plastics were first mass-produced and used in the 1950s. After steel, wood, and cement, plastic has become the fourth fundamental material due to its benefits of lightweight, low cost, and superior process ability. As of 2017, an estimated 9.2 billion metric tons of plastics were manufactured globally, with a significant amount of general-purpose plastics being thrown (Blasiak, Leander, Jouffray, & Virdin, 2021; Nxumalo, Mabaso, Mamba, & Singwane, 2020). However, plastic is not easily degraded, so the amount of plastic waste continues to increase and accumulates in environment, especially in oceans. According to Jambeck et al. (2015), 275 million metric tons (MT) of plastic waste generated by 192 coastal countries in 2010 of which 4.8–12.7 million MT entered the marine area. By 2025, it is estimated that the cumulative amount of plastic waste originating from land will increase in the oceans due to the population increase (Lebreton & Andrady, 2019; Löhr et al., 2017).

Plastics are persistent although they can be degraded into smaller particles with a diameter of 1–5 mm in water, called microplastics. Plastic particles with a diameter from 1 to 100 nm are called nanoplastic. The small pieces of plastic forming fibers, waste, or particles pollute the environment (Masura, Baker, Foster, & Arthut, 2015). Food packaging are mostly dominated microplastics source as reported by Sarinah Basri, Basri, Syaputra, and Handayani (2021), which is around 39.6% of the wasted plastics in Europe. Microplastics from food packaging material have risen in number and dispersion in the aquatic system, the terrestrial environment, and the atmosphere in recent decades. By 2060, 155–265 million tons of plastic will have accumulated in the natural environment, with microplastics accounting for 13.2% of that weight. According to European Food Safety Authority (2016), 68,500–275,000 tons of microplastics enter the sea per year. Previous research by Hiwari et al. (2019) found that the average concentration of microplastics in the waters was 0.018 ± 0.175 items/m^3 and the highest microplastic particles in the form of fragments, filaments, and films found at the Oficina station was 0.05 items/individual. The color of microplastics commonly found is black as much as 50% of the identified colors, which are identified as polyethylene polymers. The most-used packaging materials would be explained in this book chapter.

Microplastics are of particular concern due to their long-term durability in the environment their great potential for releasing plastic monomers and additives/chemicals, and their amazing vector-capacity for adsorbing or collecting other pollutants (Lim, 2021). The very small size of microplastics and their large number in the oceans also make them ubiquitous (can live/live in various places) and high bioavailability (available) for aquatic organisms. As a result, microplastics can be eaten by marine biota (Li et al., 2016). In addition, a food packaging made of plastic is composed of polymers, namely long chains of smaller units called monomers. Monomers in plastic packaging can trigger migration. Migration is the process of transferring a substance from the food packaging material to the food product. This migration affects the aroma, smell and taste of the product and has an impact on human health (Sarinah Basri et al., 2021). Heat or inappropriate use is one of the main factors causing migration (Guan et al., 2021). Consumption of foods containing migrating monomers can lead to accumulation in the body. This build-up of monomers in the body can trigger cancer. These monomers are insoluble in water so they cannot be excreted through either urine or feces. Microplastics have been found in human feces, albeit their exact toxicity to humans has yet to be determined. Microplastics have been found to elicit a localized immunological reaction once they reach our systems, through either ingestion or inhalation (Kannan & Vimalkumar, 2021). As a result, the book chapter explore and explain the possible dangers of microplastics for human health.

This book chapter focuses on the potential commercial food packaging materials, releasing the packaging materials to the food products, quantification of particles and chemicals of food packaging material, factors influencing microplastic to the food products, as well as their potential exposure and threats on human health.

2. Commercial food packaging contact materials

Types of plastic that are generally used as food packaging considered as microplastics, are specifically polyethylene terephthalate (PET), high-density polyethylene (HDPE), low-density polyethylene (LDPE), polypropylene (PP), polyvinyl chloride (PVC), polystyrene (PS). Those plastics are among the most hazardous, according to Lithner, Larsson, and Dave (2011), who categorized polymer types based on the hazard of their monomers. Thermoplastic resins and containers are often used in these plastic packaging materials due to their ability to be mechanically recycled (Geueke, Groh, & Muncke, 2018). Since each polymer type has its own set of characteristics, it is used for a variety of packaging applications (Table 1). Plastic containers are widely used and favored packaging materials all over the world, and they can be readily molded into a wide range of goods that may be utilized in a number of applications (Jadhav, Sankhla, Bhat, & Bhagat, 2021), e.g., PE, is generally said to be a single-use plastic. The advantages of plastic packaging that are light, flexible, multipurpose, strong, do not rust, can be colored and the price is low, causes people to not realize the negative impact caused by non-biodegradable plastics in the environment. Polypropylene (PP) plastic, which is used as packaging products, textiles, and stationery, is dangerous if there is a transfer of constituent substances from plastic into food. As a result, the volume of packaging plastics grows, so the amount of produced plastic garbage eventually finds up in the environment and the seas as microplastics. Table 2 shows the microplastics released from the commercial plastics, including the total microplastics particles found from packaged food products.

Table 1 Properties of commercial food packaging materials.

Plastics properties	Polypropylene (PP)	Polyethylene (PE)		Polystyrene (PS)	Polyethylene terephthalate (PET)
		LDPE	HDPE		
Monomer	Propylene	Ethylene	Ethylene	Styrene	Terephthalic acid, ethylene glycol
T_g (°C)	−10	−120	−120	74–105	73–80
T_m (°C)	160–175	105–115	128–138		245–265
WVTR g μm/m^2 day at 90% RH and 38°C	590	375–500	125	1750–3900	390–510
Tensile strength, MPa	31–42	8.2–31.4	17.3–44.8	35.8–51.7	48.2–72.3
Tensile modulus, MPa	1140–1550	172–517	620–1089		2756–4135
Elongation at break, %		100–965	10–1200	1.2–2.5	30–3000%
Tear strength, g/25 μm, film	50	200–300	20–60	4–20	30
Density, g/cm^3	0.902	0.910–0.925	0.94–0.965	1.04–1.05	1.29–1.40
O_2 permeability, cm^{3}μm/m^2 day atm.	146,000	163,000–213,000	40,000–73,000	98,000–150,000	1200–2400
CO_2 permeability, cm^{3}μm/m^2 day atm.	200,000–320,000	750,000–1,060,000	200,000–250,000	350,000	5900–9800
Water absorption, %	0.01–0.03	<0.01	<0.01	0.01–0.03	0.1–0.2

Continued

Table 1 Properties of commercial food packaging materials.—cont'd

Plastics properties	Polypropylene (PP)	Polyethylene (PE)		Polystyrene (PS)	Polyethylene terephthalate (PET)
		LDPE	HDPE		
Uses as food packaging	Yogurt, margarine tubs, plastic cups, plastic bottles, and snack wrappers	Plastic grocery bags, sandwich bags, squeezable bottles, juice and milk cartons	Plastic chairs, plastic tables, jugs, buckets, toys, shampoos and soap bottle	Egg cartons, foam cups, dairy and fishery food packaging	Water bottles, soft drink bottles, oil and peanut butter containers, and balloons
Molecular orientation in solid	Semicrystalline or amorphous	Semicrystalline	Semicrystalline	Amorphous	Semicrystalline
Code in food packaging	05 PP	04 PE-LD	02 PE-HD	06 PS	01 PET
Environmental weakness	– Prone to bending, breaking, and crushing resistance (tenacity) at frigid temperatures – Brittle at 0 °C – Crazed and small cracks form as a result of oxidation caused by heat and sunshine	Easily cracked under stress and photooxidized	Easily cracked under stress and crazed in sunlight	Brittle substance and degraded in sunlight and turns to yellow color	Hydrolyzed in water, at high humidity at >73–78 °C
Effects on environment and human	UV degradation, highly susceptible	Exposure by radiation produces methane and ethylene	Emits carbon dioxide, monoxide, nitrogen monoxide	Release chlorofluorocarbons and emits carbon monoxide	Toxic chemical antimony causing to human life such as cancer

Table 2 Microplastics released from the commercial plastics, including total microplastics particles found from packaged food products.

Microplastics	Plastic materials	Size range of microplastics	Food source	Concentration of microplastics	Reference
Fibers, fragments and films	PP and PE	38 μm to 1 mm	Fish	0.96 ± 0.08 per fish	Wootton, Reis-Santos, Dowsett, Turnbull, and Gillanders (2021)
Fibers, fragments and sheets	PE and PP	157–2785 μm	Indian white shrimp	0.04 ± 0.07 per gram	Daniel, Ashraf, and Thomas (2020)
Fibers, films, fragments, pellets, beads	PE, PP, PS	<100–1000 mm	Golden anchovy	6.78 ± 2.73 per fish	Gurjar et al. (2021)
Fibers, fragments, films and pellets	PP	0.11–4.97 mm	Seaweed nori	1.8 ± 0.7 per g	Li, Feng, Zhang, Ma, and Shi (2020)
Fibers, fragments and films	PET, PS, PP, and LDPE	Fibers: 100–8000 μm Fragments: 10–1100 μm Films: 70–1000 μm	Canned fish	1.28 ± 0.04 per g	Akhbarizadeh et al. (2020)
Microfibers and particles	PP and PE	3.3–4460 μm	Table salts	38.42 ± 24.62 per kg	Fadare, Okoffo, and Olasehinde (2021)
Fragments and fibers	PE and HDPE	1–500 μm	Vinegar	51.35 ± 20.73 per liter	Makhdoumi et al. (2021)
Fibers and fragments	PET	<1–2 mm	Beer	152 ± 50.97 per drinks	Kutralam-Muniasamy, Pérez-Guevara, Elizalde-Martínez, and Shruti (2020)

Continued

Table 2 Microplastics released from the commercial plastics, including total microplastics particles found from packaged food products.—cont'd

Microplastics	Plastic materials	Size range of microplastics	Food source	Concentration of microplastics	Reference
Particles and films	PS, PP, PE, and PET	$\leq$500–1001 μm	Unspecified food	1–41 per container	Du, Cai, Zhang, Chen, and Shi (2020)
Particles and films	HDPE	150 nm to 764.8 μm	Water	25,000 per 100 mL	Ranjan, Joseph, and Goel (2021)
Fibers, fragments, and particles	PET and HDPE	>3 μm	Water	148 ± 253 per liter	Winkler et al. (2019)
Particles and nanoparticles	PET	100 nm to 5 mm	Tea	11.6 billion per cup	Hernandez et al. (2019)
Particles	PP	1–20 μm	Infant formula	16.2 million per liter	Li, Shi, et al. (2020)

Modified from Cverenkárová, K., Valachovičová, M., Mackuľak, T., Žemlička, L., & Bírošová, L. (2021). Microplastics in the food chain. *Life*, 11(12), 1349. doi:10.3390/LIFE11121349; Jadhav, E. B., Sankhla, M. S., Bhat, R. A., & Bhagat, D. S. (2021). Microplastics from food packaging: An overview of human consumption, health threats, and alternative solutions. *Environmental Nanotechnology, Monitoring & Management*, *16*, 100608. doi:10.1016/J.ENMM.2021.100608.

2.1 Polypropylene (PP)

Polypropylene (PP) is a thermoplastic made from the combination of propylene monomers. PP manufacturing procedures include pressured injection of melted master batch into a mold cavity, which results in smooth surfaces (Du et al., 2020). A tiny quantity of ethylene (2–5%) is added to the polymerization process to generate the latter. Low density (0.89–0.92 g/cm^3) and chemical and mechanical fatigue resistance, particularly environmental stress cracking, are characteristics of thermoplastic PP polymers. Yogurt, margarine tubs, plastic cups, plastic bottles, and snack wrappers are just a few of the food packaging applications for polypropylene. PP manufacturers are always developing new grades with enhanced or altered qualities. In comparison to LDPE and HDPE, PP has a lower density, a greater melting temperature, and a higher stiffness (modulus) (Salakhov et al., 2021). PP is hard, resistant to heat and chemicals, and float in water with a molecular weight of between 200,000 and 600,000. Stronger rigidity and simplicity make PP ideal for stretch applications, while its superior heat resistance gives more possibility for PP to be sterilized in an autoclave.

Demand for polypropylene is rapidly increasing and consequently, it is one of the most common types of microplastic found in the marine environment (Barbeş, Rădulescu, & Stihi, 2014). Table 2 shows that microplastics from PP forms fibers, fragments and films with the sizes 10–2785 μm, found in seafood in marine environments (Cverenkárová, Valachovičová, Mackul'ak, Žemlička, & Bírošová, 2021). PP degraded to microplastic and can be affected by environmental conditions such as temperature and sunlight. PP has poor resistance to bending, breaking, crushing (tenacity) at cold temperatures. At temperatures below 0 °C, PP tends to be particularly brittle and thus leads to fracture. Oxidation from heat and sunlight cause fine cracks that deepen and become more severe in longer periods of time. Besides, microorganisms and fungi, such as *Pseudomonas* sp., *Vibrio* sp., and *Aspergillus niger* have degradative effects on PP (Cacciari et al., 1993; Muenmee, Chiemchaisri, & Chiemchaisri, 2015).

2.2 Polyethylene (PE)

According to density of polyethylene, PE was divided as high density of polyethylene (HDPE) and low-density polyethylene (LDPE).

2.2.1 HDPE

HDPE is the second most-used packaging material. It is made by polymerizing ethylene, but it has an almost linear structure, as opposed to the extremely branched structure of low-density polyethylene (Wypych, 2016). As a result, the capacity to crystallize improves, resulting in a tighter packing of molecules and a higher density. HDPE is a nonpolar, milky-white linear thermoplastic. It has a melting temperature of roughly 128–138 °C and a density of 0.940–0.965 g/cm^3. The average molecular weight and distribution of molecular weight, like with other polymers, influence the characteristics. Tensile strength, impact strength, and stress fracture resistance all rise as relative molecular mass (or molecular weight) increases. Table 1 shows physical properties of HDPE.

It is the second most often used plastic in the packaging sector and is one of the most versatile polymers. Containers for milk, detergent, bleach, juice, shampoo, water, and industrial chemical drums made by extrusion blow molding, as well as thin-walled dairy containers and closures made by injection molding; blown and cast films used in flexible packaging applications such as cereal, cracker, and snack food packaging, delicatessen wraps, and grocery sacks (Cherif Lahimer, Ayed, Horriche, & Belgaied, 2017; Wypych, 2016). HDPE is designed for water-based products, medium molecular weight aliphatic hydrocarbons, alcohols, acetone, ketones, dilute acids and bases are only a few of the chemical substances that is resistant for HDPE. Aromatic hydrocarbons, such as benzene, are not suitable. HDPE has a tensile strength of up to 45 MPa and strong moisture barrier capabilities but low oxygen and organic chemical barrier properties (Polyethylene, HDPE, 2012).

Table 2 shows that microplastics of HDPE are found as fragments and fibers with the size 1–500 μm from vinegar packaging (Cverenkárová et al., 2021; Makhdoumi et al., 2021). HDPE is susceptible to environmental stress cracking. It is described as the failure of a material under stress and exposed to a chemical substance in settings where neither the stress nor the chemical alone causes failure (Weinstein, Crocker, & Gray, 2016). Under mild mechanical stress, UV radiation exposure cause the HPDE to be fragmented to microplastic.

2.2.2 LDPE

LDPE offers a variety of desired properties, including clarity, flexibility, heat sealability, and processing simplicity. The physical properties of LDPE is shown in Table 1. The balance between molecular weight, distribution of molecular weight, and branching determines the actual values of these attributes. LDPE is also adaptable to blown film, cast film, extrusion coating, injection molding, and blow molding in terms of processing mode.

Food and apparel containers and bags, industrial liners, vapor barriers, agricultural films, domestic items, and shrink and stretch wrap films are all constructed of LDPE. Superb flexibility, good impact strength, fair machinability, good oil resistance, fair chemical resistance, strong heat sealing qualities, and low cost are all features of LDPE (Kumar, Maitra, Singh, & Burnwal, 2020; Loredo-Treviño, Gutiérrez-Sánchez, Rodríguez-Herrera, & Aguilar, 2011). Due to its lower percentage crystallinity, it has a higher transparency than HDPE, a decent water vapor barrier, and an even worse gas barrier than HDPE.

Microfibers of LDPE are found as fragments, films and fibers with different sizes 0.01–8 mm from canned fish (Akhbarizadeh et al., 2020; Cverenkárová et al., 2021). It indicates that the canned fish could eat the microfibers before canning process. The form of microfibers is as a result of its sensitivity under stress causing a crack and photooxidation. Khandare, Agrawal, Mehru, and Chaudhary (2022) also reported LDPE could be degraded by marine bacteria, *Marinobacter* sp. and *Bacillus* sp., which shows a tensile strength decrease and thermal weight loss. Isolates, *Aspergillus nomius and Streptomyces* sp., were observed to efficiently degrade LDPE after 90 days degradation (Abraham, Ghosh, Mukherjee, & Gajendiran, 2017).

2.3 Polystyrene (PS)

Polystyrene (PS) is a styrene addition polymer (Lithner et al., 2011). PS used in packaging is atactic, meaning it will not crystallize due to its properties as an amorphous polymer. PS is a stiff, brittle material owing to the bulkiness of the benzene ring substituent, which provides significant resistance to chain rotation. Table 1 shows its density of 1.05 g/cm^3 and a T_g range of 74–105 °C. It is exceptionally transparent due to its lack of crystallinity. PS is not appropriate for usage at high temperatures owing to liquid flow at around 100 °C (212 °F). PS does not have a definite melting point due to its amorphous properties, but softens over a wide temperature range. PS is easier to extrude and thermoform due to its ability to flow under stress at relatively increased temperatures. Polystyrene is a low-cost polymer with a weak barrier to water vapor and gases in general and has a higher chemical reactivity than PE and PP.

PS is one of the most adaptable packaging resins on the market. Thermoformed containers for dairy products are frequently made of high impact, and egg cartons and meat trays are examples of food packaging polystyrene. PS foam provides superb stress absorption and thermal insulation properties. Owing to wide packaging application, PS is widely found to be degraded in microplastics in the forms of fibers, films, pellets and beads with the sizes 100–1000 mm (Cverenkárová et al., 2021; Gurjar et al., 2021). Its degradation is caused by brittleness and sunlight causing to form yellow color. PS can also be degraded by indigenous microbial consortium under natural conditions on mangrove environment, where microplastic degradation shows 18% weight loss after treated for 90 experimental days (Auta et al., 2022).

2.4 Polyethylene terephthalate (PET)

para-Xylene and ethylene are combined to make PET. *p*-Xylene is transformed to dimethyl terephthalate or terephthalic acid, whereas ethylene is converted to ethylene glycol. Usage of terephthalic acid for a condensation process produces water as a by-product molecule. PET has a good oxygen and carbon dioxide barrier (Aguilar-Guzmán et al., 2022). The most common application of PET is in soft drink bottles although its use in nonsoft-drink "custom" bottles has grown quickly in recent years. Bottles are made using injection blow molding and stretch blow molding and PET sheet is frequently thermoformed and is used in extrusion coating. Since orientated PET deforms when exposed to high temperatures, both the film and the bottles can be heat-set to increase their stability, enabling to be used in applications like as hot filling (Shin et al., 2021). Food, distilled liquor, carbonated soft drinks, noncarbonated beverages, and hygiene are all packaged in PET (Lemos Junior, do Amaral dos Reis, de Oliveira, Lopes, & Pereira, 2019). Mustard, peanut butter, spices, edible oil, syrups, and cocktail mixers are examples of common culinary items. PET that is biaxially oriented is utilized in the packaging of meat and cheese. Ovenable board is created by covering paperboard with PET for usage in applications such as frozen meals.

Wide variety of PET usage for food packaging application due to its microorganism repulsion, high resistance, and corrosion resistance entails a big waste in the environments (Aguilar-Guzmán et al., 2022). Microplastics from bottled water were mostly PET and PP with size 5–20 μm and average contents around 118 particles per liter for PET (Kannan & Vimalkumar, 2021). It is reported that PET is degraded to microplastics forming like fibers, fragment and particles and its degradation occurs under water hydrolyzation at high humidity and at high temperatures (>73–78 °C) (Zhang et al., 2020). Auta et al. (2022) reported PET, like PS, could be degraded by indigenous microbial consortium under natural conditions on mangrove environment after treated for 90 experimental days. PET microplastics can be degraded by bacterial whole-cell biocatalysts, reducing PET particles sizes dramatically under neutral conditions (Gong et al., 2018).

2.5 Polyvinyl chloride (PVC)

Polyvinyl chloride (PVC or vinyl) contains 57% by weight of chlorine. Polyvinyl chloride is a vinyl polymer consisting of repeating vinyl groups in which one of the hydrogen atoms is replaced by a chlorine atom. PVC was first obtained in 1872 by the German chemist Eugen Bauman. However, Eugen Bauman has not registered a patent for polyvinyl chloride. The first person to register a patent for the invention of PVC is the German chemist Fritz Klatte. The method of preparation is based on the reaction of acetylene with hydrochloric acid under the action of sunlight using mercury chloride as a catalyst. This method was widely used in the 1930s and 1940s (Akovali, 2012).

PVC is used for the manufacture of food packaging, conveyor belts and pipes used in food production, gaskets for fixing metal lids on glass jars and bottles (Bueno-Ferrer, Garrigós, & Jiménez, 2010). In its pure form, PVC is a hard material to which plasticizers are added to give it flexibility and elasticity. As plasticizers, both phthalic acid derivatives and plasticizers of a nonphthalate nature can be used (ESBO, ATBC, DEHT, DINCH, DEHA and DINA) (de Anda-Flores et al., 2021). The density of polyvinyl chloride ranges from 0.5 to 1.45 g/cm^3 at 25 °C and its' elongation at tension can reach 190%. The coefficient of thermal expansion of the material is 5×10^{-5} $°C^{-1}$.

There are conflicting data on the environmental spread of PVC microplastics. While Vidyasakar et al. (2020) reported that more than 75% of microplastic particles along Silver Beach in South India consist of polyvinyl chloride, Tiwari, Rathod, Ajmal, Bhangare, and Sahu (2019) found that only 1.33% of plastic particles in the micrometer range consist of PVC. The study was conducted on sand of beaches along the Indian coast from three different locations: Gurgaon Mumbai (Arabian Sea coast), Tuticorin and Dhanushkodi (Bay of Bengal coast). Lu et al. (2020) evaluated the presence of microplastics in animal-based medicinal materials. Studies showed prevalence of microplastics in more than 90% of materials, but PVC is contained only in 5.52% of the studied samples.

It is important to note that PVC is chemically resistant to a large number of inorganic and organic chemicals, which are dilute and concentrated acids, alcohols, alkalis, aliphatic hydrocarbons, aldehydes, esters, aromatic as well as halogenated hydrocarbons and ketones (Akovali, 2012). In this regard, it is relevant to develop a nonchemical method of degradation of microplastics from PVC. Ouyang et al. (2022) found that PVC microplastics can be destroyed by ultraviolet radiation. Miao et al. (2020) proposed an EF-like technology with TiO_2/graphite cathode for degradation of polyvinyl chloride, based on simultaneous cathodic reduction of dechlorination and oxidation of hydroxyl radical. The authors reached 70% degradation efficiency of microplastics from polyvinyl chloride. Microbiological methods of degradation of microplastics from PVC are also known and have prospects in scientific developments and practical application (Liu, Xu, Ye, & Zhang, 2022).

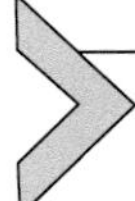

3. Release mechanisms of microplastics from packaging into foods and specific migration limits (SML)

Flexibility and other advantageous properties of commercial plastics, as explained in the previous section, trigger consumers to purchase food products packaged by the commercial plastics. However, many consumers do not realize negative effects of the plastics. The plastics of packaging products can release microparticles and nanoparticles into the packaged foods and beverages, in which the hazardous effects of the microparticles depend on the physical and chemical properties. (Bouwmeester, Hollman, & Peters, 2015; Fadare, Wan, Guo, & Zhao, 2020). During plastic production, performance-improving chemicals and additives, such as plasticizers and antioxidant are added to the polymers (Bahmid, Dekker, Fogliano, & Heising, 2021; Bahmid, Syamsu, & Maddu, 2014). These chemical substances can also be released to the environments, as reported that around 35–917 tons of the chemicals are release annually to the oceans (Suhrhoff & Scholz-Böttcher, 2016). Those known chemicals and additives are mostly bisphenol A, phthalate esters, nonylphenol (Hermabessiere et al., 2017). The release of microplastics and those chemical compounds are discussed in this section.

3.1 Release mechanisms of microplastics from packaging into foods

Release of microplastics to the foods starts from plastic degradations. The degradation mechanisms of microplastic is shown in Fig. 1A. The extended

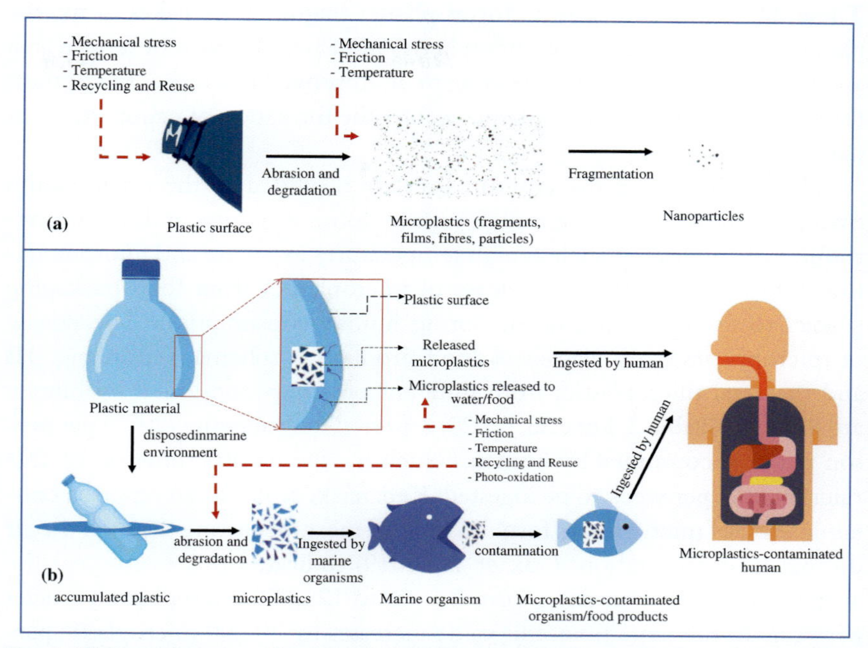

Fig. 1 (A) Degradation mechanism of plastic materials and (B) possible migration mechanisms of microplastics to food products before its consumption by human. *Panel (A): adapted from Jadhav, E. B., Sankhla, M. S., Bhat, R. A., & Bhagat, D. S. (2021). Microplastics from food packaging: An overview of human consumption, health threats, and alternative solutions. Environmental Nanotechnology, Monitoring & Management, 16, 100608. doi:10.1016/J.ENMM.2021.100608*

exposures from aforementioned factors, such as heat, light, mechanical force cause fragmentation into smaller particles and molecules (Jadhav et al., 2021). This degradation for each plastics types depends on its properties, which categorizes mechanical, photo-, biological and thermal degradation (Svedin, 2020). Degradation of plastic to microplastics categorized as:

(1) Mechanical degradation includes the physical stress, external abrasive and repetitive use;

(2) Photodegradation includes sunlight radiations exposure breaking the bond of C—H and C—C, which happens effectively by oxygen compared to UV light;

(3) Biological degradation occurs due to microorganisms or causative agents or enzyme under anaerobic or aerobic circumstances;

(4) Thermal degradation includes the exposure of high temperatures influencing the physical and chemical structure of plastic polymer.

These degradations of plastics forms plastics fragments or flakes of microplastics. The reduced size of microplastics improves the ratio of surface area and volume of microplastics, leading to an improved reactivity. In contrast, further degradation of microplastic enables the formation of nanoparticles at nanosized particles.

Plastics thrown in the environment are degraded to the microplastics owing to mechanical force, rough surface, loose structure, and other environmental conditions, such as high temperatures exposure and photooxidation (Du et al., 2020). The release of microplastics from food packaging plastics to food products occur during human consumption. The release of microplastics from plastics to food products is schematized in Fig. 1B and the total microplastics from different polymers consumed by human are shown in Table 2. Du et al. (2020) reported microplastics (2977) per person per year consumed by human, which is exposed 407 times or 12–203 microplastics per week to be ingested. Kedzierski et al. (2020) reported contamination of microplastics form PS trays to meat, occurring through the air during meat preparation. Cox et al. (2019) indicated that every year the average American consumes from 74,000 to 121,000 microplastic particles per year. At the same time, this value increases by 90,000 microplastic particles per year if a person uses bottled water and by 4000 microplastic particles per year if he consumes tap water.

Contamination of microplastics is found in bottled water, where about 93% tested bottles out 259 bottles are contaminated with microparticles (Mason, Welch, & Neratko, 2018). The release occurs owing to the mechanical stresses (opening and closing the caps). Furthermore, the release of microplastics also occurs due to temperatures. Guan et al. (2021) simulated scenarios of victuals packaging materials at different temperatures to investigate the microparticles release. At only 50 °C, PS foam containers can release various sizes and shapes of microplastics.

This plastic degradation to microplastics or even nano plastics can endanger organisms in the marine environments and can even circulate to reach food products through the food chain. Fig. 1B shows migration mechanisms of microplastics contaminated by marine organisms as a food product that would be consumed by human. Accumulation of microplastics in the marine environment cause a consumption of the microplastics by marine organisms, such as fish, zooplankton, etc. (Cole, Lindeque, Fileman, Halsband, & Galloway, 2015). Microparticles enter the body of the microorganisms through biomagnification and bioaccumulations, cause a potential negative effect for the organisms (Smith, Love, Rochman, & Neff, 2018). Microparticles can be ingested and accumulated in intestine of microorganisms. According to some studies, fish, seaweed, chicken, honey, salt contains microparticles (Bessa et al., 2018; Karami et al., 2017; Neves, Sobral, Ferreira, & Pereira, 2015; Oßmann et al., 2018). The European Food Safety Authority (EFSA) reported the microplastics found in shrimp, bivalves, and fish at concentration ranges from 0.75, 0.2–4, to 1–7 particles/g, respectively (European Food Safety Authority, 2016). In addition, according to Wakkaf et al. (2020), the average Tunisian who consumes wild mussels can get about 4200 microplastic particles per year. Moreover, other chemical from microparticles, such as plasticizers, can be accumulated in the marine environment, leading to cumulative effects to the marine organisms (Sun et al., 2021).

Plastic particles are also found in other human food products such as tea bags, honey, sugar, soft drinks, table salt, milk, beer, tap and bottled drinking water (Bessa et al., 2018; Karami et al., 2017; Li, Peng, Fu, Dai, & Wang, 2022; Neves et al., 2015; Oßmann et al., 2018; Shruti, Pérez-Guevara, Elizalde-Martínez, & Kutralam-Muniasamy, 2020; Yee et al., 2021). According to WHO, the recommended level of salt intake is 5 g per day for an adult (Alimba, Faggio, Sivanesan, Ogunkanmi, & Krishnamurthi, 2021). Iñiguez, Conesa, and Fullana (2017) found in 1 kg of salt from Spain approximately 280 plastic particles. Yang et al. (2015) claim that 1 kg of salt from China contains from 204 to 861 particles of micro and nanoplastics. These results are consistent with the research of Kosuth, Mason, and Wattenberg (2018) who declared that 1 kg of table salt can contains up to 806 particles of micro and nanoplastics. Based on the data above, it is possible to calculate the consumption of microplastics by the average person per year according to formula (1):

$$C(\text{in year}) = CP \times CR \times N \tag{1}$$

C(in year)—consumption of microplastics by the average person per year (particles);

CR—consumption rate (g/day);

CP—microplastic content in food (particles/g);

N—number of days in a year (days).

Calculations shows that consumption of microplastics from table salt by the average person ranges from 372 to 1571 particles per year. Similarly, the consumption of microplastics from other food products can be calculated using same formula (1) and data of microplastic content in foods from other studies (Diaz-Basantes, Conesa, & Fullana, 2020; Karami et al., 2017; Liebezeit & Liebezeit, 2015; Shruti et al., 2020). For example, the consumption of micro and nanoplastics ranges from 1095 to 6570 plastic per year from iced tea, from 219 to 803 plastic per year from cow's milk, etc. Thus, it can be concluded that a significant part of micro and nanoplastics enters the human body from food.

Consequently, the organisms consuming microplastics, like fish, shrimps, enter human food chain. The consumption of the seafood can result in carcinogenic effects for human body, depending on the intake rate (Sarinah Basri et al., 2021). Yudhantari, Hendrawan, and Puspitha (2019) investigated microplastics in the faces of pregnant woman and found microplastics in all faces samples of the woman. This microplastic can be found aquatic ecosystems and can be digested by organisms, including human, through consumptions (Naidoo, Rajkaran, & Sershen., 2020; Yu et al., 2020). In other cases, the microplastics can be consumed by marine microorganisms, and then further consumed by human.

3.2 Specific migration limits of microplastics to food products for human consumption

Estimation of microplastics intake in humans' body are required owing to the negative effects of the microplastics for human health. Microplastics contamination and release to the foods have been analyzed, even though little known on the major food groups. For baby, around 3 million microparticles per day can be exposed to the babies' milk prepared in baby bottles, like PP. For children, over 3 million microparticles per kg per body weight per day can be consumed using drinking bottled water (Zuccarello et al., 2019). This high concentration of microparticles can be found in the bottled water or milk due the release from the plastics of the container. Besides, daily consumption quantity of microparticles for adults is around a million microparticles/kg/body weight/day (Zuccarello et al., 2019). A meta-analysis on the microparticles concentrations in various food exposed to human found that a suggested ingestion dose is around 0.10–5 g/week (Senathirajah et al., 2021). The suggested an excretion rate of 0.03–677 mg microparticles per week was reported for microparticles of PET (Zhang, Wang, & Kannan, 2019). Ju et al. (2020) estimated the global average for humans' ingestion dose around 0.1–5 g of microplastic each week. In consideration of contamination risk for adults and children, safe fish and food for adults and children are 300 and 50 g per week, respectively (Senathirajah et al., 2021).

The different results for microplastics migration limits might be influenced by different methods of measurement in the foods and human intakes. The specific migration limits for human body are still limited. Kannan and Vimalkumar (2021) reviewed toxic dose for human exposure, but majority were investigated on microparticles of PS and limited investigation on the effects of PET, PP and other microparticles. Therefore, further research on this topic are still essential to understand the exposure limit of microplastics in food products.

4. Identification and quantification of particles and chemicals from plastic packaging in foods

Microplastics can be studied by a variety of analytical techniques, including Raman spectroscopy and Fourier-transform infrared spectroscopy (FTIR). Using Raman spectroscopy, large samples greater than 1 μm can be analyzed. While Raman spectroscopy can detect nonpolar symmetric bonds more effectively than FTIR, FTIR is more sensitive to identifying polar groups. Insights into polymers' dissociation products can be obtained using electrospray ionization coupled with mass spectrometry (ESI-MS). The technology enable to determine both the identity of the polymer and the organic filler simultaneously. In addition, low-density polyethylene (LDPE) and high-density polyethylene (HDPE) are tougher to distinguish using MS. The spectroscopy of nuclear magnetic resonance (NMR) can be used to provide information about the chemical structure of microplastic polymer chains, comprehensive information on monomers in copolymer compounds, crystallinity in semicrystalline polymers, as well as information about branching and tacticity. However, no data found regarding the NMR analysis of microplastic in food packaging. New technologies are continually being developed for complicated microplastics analysis that differ principally in the sample preparation process. Extraction using compressed liquid is one of these innovative approaches. Microplastics may also be measured using selective fluorescence tagging with the lipophilic dye Nile Red, which is a novel method. Examining polymers with dimensions less than 100 nm is a considerable challenge. Scanning electron microscopy (SEM) methods are mostly employed for particle and chemicals analysis (Cverenkárová et al., 2021). X-ray Photoelectron Spectroscopy (XPS) is an effective tool for analyzing the surface of polymers and is used frequently in many laboratories across the globe for this purpose. With the exception of hydrogen, all of the component constituents of organic polymers are observable and exhibit chemical shifts that reveal their chemical environment.

4.1 Liquid chromatography-high resolution mass spectrometry (LC-HRMS)

HRMS was sensitive enough to detect five contaminants in all pieces of beef meat that came into touch with the vacuum-plastic packing, according to the examination of the survey spectra: phthalic anhydride, stearamide, diisooctyl phthalate, and polyethylene glycol (PEG). The duplicates investigation demonstrated that the procedure maintained the same spectral profile for each of the sample types investigated (negative control meat, packed meat, and packing material), proving the method's repeatability. When comparing the spectra, no signals associated with these compounds are found in the spectrum from butcher's shop beef meat (negative control); nevertheless, they are present in the profiles for packing material and packed beef meat. Moreover, the signal strength of the discovered components is greater in the packing material spectrum than in the packed beef meat spectrum, validating the fact that these components belong to the studied material and were able to travel into meat. There were no variations in the profiles revealed for the samples from superficial and deeper cuts (Guerreiro, de Oliveira, Melo, de Oliveira Lima, & Catharino, 2018). However, these bands are present in the spectra of packaging material and packaged meat. It is important to note that the intensity of these bands decreases in samples of packaged meat, which indicates the migration of packaging material into meat.

4.2 Mass spectrometry/mass spectrometry (MS/MS)

In order to track less volatile migrants' movements, mass spectrometry-associated electrospray ionization (ESI-MS) was examined. Food simulants were heated to 80 °C, while the water was heated to 100 °C.

Two prominent peaks with high masses at m/z 851 and 867 were observed in the PP-R mass spectra arising from migration to water and 10% ethanol. In the isooctane/ethanol combination, these masses were also present, while they were reduced.

There were 44 amu difference in mass between a series of molecules ranging from 551 to 815 in the 10% ethanol extracts, this is due the repeating unit weight ($—CH_2—CH_2—O—$). Based on the distinctive peak series and mass difference of 44 amu, this chemical series has tentatively been identified as poly(ethylene glycol) (PEG) oligomers. In the manufacture of other polymers, PEG is frequently utilized as a plasticizer. Although to a lesser extent, these oligomers also migrated to water. It was difficult to identify these peaks when ethanol and isooctane were used as food simulants because other interfering peaks obscured them. Normally, heated ethanol samples, however, were not detected to contain PEG oligomers. From the PP-C samples, the most dominant ion of the migrants was identified at *m/z* 289. Different concentrations of this substance were detected in all food simulants (Alin & Hakkarainen, 2011).

4.3 Liquid chromatography electrospray ionization quadrupole time-of-flight mass spectrometry (LC-ESI-Q-TOF/MS)

To identify unexpected migrants from multilayer plastic packaging materials, the LC-ESI-Q-TOF-MS method was applied by Gómez Ramos, Lozano, and Fernández-Alba (2019). The bulk of the migrants were found in both materials, with the exception of compounds 8, 13, and 26, which were only found in the PE-based material, and compounds 20–24, which were only found in the simulants created from the LDPE + nylon composite. The only found component specified in Regulation (EU) No 10/2011 was caprolactam, a monomer of polyamide 6 (nylon) that is a likely breakdown product of compounds often used in polyurethane adhesives.

Migration between simulants B and C was found to be highly similar. Only two molecules (peaks 25 and 26) were found solely during migration to simulant B, with acetic acid accounting for 3% of the total. Both compounds were observed during the migration from the PE-based material while only peak 25 was found in the LDPE + nylon material. According to their masses, these molecules are the cyclic oligomers AA-DEG and AA-DEG-PA-DEG plus a molecule of H_2O (Gómez Ramos et al., 2019).

4.4 Raman spectrometry (RS) and Fourier-transform infrared spectroscopy (FTIR)

The researchers (Katsara, Kenanakis, Viskadourakis, & Papadakis, 2021) analyzed cheese samples on day 0 of testing to determine whether LDPE packaging as well as the characteristic Raman and ATR/FTIR peaks of cheese samples overlapped. It is evident that LDPE Raman peaks are absent in cheese's spectrum, except for one peak at 1129 cm/cm; in Kefalotyri cheese, this peak appears at 1131 cm/cm; whereas in Parmesan cheese it appears at 1134 cm/cm. The Raman peaks from the multiple cheese samples do not exactly correspond to the same LDPE peak of 1129 cm^{-1}, nevertheless this peak is regarded as the same and they are regarded as the same. LDPE migration identification was based on intensity changes, since LDPE ATR/FTIR peaks can be found in all cheese samples (Katsara et al., 2021).

In a study, the existence of four types of polymers was found by FTIR analysis, which are polyesters, polystyrene, polyamide, and polyethylene (Fig. 2). Among the polyesters, they were polyethylene terephthalate

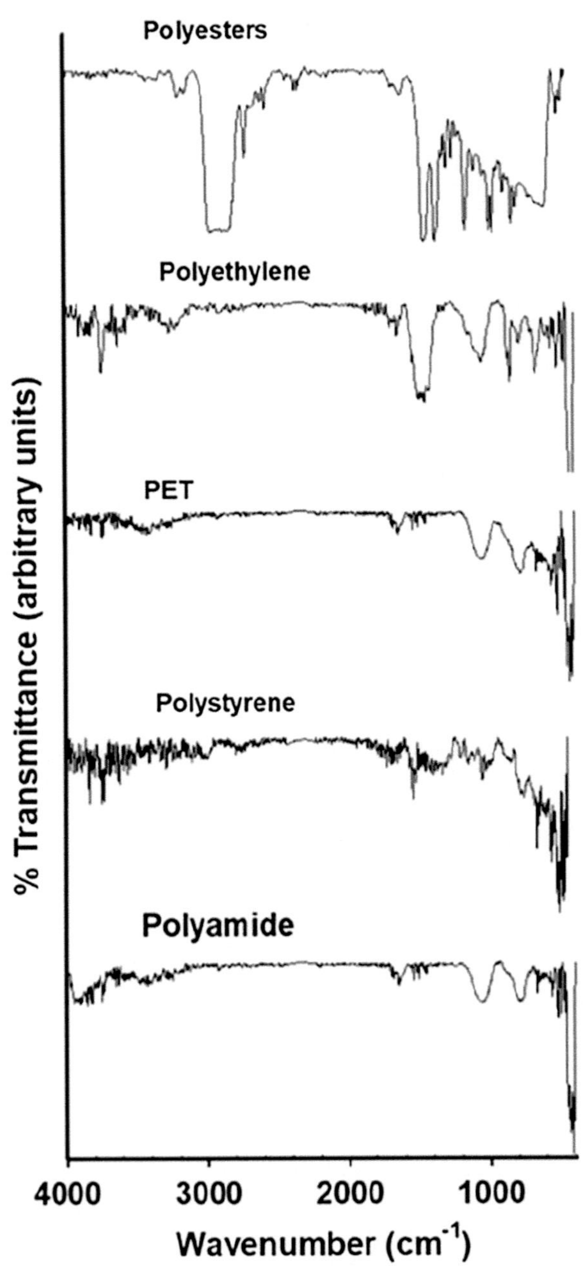

Fig. 2 Identification of the polymer type of extracted particles using FTIR.

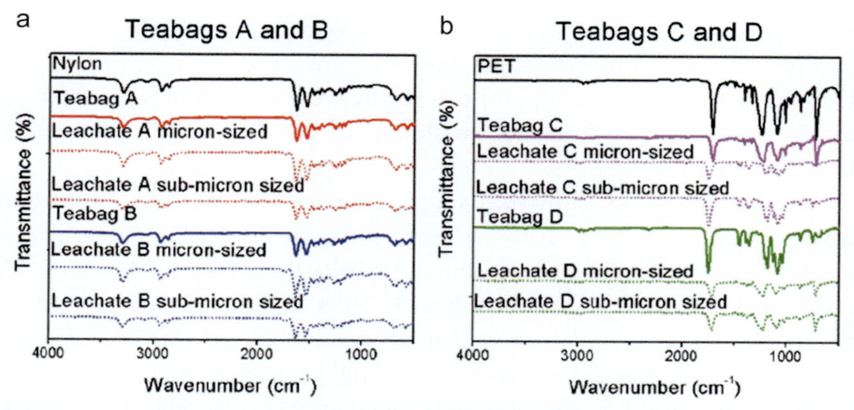

Fig. 3 Nylon-6,6, PET, the original teabag, and their matching leachates: FTIR spectra (Hernandez et al., 2019).

(PET) and others. A total of 626 microplastic particles were detected in all of the samples (Seth & Shriwastav, 2018).

All the different kinds of teabags examined had nearly identical FTIR spectra spanning from 500 to 4000 cm^{-1} (Fig. 3A and B) as their original teabags, regardless of the fractional size of the leachates. Fig. 3A shows an FTIR peak at 3289 cm^{-1} corresponding to the stretching vibration frequency of N—H groups, and the peaks observed at 2932 and 2860 cm^{-1} are related to ethylene sequences in nylon-6,6 (CH_2 asymmetric stretching). In the fingerprint region of their FTIR spectra (2000–500 cm^{-1}), teabags A and B exhibit typical nylon-6,6 peaks: 1634 cm^{-1} (amide I band, with a major contribution from C=O stretching), 1535 cm^{-1} (amide II band, bending vibration frequency of N—H), 1371 cm^{-1} (amide III band, CH_2 wagging), and 681 cm^{-1} (bending vibration frequency of N—H). The first peak is in at 1748 cm^{-1} (acid ester C=O group), the second peak at 1375 and the third peak at 1347 cm^{-1} (CH_2 wagging of glycol, sample D specifically), and the fourth peak at 1226 cm^{-1} and fifth peak 1089 cm^{-1} (broad bands, asymmetric C—C—O and O—C—C stretching, respectively), the sixth peak at 1025 cm^{-1} (in-plane vibration of benzene), and the seventh peak at 730 cm^{-1} (C—H wagging vibrations from the aromatic structure, out of plane of benzene) (Fig. 3B). There is no difference in the FTIR spectra between teabags A and B, or their respective leachates, as compared with a poly(hexamethylene adipamide) such as nylon-6,6. PET vibrations confirm the presence of PET in samples C and D (teabags and their leachates). FTIR data reveals that teabags A and B are made from nylon-6,6 and that teabags C and D are made from PET. The resultant spectra were compared with those of commercial PET and nylon-6,6 tested as validation samples, and the results were consistent with those of the teabags and their leachates, as shown in Fig. 3A and B (Hernandez et al., 2019).

The outcomes are displayed as a pie chart in Fig. 4B. The most often utilized bulk packaging materials were polyethylene (PE), polyethylene isophthalate (PEI), and polyethylene terephthalate (PETs). When Fig. 4A and B are examined, it is obvious that PP, PVA, and PE-PP were not identified in bulk packaging, meaning that a portion of the microplastics isolated from the salt samples did not come from bulk packaging. As a result, it was suspected that some of the microplastics discovered in table salt samples originated in the environment or from salt sources during manufacturing (Fadare et al., 2021).

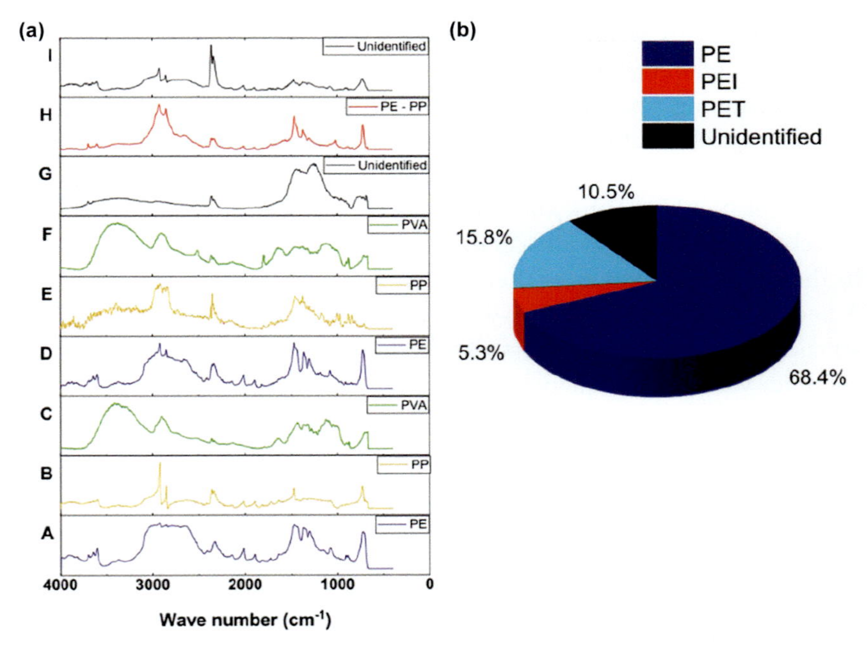

Fig. 4 (A) FTIR spectra of microplastics sampled. (B) The bulk packing materials' chemical composition. Polyethylene; PE, polypropylene; PP, polyethylene terephthalate; PET, polyethylene isophthalate; PEI and polyvinyl alcohol; PVA. NB: Unidentified means the article spectra did not match a plastic material from the library database (Fadare et al., 2021).

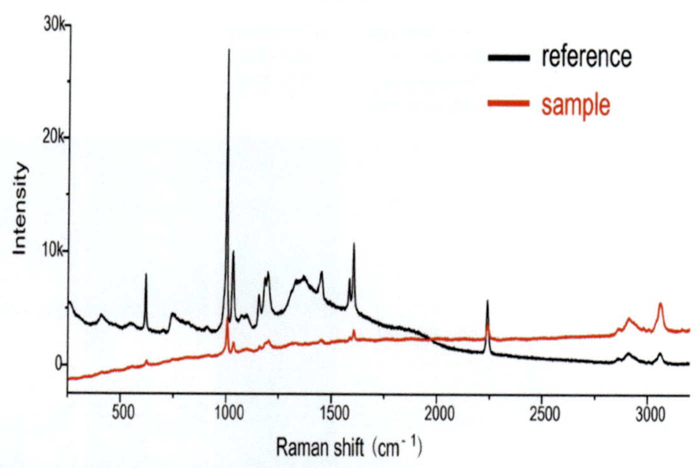

Fig. 5 Characteristic Raman spectra of SAN14 material (reference) and its microplastics particles (sample) (He et al., 2021).

In another study, in a sample of the microplastics particles or fractions in the microwavable plastic food containers demonstrated comparable characteristics to the original microwavable plastic food containers from 500 to $3500\,cm^{-1}$ in terms of the Raman spectroscopy spectrum peaks (Fig. 5) (He et al., 2021).

4.5 Nile Red-microscopic imaging (NRMI)

The number of particles in microplastics is frequently estimated using a microscopic technique. Accuracy and precision are influenced by the nature and visibility of the particles. Several techniques have been explored to improve particle detection. Staining with Nile Red is one of these techniques. This dye was chosen after trying a variety of alternatives in combination with a variety of solvents. Nile Red is solvatochromic, which means that when the polarity of the material surrounding the Nile Red molecules rises, the fluorescence emission spectrum shifts red. When utilizing a set of fluorescence filters, different kinds of microplastics exhibit distinct responses as you can see in Fig. 6 (van Raamsdonk et al., 2020). The application of Nile Red staining increases microplastic identification in this procedure and removes most nonplastic materials such as lipid, chitin, and wood (Akhbarizadeh et al., 2020).

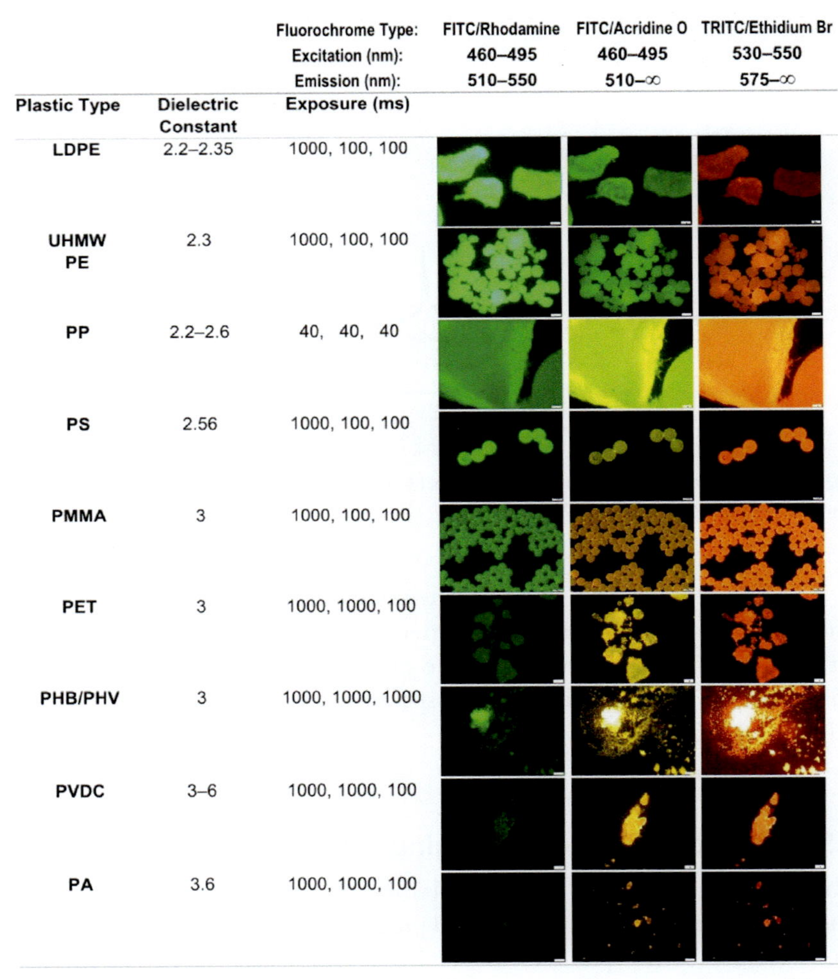

Fig. 6 UV fluorescence photos show nine different types of plastic with three different sets of excitation and emission wavelengths following Nile Red staining in hexane solvent. The photos were obtained after the solvent had completely evaporated. For accurate picture comparison, exposure in microseconds is given. Plastics are classified based on their dielectric constant. Legend: PHB/PHV biopolymer (polyhydroxybutyrate/polyhydroxyvalerate biopolymer); PVDC (polyvinylidene dichloride). Naomi Dam of Wageningen Food Safety Research provided this image (van Raamsdonk et al., 2020).

In another investigation, in a fluorescent microscope, as illustrated in Fig. 7, the NRMI of the MPs particles displayed a unique green fluorescence without background staining, giving additional evidence demonstrating their migration from the microwavable plastic food containers

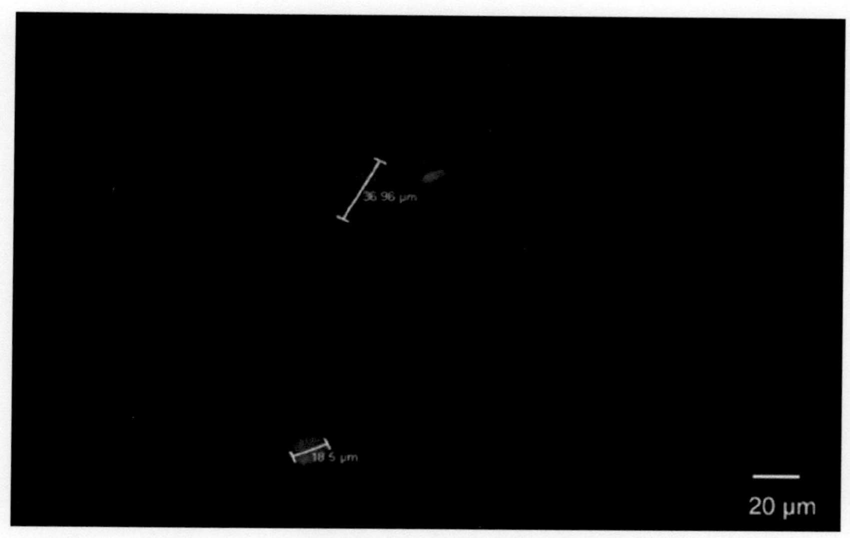

Fig. 7 Green fluorescence Nile red microscopic imaging of microplastics (He et al., 2021).

product. These qualitative evaluations demonstrated the dependability of microplastics retreatment and ruled out the influence of other contaminants. Owing to the optical limits of Nile Red microscopy, counting microplastics particles was difficult if the average size of the particles found was less than 10 µm, according to the preliminary experiment (He et al., 2021).

In a study, Fig. 8 shows various microplastics collected in canned tuna and mackerel fish samples, as well as their fluorescent pictures obtained using Nile Red (Akhbarizadeh et al., 2020).

4.6 Scanning electron microscopy (SEM): Energy dispersive X-ray spectroscopy (EDX)

The surface morphology of identified microplastics in fish muscles was studied using scanning electron microscopy (SEM). SEM scans (Fig. 9) indicated that distinct pieces had varying shapes and surface roughness. All of the fibers had a smooth surface. However, some materials, such as organic matter and/or minerals, can be found on the surface of microplastics (Akhbarizadeh et al., 2018).

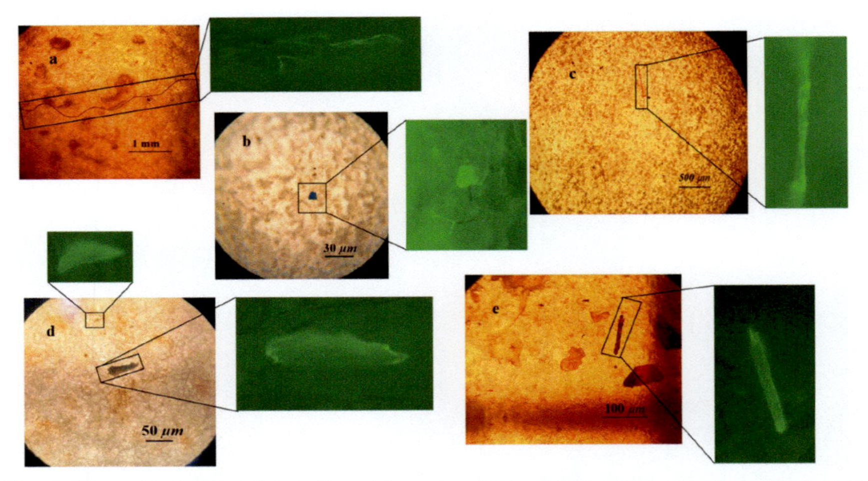

Fig. 8 Microplastics have been found in canned fish samples. The scale bars for A, B, C, D, and E indicate 1 mm, 30 m, 500 μm, 50 μm, and 100 μm, respectively (Akhbarizadeh et al., 2020).

In a study, SEM was used to identify certain plastic-like particles found in mussels. Polymers, in general, have a smooth or unregulated surface (Fig. 10A and B) (Li et al., 2016).

In an examination of the microplastics, the SEM was utilized to count the number of particles as well as to examine their morphological properties (Fig. 11). Three size groups of microplastics particles in each microwavable plastic food container have been estimated, which are 1–5, 5–50, and over 50 μm (He et al., 2021).

In a previous study, SEM-EDX of PVC reference particles produced distinct characteristics such as high chlorine peaks in the EDX spectrum (Fig. 12D and E), reasonably bright backscattered electrons (BSE) frequency, and relatively smooth and rough surfaces. Other forms of the identified plastic (i.e., PET, PS, and PP) exhibit no distinct EDX peak other than a high carbon peak and a lesser O peak (Fig. 12) (Akhbarizadeh et al., 2020).

Zhou, Wang, and Ren (2022) analyzed the presence of plastic particles in takeaway food containers. It was found that nanoscale and microsized particles of various types of plastic are present in the samples obtained. The authors reported that based on the abundance of microplastics in takeaway containers, people who order takeaway food 5–10 times a month can swallow from 145 to 5520 pieces of microplastics from containers.

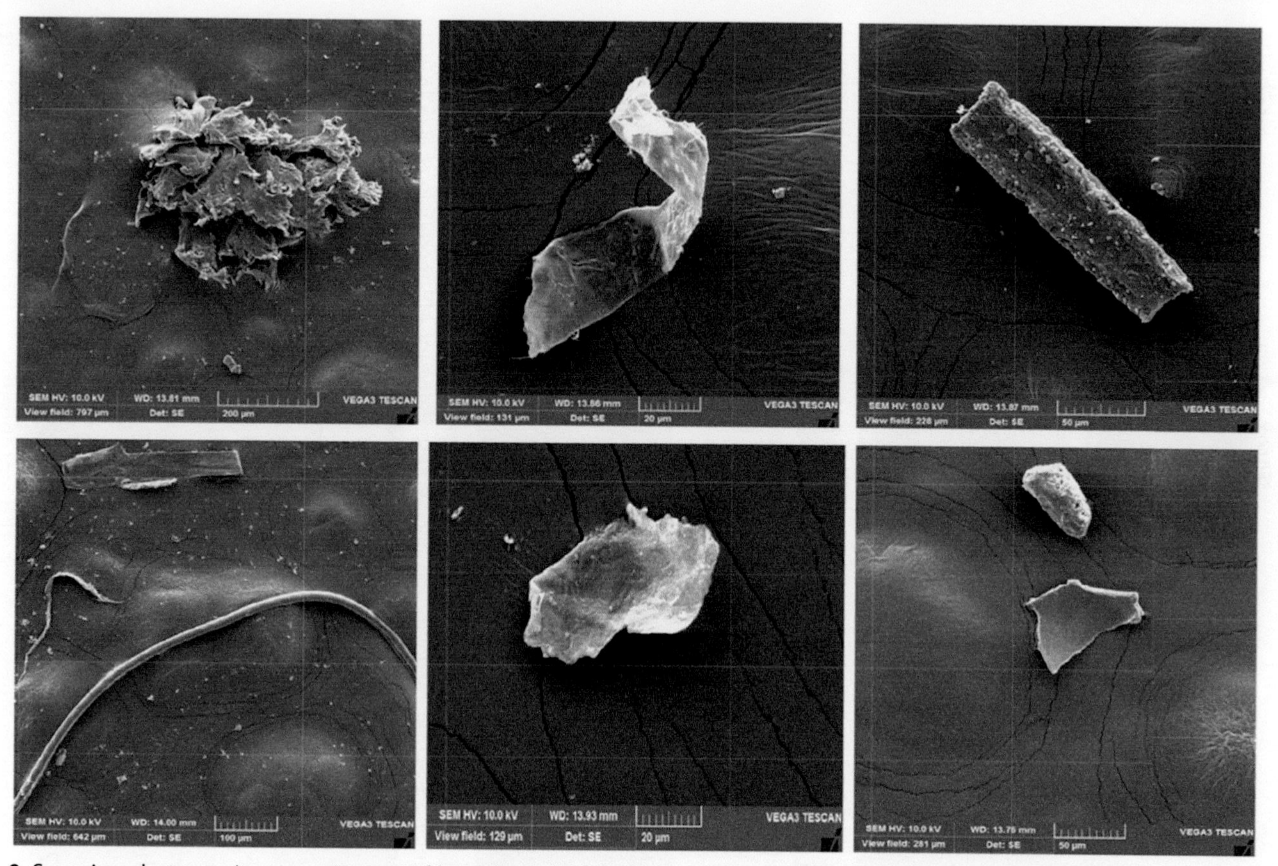

Fig. 9 Scanning electron microscope images of investigated microplastics (Akhbarizadeh, Moore, & Keshavarzi, 2018).

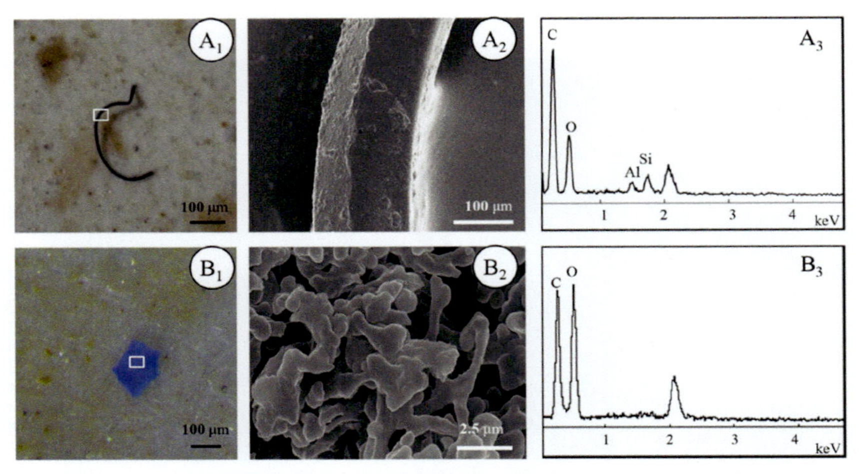

Fig. 10 Microplastic identification using SEM/EDS. The photographs on the left were obtained using microscopes, the middle photographs were taken under SEM for the white box areas of the left ones, and the right photographs were the spectra of EDS for particles in the middle photographs (Li et al., 2016).

Fig. 11 Scanning electron micrographs of blank control, released MPs particles at $\times 100$ scanning electron magnification and representative of MP morphology at $\times 18{,}000$ scanning electron magnification (He et al., 2021).

Fadare et al. (2021) investigated the presence of plastic particles in disposable tableware. The authors studied micro- and nanoparticles of plastic on the surface of containers. According to results, the particle sizes of plastic are in the range from 10 to 210 nm, while more than 50% of the particles have a diameter of less than 50 nm. The particles in the samples have cubic, spherical, rod-shaped, and irregular shapes.

4.7 X-Ray photoelectron spectroscopy (XPS)

For further confirmation of the composition of the teabags and their leachates, XPS was used by Hernandez et al. (2019) to determine their

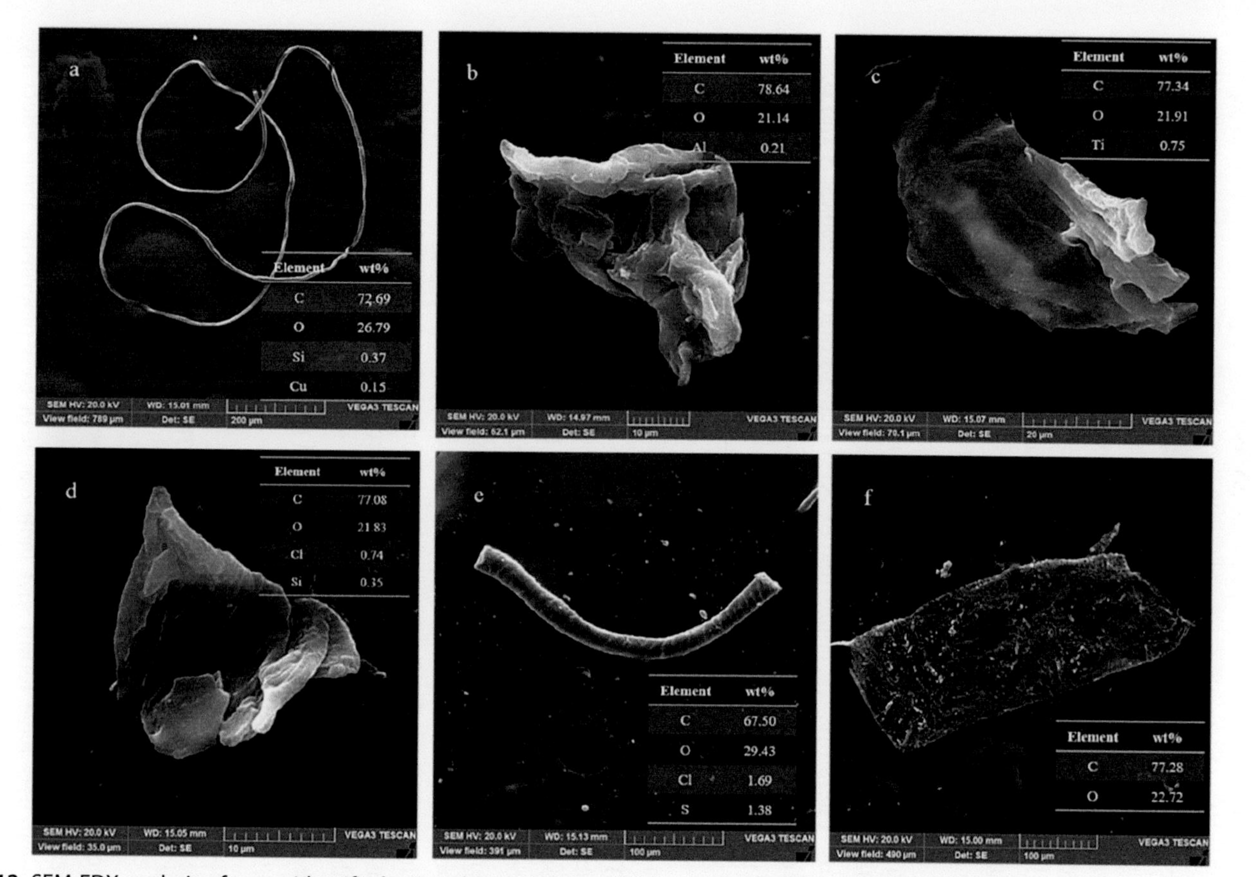

Fig. 12 SEM-EDX analysis of some identified microplastics in canned fish samples (Akhbarizadeh et al., 2020).

elemental content and electronic arrangement. Samples A and B have distinct peaks in the C1s spectrum. A significant peak at 285–286 eV can be found, which corresponds to three carbon-containing groups: C—C, C—N, and C—O/C—OH. In the binding energy range of 287–288 eV, there is a minor peak corresponding to the fourth carbon-containing group, namely CONH, which is characteristic of nylon-6,6. There is a monomodal peak detected with a maximum value at 532 eV for the O1s spectra of teabags and leachates A and B. Oxygen-containing groups CONH/COOH are probably important contributors to this peak. C—O/C—OH groups are associated with a tailing to the left of the peak. It was also explored the N1s spectral region, finding a peak in the range of 399–400 eV, thus indicating the presence of nitrogen-containing groups (NH). XPS analysis of a nylon-6,6 sample led to extremely similar spectra, supporting these designations. Therefore, it is confirmed by XPS analysis that the teabags A and B, as well as their leachates, contain nylon-6,6. Teabags C and D and their leachates were found to contain PET. Three distinct peaks can be seen in the C1s spectrum: the intense peak at 284 eV (C—(CH)), the smaller peak at 287 eV (C—O) and the peak at 290 eV (carbon ester). Two significant components can be observed in the O1s spectra of teabags C and D. The two forms of oxygen present in teabags and leachates C and D are evidenced by a bimodal distribution with peaks at 534 eV (C—O—C groups) and 532 eV (COO groups). The spectra of a commercial PET sample were compared with those of the teabags and their leachates after XPS analysis. The XPS measurements demonstrate that the plastic teabag's parent material was also used to manufacture the microsized and submicron particles in the leachate. It was estimated that an individual could consume 13–16 μg of plastic micro- and nanoparticles when drinking a single cup of tea prepared with a plastic teabag, based upon the density of PET and nylon, the size and density of the particles observed, and the estimated particle count per cup (Hernandez et al., 2019).

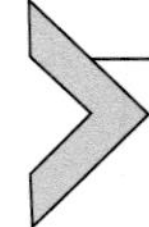

5. Factors influencing the microplastics into foods

5.1 Ultraviolet (UV)

Microplastic migration into food can be prevented by understanding the effects of ultraviolet on plastic fragmentation process and mechanism. Researchers have found that light can lead to a synergistic aging effect by increasing the chemical oxidation of microplastics. Polymer photooxidation includes chain initiation, chain growth, chain creation, and chain termination. The combined action of UV irradiation and oxygen will accelerate the aging process of the polymer. Polymers photodegrade as a result of unsaturated structures or additives in the polymer absorbing ultraviolet light and creating free radicals that interact with oxygen to generate further free radicals, and speed up aging and subsequent polymer chain breakdown (Cheng, Zhang, Liu, Zhang, & Qu, 2021). The combined effects of UV exposure duration and mechanical abrasion on microplastics breaking were investigated in a research. It was revealed that the combined impact of 6 months of UV exposure and 2 months of mechanical abrasion may yield $12,152 \pm 3276$ plastic particles, with a percentage of the particles breaking down into undetectable submicron particles at the same time (Song et al., 2017). PS, PF, PE, and PVC photooxidize in the presence of simulated sunlight, which leads to the breakdown of C—C and C—H bonds (Zhu et al., 2019). During 60 days of UV exposure, the low-density polymer surfaces of PE, PP, and PS were damaged, and new carbonyl groups, vinyl groups, and hydroxyl groups were produced (Ainali, Bikiaris, & Lambropoulou, 2021).

5.2 High temperature

In Fig. 13, it can be seen how temperature affects the rate of leaching of the compounds. Chemical leaching from the PAH and PAE groups takes place more efficiently at higher temperatures, but only for the latter group, whereas it is not as efficient for the former. For example, PAHs leaching at higher temperatures (60 °C) are not significantly increased, while DEPs leaching from PVC and PP at lower temperatures (60 °C) were not detected. Physicochemical properties of the additive used in addition to the kind of plastic used may influence the potential for increased migration as well as the onset time. Based on statistical analysis, significant differences were observed in the temperature at which PVC components leached between DEP ($P < \alpha$, $P = 0.0221$), DBP ($P < \alpha$, $P = 0.0395$), DEHP ($P < \alpha$, $P = 0.0265$), PHE ($P < \alpha$, $P = 0.0379$) and PYE ($P < \alpha$, $P = 0.0390$). Statistically significant differences in DEHP migration from phthalate foils were observed at 60 and 20 °C ($P < \alpha$, $P = 0.0273$). Comparatively, the DEHP levels leached from PVC ($P < \alpha$, $P = 0.0265$) and PP ($P < \alpha$, $P = 0.0390$) were almost twice as high, and those leached from PS ($P < \alpha$, $P = 0.0273$) and rubber ($P < \alpha$, $P = 0.0390$) were almost twice times as high (Kida & Koszelnik, 2021).

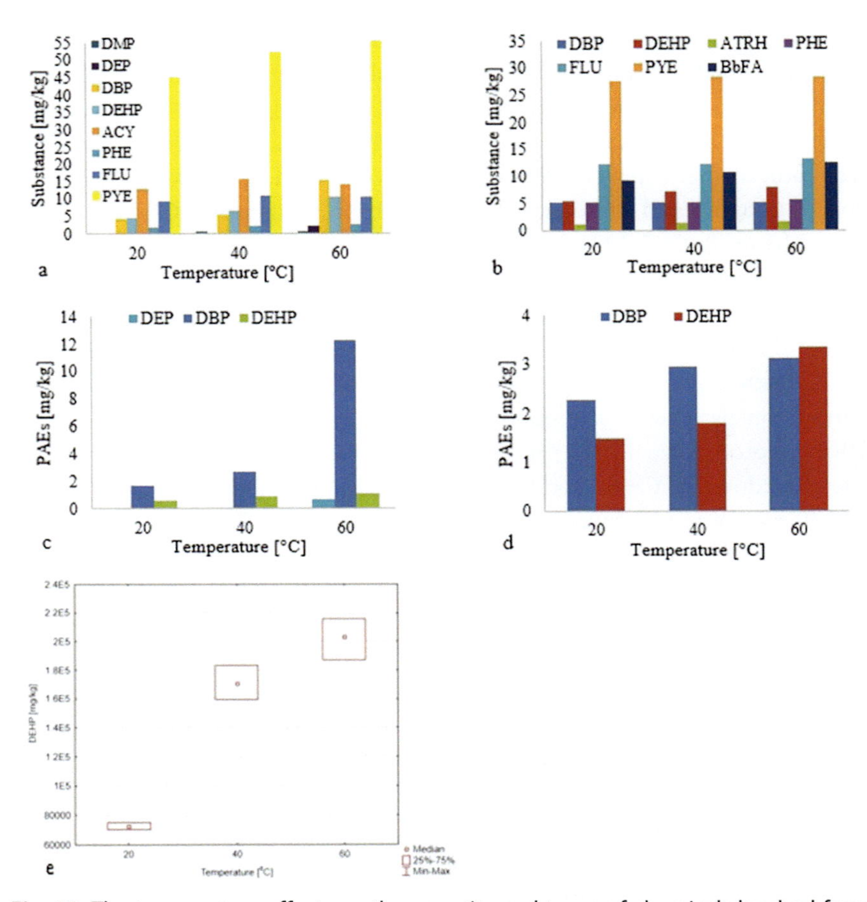

Fig. 13 The temperature effects on the quantity and types of chemicals leached from the following materials: A: PVC, B: rubber, C: PS, D: PP, and E: phthalate foil at 60 °C for 2 h (Kida & Koszelnik, 2021).

5.3 Bacteria

In both aerobic and anaerobic conditions, synthetic polymers can degrade. It is possible to decompose a polymer into CO_2, H_2O, N_2, H_2, CH_4, salts, minerals, as well as biomass (mineralization). The polymer chain is degraded partially or entirely, yielding stable or temporarily stable transition products. Three important criteria are associated with biodegradation, which are:

1. The presence of microorganisms capable of depolymerizing the target material and mineralizing the monomeric molecules through an appropriate metabolic route is required.

2. Temperature, pH, moisture, and salinity are all environmental characteristics that must be satisfied in order for biodegradation to occur.
3. Biofilm formation requires the morphology of polymer particles to allow microorganisms to adhere to them, while the structure of the polymeric substrate, such as chemical bonding, degree of polymerization, branching, and parameters such as hydrophobicity and crystallinity, must not interfere with microbial growth.

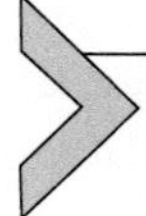

6. Potential exposure risk of the migrated microplastics of foods on human health

In humans, the specific health effects of microplastics intake remain unclear. For instance, several sublethal consequences including inflammatory, immunological, and metabolic abnormalities have been observed in other organisms like algae, zooplankton, fish, and mice, indicating potential human health risks (Hernandez et al., 2019; Jadhav et al., 2021; Lim, 2021; Revel, Châtel, & Mouneyrac, 2018). Some works revealed the influence of micro and nanoplastic on the growth and development of animals and plants. Mao et al. (2018) claim that the addition of polystyrene with a particle diameter from 0.1 to 1 μm leads to morphological changes and a decrease in photosynthetic activity of microalgae *Chlorella pyrenoidosa*, which is explained by physical damage and oxidative stress caused by microplastics. The addition of HDPE with a particle diameter from 1 to 50 μm led to the formation of stress-related proteins and to a decrease in immunity in *Mytilus* spp. mussels (Détrée & Gallardo-Escárate, 2018). Barboza, Vieira, and Guilhermino (2018) found that microplastics with a particle diameter from 1 to 5 μm exhibit neurotoxic properties, induce oxidative stress and lipid peroxidation. The study was carried out on juvenile European sea bass (*Dicentrarchus labrax*). Amereh, Babaei, Eslami, Fazelipour, and Rafiee (2020) evaluated the reproductive toxicity of microplastics in laboratory animals. The authors detected increase in the dosage of plastic nanoparticles (diameter = 38.92 nm), a proportional decrease in the concentrations of testosterone, luteinizing hormone (LH) and follicle-stimulating hormone (FSH) in the blood serum of laboratory animals. At the lowest tested dosage (1 mg/kg-day), changes in tissue and cellular structures were observed, which significantly increased with an increase in the dosage of plastic nanoparticles.

Some works also claim that microplastics can accumulate in the lungs, migrate to the lymph nodes or are transported to other tissues and organs through the circulatory system (Song, Murphy, Narayan, & Davies, 2009). This fact is confirmed by a large percentage of patients with lung, stomach and esophageal cancer among workers in the field of synthetic textiles, sheep breeding and plastics production. The occurrence of such diseases is associated with exposure to a high content of microplastics for a long time (Prata, 2018). Ju et al. (2020) investigated the effect of polyvinyl chloride microplastics on human serum albumin. They found that microplastic particles with a diameter of 5 μm can quench fluorescence of human serum albumin and cause changes in its secondary structure and a decrease in the α-helix. The introduction of microparticles of plastic leads to oxidative stress of human cerebral and epithelial cells (Schirinzi et al., 2017). Hwang, Choi, Han, Choi, and Hong (2019) found that direct contact of polypropylene particles with human cells leads to the release of proteins characteristic of the inflammatory process.

Considering that plastic particles in the micro and nanometer range have a high value of specific surface area, it is possible to assume that the presence of microplastics can lead to the development of potentially pathogenic microorganisms (*Vibrio* spp. and *Aeromonas salmonicida*) and the formation of biological films inside the animal and human body (Kirstein et al., 2016). The authors (Jin, Lu, Tu, Luo, & Fu, 2019) evaluated the effect of polystyrene microparticles on the microbiota of laboratory animals. Studies have shown that microplastics induce intestinal microbiota and cause dysbiosis and dysfunction of the intestinal barrier, as well as metabolic disorders in the body of laboratory animals. Wang, Peng, Li, Zhang, and Liu (2021) reported that the presence of microplastic particles could lead to changes in the intestinal flora and the composition of microorganisms. Ahrendt et al. (2020) studied the development of the intestines of *Girella laevifrons* fish under the influence of various amounts of microplastics. The authors reported that in the group of fish feeding on microplastics, infiltration and hyperemia of leukocytes were more serious compared to the control group. Loss of crypt cells and loss of villi cells were observed in groups whose diet included microplastics.

Microparticles of plastic are able to leave the body of animals naturally. D'Souza, Windsor, Santillo, and Ormerod (2020) and Nelms, Galloway, Godley, Jarvis, and Lindeque (2018) declare that microplastic can be freely excreted from the animal's body, as evidenced by its presence in the fecal masses of wild Eurasian bears (*Cinclus cinclus*), gray seals (*Halichoerus grypus*) and mackerel (*Scomber scombrus*). Studies on mussels (*Mytilus edulis*) have shown that microparticles and nanoparticles of plastic are able to move freely inside the animal's body, which significantly increases the danger of this material (Barboza, Vethaak, Lavorante, Lundebye, & Guilhermino, 2018).

Translocation of various kinds and sizes of microparticles (between 0.1 and 150 mm) through the mammalian stomach into the lymphatic system has been established in human investigations (0.2 and 150 mm). In an in vitro research, just 0.2% of polylactide-*co*-glycolide microparticles (3 mm) were absorbed in human mucosal colon tissue. The mucosal colon tissue of patients with inflammatory bowel illness revealed higher transport (0.45% vs 0.2% in healthy controls) in response to increased gut permeability (Revel et al., 2018). Since this size of microplastics were found in lymph in animal investigations, absorption of microplastics by the intestinal epithelium is most likely limited to microplastics up to 150 μm in size. Exposure to these microplastics leads in systematic exposure, whereas bigger microplastics can only have a limited impact on the immune system (e.g., inflammation of the intestine). Even the tiniest amount of microplastic (<1.5 μm) can penetrate the organs and cause harm. For the first time, the existence of microplastics in human placenta was shown in a 2021 study. These were 12 pieces ranging in size from 5 to 10 μm that were discovered in four placentas. The technique by which they enter the placenta remains unclear, as are the potential consequences on pregnancy and the fetus. Microplastics were found in human feces. The microplastics were found in the stool samples of eight people, ranging from 50 to 500 μm. In the study, PP and PET were the most common materials, while PP and fibers were the most common fragments and fibers. Twenty-three out of 24 stool samples taken from young Chinese men were found to contain microplastics. Microplastics with diameters ranging from 20 to 800 μm were most common, with the most common types being PP, PET, and PS. There is still no clear evidence of long-term health effects due to microplastics. Because of reactive oxygen species produced during an inflammatory response, their adverse effects on the body may include oxidative stress, which may have cytotoxic consequences. Energy balance, metabolism, and the immune system may be affected by microplastics. In addition to their absorption into food, microplastics can also interact with microorganisms on their surface. Microplastics contain a variety of harmful organisms, and seafood consumption increases human exposure to them. Hazardous chemicals such as bisphenol A, PCBs, PAHs, chlorinated insecticides, BFRs, and antibiotics can leach into food from microplastics, leading to cancer and mutagenesis as well as endocrine disruption. Several studies suggest that persistent organic pollutants present a modest level of health hazards for humans when consumed with microplastics. As an example, a typical meal has a projected daily dose of bisphenol A, 40 million times greater than contaminated seafood. The exposure to PCBs and PAHs resulting from contaminated microplastics in mussels would be <0.006% and <0.004%, respectively (Cverenkárová et al., 2021).

There is evidence that the dyes and plasticizers included in plastics are toxic and carcinogenic. These substances can leach from microplastics and accumulate on the surface of the particles, and then be absorbed by the body when microplastic particles enter the human body. The most common plasticizers are phthalates. It is known that an increase in phthalate levels in the blood can lead to asthma and allergies in children (Bamai et al., 2014) and premature pregnancy (Latini et al., 2003). Another plasticizer Bisphenol-A has also been studied similarly to phthalates. Peretz et al. (2014) found that Bisphenol-A has a toxic effect on the reproductive system. The composition of polyvinyl chloride contains a carcinogenic monomer and several dangerous additives, which makes it one of the most dangerous plastics in terms of toxicity. Lithner, Damberg, Dave, and Larsson (2009) found that 9 out of 32 polyvinyl chloride products and filtrates are toxic.

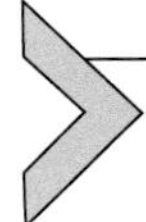

7. Future assessment to reduce the migration and conclusion

Size, quality, and amount of microplastics found in ordinary meals originating from regularly used plastic food packaging and packaging materials for fast food delivery, water, and food packing throughout the world is presented in this chapter. Several studies prove that plastic food packaging poses a threat to the environment and to consumers' health. The purpose of food packaging is to protect, contain, make it convenient for the consumer, and communicate with them. Therefore, packaging should also be nontoxic to the body in order to protect the health of consumers (Jadhav et al., 2021). Some suggestions to reduce plastic footprint, and thus limit the release of microplastics, involve using glass or ceramic dishes to heat food in the microwave. Microwave food should never be heated in plastic containers. Allow food to cool to room temperature before placing it in plastic storage containers. Food and drinking water should be stored in glass or stainless/ceramic/wood/steel/earthenware containers. Consume as many fresh meals as possible. Reduce your intake of fast food and packaged or processed meals. For shopping, use cotton or canvas bags. Avoid placing plastic containers in the dishwasher since they leach toxins onto other dishes. Plastic containers should be hand-washed. Consider reusable containers and dispose of recyclable plastics correctly. This varies by place, hence, know the recycling regulations in your area. There are also filters in residential washing machines that remove microplastic from water after washing, reducing the microplastic load entering the wastewater treatment system. Consider avoiding goods containing microbeads, such as some toothpastes, face and body washes, and cosmetics, to reduce further the exposure to and release of microplastics (Kuna & Sreedhar, 2022). Hazard evaluation and the resultant implementation of dietary pointers for high-danger foods with a better microplastic content would be every other useful tool for mitigating the viable unfavorable effects of microplastics in foods. To deal with the difficulty of microplastic contamination, new techniques for their degradation within side the environment are required. Likewise, it is important to improve public awareness and waste management regarding microplastics. Introducing regulations on the use of significant microplastics and their release into the environment would be a good first step to reduce the load of microplastics in the food chain and ecosystem. The maximum severe hassle in figuring out microplastic contamination in foods is the shortage of a standardized approach. Since the approach hired in numerous investigations varies, the assessment of infection is complex and hard to evaluate. The conclusions of the researches show that the impact of microplastics in the food chain, and in particular the impact of microplastics on human health, needs to be addressed much more aggressively (Cverenkárová et al., 2021).

References

Abraham, J., Ghosh, E., Mukherjee, P., & Gajendiran, A. (2017). Microbial degradation of low density polyethylene. *Environmental Progress & Sustainable Energy*, *36*(1), 147–154. https://doi.org/10.1002/EP.12467.

Aguilar-Guzmán, J. C., Bejtka, K., Fontana, M., Valsami-Jones, E., Villezcas, A. M., Vazquez-Duhalt, R., et al. (2022). Polyethylene terephthalate nanoparticles effect on RAW 264.7 macrophage cells. *Microplastics and Nanoplastics*, *2*(1), 1–15. https://doi.org/10.1186/S43591-022-00027-1.

Ahrendt, C., Perez-Venegas, D. J., Urbina, M., Gonzalez, C., Echeveste, P., Aldana, M., et al. (2020). Microplastic ingestion cause intestinal lesions in the intertidal fish Girella laevifrons. *Marine Pollution Bulletin*, *151*, 110795. https://doi.org/10.1016/j.marpolbul.2019.110795.

Ainali, N. M., Bikiaris, D. N., & Lambropoulou, D. A. (2021). Aging effects on low- and high-density polyethylene, polypropylene and polystyrene under UV irradiation: An insight into decomposition mechanism by Py-GC/MS for microplastic analysis. *Journal of Analytical and Applied Pyrolysis*, *158*, 105207. https://doi.org/10.1016/J.JAAP.2021.105207.

Akhbarizadeh, R., Dobaradaran, S., Nabipour, I., Tajbakhsh, S., Darabi, A. H., & Spitz, J. (2020). Abundance, composition, and potential intake of microplastics in canned fish. *Marine Pollution Bulletin, 160*, 111633. https://doi.org/10.1016/J.MARPOLBUL.2020.111633.

Akhbarizadeh, R., Moore, F., & Keshavarzi, B. (2018). Investigating a probable relationship between microplastics and potentially toxic elements in fish muscles from northeast of Persian gulf. *Environmental Pollution (Barking, Essex: 1987), 232*, 154–163. https://doi.org/10.1016/J.ENVPOL.2017.09.028.

Akovali, G. (2012). Plastic materials: Polyvinyl chloride (PVC). In *Toxicity of building materials* (pp. 23–53). Woodhead Publishing. https://doi.org/10.1533/9780857096357.23.

Alimba, C. G., Faggio, C., Sivanesan, S., Ogunkanmi, A. L., & Krishnamurthi, K. (2021). Micro (nano)-plastics in the environment and risk of carcinogenesis: Insight into possible mechanisms. *Journal of Hazardous Materials, 416*, 126143. https://doi.org/10.1016/j.jhazmat.2021.126143.

Alin, J., & Hakkarainen, M. (2011). Microwave heating causes rapid degradation of antioxidants in polypropylene packaging, leading to greatly increased specific migration to food simulants as shown by ESI-MS and GC-MS. *Journal of Agricultural and Food Chemistry, 59*(10), 5418–5427. https://doi.org/10.1021/JF1048639.

Amereh, F., Babaei, M., Eslami, A., Fazelipour, S., & Rafiee, M. (2020). The emerging risk of exposure to nano (micro) plastics on endocrine disturbance and reproductive toxicity: From a hypothetical scenario to a global public health challenge. *Environmental Pollution, 261*, 114158. https://doi.org/10.1016/j.envpol.2020.114158.

Auta, H. S., Abioye, O. P., Aransiola, S. A., Bala, J. D., Chukwuemeka, V. I., Hassan, A., et al. (2022). Enhanced microbial degradation of PET and PS microplastics under natural conditions in mangrove environment. *Journal of Environmental Management, 304*, 114273. https://doi.org/10.1016/J.JENVMAN.2021.114273.

Bahmid, N. A., Dekker, M., Fogliano, V., & Heising, J. (2021). Development of a moisture-activated antimicrobial film containing ground mustard seeds and its application on meat in active packaging system. *Food Packaging and Shelf Life, 30*, 100753. https://doi.org/10.1016/J.FPSL.2021.100753.

Bahmid, N. A., Syamsu, K., & Maddu, A. (2014). Pengaruh ukuran serat selulosa asetat dan penambahan dietilen glikol (DEG) terhadap sifat fisik dan mekanik bioplastik. *Jurnal Teknologi Industri Pertanian, 24*(3), 226–234. https://journal.ipb.ac.id/index.php/jurnaltin/article/view/9125.

Bamai, Y. A., Shibata, E., Saito, I., Araki, A., Kanazawa, A., Morimoto, K., et al. (2014). Exposure to house dust phthalates in relation to asthma and allergies in both children and adults. *Science of the Total Environment, 485*, 153–163. https://doi.org/10.1016/j.scitotenv.2014.03.059.

Barbeş, L., Rădulescu, C., & Stihi, C. (2014). ATR-FTIR spectrometry characterisation of polymeric materials. *Romanian Reports in Physics, 66*(3), 765–777.

Barboza, L. G. A., Vethaak, A. D., Lavorante, B. R., Lundebye, A. K., & Guilhermino, L. (2018). Marine microplastic debris: An emerging issue for food security, food safety and human health. *Marine Pollution Bulletin, 133*, 336–348. https://doi.org/10.1016/j.marpolbul.2018.05.047.

Barboza, L. G. A., Vieira, L. R., & Guilhermino, L. (2018). Single and combined effects of microplastics and mercury on juveniles of the European seabass (Dicentrarchus labrax): Changes in behavioural responses and reduction of swimming velocity and resistance time. *Environmental Pollution, 236*, 1014–1019. https://doi.org/10.1016/j.envpol.2017.12.082.

Bessa, F., Barría, P., Neto, J. M., Frias, J. P. G. L., Otero, V., Sobral, P., et al. (2018). Occurrence of microplastics in commercial fish from a natural estuarine environment. *Marine Pollution Bulletin, 128*, 575–584. https://doi.org/10.1016/J.MARPOLBUL.2018.01.044.

Blasiak, R., Leander, E., Jouffray, J. B., & Virdin, J. (2021). Corporations and plastic pollution: Trends in reporting. *Sustainable Futures*, *3*, 100061. https://doi.org/10.1016/J.SFTR.2021.100061.

Bouwmeester, H., Hollman, P. C. H., & Peters, R. J. B. (2015). Potential health impact of environmentally released micro- and Nanoplastics in the human food production chain: Experiences from Nanotoxicology. *Environmental Science and Technology*, *49*(15), 8932–8947. https://doi.org/10.1021/ACS.EST.5B01090.

Bueno-Ferrer, C., Garrigós, M. C., & Jiménez, A. (2010). Characterization and thermal stability of poly (vinyl chloride) plasticized with epoxidized soybean oil for food packaging. *Polymer Degradation and Stability*, *95*(11), 2207–2212. https://doi.org/10.1016/j.polymdegradstab.2010.01.027.

Cacciari, I., Quatrini, P., Zirletta, G., Mincione, E., Vinciguerra, V., Lupattelli, P., et al. (1993). Isotactic polypropylene biodegradation by a microbial community: Physicochemical characterization of metabolites produced. *Applied and Environmental Microbiology*, *59*(11), 3695–3700. https://doi.org/10.1128/AEM.59.11.3695-3700.1993.

Cheng, F., Zhang, T., Liu, Y., Zhang, Y., & Qu, J. (2021). Non-negligible effects of UV irradiation on transformation and environmental risks of microplastics in the water environment. *Journal of Xenobiotics*, *12*(1), 1–12. https://doi.org/10.3390/JOX12010001.

Cherif Lahimer, M., Ayed, N., Horriche, J., & Belgaied, S. (2017). Characterization of plastic packaging additives: Food contact, stability and toxicity. *Arabian Journal of Chemistry*, *10*, S1938–S1954. https://doi.org/10.1016/J.ARABJC.2013.07.022.

Cole, M., Lindeque, P., Fileman, E., Halsband, C., & Galloway, T. S. (2015). The impact of polystyrene microplastics on feeding, function and fecundity in the marine copepod Calanus helgolandicus. *Environmental Science and Technology*, *49*(2), 1130–1137. https://doi.org/10.1021/ES504525U.

Cox, K. D., Covernton, G. A., Davies, H. L., Dower, J. F., Juanes, F., & Dudas, S. E. (2019). Human consumption of microplastics. *Environmental Science & Technology*, *53*(12), 7068–7074. https://doi.org/10.1021/acs.est.9b01517.

Cverenkárová, K., Valachovičová, M., Mackul'ak, T., Žemlička, L., & Bírošová, L. (2021). Microplastics in the food chain. *Life*, *11*(12), 1349. https://doi.org/10.3390/LIFE11121349.

Daniel, D. B., Ashraf, P. M., & Thomas, S. N. (2020). Abundance, characteristics and seasonal variation of microplastics in Indian white shrimps (Fenneropenaeus indicus) from coastal waters off Cochin, Kerala, India. *Science of the Total Environment*, *737*, 139839. https://doi.org/10.1016/J.SCITOTENV.2020.139839.

de Anda-Flores, Y. B., Cordón-Cardona, B. A., González-León, A., Valenzuela-Quintanar, A. I., Peralta, E., & Soto-Valdez, H. (2021). Effect of assay conditions on the migration of phthalates from polyvinyl chloride cling films used for food packaging in México. *Food Packaging and Shelf Life*, *29*, 100684. https://doi.org/10.1016/j.fpsl.2021.100684.

Détrée, C., & Gallardo-Escárate, C. (2018). Single and repetitive microplastics exposures induce immune system modulation and homeostasis alteration in the edible mussel Mytilus galloprovincialis. *Fish & Shellfish Immunology*, *83*, 52–60. https://doi.org/10.1016/j.fsi.2018.09.018.

Diaz-Basantes, M. F., Conesa, J. A., & Fullana, A. (2020). Microplastics in honey, beer, milk and refreshments in Ecuador as emerging contaminants. *Sustainability*, *12*(14), 5514. https://doi.org/10.3390/su12145514.

D'Souza, J. M., Windsor, F. M., Santillo, D., & Ormerod, S. J. (2020). Food web transfer of plastics to an apex riverine predator. *Global Change Biology*, *26*(7), 3846–3857. https://doi.org/10.1111/gcb.15139.

Du, F., Cai, H., Zhang, Q., Chen, Q., & Shi, H. (2020). Microplastics in take-out food containers. *Journal of Hazardous Materials*, *399*, 122969. https://doi.org/10.1016/J.JHAZMAT.2020.122969.

European Food Safety Authority. (2016). Presence of microplastics and nanoplastics in food, with particular focus on seafood. *EFSA Journal*, *14*(6), e04501. https://doi.org/10.2903/J.EFSA.2016.4501.

Fadare, O. O., Okoffo, E. D., & Olasehinde, E. F. (2021). Microparticles and microplastics contamination in African table salts. *Marine Pollution Bulletin*, *164*, 112006. https://doi.org/10.1016/J.MARPOLBUL.2021.112006.

Fadare, O. O., Wan, B., Guo, L. H., & Zhao, L. (2020). Microplastics from consumer plastic food containers: Are we consuming it? *Chemosphere*, *253*, 126787. https://doi.org/10.1016/J.CHEMOSPHERE.2020.126787.

Geueke, B., Groh, K., & Muncke, J. (2018). Food packaging in the circular economy: Overview of chemical safety aspects for commonly used materials. *Journal of Cleaner Production*, *193*, 491–505. https://doi.org/10.1016/j.jclepro.2018.05.005.

Gómez Ramos, M. J., Lozano, A., & Fernández-Alba, A. R. (2019). High-resolution mass spectrometry with data independent acquisition for the comprehensive non-targeted analysis of migrating chemicals coming from multilayer plastic packaging materials used for fruit purée and juice. *Talanta*, *191*, 180–192. https://doi.org/10.1016/J.TALANTA.2018.08.023.

Gong, J., Kong, T., Li, Y., Li, Q., Li, Z., & Zhang, J. (2018). Biodegradation of microplastic derived from poly(ethylene terephthalate) with bacterial whole-cell biocatalysts. *Polymers*, *10*(12), 1326. https://doi.org/10.3390/POLYM10121326.

Guan, Q.-F., Yang, H.-B., Zhao, Y.-X., Han, Z.-M., Ling, Z.-C., Yang, K.-P., et al. (2021). Microplastics release from victuals packaging materials during daily usage. *EcoMat*, *3*(3), e12107. https://doi.org/10.1002/EOM2.12107.

Guerreiro, T. M., de Oliveira, D. N., Melo, C. F. O. R., de Oliveira Lima, E., & Catharino, R. R. (2018). Migration from plastic packaging into meat. *Food Research International*, *109*, 320–324. https://doi.org/10.1016/J.FOODRES.2018.04.026.

Gurjar, U. R., Xavier, K. A. M., Shukla, S. P., Deshmukhe, G., Jaiswar, A. K., & Nayak, B. B. (2021). Incidence of microplastics in gastrointestinal tract of golden anchovy (Coilia dussumieri) from north east coast of Arabian Sea: The ecological perspective. *Marine Pollution Bulletin*, *169*, 112518. https://doi.org/10.1016/J.MARPOLBUL.2021.112518.

He, Y. J., Qin, Y., Zhang, T. L., Zhu, Y. Y., Wang, Z. J., Zhou, Z. S., et al. (2021). Migration of (non-) intentionally added substances and microplastics from microwavable plastic food containers. *Journal of Hazardous Materials*, *417*, 126074. https://doi.org/10.1016/J.JHAZMAT.2021.126074.

Hermabessiere, L., Dehaut, A., Paul-Pont, I., Lacroix, C., Jezequel, R., Soudant, P., et al. (2017). Occurrence and effects of plastic additives on marine environments and organisms: A review. *Chemosphere*, *182*, 781–793. https://doi.org/10.1016/J.CHEMOSPHERE.2017.05.096.

Hernandez, L. M., Xu, E. G., Larsson, H. C. E., Tahara, R., Maisuria, V. B., & Tufenkji, N. (2019). Plastic teabags release billions of microparticles and nanoparticles into tea. *Environmental Science and Technology*, *53*(21), 12300–12310. https://doi.org/10.1021/ACS.EST.9B02540/SUPPL_FILE/ES9B02540_SI_001.PDF.

Hiwari, H., Purba, N. P., Ihsan, Y. N., Yuliadi, L. P. S., Mulyani, P. G., Studi Ilmu Kelautan, P., et al. (2019). Condition of microplastic garbage in sea surface water at around Kupang and rote, East Nusa Tenggara Province. *Prosiding Seminar Nasional Masyarakat Biodiversitas Indonesia*, *5*(2), 165–171. https://doi.org/10.13057/PSNMBI/M050204.

Hwang, J., Choi, D., Han, S., Choi, J., & Hong, J. (2019). An assessment of the toxicity of polypropylene microplastics in human derived cells. *Science of the Total Environment*, *684*, 657–669. https://doi.org/10.1016/J.SCITOTENV.2019.05.071.

Iñiguez, M. E., Conesa, J. A., & Fullana, A. (2017). Microplastics in Spanish table salt. *Scientific Reports*, 7(1), 1–7. https://doi.org/10.1038/s41598-017-09128-x.

Jadhav, E. B., Sankhla, M. S., Bhat, R. A., & Bhagat, D. S. (2021). Microplastics from food packaging: An overview of human consumption, health threats, and alternative solutions. *Environmental Nanotechnology, Monitoring & Management, 16*, 100608. https://doi.org/10.1016/J.ENMM.2021.100608.

Jambeck, J. R., Geyer, R., Wilcox, C., Siegler, T. R., Perryman, M., Andrady, A., et al. (2015). Plastic waste inputs from land into the ocean. *Science, 347*(6223), 768–771. https://doi.org/10.1126/SCIENCE.1260352/SUPPL_FILE/JAMBECK.SM.PDF.

Jin, Y., Lu, L., Tu, W., Luo, T., & Fu, Z. (2019). Impacts of polystyrene microplastic on the gut barrier, microbiota and metabolism of mice. *Science of the Total Environment, 649*, 308–317. https://doi.org/10.1016/j.scitotenv.2018.08.353.

Ju, P., Zhang, Y., Zheng, Y., Gao, F., Jiang, F., Li, J., et al. (2020). Probing the toxic interactions between polyvinyl chloride microplastics and human serum albumin by multispectroscopic techniques. *Science of the Total Environment, 734*, 139219. https://doi.org/10.1016/J.SCITOTENV.2020.139219.

Kannan, K., & Vimalkumar, K. (2021). A review of human exposure to microplastics and insights into microplastics as obesogens. *Frontiers in Endocrinology, 12*, 978. https://doi.org/10.3389/FENDO.2021.724989/BIBTEX.

Karami, A., Golieskardi, A., Keong Choo, C., Larat, V., Galloway, T. S., & Salamatinia, B. (2017). The presence of microplastics in commercial salts from different countries. *Scientific Reports*, 7(1), 1–11. https://doi.org/10.1038/srep46173.

Katsara, K., Kenanakis, G., Viskadourakis, Z., & Papadakis, V. M. (2021). Polyethylene migration from food packaging on cheese detected by Raman and infrared (ATR/FT-IR) spectroscopy. *Materials, 14*(14), 3872. https://doi.org/10.3390/MA14143872.

Kedzierski, M., Lechat, B., Sire, O., le Maguer, G., le Tilly, V., & Bruzaud, S. (2020). Microplastic contamination of packaged meat: Occurrence and associated risks. *Food Packaging and Shelf Life, 24*, 100489. https://doi.org/10.1016/J.FPSL.2020.100489.

Khandare, S. D., Agrawal, D., Mehru, N., & Chaudhary, D. R. (2022). Marine bacterial based enzymatic degradation of low-density polyethylene (LDPE) plastic. *Journal of Environmental Chemical Engineering, 10*(3), 107437. https://doi.org/10.1016/J.JECE.2022.107437.

Kida, M., & Koszelnik, P. (2021). Investigation of the presence and possible migration from microplastics of phthalic acid esters and polycyclic aromatic hydrocarbons. *Journal of Polymers and the Environment, 29*(2), 599–611. https://doi.org/10.1007/S10924-020-01899-1/FIGURES/6.

Kirstein, I. V., Kirmizi, S., Wichels, A., Garin-Fernandez, A., Erler, R., Löder, M., et al. (2016). Dangerous hitchhikers? Evidence for potentially pathogenic vibrio spp. on microplastic particles. *Marine Environmental Research, 120*, 1–8. https://doi.org/10.1016/j.marenvres.2016.07.004.

Kosuth, M., Mason, S. A., & Wattenberg, E. V. (2018). Anthropogenic contamination of tap water, beer, and sea salt. *PLoS One, 13*(4), e0194970. https://doi.org/10.1371/journal.pone.0194970.

Kumar, V., Maitra, S. S., Singh, R., & Burnwal, D. K. (2020). Acclimatization of a newly isolated bacteria in monomer tere-phthalic acid (TPA) may enable it to attack the polymer poly-ethylene tere-phthalate(PET). *Journal of Environmental Chemical Engineering, 8*(4), 103977. https://doi.org/10.1016/J.JECE.2020.103977.

Kuna, A., & Sreedhar, M. (2022). *Microplastics in food chain*. Retrieved March 13, 2022, from. https://www.researchgate.net/publication/333719200.

Kutralam-Muniasamy, G., Pérez-Guevara, F., Elizalde-Martínez, I., & Shruti, V. C. (2020). Branded milks – Are they immune from microplastics contamination? *Science of the Total Environment, 714*, 136823. https://doi.org/10.1016/J.SCITOTENV.2020.136823.

Latini, G., De Felice, C., Presta, G., Del Vecchio, A., Paris, I., Ruggieri, F., et al. (2003). In utero exposure to di-(2-ethylhexyl) phthalate and duration of human pregnancy. *Environmental Health Perspectives, 111*(14), 1783–1785. https://doi.org/10.1289/ehp.6202.

Lebreton, L., & Andrady, A. (2019). Future scenarios of global plastic waste generation and disposal. *Palgrave Communications*, *5*(1), 1–11. https://doi.org/10.1057/s41599-018-0212-7.

Lemos Junior, W. J. F., do Amaral dos Reis, L. P., de Oliveira, V. S., Lopes, L. O., & Pereira, K. S. (2019). Reuse of refillable PET packaging: Approaches to safety and quality in soft drink processing. *Food Control*, *100*, 329–334. https://doi.org/10.1016/J.FOODCONT.2019.02.008.

Li, Q., Feng, Z., Zhang, T., Ma, C., & Shi, H. (2020). Microplastics in the commercial seaweed nori. *Journal of Hazardous Materials*, *388*, 122060. https://doi.org/10.1016/J.JHAZMAT.2020.122060.

Li, Y., Peng, L., Fu, J., Dai, X., & Wang, G. (2022). A microscopic survey on microplastics in beverages: The case of beer, mineral water and tea. *Analyst*, *147*(6), 1099–1105. https://doi.org/10.1039/D2AN00083K.

Li, J., Qu, X., Su, L., Zhang, W., Yang, D., Kolandhasamy, P., et al. (2016). Microplastics in mussels along the coastal waters of China. *Environmental Pollution*, *214*, 177–184. https://doi.org/10.1016/J.ENVPOL.2016.04.012.

Li, D., Shi, Y., Yang, L., Xiao, L., Kehoe, D. K., Gun'ko, Y. K., et al. (2020). Microplastic release from the degradation of polypropylene feeding bottles during infant formula preparation. *Nature Food*, *1*(11), 746–754. https://doi.org/10.1038/s43016-020-00171-y.

Liebezeit, G., & Liebezeit, E. (2015). Origin of synthetic particles in honeys. *Polish Journal of Food and Nutrition Sciences*, *65*(2), 143–147. https://doi.org/10.1515/pjfns-2015-0025.

Lim, X. Z. (2021). Microplastics are everywhere - but are they harmful? *Nature*, *593*(7857), 22–25. https://doi.org/10.1038/D41586-021-01143-3.

Lithner, D., Damberg, J., Dave, G., & Larsson, Å. (2009). Leachates from plastic consumer products–screening for toxicity with Daphnia magna. *Chemosphere*, *74*(9), 1195–1200. https://doi.org/10.1016/j.chemosphere.2008.11.022.

Lithner, D., Larsson, A., & Dave, G. (2011). Environmental and health hazard ranking and assessment of plastic polymers based on chemical composition. *Science of the Total Environment*, *409*(18), 3309–3324. https://doi.org/10.1016/J.SCITOTENV.2011.04.038.

Liu, L., Xu, M., Ye, Y., & Zhang, B. (2022). On the degradation of (micro) plastics: Degradation methods, influencing factors, environmental impacts. *Science of the Total Environment*, *806*, 151312. https://doi.org/10.1016/j.scitotenv.2021.151312.

Löhr, A., Savelli, H., Beunen, R., Kalz, M., Ragas, A., & van Belleghem, F. (2017). Solutions for global marine litter pollution. *Current Opinion in Environmental Sustainability*, *28*, 90–99. https://doi.org/10.1016/J.COSUST.2017.08.009.

Loredo-Treviño, A., Gutiérrez-Sánchez, G., Rodríguez-Herrera, R., & Aguilar, C. N. (2011). Microbial enzymes involved in polyurethane biodegradation: A review. *Journal of Polymers and the Environment*, *20*(1), 258–265. https://doi.org/10.1007/S10924-011-0390-5.

Lu, S., Qiu, R., Hu, J., Li, X., Chen, Y., Zhang, X., et al. (2020). Prevalence of microplastics in animal-based traditional medicinal materials: Widespread pollution in terrestrial environments. *Science of the Total Environment*, *709*, 136214. https://doi.org/10.1016/j.scitotenv.2019.136214.

Makhdoumi, P., Naghshbandi, M., Ghaderzadeh, K., Mirzabeigi, M., Yazdanbakhsh, A., & Hossini, H. (2021). Micro-plastic occurrence in bottled vinegar: Qualification, quantification and human risk exposure. *Process Safety and Environmental Protection*, *152*, 404–413. https://doi.org/10.1016/J.PSEP.2021.06.022.

Mao, Y., Ai, H., Chen, Y., Zhang, Z., Zeng, P., Kang, L., et al. (2018). Phytoplankton response to polystyrene microplastics: Perspective from an entire growth period. *Chemosphere*, *208*, 59–68. https://doi.org/10.1016/j.chemosphere.2018.05.170.

Mason, S. A., Welch, V. G., & Neratko, J. (2018). Synthetic polymer contamination in bottled water. *Frontiers in Chemistry*, *6*, 407. https://doi.org/10.3389/FCHEM.2018.00407/BIBTEX.

Masura, J., Baker, J., Foster, G., & Arthut, C. (2015). *Laboratory methods for the analysis of microplastics in the marine environment: Recommendations for quantifying synthetic particles in waters and sediments*. https://doi.org/10.25607/OBP-604.

Miao, F., Liu, Y., Gao, M., Yu, X., Xiao, P., Wang, M., et al. (2020). Degradation of polyvinyl chloride microplastics via an electro-Fenton-like system with a TiO_2/graphite cathode. *Journal of Hazardous Materials*, *399*, 123023. https://doi.org/10.1016/j.jhazmat.2020.123023.

Muenmee, S., Chiemchaisri, W., & Chiemchaisri, C. (2015). Microbial consortium involving biological methane oxidation in relation to the biodegradation of waste plastics in a solid waste disposal open dump site. *International Biodeterioration & Biodegradation*, *102*, 172–181. https://doi.org/10.1016/J.IBIOD.2015.03.015.

Naidoo, T., Rajkaran, A., & Sershen. (2020). Impacts of plastic debris on biota and implications for human health: A south African perspective. *South African Journal of Science*, *116*(5/6). https://doi.org/10.17159/SAJS.2020/7693.

Nelms, S. E., Galloway, T. S., Godley, B. J., Jarvis, D. S., & Lindeque, P. K. (2018). Investigating microplastic trophic transfer in marine top predators. *Environmental Pollution*, *238*, 999–1007. https://doi.org/10.1016/j.envpol.2018.02.016.

Neves, D., Sobral, P., Ferreira, J. L., & Pereira, T. (2015). Ingestion of microplastics by commercial fish off the Portuguese coast. *Marine Pollution Bulletin*, *101*(1), 119–126. https://doi.org/10.1016/J.MARPOLBUL.2015.11.008.

Nxumalo, S. M., Mabaso, S. D., Mamba, S. F., & Singwane, S. S. (2020). Plastic waste management practices in the rural areas of Eswatini. *Social Sciences & Humanities Open*, *2*(1), 100066. https://doi.org/10.1016/J.SSAHO.2020.100066.

Oßmann, B. E., Sarau, G., Holtmannspötter, H., Pischetsrieder, M., Christiansen, S. H., & Dicke, W. (2018). Small-sized microplastics and pigmented particles in bottled mineral water. *Water Research*, *141*, 307–316. https://doi.org/10.1016/J.WATRES.2018.05.027.

Ouyang, Z., Li, S., Zhao, M., Wangmu, Q., Ding, R., Xiao, C., et al. (2022). The aging behavior of polyvinyl chloride microplastics promoted by UV-activated persulfate process. *Journal of Hazardous Materials*, *424*, 127461. https://doi.org/10.1016/j.jhazmat.2021.127461.

Peretz, J., Vrooman, L., Ricke, W. A., Hunt, P. A., Ehrlich, S., Hauser, R., et al. (2014). Bisphenol A and reproductive health: Update of experimental and human evidence, 2007–2013. *Environmental Health Perspectives*, *122*(8), 775–786. https://doi.org/10.1289/ehp.1307728.

Polyethylene, HDPE. (2012). *Chemical resistance of thermoplastics* (pp. 582–1295). https://doi.org/10.1016/B978-1-4557-7896-6.00007-8.

Prata, J. C. (2018). Microplastics in wastewater: State of the knowledge on sources, fate and solutions. *Marine Pollution Bulletin*, *129*(1), 262–265. https://doi.org/10.1016/j.marpolbul.2018.02.046.

Ranjan, V. P., Joseph, A., & Goel, S. (2021). Microplastics and other harmful substances released from disposable paper cups into hot water. *Journal of Hazardous Materials*, *404*, 124118. https://doi.org/10.1016/J.JHAZMAT.2020.124118.

Revel, M., Châtel, A., & Mouneyrac, C. (2018). Micro(nano)plastics: A threat to human health? *Current Opinion in Environmental Science & Health*, *1*, 17–23. https://doi.org/10.1016/J.COESH.2017.10.003.

Salakhov, I. I., Shaidullin, N. M., Chalykh, A. E., Matsko, M. A., Shapagin, A. V., Batyrshin, A. Z., et al. (2021). Low-temperature mechanical properties of high-density and low-density polyethylene and their blends. *Polymers*, *13*(11), 1821. https://doi.org/10.3390/POLYM13111821.

Sarinah Basri, K., Basri, K., Syaputra, E. M., & Handayani, S. (2021). Microplastic pollution in waters and its impact on health and environment in Indonesia: A review. *Journal of Public Health for Tropical and Coastal Region*, *4*(2), 63–77. https://doi.org/10.14710/JPHTCR.V4I2.10809.

Schirinzi, G. F., Pérez-Pomeda, I., Sanchís, J., Rossini, C., Farré, M., & Barceló, D. (2017). Cytotoxic effects of commonly used nanomaterials and microplastics on cerebral and epithelial human cells. *Environmental Research*, *159*, 579–587. https://doi.org/10.1016/J.ENVRES.2017.08.043.

Senathirajah, K., Attwood, S., Bhagwat, G., Carbery, M., Wilson, S., & Palanisami, T. (2021). Estimation of the mass of microplastics ingested—A pivotal first step towards human health risk assessment. *Journal of Hazardous Materials*, *404*, 124004. https://doi.org/10.1016/J.JHAZMAT.2020.124004.

Seth, C. K., & Shriwastav, A. (2018). Contamination of Indian sea salts with microplastics and a potential prevention strategy. *Environmental Science and Pollution Research*, *25*(30), 30122–30131. https://doi.org/10.1007/S11356-018-3028-5.

Shin, H., Park, S., Thanakkasaranee, S., Sadeghi, K., Lee, Y., Tak, G., et al. (2021). Applicability of newly developed PET/bio-based polyester blends for hot-filling bottle. *Food Packaging and Shelf Life*, *30*, 100757. https://doi.org/10.1016/J.FPSL.2021.100757.

Shruti, V. C., Pérez-Guevara, F., Elizalde-Martínez, I., & Kutralam-Muniasamy, G. (2020). First study of its kind on the microplastic contamination of soft drinks, cold tea and energy drinks-future research and environmental considerations. *Science of the Total Environment*, *726*, 138580. https://doi.org/10.1016/j.scitotenv.2020.138580.

Smith, M., Love, D. C., Rochman, C. M., & Neff, R. A. (2018). Microplastics in seafood and the implications for human health. *Current Environmental Health Reports*, *5*(3), 375–386. https://doi.org/10.1007/S40572-018-0206-Z/TABLES/4.

Song, Y. K., Hong, S. H., Jang, M., Han, G. M., Jung, S. W., & Shim, W. J. (2017). Combined effects of UV exposure duration and mechanical abrasion on microplastic fragmentation by polymer type. *Environmental Science and Technology*, *51*(8), 4368–4376. https://doi.org/10.1021/ACS.EST.6B06155/SUPPL_FILE/ES6B06155_SI_001.PDF.

Song, J. H., Murphy, R. J., Narayan, R., & Davies, G. B. H. (2009). Biodegradable and compostable alternatives to conventional plastics. *Philosophical Transactions of the Royal Society B: Biological Sciences*, *364*(1526), 2127–2139. https://doi.org/10.1098/rstb.2008.0289.

Suhrhoff, T. J., & Scholz-Böttcher, B. M. (2016). Qualitative impact of salinity, UV radiation and turbulence on leaching of organic plastic additives from four common plastics — A lab experiment. *Marine Pollution Bulletin*, *102*(1), 84–94. https://doi.org/10.1016/J.MARPOLBUL.2015.11.054.

Sun, J., Yang, S., Zhou, G. J., Zhang, K., Lu, Y., Jin, Q., et al. (2021). Release of microplastics from discarded surgical masks and their adverse impacts on the marine copepod Tigriopus japonicus. *Environmental Science and Technology Letters*, *8*(12), 1065–1070. https://doi.org/10.1021/ACS.ESTLETT.1C00748/SUPPL_FILE/EZ1C00748_SI_001.PDF.

Svedin, J. (2020). *Photodegradation of macroplastics to microplastics: A laboratory study on common litter found in urban areas.* https://www.diva-portal.org/smash/record.jsf?pid=diva2:1462307.

Tiwari, M., Rathod, T. D., Ajmal, P. Y., Bhangare, R. C., & Sahu, S. K. (2019). Distribution and characterization of microplastics in beach sand from three different Indian coastal environments. *Marine Pollution Bulletin*, *140*, 262–273. https://doi.org/10.1016/j.marpolbul.2019.01.055.

van Raamsdonk, L. W. D., van der Zande, M., Koelmans, A. A., Hoogenboom, P. L. A., Peters, R. J. B., Groot, M. J., et al. (2020). Current insights into monitoring, bioaccumulation, and potential health effects of microplastics present in the food chain. *Foods*, *9*(1), 72. https://doi.org/10.3390/FOODS9010072.

Vidyasakar, A., Krishnakumar, S., Kasilingam, K., Neelavannan, K., Bharathi, V. A., Godson, P. S., et al. (2020). Characterization and distribution of microplastics and plastic debris along Silver Beach, Southern India. *Marine Pollution Bulletin*, *158*, 111421. https://doi.org/10.1016/j.marpolbul.2020.111421.

Wakkaf, T., El Zrelli, R., Kedzierski, M., Balti, R., Shaiek, M., Mansour, L., et al. (2020). Microplastics in edible mussels from a southern Mediterranean lagoon: Preliminary results on seawater-mussel transfer and implications for environmental protection and seafood safety. *Marine Pollution Bulletin*, *158*, 111355. https://doi.org/10.1016/j.marpolbul.2020.111355.

Wang, J., Peng, C., Li, H., Zhang, P., & Liu, X. (2021). The impact of microplastic-microbe interactions on animal health and biogeochemical cycles: A mini-review. *Science of the Total Environment*, *773*, 145697. https://doi.org/10.1016/j.scitotenv.2021.145697.

Weinstein, J. E., Crocker, B. K., & Gray, A. D. (2016). From macroplastic to microplastic: Degradation of high-density polyethylene, polypropylene, and polystyrene in a salt marsh habitat. *Environmental Toxicology and Chemistry*, *35*(7), 1632–1640. https://doi.org/10.1002/ETC.3432.

Winkler, A., Santo, N., Ortenzi, M. A., Bolzoni, E., Bacchetta, R., & Tremolada, P. (2019). Does mechanical stress cause microplastic release from plastic water bottles? *Water Research*, *166*, 115082. https://doi.org/10.1016/J.WATRES.2019.115082.

Wootton, N., Reis-Santos, P., Dowsett, N., Turnbull, A., & Gillanders, B. M. (2021). Low abundance of microplastics in commercially caught fish across southern Australia. *Environmental Pollution*, *290*, 118030. https://doi.org/10.1016/J.ENVPOL.2021.118030.

Wypych, G. (2016). HDPE high density polyethylene. *European Chemical News*, *72*(1900), 156–163. https://doi.org/10.1016/B978-1-895198-92-8.50053-7.

Yang, D., Shi, H., Li, L., Li, J., Jabeen, K., & Kolandhasamy, P. (2015). Microplastic pollution in table salts from China. *Environmental Science & Technology*, *49*(22), 13622–13627. https://doi.org/10.1021/acs.est.5b03163.

Yee, M. S. L., Hii, L. W., Looi, C. K., Lim, W. M., Wong, S. F., Kok, Y. Y., et al. (2021). Impact of microplastics and nanoplastics on human health. *Nanomaterials*, *11*(2), 496. https://doi.org/10.3390/nano11020496.

Yu, Q., Hu, X., Yang, B., Zhang, G., Wang, J., & Ling, W. (2020). Distribution, abundance and risks of microplastics in the environment. *Chemosphere*, *249*, 126059. https://doi.org/10.1016/J.CHEMOSPHERE.2020.126059.

Yudhantari, C. I., Hendrawan, I. G., & Puspitha, N. L. P. R. (2019). Kandungan Mikroplastik pada Saluran Pencernaan Ikan Lemuru Protolan (Sardinella Lemuru) Hasil Tangkapan di Selat Bali. *Journal of Marine Research and Technology*, *2*(2), 48–52. https://doi.org/10.24843/JMRT.2019.V02.I02.P10.

Zhang, J., Wang, L., & Kannan, K. (2019). Polyethylene terephthalate and polycarbonate microplastics in pet food and Feces from the United States. *Environmental Science and Technology*, *53*(20), 12035–12042. https://doi.org/10.1021/ACS.EST.9B03912/SUPPL_FILE/ES9B03912_SI_001.PDF.

Zhang, Q., Zhao, Y., Du, F., Cai, H., Wang, G., & Shi, H. (2020). Microplastic fallout in different indoor environments. *Environmental Science and Technology*, *54*(11), 6530–6539. https://doi.org/10.1021/ACS.EST.0C00087/SUPPL_FILE/ES0C00087_SI_001.PDF.

Zhou, X., Wang, J., & Ren, J. (2022). Analysis of microplastics in takeaway food Containers in China Using FPA-FTIR whole filter analysis. *Molecules*, *27*(9), 2646. https://doi.org/10.3390/molecules27092646.

Zhu, K., Jia, H., Zhao, S., Xia, T., Guo, X., Wang, T., et al. (2019). Formation of environmentally persistent free radicals on microplastics under light irradiation. *Environmental Science and Technology*, *53*(14), 8177–8186. https://doi.org/10.1021/ACS.EST.9B01474/SUPPL_FILE/ES9B01474_SI_001.PDF.

Zuccarello, P., Ferrante, M., Cristaldi, A., Copat, C., Grasso, A., Sangregorio, D., et al. (2019). Exposure to microplastics (<10 μm) associated to plastic bottles mineral water consumption: The first quantitative study. *Water Research*, *157*, 365–371. https://doi.org/10.1016/J.WATRES.2019.03.091.

Potential risk assessment and toxicological impacts of nano/micro-plastics on human health through food products

Shahida Anusha Siddiqui[a,b,*], Sipper Khan[c], Tayyaba Tariq[d], Aysha Sameen[d], Asad Nawaz[e,f,g], Noman Walayat[h], Natalya Pavlovna Oboturova[i], Tigran Garrievich Ambartsumov[i], and Andrey Ashotovich Nagdalian[i]

[a]Technical University of Munich Campus Straubing for Biotechnology and Sustainability, Straubing, Germany
[b]German Institute of Food Technologies (DIL e.V.), Quakenbrück, Germany
[c]Institute of Agricultural Engineering Tropics and Subtropics Group, University of Hohenheim, Stuttgart, Germany
[d]National Institute of Food Science and Technology, University of Agriculture, Faisalabad, Pakistan
[e]College of Civil and Transportation Engineering, Shenzhen University, Shenzhen, China
[f]Shenzhen Key Laboratory of Marine Microbiome Engineering, Institute for Advanced Study, Shenzhen University, Shenzhen, China
[g]Institute for Innovative Development of Food Industry, Shenzhen University, Shenzhen, China
[h]College of Food Science and Technology, Zhejiang University of Technology, Hangzhou, China
[i]Food Technology and Engineering Department, North Caucasus Federal University, Stavropol, Russia
*Corresponding author: e-mail address: s.siddiqui@dil-ev.de

Contents

Advances in Food and Nutrition Research, Volume 103
ISSN 1043-4526
https://doi.org/10.1016/bs.afnr.2022.07.006

Abstract

The problem of environmental pollution with plastic is becoming more and more acute every year. Due to the low rate of decomposition of plastic, its particles get into food and harm the human body. This chapter focuses on the potential risks and toxicological effects of both nano and microplastics on human health. The main places of distribution of various toxicants along with the food chain have been established. The effects of some examples of the main sources of micro/nanoplastics on the human body are also emphasised. The processes of entry and accumulation of micro/nanoplastics are described, and the mechanism of accumulation that occurs inside the body is briefly explained. Potential toxic effects reported from studies on various organisms are highlighted as well.

1. Introduction

The commercial production of polypropylene, polyethylene, and polyolefin in the 1950s marked the beginning of the modern era of plastic manufacture. While worldwide plastic utilisation has risen, rapid development in manufacturing and distribution has resulted in severe ecological effects (Statista, 2021). A combination of factors, such as the great strength and durability of the disintegration of plastic polymers, along with high utilisation and poor recycling practices, has contributed to the ongoing growth of plastics in the environments. With 335 million tonnes of plastic produced in 2016, the world's plastic output has increased by around 9% each year. The open ocean is the most significant known basin for microplastics. According to estimates, between 4 and 12 million metric tonnes of plastic trash enters the marine environment each year which is a significant amount. Plastic waste can be carried thousands of kilometres and contaminated sites that are quite far away. It may also gather along strandlines, in open water, and on the seabed (Jambeck et al., 2015).

Most plastics are resilient to biodegradation although they will deteriorate over time due to mechanical stress. UVB light, oxidative characteristics of the atmosphere and hydrolytic properties of seawater cause plastics to become fragile and shatter into microplastics (0.1–5000 m) or nano plastics (0.1–0.1 m). Second- and third-generation contributors of microplastics comprise industry, such as cosmetics and cleaning goods, tyre wear, as well as microfibers from machine-washed textiles (Lambert & Wagner, 2016).

The manufacturing of nano plastics is also becoming more prominent. Paints, cosmetics adhesives, medication delivery vehicles, and electronics are only a few examples (Koelmans, Besseling, & Shim, 2015). Whether by design or as a result of environmental deterioration, the reduction in particle size can result in the development of distinct particle properties that might impact their potential toxicity (Wright & Kelly, 2017). Additionally, marine creatures are damaged by microplastics, which may lead to some possible outcomes including physical harm, reduced neuro-functional activity as well as activation of oxidative stress, reproductive consequences, oxidative damage, pathology development, and even death (de Sá, Oliveira, Ribeiro, Rocha, & Futter, 2018).

Microplastics may be coated with biomolecules that react with biological systems and/or serve as a conduit for the transfer of organic pollutants (POPs) into animals' tissues (Galloway, Cole, & Lewis, 2017). Moreover, microplastics may have physical consequences on creatures when they are ingested. This dynamic nature of microplastics is due to their high surface-to-volume ratio, reactivity, curvature, and microscopic size, which allow for varying absorption rates and bio-distribution in the environment (Mattsson, Jocic, Doverbratt, & Hansson, 2018).

Plastic pollution of land and aquatic environments ranks first among the most critical worldwide environmental pollution and toxicological problems. A variety of toxicological hazards to lower and higher organisms are caused by plastics and their accompanying chemicals, which are persistent and pervasive pollutants (Alimba & Faggio, 2019; Barletta, Lima, & Costa, 2019; Chang, Xue, Li, Zou, & Tang, 2020; Prüst, Meijer, & Westerink, 2020). United States Resource Conservation and Recovery Act of 1987 categorised plastics as harmful materials, and their usage in the corporate sector were discouraged (Bean, 1987). Although a global ban on plastic production has been proposed, still, it has not been implemented strictly owing to the wide variety and high competence of plastic materials used in construction, transportation, renewable energy, packaging, medical and scientific equipment, and a wide range of other applications, as well as the rapid growth in human population and industrialisation. As a result, the level of plastic pollution in the environment has reached unprecedented levels. The production of plastic has amplified greatly from 1.5 metric million tonnes in 1950 to 368 metric million tonnes in 2019 (Statista, 2021).

Around 11% of global plastic production in 2016 found its way into freshwater environments which is expected to rise to 53 million metric tonnes yearly by 2030 (Borrelle et al., 2020). A greater proportion of processed plastics end up in the environment and aggregate in nearly all components (Akindele & Alimba, 2021; Barletta et al., 2019; Guzzetti, Sureda, Tejada, & Faggio, 2018). It exerts a larger threat to the environment and the health of wildlife due to its limited degradability (Amereh, Babaei, Eslami, Fazelipour, & Rafiee, 2020). After persistent exposure to physical erosion and ultraviolet light in the environment, plastic fragments into smaller units that can be categorised as mega-plastics/macro-plastics (large plastic pieces), meso-plastics/small plastic pieces (small plastic pieces), or nano plastics/small plastic pieces (small plastic pieces) (Song et al., 2017). Although nano plastics (NPs) and microplastics (MPs) with the size limit of 0.1–100 nm and <50 to >1000 μm are ubiquitous at all trophic levels of the environment and human food chain. Nano and micro plastics concentrations in the environment have risen as a result of their widespread use as microbeads in cosmetic formulations, plastic powders in mould-making, and other industrial applications (Cheung & Fok, 2016).

The unique properties of nano and micro plastics of being small sizes and hydrophobic properties make them able to increase their interaction ability with microorganisms which speedily absorb and waterborne harmful contaminants including polycyclic aromatic hydrocarbon, drugs, metals (Brennecke, Duarte, Paiva, Caçador, & Canning-Clode, 2016; Sharma, Elanjickal, Mankar, & Krupadam, 2020; Zhang et al., 2018). In this way, they transfer the harmful chemicals to the biological systems by acting as vectors (Caruso, 2019). In several *in vivo* and *in vitro* studies, the development of DNA damage, neurotoxicity, inflammation, changes in lipid/energy metabolism, developmental and reproductive abnormalities, and a variety of other pathophysiological malfunctions have all been documented (Jakubowska et al., 2020; Patil, Bafana, Naoghare, Krishnamurthi, & Sivanesan, 2021).

2. Human exposure to nano/micro-plastics

Plastic particles interact with their surrounding environment in saltwater, generating an "eco-corona" of organic materials, hydrophobic contaminants, nutrients, and bacteria from the water column and residues (Galloway et al., 2017). Natural organic material influences particle aggregation and the frequency at which proteins bind to one another, as well as the development of a corona (Mattsson et al., 2018; Siddiqui et al., 2022). Microplastics' volume/surface ratio, reactivity, small size, and curvature make them particularly vibrant in the environment, with varying bioavailability.

The kind of polymer, its surface chemistry, and the extent to which it is befouling by microbial rafting species and biofilms all have an effect on its bioavailability in the aquatic environment (Turner, 2015). Particles of organic matter (POM), which include zooplankton and fish faecal pellets, may also agglomerate microplastics.

Until far, the majority of research on plastic particle interactions with the environment has focused on polystyrene (PS) microparticles. In artificial seawater, nano plastics (30 nm) and microplastics (20 μm) created millimetre-long aggregates. It was discovered that $FeCl_3$ solutions aggregated as the ionic strength increased. Strangely, natural organic matter (NOM) did not affect the aggregation of nano plastics (Cai et al., 2018).

There are three main routes of exposure to nano and micro plastics including inhalation, ingestion, and dermal exposure (Fig. 1). Alimba and Faggio (2019) discovered that nearly every animal species absorbed substantial quantities and kinds of nano and micro plastics. All vertebrates and invertebrates (including bivalves, fish, crabs, turtles, marine mammals, and shorebirds) have different quantities of nano and micro plastics in their stomachs and tissues (Patil et al., 2021; Triebskorn et al., 2019). According to Dowarah, Patchaiyappan, Thirunavukkarasu, Jayakumar, and Devipriya (2020), a typical Indian might ingest 3918 MPs each year. Furthermore, a Tunisian who eats wild mussels might absorb up to 4.2 MPs yearly (Wakkaf et al., 2020) and 2757 MPs per person yearly

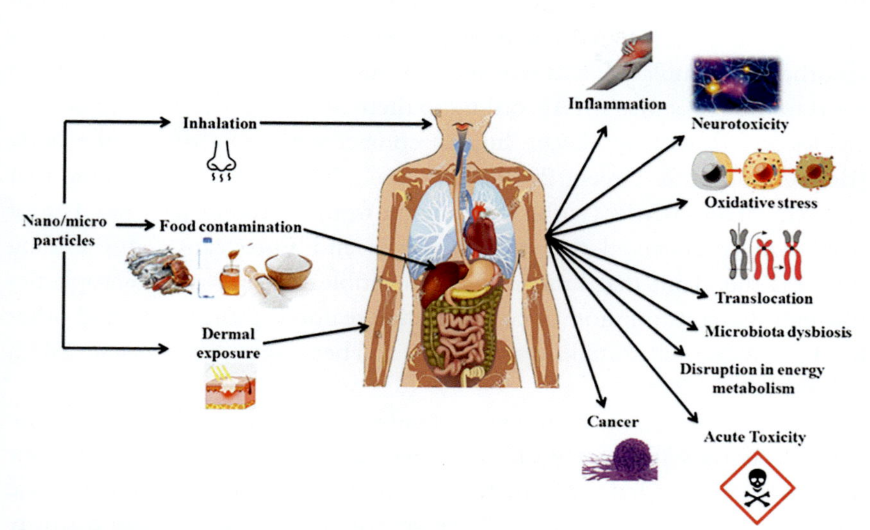

Fig. 1 Human exposure to nano and micro plastics.

(Abidli, Lahbib, & el Menif, 2019). By considering the diverse dietary sources and their consumption, Cox and colleagues determined the average American plastics consumption yearly as the higher value between 39,000 and 52,000 MPs (Cox et al., 2019). Fish intake was unambiguously linked in these findings as a key pathway of human exposure to plastic particles (Triebskorn et al., 2019).

Nano and micro plastics get into contact with humans either by environmental, occupational exposure, or through contaminated foods. Honey, beverages, sugar, milk, teabags, tap and bottled drinking water, beer, and commercial salts include nano and micro plastics with a range of 100 nm-5000 m. According to the WHO's recommendation, an adult consumes 5 g of table salts per day and the fact that NP/MPs are found in salts from different countries. Therefore, this way, ingestion of polluted salts is a main oral route of human exposure to plastic particles. Contamination of salts with nano and microplastics varies from region to region as well (Peixoto et al., 2019).

The use of tap water and milk in beverage manufacturing and their consumption with bear enhances human exposure to nano and micro plastics. Another vital exposure pathway of plastic particles to humans is *via* plant-based foods intake, which appears to be neglected. There have already been several scientific studies about the beneficial role of edible plants such as lettuce (Li, Zhou, Yin, Tu, & Luo, 2019), and seeds (Bosker, Bouwman, Brun, Behrens, & Vijver, 2019; Siddiqui et al., 2021) on human health (Li et al., 2020). NP/MPs with sizes ranging from 50 to 4800 nm were absorbed, accumulated, and translocated through the vascular system from roots to the shoot system. According to these researchers, the intake of vegetables and fruits may elevate human contact with NP/MPs in the body (Boots, Russell, & Green, 2019; Guo et al., 2020). One study found that approximately 20 MPs (with sizes ranging from 50 to 500 μm) per 10 g of stool sample collected from some Asians and Europeans. This finding supports the claim that humans are susceptible to nano and microplastics through the consumption of contaminated seafood, plant foods, and other food products such as milk, water, beers, and beverages (Blinov et al., 2022; Schwabl et al., 2019).

According to research studies, individuals unintentionally inhale plastic particles along with their meals and beverages, which accumulate in their digestive systems. Oral exposure to NP/MPs through the ingestion of plastic particle-contaminated foods and beverages appears to be the main pathway of human exposure to plastic particles that have been extensively researched although inhalation of air polluted with NPs and MPs cannot be overlooked. The detection of MPs in air samples (Dris et al., 2015) highlighted concerns that inhalation might be a considerable source of NP/MP contamination in addition to the skin. Due to the detection of 33% microfibers of plastic in both interior and outdoor air (including dust and on the floor surrounding buildings), it has been determined that people inhale NP/MPs at a frequency of 11 micro plastics hourly on average (Dris et al., 2017; Dris, Gasperi, Saad, Mirande, & Tassin, 2016). This concludes higher household inhalation of plastics than from mussel consumption. NP/MPs in indoor or residential air are likely to be placed on food and beverages, increasing microplastic exposure and intake (Zhang et al., 2020). In accordance with the studies on the frequency of microplastics acquired from several sample locations in Iran, the range of MPs consumed by adults through dust consumption is estimated to be 100 mg/day on average by Dehghani, Moore, and Akhbarizadeh (2017). Topical use of cosmetics containing nano and micro plastics microbeads may also expose people to NP/MPs through the skin (dermal route of exposure.

According to research, microplastic beads are utilised as components in a variety of personal use products, including shaving creams, facial wash, toothpaste, wrinkle creams, facial scrubs, moisturisers, sunscreen lotions, deodorants, soaps, shampoos, facemasks, eye shadows, and lipsticks (Cheung & Fok, 2016). Micro plastics absorb through the skin since their size is greater than the size of the skin pore. Nano plastics, in contrast, just require particles less than 100 nm in size to get through the cornified layer of the epithelia (Revel, Châtel, & Mouneyrac, 2018). In the animal body, NPs are transported to various viscera by the proper blood vessels, including those in the liver, heart, spleen, lung, placenta, reproductive organs, kidney, thymus and even the brain (they can pass through the blood-brain barrier) (Prüst et al., 2020; Ragusa et al., 2021).

2.1 Ingestion

Humans are most exposed to MPs *via* contaminated food staff and drink. Consumption amounts varied among age groups and sexes owing to dietary and lifestyle factors. Additionally, it was discovered that persons who drank just bottled water swallowed 90,000 more particles than those who drank only tap water (Galloway, 2015).

Microplastics have been found in a variety of food products, including bivalves (Li et al., 2016), crustaceans and commercial fish, sugar, salt, and bottled water. Bivalves alone contained 11,000 MP particles each year. In Europe, 37 MPs per person are ingested yearly through table salt (Karami, Golieskardi, Ho, Larat, & Salamatinia, 2017), although per person consumes 100 MPs in China. Additionally, dust that accumulates on serving dishes, food containers, and packaging contains additional MPs. PET particles have been discovered in food and the environment (Galloway, 2015; Karbalaei, Hanachi, Walker, & Cole, 2018; Toussaint et al., 2019).

Numerous entrance paths for MPs have been suggested. MPs may be digested by intestinal lymphoid M-cells or absorbed directly, dependent on their adherence to the gastrointestinal mucus membrane. Another research team discovered that when MPs are mixed with other intestinal components, they might infiltrate the intestinal membrane directly. It was determined that nano plastics transit transcellular and paracellularly through the rat intestinal mucosa. The researchers found that oral nanoparticles passed through rats' lymphatic and circulatory systems (Prata, da Costa, Lopes, Duarte, & Rocha-Santos, 2020). Additionally, Fiorentino et al. (2015) discovered that 44-nm polystyrene particles were secreted passively by human colon fibroblasts. MPs may induce an inflammatory response, inhibit mucus production, affect intestinal barrier function, increase mucosal permeability, and alter metabolism (lipogenesis, triglyceride synthesis, *etc.*) (Jin, Lu, Tu, & Luo, 2019; Lu, Wan, Luo, Fu, & Jin, 2018). Plastic particles (44 nm or smaller) have been demonstrated to impact gene expression after absorption across the stomach mucosa pro-inflammatory responses are activated by them (IL6, IL8, and IL1 genes), and decrease cell viability (Forte et al., 2016).

2.2 Inhalation

MPs enter the environment by several sources, including synthetic garments and textiles (Dris et al., 2015), shedding from construction materials (Prata, 2018), waste landfilling and incineration and waste incineration and landfilling (Dris et al., 2016). Dris et al. (2017) determined inhalable MPs concentrations both outdoors (0.3–1.5 particles per cubic metre) and indoors (0.4–56.5 particles per cubic metre).

According to Prata (2018), an average individual inhales between 26 and 130 MP particles per day although Vianello, Jensen, Liu, and Vollertsen (2019) discovered that up to 272 MP particles are inhaled by a sedentary man per day. Since the estimations were based on sampling technique, particle properties, and other considerations such as seasonal changes, furniture materials, surface cleaning schedules, and overall air quality, these disparities are unsurprising (Prata et al., 2020). The density, size, surface charge, and hydrophobicity of particles may all affect their absorption and deposition in the respiratory system, with lighter and smaller particles reaching critically in the lungs (Rist, Almroth, Hartmann, & Karlsson, 2018).

The inhalation of MPs is harmful to human health owing to inflammation, chemical toxicity, and the spread of pathogens (which are transmitted through MPs as vectors) (Rist et al., 2018; Vethaak & Leslie, 2016; Wright & Kelly, 2017). Due the large alveolation surface area of 150 m^2 and the low tissue barrier of less than 1 μm, nanoparticles can pass through the human lungs with little difficulty (Lehner, Weder, Petri-Fink, & Rothen-Rutishauser, 2019). MPs are absorbed by macrophages and translocated to the circulatory and lymphatic systems after they entered the respiratory system (Vethaak & Leslie, 2016). Certain inflammatory responses are being inserted by MPs in lung tissue. According to Prata (2018), MPs may induce chronic inflammation in the lungs and contribute to the development of lung cancer. Accumulation of particles in the respiratory system may contribute to excessive secretion of pro-inflammatory chemotactic factors, which can lead to chronic inflammation, a condition recognised as "dust overload" (Prata, 2018; Wright & Kelly, 2017). In a cultured environment, Xu et al. (2019) showed that polystyrene particles with a diameter of 50 nm may be genotoxic and cytotoxic to the macrophages and respiratory lining epithelium.

Interstitial lung diseases and inflammatory airways have been connected to severe respiratory symptoms in workers of synthetic textile mills exposed to airborne MPs. *In vivo* study results verified these outcomes. Prata (2018) reported that chronic and acute inhalation of MPs may cause immediate asthmatic-like bronchial reactions as well as granulomatous and inflammatory changes in bronchial tissues, which may result in extrinsic allergic alveolitis and chronic pneumonia depending on the individual susceptibility and exposure level to the substance. As a result, a high concentration of MPs in the air may have a detrimental impact on the respiratory system of susceptible individuals who inhale the MPs.

2.3 Dermal contact

The adverse consequences of cutaneous contact with microplastics are unclear at this time. According to Revel et al. (2018), although particles size greater than 100 nm cannot be absorbed through the skin, nano size plastics can. The German Federal Institute for Risk Assessment assessed the health hazards associated with the use of face washes, facemasks, hand cleansers, and toothpaste. They established that MPs in cosmetics induced local inflammation and cytotoxicity, resulting in skin damage (Sharma & Chatterjee, 2017). Microplastics derived from personal care products and toothpaste may be absorbed by the skin, gastrointestinal mucosa, and mucosal membranes (Lassen et al., 2012).

It is well established that plastic used in surgery and prosthetic body parts causes foreign body responses and inflammation. Subcutaneous MP injections *in vivo* induced inflammation in mice. Schirinzi et al. (2017) discovered that cutaneous exposure to NPs/MPs resulted in the induction of oxidative stress in human epithelial cells. NPs/MPs have been shown to induce foreign body reactions, while the characteristics of MPs differ, prompting more research.

3. Occurrence of nano/micro-plastics in the food chain

The presence of NPs/MPs in the food chain is becoming a health concern due to the increase in plastic waste (Kolandhasamy et al., 2018; Wang et al., 2020). Nano and micro plastics are highly likely to be found in numerous food products due to their widespread ubiquity and bioavailability in both terrestrial and aquatic environments (Table 1). They penetrate the human food chain in a variety of ways, according to several research findings: animals taking them in their natural environment (Santillo, Miller, & Johnston, 2017), contaminating them during food preparation operations

Table 1 Average levels of nano/microplastics in the food chain (Yee et al., 2021).

Food commodity	Nano/micro particles
Seafood	1.48 particles/g
Sugar	0.44 particles/g
Honey	0.10 particles/g
Salt	0.11 particles/g
Alcohol	32.27 particles/L
Bottled water	94.37 particles/L
Tap water	4.23 particles/L

(Karami et al., 2017), and/or leaching from food and drink containers/ package (Mason, Welch, & Neratko, 2018). They were found in tap, bottled, and spring water samples analysed using Fourier-Transform Infrared Spectroscopy (FTIR). Micro plastic particles measuring less than 5 mm in size were found in 81% of tap water samples collected from 159 different sources throughout the world (Kosuth, Mason, & Wattenberg, 2018).

Microplastic particles were identified in 93% of the 259 bottles of water tested, which were from 11 different brands and 27 different batches (Mason et al., 2018). Based on these numbers, the average human consumes roughly 39,000 to 52,000 microplastic particles each year, with gender and age having an impact on the total amount. If inhalation of plastic particles is factored in, the total number of particles rises from 74,000 to 121,000 per year. Besides, compared to people drinking tap water only, they have a chance to ingest only 4000 extra particles, as compared to the 90,000 particles by people with bottled water (Cox et al., 2019). These findings suggest that microplastic ingestion by humans is mostly sourced from the human food chain. There are presently no data on the prevalence of nanoplastics in food due to a lack of analytical instruments. Nanoplastics were generated overtime when the polystyrene drinking cup lids degraded (Lambert & Wagner, 2016).

Heavy metals and chemicals used in the production of MP have been found to seep into the water, prompting worries about the safety of food produced consequently. In addition, the existence of non-pathogenic and pathogenic bacteria in MPs has been emphasised in different studies. Moreover, it is considered that organism shape forms, such as pieces, fibres, beads, foams, and pellets, have a significant influence on the variety of organisms researched (Lusher, Hollman, & Mendoza-Hill, 2017). It is commonly known that the build-up of these various germs and chemicals causes significant damage to consumers.

According to the literature, MPs have been discovered in almost all abiotic products and marine organisms (Hantoro, Löhr, van Belleghem, Widianarko, & Ragas, 2019). All the trophic levels of the marine environment are affected by the transfer of MPs starting with plankton and proceeding through fish (Tanaka & Takada, 2016), bivalves (Leslie, Brandsma, van Velzen, & Vethaak, 2017), crustaceans, and may harm the health, feeding, development, and marine species' survival (Tanaka & Takada, 2016). MPs may be swallowed purposefully (owing to their likeness to a filter-feeding or natural food source) or accidentally (due to feeding behaviour or clinging to external appendages) (Lusher et al., 2017).

3.1 Seafood

Consumption of adulterated seafood leads to the build-up of the pollutant, which is then passed to other consumers through trophic transfer, resulting in negative health effects in consumers, especially the children, elderly, and pregnant women (Lusher et al., 2017). Seafood eating is vital for human nutrition and health. It is recommended for the consumption of humans at all stages of development owing to the excellent quality of the fats (PUFAS) and protein it contains, as well as the micronutrient content proven to provide significant health benefits. More cases of seawater body pollution by MPs pose a potential health hazard to consumers and increase public awareness of the pollution problem. It was reported that some of the MPs (1.5 m) may enter the circulatory system and accumulate in shellfish (Grigorakis, Mason, & Drouillard, 2017; Gündoğdu, Eroldoğan, Evliyaoğlu, Turchini, & Wu, 2021).

Primary MPs have at least one micro dimension and are manufactured using several industrial procedures. They may be found in a variety of goods, including powder plastic, resin pellets, personal care products, microbeads, and textile. Secondary MPs are formed due to the breakdown and fragmentation of large plastic products, which may occur as a result of many chemical and physical processes (Hale, Seeley, la Guardia, Mai, & Zeng, 2020).

MPs have been found in a variety of important seafood, and they have become a serious cause of concern. MPs may be found in a broad spectrum of fish, with a sizes range of $<100\,\mu m$ to $>1000\,\mu m$ (PMMA). Shrimp, lobster, and crabs all possessed microparticles ranging in size from 1 to 1000 μm, with the great majority of them being fibrous (Abbasi et al., 2018). Some recent studies have shown that MPs have been detected in canned and dried fish (Akhbarizadeh et al., 2020; Karami, Golieskardi, Ho, et al., 2017).

After being degraded by MPs, these contaminants are circulated and deposited in the sea. A significant amount of plastic waste is present in the marine environment, either in secondary or primary forms and has been devoured as food by shellfish or passed through the gills of fish (Tanaka & Takada, 2016). Farrell and Nelson released an article explaining how MP is transmitted trophically across the marine food chain. At the lowest water body level, microplastic pollution originates at the sea surface and travels downward through sedimentation and ingestion by phytoplankton and crustaceans (Farrell & Nelson, 2013; Murphy & Margaret, 2017; SAPEA, 2019). MPs are often considered the single largest source of marine pollution. The number of zooplankton that follows depends on the size and concentration of the MP. Zooplankton is subsequently eaten by a variety of predatory fish, including sharks and rays (EFSA, 2016). MPs are absorbed through prey consumption, the mouth is at a higher trophic level or when misguided for food due to bivalves' suspension-feeding behaviour. MPs are also eaten by higher-trophic level fish by their gills (Setälä, Magnusson, Lehtiniemi, & Norén, 2016). According to recent research, seaweed, which is eaten as a traditional meal across the globe owing to its renowned nutraceutical worth, can absorb MPs smaller than or similar to 20 m in size (Sundbæk et al., 2018), and trophic transfer of MPs, as well as trophic levels (Gutow, Eckerlebe, Giménez, & Saborowski, 2016).

Plastics are particularly resilient in the marine environment because they do not go through the mineralisation process and instead break down into ever-smaller pieces, ultimately becoming NPs/MPs. Researchers are currently working on finding novel strategies to harvest MPs from aquatic ecosystems since MPs are almost hard to recycle or dispose of (Guo & Wang, 2019).

In aquatic environments, small and large quantities of MPs unquestionably reach the food chain. Microbial contaminants may be absorbed by marine organisms and through the biological accumulation process are transferred to the highest trophic level. Contamination of the food chain with MPs has the potential to have serious effects on human health. According to the Environmental Protection Agency, MPs present in seafood may cause the leaching of adsorbed pollutants or plastic additives (such as pesticides, heavy metals, and other poisons) from aquatic habitats. As a result, one of the major ways of human exposure to MPs and related pollutants is through seafood intake (Wei et al., 2022; Zazouli et al., 2022).

MPs are getting common in the aquatic environment, thus choosing the right species as an indicator is important to assess the human health and ecological risks they pose. Mussels, for example, might be regarded as one of the major species of indicators since they are filter-feeding organisms with a global distribution. As mussels are exposed to direct MP pollution in their capacity as filter feeders, there is a positive relationship between MPs existence in the adjacent water and their tissues. Mussels are swallowed whole, which means they contain high quantities of MPs, and they are an important way of human exposure to MPs through food. They may be utilised as indicators of MP contamination and food quality in this approach. However, several other forms of seafood serve as key entrance sites for MPs into the human food chain. Because their gastrointestinal system is digested as part of a complete animal, and since MP concentrations in the gastrointestinal tract are generally higher, microscopic fish (*e.g.*, anchovies) can cause direct human exposure through food. The microplastics determined that since the gastrointestinal tract of larger fish is not regularly removed, there may be no direct harm to human health in general. Besides, it is unknown if MPs remain in the gut lumen after consumption of the drug. Moreover, it is unknown if MPs can cross the intestinal epithelium. Gut cells, however, can absorb MPs as tiny as a few microns, while Peyer's patch cells in the ileum can identify MPs as small as 10 μm (Cox et al., 2019). Therefore, MP-infected fish may be considered a source of human exposure to MPs in the environment, whether consumed whole or without the gastrointestinal tracts.

MPs have been found in a range of seafood, including seaweed (Li, Luo, et al., 2020), scallops (Akoueson et al., 2020), shrimp, crab (Daniel, Ashraf, Thomas, & Thomson, 2021), mussels (Vinay Kumar, Löschel, Imhof, Löder, & Laforsch, 2021), fish. MPs in seafood do not always arise from the aquatic as MPs are abundant in the atmospheric and terrestrial environment, they may arise from a range of sources that are both varied and intricate. According to the Environmental Protection Agency, food processing and packaging are additional causes of MP contamination in seafood, with air pollution and terrestrial pollution. MPs have been found in canned and fresh fish (Akhbarizadeh et al., 2020; Karami et al., 2018) and stuffed mussels (Gündoğdu, Çevik, & Ataş, 2020), implying that contamination with MPs can occur during catching, storage, transportation, processing, and canning, as well as after addition to other foods like oil, rice, and other ingredients. Ferrante et al. (2022) recently discovered a high-assessed daily consumption of microplastic in six Mediterranean marine species that has edible flesh, indicating that these species may constitute a health risk to people.

Aquaculture feeds such as fish meals that are formed from fish waste consisting the parts of the digestive system and are a key source of MP deposition for fish, are a popular source of lipids, vitamins, minerals and protein for aquaculture, particularly shrimp and fish (Promthale, Pongtippatee, Withyachumnarnkul, & Wongprasert, 2019). In this approach, feeding MP-containing fish meals to aquaculture may result in seafood consumption (Gündoğdu et al., 2021), ascribing the presence of microplastics in water sources where fish were reared had an influence on the fish to the feeding of MP-containing fish meals to aquaculture (Zazouli et al., 2022). Similarly, when it comes to small, dry fish, cutting and cleaning might be difficult or impossible. They are conserved dry or whole, with the stomach and head remaining complete, according to Karami, Golieskardi, Ho, et al. (2017), indicating that they might be another source of MPs (Feng et al., 2019; Mistri et al., 2022). Microplastic (polyethylene, polystyrene, polypropylene, polyethylene terephthalate, and polyvinyl chloride) can be found in commercially available dried fish widely consumed across Asian countries (Piyawardhana et al., 2022).

Filter feeders like mussels and bivalves have been discovered to be subtle to a broad variety of pollutants, including microplastics due to their feeding habits. Bivalves have been presented as a feasible answer as biological monitors of pollution of MP in ecosystems. Therefore, bivalves, especially mussels, have been widely investigated in recent years for contamination of MP (Klasios, de Frond, Miller, Sedlak, & Rochman, 2021). Researchers have started to look at edible mussel species, since they may be a substantial source of human exposure to MPs due to their high capacity to absorb them. The contamination of cultured and wild mussels with MP has been investigated by a significant number of researchers on a global scale. However, the presence of MP in processed mussels has only been investigated once (Gündoğdu et al., 2020), with inconsistent findings.

Stuffed mussels are available for purchase in numerous Turkish towns where *Mytilus galloprovincialis* is eaten. It is self-evident that MP contamination may occur throughout the production process, irrespective of the fact whether or not hygiene standards are obeyed. Therefore, the filter-feeding mussels' load of MPs may significantly rise throughout the processing process. The MP ratio of ready-to-eat filled mussels bought from 41 different retailers in five different Turkish towns was investigated by Gündoğdu et al. (2020). During their investigation, they noticed that 0.6 items were the average number of MPs per mussel and that polyethylene (35%) and polypropylene (35%) were the most common polymer types (15%). The average quantity of MP discovered in stuffed mussels meant for human consumption was also used by this group of researchers to determine the risk of MP exposure. According to their findings, if a typical customer eats 100 g of filled mussels per serving, they risk consuming 5.8% of their daily-recommended MP consumption.

3.2 Seaweed

Macroalgae dominate primary production in coastal oceans. The high levels of dietary fibre, vitamins, minerals, and proteins in macroalgae make them ideal marine vegetables. It is vital to remember that macroalgae, like all other marine organisms, are affected directly by pollution in the ocean. Due to their eating, they might spread the contamination to others through their food. It was reported that some macroalgae have been shown to enhance the number of MPs in seawater by tricking and concentrating them (Seng et al., 2020). Despite the extensive presence of MPs in the environment, little study has been done on their impact on marine cultures and macroalgae (Feng et al., 2019; Gao, Li, Hu, Li, & Sun, 2020; Seng et al., 2020). Apart from that, just one research has examined the MP contamination caused by macroalgae utilised in processed seafood (Li, Feng, Zhang, Ma, & Shi, 2020). These researchers tested the existence of MP contamination using nori, an edible seaweed (Pyropia spp.). This study looked at the final commercial form of the product as well as intermediate products created at different stages of the manufacturing process. Between 0.9 and 3.0 MPs/g (dry weight) were found in 24 brands of commercially packaged nori samples from the Chinese market, with 0.9 being the lowest concentration. The researchers' results determine that MPs were found in abundance in commercially accessible seaweed nori samples although their quantities lingered relatively low.

The oceans have been severely contaminated due to human activity. Abiotic marine materials, such as salts, also include MPs. Salt is usually produced by drying saltwater in the sun and wind. Table salt's MP load is estimated to be (0–7.3) 104 pieces per person per year (Zhang et al., 2020). Salt is extracted from brine wells, salt lakes, and salt rocks. MP is expected to be present in many anthropogenic areas where salt is produced (Peixoto et al., 2019). MPs were found in ten commercial sea salts from various sources (Sivagami et al., 2021). There are 700 MP/kg of sea salts, with particle sizes ranging from 3.88 M to 5.2 mM (PAR). In some studies, it was shown that the salt included MPs, which were damaging to human cells. Refined sea salt (1400–1900 particles/kg), unprocessed sea salt (1900–2300 particles/kg), and rock salt all include MPs (200–400 particles kg). MPs are higher in sea salts than in rock salts. MPs were most often between the sizes of 1–4 m. In addition, a technique for separating and extracting MPs from table salt was developed. Karami, Golieskardi, Ho, et al. (2017) tested 17 different salt brands from eight different nations. Only one brand of salt contained no MPs, whereas the others had 1–10 items/kg with a particle size of 515 m. PP (40%) and PE (40%) were the most common polymers discovered in the samples (33.3%).

The results illustrate how the distribution of microplastics in sea salts changes from place to place. However, due to a shortage of research, these findings cannot be considered general. Recently, there has been an increase in the amount of study on this issue. Local environmental factors and plastic garbage contamination in coastal regions have a significant role in the wide range of sea salt varieties seen in the region (Parvin, Nath, Hannan, & Tareq, 2022). Sea salts from Croatia, Indonesia, and Bangladesh provide the greatest quantities of MPs/kg. Due to human populations and activities, coastal landscapes are becoming overexploited (Kim, Lee, Kim, & Kim, 2018; Parvin et al., 2022; Renzi & Blašković, 2018).

4. Intake and accumulation of nano/micro-plastics inside the body

The environment, human health, and animals may be negatively affected by NP and MP pollution. MP contamination in aquatic environment is expected to rise rapidly over the next several years although the toxicity of this pollution to humans and ecosystems remains a mystery. Because MPs are not water-soluble, there is minimal possibility that they will be absorbed by the gastrointestinal tract or interact with mammalian biological systems. Anbumani and Kakkar (2018) claim that MPs are currently considered potentially harmful pollutants to organisms; nonetheless, the impact of MPs on human health remains a matter of debate in the scientific community (Barboza, Dick Vethaak, Lavorante, Lundebye, & Guilhermino, 2018; Paul et al., 2020).

MP pollution is inevitable as a result of the growing production and use of things packaged in plastic. MPs may be swallowed by food or beverages, or they can be breathed through the air. MPs may also be swallowed or released through the eyes, and sandy beaches have been shown to have MP particles in their sediment (Lee et al., 2015). MP particles may be discovered in sand, silt, and water on the beach (Suteja et al., 2021; Wilson, Godley, Haggar, Santillo, & Sheen, 2021). When compared to the sea surface and deep soil, beaches provide macro- and MPs with much more UV radiation, oxygen, mechanical stressors, and high temperatures.

Toxicities caused by nano and micro plastics in organisms, such as oxidative stress, inflammatory immunological responses, and necrosis are because by the passage and build-up of NP/MPs through the cellular compartments (Barletta et al., 2019; Chang et al., 2020; Prüst et al., 2020; Strungaru, Jijie, Nicoara, Plavan, & Faggio, 2019). Uptake and translocation of NP/MPs refer to the capacity of plastic materials to permeate through cell membranes and epithelial linings into tissue and cells. However, they are in the tissue or cell, plastic materials either collected or were sought to be eradicated by the organism's cellular defence system (Deng, Yan, Zhu, & Zhang, 2020). Even though the size and form of plastics affect accumulation and membrane translocation (Park et al., 2020; Triebskorn et al., 2019), sufficient evidence supports that NP/MPs translocate tissues and cells in numerous species of vertebrates, and invertebrates such as humans (*in vitro* studies) (Deng et al., 2020; Ragusa et al., 2021; Triebskorn et al., 2019). After being consumed or inhaled, they enter the body through the alimentary canal leading to the tracheal system (Chang et al., 2020; Deng et al., 2020).

Regardless of their structure, nano and microplastics can pass through the epithelial linings of the alveoli and stomach, as they possess a size range of 20–20,000 nm, and then enter the circulatory system, where they are delivered to different organs throughout the body. For example, Peda et al. (2016) conducted a study on European sea bass (*Dicentrarchus labrax*, bodyweight 140 8.42 g), which were fed PVC plastic particles with a 0.3 mm particle size for 30, 60, and 90 days. They detected a wide range of pathological anomalies in the intestinal segments when they examined the histological sections. According to the observed lesions, PVC translocated into the gut and generated detrimental effects on the submucosa of the three regions of the intestine, and muscular mucus associated with circulatory, structural, and inflammatory problems.

Lu et al. (2016) administered polystyrene (PS) to zebrafish (*Danio rerio*) for 7 days and assessed the amount of PS absorbed and accumulated in the intestines, liver, and gills. According to the findings, PS with a size of 5 m accumulated in the stomach, liver, and gills, but PS with a size of 20 m accumulated primarily in the gut and gills, as previously reported. They performed a histological examination of the liver and discovered 5 m and 70 nm PS in the tissues, as well as a variety of pathological problems such as lipid accumulation, inflammation, and alterations in oxidative stress protein levels.

Nano/micro-plastics easily diffuse and deposit in the cells and tissues of the gastrointestinal tracts, reproductive system, respiratory system, neurological system, embryos, and hemolymph as well as the kidneys, spleen, and liver, according to Triebskorn et al. (2019) and Deng et al. (2020). In light of the discovery of MPs in human faeces samples (Schwabl et al., 2019), it is undeniable that NP/MPs are able to controllably transport tissue membranes into circulation and deliver them to various regions of the body, such as the human placenta. The nano-scale size of plastics is responsible for the reduction in surface area to hydrophobicity, surface functionalization, volume ratio, and surface charges seen in the experiments. Due to these characteristics, NPs can pass through tissue epithelia and membrane barriers (transmucosal transit) into the circulatory and lymphatic systems, where they can be transported to various organs including the brain, liver, reproductive tissues, embryo, and other body tissues (Chang et al., 2020; Ragusa et al., 2021). MPs smaller than 150 μm in diameter can be consumed; nevertheless, MPs exceeding 150 μm in diameter are stored by intestinal mucus when they make contact with the apical region of epithelial cells.

Patches made by Peyer microplastics are transported into the circulatory system through the intestinal epithelium (at the ends of the villi) and into the bloodstream by endocytosis or paracellular diffusion pathways mediated by macrophages (M cells) (gut perception). The distal section of the gut, which is a key location of endocytosis (uptake and transport) in the gut, appears to be the principal pathway for NPs/MPs uptake from the gut (Arumugasaamy, Navarro, Kent Leach, Kim, & Fisher, 2019; Ragusa et al., 2021). Plastic particles are transferred into the blood and lymphatic circulation by inhalation of the particles (Prata, 2018). Certain factors influence the sorption of chemical contaminants to microplastics (Table 2).

In contrast to inhaled MP particle cytotoxicity, research on ingested MP particle toxicity is poorly recognised. In the end, MPs are mostly eaten by

Table 2 Factors affecting the sorption of nano and micro plastics.

Physical and chemical properties of the sorbent	Chemical characteristics of medium	Chemical characteristics of the sorbate	References
Temperature	pH	Hydrophobicity	Gopinath, Saranya, Vijayakumar, et al., 2019 Lionetto & Esposito, 2021
Polarity	Salinity	Ionic property	Velzeboer, Kwadijk, & Koelmans, 2014 Agboola & Benson, 2021
Polymer type (Crystalline, semi-crystalline, or amorphous)	Ionic strength	Functional groups	Reichel, Graßmann, Knoop, Drewes, & Letzel, 2021 Wang, Zhang, et al., 2020

the general public. Annual MP consumption of an individual is predicted to be 39,000 to 52,000 particles (Cox et al., 2019). There are several methods by which MP particles might enter food systems. Water, honey, sweets, and milk are among the other foods that contain MPs (Danopoulos, Jenner, Twiddy, & Rotchell, 2020; Gündoğdu et al., 2020). New-born contact with MP from PP-based materials utilised in formula preparation is far higher than previously assumed, according to recent research (Li, Luo, et al., 2020).

After consuming shellfish, researchers discovered PS, PE glycol, and PU in the faeces of eight participants from various European countries (van Cauwenberghe & Janssen, 2014). Despite the fact that MPs may be found in food and excreted in faeces, no studies have examined whether MPs are systemically bioavailable in people (until a recent Italian study). For the first time, researchers have detected 12MP particles in four out of six human placentas in Italy. Poisoning by eating fish that has been contaminated with MPs, additives, and other dangerous chemicals is possible. Because marine biosystems may amass and bioamplify a wide range of potentially hazardous substances linked to marine pollutants. With an average of 1800 particles per year, European nations that eat a lot of shellfish are expected to have greater annual MP particle intakes (up to 11,000 particles) (van Cauwenberghe & Janssen, 2014).

A food supply contaminated by plastic additives and organic pollutants may pose a risk to human health. PVC is a very heat- and abrasion-resistant polymer. Sixty four out of 400 plastic additives are detrimental to human health according to a new study. Prior animal research has shown that bisphenol and phthalates exert a significant risk to human health (Liu, Shi, Xie, Dionysiou, & Zhao, 2019).

Intestinal epithelium only allows for particles smaller than 150 m in diameter to penetrate the systemic availability of MPs (less than 0.3%) (EFSA, 2016). Particle size and systemic bioavailability were shown to have an unfavourable relationship in a previous study. Particles less than 100 nm were separated from those larger than 100 nm. It was revealed in the bone marrow and blood that the latter were present. There are macrophages in the lymph nodes and the liver can take up particles bigger than 10 mm (Tomazic-Jezic, Merritt, & Umbreit, 2001). In fish intestines and livers, plastic particles with a diameter of 200 to 600 m were found to be present.

In 13 healthy adults, the gastrointestinal transit time of PEMPs (15 g, 1-2 mm) was examined utilising a meal test. Mechanical stimulation of mucosal receptors was thought to be the cause of the longer transit time through the digestive tract. There is an urgent need for research on the health impacts of MPs and other chemicals found in seafood (Rist et al., 2018).

There is a shortage of literature on NP particles because of difficulties in detecting and analysing them. Additional damage might be caused if these particles penetrate the animal's cellular membranes (Koelmans et al., 2015). Plastic particles might enter the liver, kidneys, brain, and muscles (Wright & Kelly, 2017). Immunotoxicity is a possibility because of MP-immune system interactions. The human brain and epithelial cells were used in an *in vitro* study to examine the effects of NPs/MPs on oxidative stress, which produced negative results (Schirinzi et al., 2017).

Diseases including obesity, infertility, and cancer may arise when MPs are consumed by humans (Sharma & Chatterjee, 2017). Due to their resistance to artificial digestion fluids, intestinal plastic particles may not break down or alter in form or size (Stock et al., 2020). An MP-free gastrointestinal system may therefore be possible. If an individual has a problem with its digestion, the treatment options might vary. Inflammatory gastrointestinal mucosal illnesses such as ulcerative colitis and Crohn's disease may be affected by nano/microparticles, according to preliminary investigations. Nanoparticles were found in ulcerous lesions, although no biodegradable polymers were found. In contrast, NP translocation to the serosa was significantly increased in the human ulcerative intestinal mucosa. As a consequence, the danger of systemic spread has risen. Further investigation into the role of NPs/MPs in disorders of the digestive mucosa is thus required to change their association with certain ailments and impact on human health (Paul et al., 2020) (Table 3).

During experimental study on animals, the gastrointestinal tract was discovered as a possible target region. Lu et al. (2018) found that oral contact to polystyrene MP (0.5 and 50 m) at doses of 100 and 1000 g/L in drinking water for 35 days decreased mucus production in the gut, induced gut microbiota dysbiosis, and altered hepatic lipid metabolism in mice. Similarly, after 6 weeks of exposure to polystyrene MPs (5 m) at concentrations of 100–1000 mg/L in drinking water, an increase in bile acid secretion in the liver and a decrease in mucus secretion in the colon were seen in mice (Jin et al., 2019). Zheng, Wang, Wei, Chang, and Liu (2021) reported that when polystyrene (5 m) was added to drinking water over a 28-day period, it increased intestinal permeability, resulting in acute colitis and lipid disorders in the livers of mice.

Table 3 Threats to human health due to nano and micro plastic intake.

Physical threats	Biological threats	Chemical threats	References
Obstruction and physical damage to the digestive system	Act as a vector in spreading different pathogens for diseases	Endocrine disruption leading to male infertility	Jiang, Kauffman, & Li, 2020 Galloway, 2015
Reduction in the digestive ability of the system leading to starvation	Air pollution	Inflammation and oxidative stress	Yee et al., 2021 Ojinnaka & Mame Marie, 2020
Animal and net entanglement by plastic debris	Contaminated aquatic bodies	Microbiota dysbiosis	Lu et al., 2018
Stenosis in respiratory and gastrointestinal tracts	Disturbance and destruction of natural habitats	Dermatitis due to dermal exposure	Brouwer et al., 2016

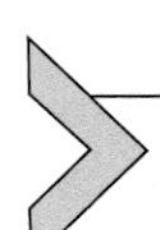

5. Potential toxic impacts of plastic intake on the human health

Owing to the wide scale presence of toxicants in the environment, public concerns are visible over its consumption by aquatic organisms, resulting in overall food chain contamination. In the case of gastrointestinal toxicity, microplastic accumulation by ingestion was studied in fish, scallops, lobsters, mussels, and oysters, resulting in microplastic gastrointestinal toxicities. This accumulation in marine environment is evident in studies although a lag in studies conducted on humans and mammals is also visible. Histological data exhibited a strong inflammatory response on polyethylene ingestion by blue mussel (*Mytilus edulis*). *In vivo* evidence was confirmed regarding inflammatory function, additionally, an increase in the intestinal mucus volume is associated with a reduction in bacteroidetes and proteobacteria at 1000 mg/L, in male (adult) zebrafish inspected for gut microbiota when exposed to 100–1000 mg/L for 14 consecutive days (Chang et al., 2020).

Jin et al. (2018) reported the increased expression of interleukin-1a (IL-1α), IL-1β, and interferons in the gut due to the presence of 0.5 μm polystyrene microplastics presence. This also highlighted the inflammation and microbiota dysbiosis in adult zebrafish gut. Another research explained the intestinal damages in terms of damaged and cracking villi accompanied with enterocyte splitting after exposure to polystyrene particle, polyvinylchloride, polyethylene, and polypropylene (70 μm polyamine) (Jin et al., 2019). Similar damages are also reported in rodents, where male mice on exposure to 5 μm polystyrene for 6 consecutive weeks resulted in its accumulation in the gut resulting in a reduction in intestinal mucus along with damages to the gut barrier function (Esnouf, Latrille, Steyer, & Helias, 2018). Further metabolic disorders are also associated with its consumption in mice. Polystyrene is also linked with gastric adenocarcinoma viability in the cells resulting in induction of inflammatory cell expression including IL-6 and IL-8 with gastric pathology cytokines in gastric adenocarcinoma cells. However, limited evidence is available for *in vitro* mammalian studies (Forte et al., 2016).

After the oral consumption of micro and nano plastics, they are distributed through the blood and lymphatic system to reach the liver. Zebrafish were also identified to have 5 and 70 nm polystyrene accumulated in the gills, liver, and gut structure (Lu et al., 2016). Another study of the liver of freshwater fish in Paris also showed a microplastic average length of 2.41 mm. (Collard et al., 2018). This phenomenon is associated with oxidative stress induction and changes in the metabolic profiles resulting in a disruption in both the lipid framework and energy metabolism (Lu et al., 2016). This oxidative triggering was observed in *Eriocheir sinensis*'s liver. This was also accompanied by acetylcholinesterase (AChE) reduction in the activity along with catalase and alanine aminotransferase in this crab's liver (Yu et al., 2018). Additionally, these plastics reduce the genetic expressions along with antioxidants concentration including catalase, superoxide dismutase, glutathione peroxidase, and glutathione S-transferase. On 5-week consecutive oral exposure to 1000 μg/L polystyrene in mice, hepatic triglyceride and total cholesterol levels reportedly decreased. These research studies explain small bits of assessment and evaluation of microplastic toxicity in the existing beings (Chang et al., 2020).

Since the nanoparticles easily translocate through the gut barrier, with variations depending on their surface charge and size proportions. Studies show that micro- and nano-plastics as an entry into the bloodstream cause nervous and reproductive system dysfunction. Freshwater zebra mussels (*Dreissena polymorpha*) an exposure to (10 and 1 μm) polystyrene microbeads produced a higher concentration of dopamine, concluding that neurotransmitters can eliminate some microplastics (Magni et al., 2018). For *in vitro* studies, polystyrene is associated with increased reactive oxygen species causing oxidative stress in T98G cells (Schirinzi et al., 2017). Similarly, reproductive toxicity by microplastics has also been studied both *in vivo* and *in vitro*. Freshwater *Hydra attenuates* exhibited significant morphological and reproductive system changes on exposure to microplastics. *Daphnia magna* also exhibited severe reproductive changes after treatment with 70 nm polystyrene beads for 21 consecutive days. In oysters, 0.023 mg/L polystyrene exposure significantly reduced the D-larval yield, oocyte number, diameter, and sperm velocity, concluding reproductive disruptions (Sussarellu et al., 2016). Conclusively micro and nano plastics toxicity mainly impacts inflammatory systems, oxidative stress, and disruption in energy metabolism among humans. Further studies are needed to understand the toxic mechanisms (Chang et al., 2020).

Leaching additives of microplastics can also cause toxicity in aquatic animals such as invertebrates, fish, amphibians, carp, zebrafish, and crayfish. This toxic aquatic ecosystem disturbance is a sensitive indicator of water pollution, specifically behaviour malformations changes in the reproductive system, alterations in haematological parameters, instances of acute embryonic toxicity, oxidative stress as well as variations in mitochondrial activities (Sehonova, Svobodova, Dolezelova, Vosmerova, & Faggio, 2018; Stara et al., 2019).

6. Legislations of nano/micro-plastics

In the global landscape of modern society invention of plastic is considered a success however the negative impacts are somewhat neglected threatening the environment, therefore, resulting in a complex societal challenge (Allan et al., 2021). Approximately 15% of the plastic waste is recycled while the majority ends up in a landfill. This demands a regulatory framework aimed at improving the recyclability of plastics. For the European Union safe-by-design: creating nanomaterials of tomorrow concept mainly revolves around the safe usage of nanotechnology across the complete lifecycle, from production to recycling and re-usage. Recent ambitious road map, the European Commission's Green Deal aimed for climate neutral, zero pollution, sustainable, circular, and inclusive economic standards, it was also aimed to develop the new industrial strategy for Europe (Deal, 2018). Additionally, it plays an essential role in implementing United Nations 2030's agenda for sustainable development. This agenda follows up on the 2018 plastic strategy, tackling microplastics, plastics unintentional release from textiles and tires, and the EU's regulatory framework for the development of biodegradable and bio-based plastics. From January 2018, the European strategy for plastics was adopted in the product usage and recycling of plastics in the EU. Higher recycling rates, better recycle its quality were also addressed. The European Commission has restricted microplastics usage for consumer and professional products. Only this restriction can reduce the microplastics amount by about 400,000 tons over 20 years in the EU. Moreover, the regulatory framework for biodegradable properties is encompassed in the plastics strategy that will further minimise the usage of micro and nano plastics (Allan et al., 2021).

In the United States, activities on nano plastics are being monitored by an informal nano plastic interest group comprised of over 20 US agencies. They leverage the lessons learned from nanotechnology and the mechanisms used for collaboration to address the current concerns and prevent any toxicity to the environment and food chain. Major work is being carried out on nanotechnology core facilities in the FDA although there is a need to increase the research on the impact of environmental contamination due to nano plastics. In Canada, several government departments and agencies are collaborating not only on the safe usage of nanotechnology but also on the nano plastic risk assessment (Allan et al., 2021). These agencies include Health Canada, Environment and Climate Change Canada, Agriculture and Agri-food Canada, and the Canadian Food Inspection Agency. The regulatory approach in Canada follows the OECD (Organisation for Economic Cooperation and Development) council recommendation on safety testing and assessment of manufactured nanomaterials. They are regulated under Canadian Environmental Act. In close collaboration with the OECD framework, Canadian agencies are working in a similar regulatory framework. In Chile, researchers have investigated quite a several patents related to plastics microplastics and their worldwide regulation. Circular economic models are sought out for recycling purposes and banning plastic bags. Chile is the first South American country to initiate this ban in 2018. It is also planning to generate a baseline of microplastics and nano plastics in the marine environment in the southern part of Chile (Allan et al., 2021).

7. Future perspectives of nano/micro-plastics

The toxicity from NPs/MPs is mainly influencing the inflammatory system, oxidative stress, and disruption in the energy metabolism of humans however, more studies are needed to exactly understand the toxic mechanisms associated with all these processes. The oral intake of marine products or microplastic contaminated water is the primary route for human exposure. The indirect toxicity in the form of leachable additives and adsorbents also need further research to identify toxicological mechanisms on human health. In addition, collaborative assessment studies should be conducted to identify bioaccumulation of both microplastics and nanoplastics in combination with combined toxicants, including heavy metals. These studies can magnify the impact of toxicants on human health thus, further research is needed to identify the routes.

8. Conclusion

In this chapter, it has been identified potential risk assessments and the toxicological impact of both nano and microplastics on human health. Basic pathways through which humans get exposed to the micro and the nano plastics are also identified and enlisted. Similarly, it has also been narrowed down the basic occurrence of different toxicants by their food chain. Therefore intake and the accumulation of NPs/MPs are summarised and how the accumulation mechanism happens inside the body is briefly explained. The potential toxic impacts recorded by research in different organisms have been reported in the chapter although more research is needed to identify exactly the toxicological mechanisms on human health. Moreover, in the human body, the primary influenced systems include the inflammatory system, oxidative stress management system and disruption in the energy metabolism system owing to the different toxicities reported. Furthermore, the impact of different additives and exorbitance have another indirect toxicological status. Therefore, more studies are needed to consolidate the effect of microplastics, nano plastics, and the joint bioaccumulation with different heavy metals.

Conflict of interest

The authors declare no conflict of interest.

References

Abbasi, S., Soltani, N., Keshavarzi, B., Moore, F., Turner, A., & Hassanaghaei, M. (2018). Microplastics in different tissues of fish and prawn from the Musa estuary, Persian gulf. *Chemosphere*, *205*, 80–87.

Abidli, S., Lahbib, Y., & el Menif, N. T. (2019). Microplastics in commercial molluscs from the lagoon of Bizerte (Northern Tunisia). *Marine Pollution Bulletin*, *142*, 243–252.

Agboola, O. D., & Benson, N. U. (2021). Physisorption and chemisorption mechanisms influencing Micro (Nano) plastics-organic chemical contaminants interactions: A review. *Frontiers in Environmental Science*, *9*, 678574. https://doi.org/10.3389/fenvs.2021.678574.

Akhbarizadeh, R., Dobaradaran, S., Nabipour, I., Tajbakhsh, S., Darabi, A. H., & Spitz, J. (2020). Abundance, composition, and potential intake of microplastics in canned fish. *Marine Pollution Bulletin*, *160*, 111633.

Akindele, E. O., & Alimba, C. G. (2021). Plastic pollution threat in Africa: Current status and implications for aquatic ecosystem health. *Environmental Science and Pollution Research*, *28*(7), 7636–7651.

Akoueson, F., Sheldon, L. M., Danopoulos, E., Morris, S., Hotten, J., Chapman, E., et al. (2020). A preliminary analysis of microplastics in edible versus non-edible tissues from seafood samples. *Environmental Pollution*, *263*, 114452. https://doi.org/10.1016/j.envpol.2020.114452.

Alimba, C. G., & Faggio, C. (2019). Microplastics in the marine environment: Current trends in environmental pollution and mechanisms of toxicological profile. *Environmental Toxicology and Pharmacology*, *68*, 61–74.

Allan, J., Belz, S., Hoeveler, A., Hugas, M., Okuda, H., Patri, A., et al. (2021). Regulatory landscape of nanotechnology and nanoplastics from a global perspective. *Regulatory Toxicology and Pharmacology*, *122*, 104885.

Amereh, F., Babaei, M., Eslami, A., Fazelipour, S., & Rafiee, M. (2020). The emerging risk of exposure to nano (micro) plastics on endocrine disturbance and reproductive toxicity: From a hypothetical scenario to a global public health challenge. *Environmental Pollution*, *261*, 114158.

Anbumani, S., & Kakkar, P. (2018). Ecotoxicological effects of microplastics on biota: A review. *Environmental Science and Pollution Research*, *25*(15), 14373–14396. https://doi.org/10.1007/s11356-018-1999-x.

Arumugasaamy, N., Navarro, J., Kent Leach, J., Kim, P. C. W., & Fisher, J. P. (2019). In vitro models for studying transport across epithelial tissue barriers. *Annals of Biomedical Engineering*, *47*(1), 1–21.

Barboza, L. G. A., Dick Vethaak, A., Lavorante, B. R. B. O., Lundebye, A.-K., & Guilhermino, L. (2018). Marine microplastic debris: An emerging issue for food security, food safety and human health. *Marine Pollution Bulletin*, *133*, 336–348. https://doi.org/10.1016/j.marpolbul.2018.05.047.

Barletta, M., Lima, A. R. A., & Costa, M. F. (2019). Distribution, sources and consequences of nutrients, persistent organic pollutants, metals and microplastics in south American estuaries. *Science of the Total Environment*, *651*, 1199–1218.

Bean, M. J. (1987). Legal strategies for reducing persistent plastics in the marine environment. *Marine Pollution Bulletin*, *18*(6), 357–360.

Blinov, A. V., Siddiqui, S. A., Blinova, A. A., Khramtsov, A. G., Oboturova, N. P., Nagdalian, A.A., et al. (2022). Analysis of the dispersed composition of milk using photon correlation spectroscopy. *Journal of Food Composition and Analysis*, *108*, 104414. https://doi.org/10.1016/j.jfca.2022.104414.

Boots, B., Russell, C. W., & Green, D. S. (2019). Effects of microplastics in soil ecosystems: Above and below ground. *Environmental Science & Technology*, *53*(19), 11496–11506.

Borrelle, S. B., Ringma, J., Law, K. L., Monnahan, C. C., Lebreton, L., McGivern, A., et al. (2020). Predicted growth in plastic waste exceeds efforts to mitigate plastic pollution. *Science*, *369*(6510), 1515–1518.

Bosker, T., Bouwman, L. J., Brun, N. R., Behrens, P., & Vijver, M. G. (2019). Microplastics accumulate on pores in seed capsule and delay germination and root growth of the terrestrial vascular plant Lepidium sativum. *Chemosphere*, *226*, 774–781.

Brennecke, D., Duarte, B., Paiva, F., Caçador, I., & Canning-Clode, J. (2016). Microplastics as vector for heavy metal contamination from the marine environment. *Estuarine, Coastal and Shelf Science*, *178*, 189–195.

Brouwer, D., Spaan, S., Roff, M., Sleeuwenhoek, A. J., Tuinman, I. L., Goede, H., et al. (2016). Occupational dermal exposure to nanoparticles and nano-enabled products: Part 2, exploration of exposure processes and methods of assessment. *International Journal of Hygiene and Environmental Health*, *219*(6), 503–512.

Cai, L., Hu, L., Shi, H., Ye, J., Zhang, Y., & Kim, H. (2018). Effects of inorganic ions and natural organic matter on the aggregation of nanoplastics. *Chemosphere*, *197*, 142–151.

Caruso, G. (2019). Microplastics as vectors of contaminants. *Marine Pollution Bulletin*, *146*, 921–924.

Chang, X., Xue, Y., Li, J., Zou, L., & Tang, M. (2020). Potential health impact of environmental micro-and nanoplastics pollution. *Journal of Applied Toxicology*, *40*(1), 4–15.

Cheung, P. K., & Fok, L. (2016). Evidence of microbeads from personal care product contaminating the sea. *Marine Pollution Bulletin*, *109*(1), 582–585.

Collard, F., Gasperi, J., Gilbert, B., Eppe, G., Azimi, S., Rocher, V., et al. (2018). Anthropogenic particles in the stomach contents and liver of the freshwater fish Squalius cephalus. *Science of the Total Environment*, *643*, 1257–1264.

Cox, K. D., Covernton, G. A., Davies, H. L., Dower, J. F., Juanes, F., & Dudas, S. E. (2019). Human consumption of microplastics. *Environmental Science & Technology*, *53*(12), 7068–7074.

Daniel, D. B., Ashraf, P. M., Thomas, S. N., & Thomson, K. T. (2021). Microplastics in the edible tissues of shellfishes sold for human consumption. *Chemosphere*, *264*, 128554. https://doi.org/10.1016/j.chemosphere.2020.128554.

Danopoulos, E., Jenner, L., Twiddy, M., & Rotchell, J. M. (2020). Microplastic contamination of salt intended for human consumption: A systematic review and meta-analysis. *SN Applied Sciences*, *2*(12), 1950. https://doi.org/10.1007/s42452-020-03749-0.

de Sá, L. C., Oliveira, M., Ribeiro, F., Rocha, T. L., & Futter, M. N. (2018). Studies of the effects of microplastics on aquatic organisms: What do we know and where should we focus our efforts in the future? *Science of the Total Environment*, *645*, 1029–1039.

Deal, G. (2018). *Communication from the commission to the European Parliament, the European council, the council*. the European Economic and Social Committee and the Committee of the Regions.

Dehghani, S., Moore, F., & Akhbarizadeh, R. (2017). Microplastic pollution in deposited urban dust, Tehran metropolis, Iran. *Environmental Science and Pollution Research*, *24*(25), 20360–20371.

Deng, Y., Yan, Z., Zhu, Q., & Zhang, Y. (2020). Tissue accumulation of microplastics and toxic effects: Widespread health risks of microplastics exposure. In *Microplastics in terrestrial environments* (pp. 321–341). Springer.

Dowarah, K., Patchaiyappan, A., Thirunavukkarasu, C., Jayakumar, S., & Devipriya, S. P. (2020). Quantification of microplastics using Nile red in two bivalve species Perna viridis and Meretrix meretrix from three estuaries in Pondicherry, India and microplastic uptake by local communities through bivalve diet. *Marine Pollution Bulletin*, *153*, 110982.

Dris, R., Gasperi, J., Mirande, C., Mandin, C., Guerrouache, M., Langlois, V., et al. (2017). A first overview of textile fibers, including microplastics, in indoor and outdoor environments. *Environmental Pollution*, *221*, 453–458.

Dris, R., Gasperi, J., Rocher, V., Saad, M., Renault, N., & Tassin, B. (2015). Microplastic contamination in an urban area: A case study in greater Paris. *Environmental Chemistry*, *12*(5), 592–599.

Dris, R., Gasperi, J., Saad, M., Mirande, C., & Tassin, B. (2016). Synthetic fibers in atmospheric fallout: A source of microplastics in the environment? *Marine Pollution Bulletin*, *104*(1–2), 290–293.

EFSA Panel on Contaminants in the Food Chain (CONTAM). (2016). Presence of microplastics and nanoplastics in food, with particular focus on seafood. *EFSA Journal*, *14*(6). https://doi.org/10.2903/j.efsa.2016.4501.

Esnouf, A., Latrille, É., Steyer, J.-P., & Helias, A. (2018). Representativeness of environmental impact assessment methods regarding life cycle inventories. *Science of the Total Environment*, *621*, 1264–1271.

Farrell, P., & Nelson, K. (2013). Trophic level transfer of microplastic: Mytilus edulis (L.) to Carcinus maenas (L.). *Environmental Pollution*, *177*. https://doi.org/10.1016/j.envpol.2013.01.046.

Feng, Z., Zhang, T., Li, Y., He, X., Wang, R., Xu, J., et al. (2019). The accumulation of microplastics in fish from an important fish farm and mariculture area, Haizhou Bay, China. *Science of the Total Environment*, *696*, 133948. https://doi.org/10.1016/j.scitotenv.2019.133948.

Ferrante, M., Pietro, Z., Allegui, C., Maria, F., Antonio, C., Pulvirenti, E., et al. (2022). Microplastics in fillets of Mediterranean seafood. A risk assessment study. *Environmental Research*, *204*, 112247. https://doi.org/10.1016/j.envres.2021.112247.

Fiorentino, I., Gualtieri, R., Barbato, V., Mollo, V., Braun, S., Angrisani, A., et al. (2015). Energy independent uptake and release of polystyrene nanoparticles in primary mammalian cell cultures. *Experimental Cell Research*, *330*(2), 240–247.

Forte, M., Iachetta, G., Tussellino, M., Carotenuto, R., Prisco, M., de Falco, M., et al. (2016). Polystyrene nanoparticles internalization in human gastric adenocarcinoma cells. *Toxicology In Vitro*, *31*, 126–136.

Galloway, T. S. (2015). Micro-and nano-plastics and human health. In *Marine anthropogenic litter* (pp. 343–366). Cham: Springer.

Galloway, T. S., Cole, M., & Lewis, C. (2017). Interactions of microplastic debris throughout the marine ecosystem. *Nature Ecology & Evolution*, *1*(5), 1–8.

Gao, F., Li, J., Hu, J., Li, X., & Sun, C. (2020). Occurrence of microplastics carried on Ulva prolifera from the Yellow Sea, China. *Case Studies in Chemical and Environmental Engineering*, *2*, 100054. https://doi.org/10.1016/j.cscee.2020.100054.

Gopinath, P. M., Saranya, V., Vijayakumar, S., et al. (2019). Assessment on interactive prospectives of nanoplastics with plasma proteins and the toxicological impacts of virgin, coronated and environmentally released-nanoplastics. *Scientific Reports*, *9*, 8860. https://doi.org/10.1038/s41598-019-45139-6.

Grigorakis, S., Mason, S. A., & Drouillard, K. G. (2017). Determination of the gut retention of plastic microbeads and microfibers in goldfish (Carassius auratus). *Chemosphere*, *169*, 233–238.

Gündoğdu, S., Çevik, C., & Ataş, N. T. (2020). Stuffed with microplastics: Microplastic occurrence in traditional stuffed mussels sold in the Turkish market. *Food Bioscience*, *37*, 100715. https://doi.org/10.1016/j.fbio.2020.100715.

Gündoğdu, S., Eroldoğan, O. T., Evliyaoğlu, E., Turchini, G. M., & Wu, X. G. (2021). Fish out, plastic in: Global pattern of plastics in commercial fishmeal. *Aquaculture*, *534*, 736316.

Guo, J.-J., Huang, X.-P., Xiang, L., Wang, Y.-Z., Li, Y.-W., Li, H., et al. (2020). Source, migration and toxicology of microplastics in soil. *Environment International*, *137*, 105263.

Guo, X., & Wang, J. (2019). The chemical behaviors of microplastics in marine environment: A review. *Marine Pollution Bulletin*, *142*, 1–14. https://doi.org/10.1016/j.marpolbul.2019.03.019.

Gutow, L., Eckerlebe, A., Giménez, L., & Saborowski, R. (2016). Experimental evaluation of seaweeds as a vector for microplastics into marine food webs. *Environmental Science & Technology*, *50*(2), 915–923. https://doi.org/10.1021/acs.est.5b02431.

Guzzetti, E., Sureda, A., Tejada, S., & Faggio, C. (2018). Microplastic in marine organism: Environmental and toxicological effects. *Environmental Toxicology and Pharmacology, 64*, 164–171.

Hale, R. C., Seeley, M. E., la Guardia, M. J., Mai, L., & Zeng, E. Y. (2020). A global perspective on microplastics. *Journal of Geophysical Research, Oceans, 125*(1), e2018JC014719.

Hantoro, I., Löhr, A. J., van Belleghem, F. G. A. J., Widianarko, B., & Ragas, A. M. J. (2019). Microplastics in coastal areas and seafood: Implications for food safety. *Food Additives & Contaminants: Part A, 36*(5), 674–711.

Jakubowska, M., Białowąs, M., Stankevičiūtė, M., Chomiczewska, A., Pažusienė, J., Jonko-Sobuś, K., et al. (2020). Effects of chronic exposure to microplastics of different polymer types on early life stages of sea trout Salmo trutta. *Science of the Total Environment, 740*, 139922.

Jambeck, J., Geyer, R., Wilcox, C., Siegler, T. R., Perryman, M., Andrady, A., et al. (2015). Marine pollution. Plastic waste inputs from land into the ocean. *Marine Pollution, 347*(6223).

Jiang, B., Kauffman, A., & Li, L. (2020). Health impacts of environmental contamination of micro- and nanoplastics: A review. *Environmental Health and Preventive Medicine, 25*, 29 (2020). https://doi.org/10.1186/s12199-020-00870-9.

Jin, Y., Lu, L., Tu, W., Luo, T., & Fu,Z. (2019). Impacts of polystyrene microplastic on the gut barrier, microbiota and metabolism of mice. Science of the Total Environment, 649, 308–317.

Jin, Y., Xia, J., Pan, Z., Yang, J., Wang, W., & Fu, Z. (2018). Polystyrene microplastics induce microbiota dysbiosis and inflammation in the gut of adult zebrafish. *Environmental Pollution, 235*, 322–329.

Karami, A., Golieskardi, A., Choo, C. K., Larat, V., Karbalaei, S., & Salamatinia, B. (2018). Microplastic and mesoplastic contamination in canned sardines and sprats. *Science of the Total Environment, 612*, 1380–1386. https://doi.org/10.1016/j.scitotenv.2017.09.005.

Karami, A., Golieskardi, A., Ho, Y.b., Larat, V., & Salamatinia, B. (2017). Microplastics in eviscerated flesh and excised organs of dried fish. *Scientific Reports*, 7(1), 1–9.

Karami, A., Golieskardi, A., Keong Choo, C., Larat, V., Galloway, T. S., & Salamatinia, B. (2017). The presence of microplastics in commercial salts from different countries. *Scientific Reports*, 7(1), 1–11.

Karbalaei, S., Hanachi, P., Walker, T. R., & Cole, M. (2018). Occurrence, sources, human health impacts and mitigation of microplastic pollution. *Environmental Science and Pollution Research, 25*(36), 36046–36063.

Kim, J.-S., Lee, H.-J., Kim, S.-K., & Kim, H.-J. (2018). Global pattern of microplastics (MPs) in commercial food-grade salts: Sea salt as an Indicator of seawater MP pollution. *Environmental Science & Technology, 52*(21), 12819–12828. https://doi.org/10.1021/acs.est.8b04180.

Klasios, N., de Frond, H., Miller, E., Sedlak, M., & Rochman, C. M. (2021). Microplastics and other anthropogenic particles are prevalent in mussels from San Francisco Bay, and show no correlation with PAHs. *Environmental Pollution, 271*, 116260. https://doi.org/10.1016/j.envpol.2020.116260.

Koelmans, A. A., Besseling, E., & Shim, W. J. (2015). Nanoplastics in the aquatic environment. Critical review. *Marine Anthropogenic Litter*, 325–340.

Kolandhasamy, P., Su, L., Li, J., Qu, X., Jabeen, K., & Shi, H. (2018). Adherence of microplastics to soft tissue of mussels: A novel way to uptake microplastics beyond ingestion. *Science of the Total Environment, 610*, 635–640.

Kosuth, M., Mason, S. A., & Wattenberg, E.v. (2018). Anthropogenic contamination of tap water, beer, and sea salt. *PLoS One, 13*(4), e0194970.

Lambert, S., & Wagner, M. (2016). Characterisation of nanoplastics during the degradation of polystyrene. *Chemosphere, 145*, 265–268.

Lassen, C., Hansen, S. F., Magnusson, K., Norén, F., Hartmann, N. I. B., Jensen, P. R., et al. (2012). Microplastics-occurrence, effects and sources of. *Significance*, *2*, 2.

Lee, J., Lee, J. S., Jang, Y. C., Hong, S. Y., Shim, W. J., Song, Y. K., et al. (2015). Distribution and size relationships of plastic marine debris on beaches in South Korea. *Archives of Environmental Contamination and Toxicology*, *69*(3), 288–298. https://doi.org/10.1007/s00244-015-0208-x.

Lehner, R., Weder, C., Petri-Fink, A., & Rothen-Rutishauser, B. (2019). Emergence of nanoplastic in the environment and possible impact on human health. *Environmental Science & Technology*, *53*(4), 1748–1765.

Leslie, H. A., Brandsma, S. H., van Velzen, M. J. M., & Vethaak, A. D. (2017). Microplastics en route: Field measurements in the Dutch river delta and Amsterdam canals, wastewater treatment plants, North Sea sediments and biota. *Environment International*, *101*, 133–142.

Li, Q., Feng, Z., Zhang, T., Ma, C., & Shi, H. (2020). Microplastics in the commercial seaweed nori. *Journal of Hazardous Materials*, *388*, 122060. https://doi.org/10.1016/j.jhazmat.2020.122060.

Li, L., Luo, Y., Li, R., Zhou, Q., Peijnenburg, W. J. G. M., Yin, N., et al. (2020). Effective uptake of submicrometre plastics by crop plants via a crack-entry mode. *Nature Sustainability*, *3*(11), 929–937.

Li, J., Qu, X., Su, L., Zhang, W., Yang, D., Kolandhasamy, P., et al. (2016). Microplastics in mussels along the coastal waters of China. *Environmental Pollution*, *214*, 177–184.

Li, L., Zhou, Q., Yin, N., Tu, C., & Luo, Y. (2019). Uptake and accumulation of microplastics in an edible plant. *Chinese Science Bulletin*, *64*(9), 928–934.

Lionetto, F., & Esposito, C. C. (2021). An overview of the sorption studies of contaminants on poly(ethylene terephthalate) microplastics in the marine environment. *Journal of Marine Science and Engineering.*, *9*(4), 445. https://doi.org/10.3390/jmse9040445.

Liu, X., Shi, H., Xie, B., Dionysiou, D. D., & Zhao, Y. (2019). Microplastics as both a sink and a source of bisphenol A in the marine environment. *Environmental Science & Technology*, *53*(17), 10188–10196. https://doi.org/10.1021/acs.est.9b02834.

Lu, L., Wan, Z., Luo, T., Fu, Z., & Jin, Y. (2018). Polystyrene microplastics induce gut microbiota dysbiosis and hepatic lipid metabolism disorder in mice. *Science of the Total Environment*, *631*, 449–458.

Lu, Y., Zhang, Y., Deng, Y., Jiang, W., Zhao, Y., Geng, J., et al. (2016). Uptake and accumulation of polystyrene microplastics in zebrafish (Danio rerio) and toxic effects in liver. *Environmental Science & Technology*, *50*(7), 4054–4060.

Lusher, A., Hollman, P., & Mendoza-Hill, J. (2017). *Microplastics in fisheries and aquaculture: Status of knowledge on their occurrence and implications for aquatic organisms and food safety*. FAO.

Magni, S., Gagné, F., André, C., della Torre, C., Auclair, J., Hanana, H., et al. (2018). Evaluation of uptake and chronic toxicity of virgin polystyrene microbeads in freshwater zebra mussel *Dreissena polymorpha* (Mollusca: Bivalvia). *Science of the Total Environment*, *631*, 778–788.

Mason, S. A., Welch, V. G., & Neratko, J. (2018). Synthetic polymer contamination in bottled water. *Frontiers in Chemistry*, 407.

Mattsson, K., Jocic, S., Doverbratt, I., & Hansson, L.-A. (2018). Nanoplastics in the aquatic environment. *Microplastic Contamination in Aquatic Environments*, 379–399.

Mistri, M., Sfriso, A. A., Casoni, E., Nicoli, M., Vaccaro, C., & Munari, C. (2022). Microplastic accumulation in commercial fish from the Adriatic Sea. *Marine Pollution Bulletin*, *174*, 113279. https://doi.org/10.1016/j.marpolbul.2021.113279.

Murphy, & Margaret. (2017). *Microplastics Expert Workshop Report*.

Ojinnaka, D., & Mame Marie, A. W. (2020). Micro and Nano Plastics: A Consumer Perception Study on the Environment, Food Safety Threat and Control Systems. *Biomedical Journal of Scientific & Technical Research*, *31*(2). BJSTR.MS.ID.005064.

Park, E.-J., Han, J.-S., Park, E.-J., Seong, E., Lee, G.-H., Kim, D.-W., et al. (2020). Repeated-oral dose toxicity of polyethylene microplastics and the possible implications on reproduction and development of the next generation. *Toxicology Letters*, *324*, 75–85.

Parvin, F., Nath, J., Hannan, T., & Tareq, S. M. (2022). Proliferation of microplastics in commercial sea salts from the world longest sea beach of Bangladesh. *Environmental Advances*, 7, 100173. https://doi.org/10.1016/j.envadv.2022.100173.

Patil, S., Bafana, A., Naoghare, P. K., Krishnamurthi, K., & Sivanesan, S. (2021). Environmental prevalence, fate, impacts, and mitigation of microplastics—A critical review on present understanding and future research scope. *Environmental Science and Pollution Research*, *28*(5), 4951–4974.

Paul, M. B., Stock, V., Cara-Carmona, J., Lisicki, E., Shopova, S., Fessard, V., et al. (2020). Micro- and nanoplastics-current state of knowledge with the focus on oral uptake and toxicity. *Nanoscale Advances*, *2*(10), 4350–4367. Royal Society of Chemistry https://doi.org/10.1039/d0na00539h.

Peda, C., Caccamo, L., Fossi, M. C., Gai, F., Andaloro, F., Genovese, L., et al. (2016). Intestinal alterations in European sea bass Dicentrarchus labrax (Linnaeus, 1758) exposed to microplastics: Preliminary results. *Environmental Pollution*, *212*, 251–256.

Peixoto, D., Pinheiro, C., Amorim, J., Oliva-Teles, L., Guilhermino, L., & Vieira, M. N. (2019). Microplastic pollution in commercial salt for human consumption: A review. *Estuarine, Coastal and Shelf Science*, *219*, 161–168.

Piyawardhana, N., Weerathunga, V., Chen, H.-S., Guo, L., Huang, P.-J., Ranatunga, R. R. M. K. P., et al. (2022). Occurrence of microplastics in commercial marine dried fish in Asian countries. *Journal of Hazardous Materials*, *423*, 127093. https://doi.org/10.1016/j.jhazmat.2021.127093.

Prata, J. C. (2018). Airborne microplastics: Consequences to human health? *Environmental Pollution*, *234*, 115–126.

Prata, J. C., da Costa, J. P., Lopes, I., Duarte, A. C., & Rocha-Santos, T. (2020). Environmental exposure to microplastics: An overview on possible human health effects. *Science of the Total Environment*, *702*, 134455.

Promthale, P., Pongtippatee, P., Withyachumnarnkul, B., & Wongprasert, K. (2019). Bioflocs substituted fishmeal feed stimulates immune response and protects shrimp from Vibrio parahaemolyticus infection. *Fish & Shellfish Immunology*, *93*, 1067–1075. https://doi.org/10.1016/j.fsi.2019.07.084.

Prüst, M., Meijer, J., & Westerink, R. H. S. (2020). The plastic brain: Neurotoxicity of micro-and nanoplastics. *Particle and Fibre Toxicology*, *17*(1), 1–16.

Ragusa, A., Svelato, A., Santacroce, C., Catalano, P., Notarstefano, V., Carnevali, O., et al. (2021). Plasticenta: First evidence of microplastics in human placenta. *Environment International*, *146*, 106274.

Reichel, J., Graßmann, J., Knoop, O., Drewes, J. E., & Letzel, T. (2021). Organic contaminants and interactions with Micro- and Nano-plastics in the aqueous environment: Review of analytical methods. *Molecules*, *26*(4), 1164. https://doi.org/10.3390/molecules26041164.

Renzi, M., & Blašković, A. (2018). Litter & microplastics features in table salts from marine origin: Italian versus Croatian brands. *Marine Pollution Bulletin*, *135*, 62–68. https://doi.org/10.1016/j.marpolbul.2018.06.065.

Revel, M., Châtel, A., & Mouneyrac, C. (2018). Micro (nano) plastics: A threat to human health? *Current Opinion in Environmental Science & Health*, *1*, 17–23.

Rist, S., Almroth, B. C., Hartmann, N. B., & Karlsson, T. M. (2018). A critical perspective on early communications concerning human health aspects of microplastics. *Science of the Total Environment*, *626*, 720–726.

Statista. (2021). *Production of plastics worldwide from 1950 to 2019 (in million metric tons)*. https://www.statista.com/statistics/282732/global-production-of-plasticssince-1950/ (accessed 27 February 2022).

Santillo, D., Miller, K., & Johnston, P. (2017). Microplastics as contaminants in commercially important seafood species. *Integrated Environmental Assessment and Management, 13*(3), 516–521.

SAPEA. (2019). A scientific perspective on microplastics in nature and society. In *Evidence review report* (issue 4).

Schirinzi, G. F., Pérez-Pomeda, I., Sanchís, J., Rossini, C., Farré, M., & Barceló, D. (2017). Cytotoxic effects of commonly used nanomaterials and microplastics on cerebral and epithelial human cells. *Environmental Research, 159*, 579–587.

Schwabl, P., Köppel, S., Königshofer, P., Bucsics, T., Trauner, M., Reiberger, T., et al. (2019). Detection of various microplastics in human stool: A prospective case series. *Annals of Internal Medicine, 171*(7), 453–457.

Sehonova, P., Svobodova, Z., Dolezelova, P., Vosmerova, P., & Faggio, C. (2018). Effects of waterborne antidepressants on non-target animals living in the aquatic environment: A review. *Science of the Total Environment, 631*, 789–794.

Seng, N., Lai, S., Fong, J., Saleh, M. F., Cheng, C., Cheok, Z. Y., et al. (2020). Early evidence of microplastics on seagrass and macroalgae. *Marine and Freshwater Research, 71*(8), 922–928. https://doi.org/10.1071/MF19177.

Setälä, O., Magnusson, K., Lehtiniemi, M., & Norén, F. (2016). Distribution and abundance of surface water microlitter in the Baltic Sea: A comparison of two sampling methods. *Marine Pollution Bulletin, 110*(1), 177–183. https://doi.org/10.1016/j.marpolbul.2016.06.065.

Sharma, S., & Chatterjee, S. (2017). Microplastic pollution, a threat to marine ecosystem and human health: A short review. *Environmental Science and Pollution Research, 24*(27), 21530–21547.

Sharma, M. D., Elanjickal, A. I., Mankar, J. S., & Krupadam, R. J. (2020). Assessment of cancer risk of microplastics enriched with polycyclic aromatic hydrocarbons. *Journal of Hazardous Materials, 398*, 122994.

Siddiqui, S. A., Blinov, A. V., Serov, A. V., Gvozdenko, A. A., Kravtsov, A. A., Nagdalian, A. A., et al. (2021). Effect of selenium nanoparticles on germination of Hordéum Vulgáre barley seeds. *Coatings, 11*(7), 862. https://doi.org/10.3390/coatings11070862.

Siddiqui, S. A., Ristow, B., Rahayu, T., Putra, N. S., Widya Yuwono, N., Nisa, K., et al. (2022). Black soldier fly larvae (BSFL) and their affinity for organic waste processing. *Waste Management, 140*, 1–13. https://doi.org/10.1016/j.wasman.2021.12.044.

Sivagami, M., Selvambigai, M., Devan, U., Velangani, A. A. J., Karmegam, N., Biruntha, M., et al. (2021). Extraction of microplastics from commonly used sea salts in India and their toxicological evaluation. *Chemosphere, 263*, 128181. https://doi.org/10.1016/j.chemosphere.2020.128181.

Song, Y. K., Hong, S. H., Jang, M., Han, G. M., Jung, S. W., & Shim, W. J. (2017). Combined effects of UV exposure duration and mechanical abrasion on microplastic fragmentation by polymer type. *Environmental Science & Technology, 51*(8), 4368–4376.

Stara, A., Bellinvia, R., Velisek, J., Strouhova, A., Kouba, A., & Faggio, C. (2019). Acute exposure of common yabby (Cherax destructor) to the neonicotinoid pesticide. *Science of the Total Environment, 665*, 718–723.

Stock, V., Fahrenson, C., Thuenemann, A., Dönmez, M. H., Voss, L., Böhmert, L., et al. (2020). Impact of artificial digestion on the sizes and shapes of microplastic particles. *Food and Chemical Toxicology, 135*, 111010. https://doi.org/10.1016/j.fct.2019.111010.

Strungaru, S.-A., Jijie, R., Nicoara, M., Plavan, G., & Faggio, C. (2019). Micro-(nano) plastics in freshwater ecosystems: Abundance, toxicological impact and quantification methodology. *TrAC Trends in Analytical Chemistry, 110*, 116–128.

Sundbæk, K. B., Koch, I. D. W., Villaro, C. G., Rasmussen, N. S., Holdt, S. L., & Hartmann, N. B. (2018). Sorption of fluorescent polystyrene microplastic particles to edible seaweed Fucus vesiculosus. *Journal of Applied Phycology, 30*(5), 2923–2927. https://doi.org/10.1007/s10811-018-1472-8.

Sussarellu, R., Suquet, M., Thomas, Y., Lambert, C., Fabioux, C., Pernet, M. E. J., et al. (2016). Oyster reproduction is affected by exposure to polystyrene microplastics. *Proceedings of the National Academy of Sciences*, *113*(9), 2430–2435.

Suteja, Y., Atmadipoera, A. S., Riani, E., Nurjaya, I. W., Nugroho, D., & Cordova, M. R. (2021). Spatial and temporal distribution of microplastic in surface water of tropical estuary: Case study in Benoa Bay, Bali, Indonesia. *Marine Pollution Bulletin*, *163*, 111979. https://doi.org/10.1016/j.marpolbul.2021.111979.

Tanaka, K., & Takada, H. (2016). Microplastic fragments and microbeads in digestive tracts of planktivorous fish from urban coastal waters. *Scientific Reports*, *6*(1), 1–8.

Tomazic-Jezic, V. J., Merritt, K., & Umbreit, T. H. (2001). Significance of the type and the size of biomaterial particles on phagocytosis and tissue distribution. *Journal of Biomedical Materials Research*, *55*(4), 523–529. https://doi.org/10.1002/1097-4636(20010615)55:4<523::aid-jbm1045>3.0.co;2-g.

Toussaint, B., Raffael, B., Angers-Loustau, A., Gilliland, D., Kestens, V., Petrillo, M., et al. (2019). Review of micro-and nanoplastic contamination in the food chain. *Food Additives & Contaminants: Part A*, *36*(5), 639–673.

Triebskorn, R., Braunbeck, T., Grummt, T., Hanslik, L., Huppertsberg, S., Jekel, M., et al. (2019). Relevance of nano-and microplastics for freshwater ecosystems: A critical review. *TrAC Trends in Analytical Chemistry*, *110*, 375–392.

Turner, J. T. (2015). Zooplankton fecal pellets, marine snow, phytodetritus and the ocean's biological pump. *Progress in Oceanography*, *130*, 205–248.

van Cauwenberghe, L., & Janssen, C. R. (2014). Microplastics in bivalves cultured for human consumption. *Environmental Pollution*, *193*, 65–70. https://doi.org/10.1016/j.envpol.2014.06.010.

Velzeboer, I., Kwadijk, C. J., & Koelmans, A. A. (2014). Strong sorption of PCBs to nanoplastics, microplastics, carbon nanotubes, and fullerenes. *Environmental Science & Technology*, *48*(9), 4869–4876. https://doi.org/10.1021/es405721v.

Vethaak, A. D., & Leslie, H. A. (2016). *Plastic debris is a human health issue*. ACS Publications.

Vianello, A., Jensen, R. L., Liu, L., & Vollertsen, J. (2019). Simulating human exposure to indoor airborne microplastics using a breathing thermal manikin. *Scientific Reports*, *9*(1), 1–11.

Vinay Kumar, B. N., Löschel, L. A., Imhof, H. K., Löder, M. G. J., & Laforsch, C. (2021). Analysis of microplastics of a broad size range in commercially important mussels by combining FTIR and Raman spectroscopy approaches. *Environmental Pollution*, *269*, 116147. https://doi.org/10.1016/j.envpol.2020.116147.

Wakkaf, T., el Zrelli, R., Kedzierski, M., Balti, R., Shaiek, M., Mansour, L., et al. (2020). Microplastics in edible mussels from a southern Mediterranean lagoon: Preliminary results on seawater-mussel transfer and implications for environmental protection and seafood safety. *Marine Pollution Bulletin*, *158*, 111355.

Wang, F., Zhang, M., Sha, W., Wang, Y., Hao, H., Dou, Y., et al. (2020). Sorption behavior and mechanisms of organic contaminants to Nano and Microplastics. *Molecules*, *25*(8), 1827. https://doi.org/10.3390/molecules25081827.

Wei, L., Wang, D., Aierken, R., Wu, F., Dai, Y., Wang, X., et al. (2022). The prevalence and potential implications of microplastic contamination in marine fishes from Xiamen Bay, China. *Marine Pollution Bulletin*, *174*, 113306. https://doi.org/10.1016/j.marpolbul.2021.113306.

Wilson, D. R., Godley, B. J., Haggar, G. L., Santillo, D., & Sheen, K. L. (2021). The influence of depositional environment on the abundance of microplastic pollution on beaches in the Bristol Channel, UK. *Marine Pollution Bulletin*, *164*, 111997. https://doi.org/10.1016/j.marpolbul.2021.111997.

Wright, S. L., & Kelly, F. J. (2017). Plastic and human health: A micro issue? *Environmental Science & Technology*, *51*(12), 6634–6647.

Xu, M., Halimu, G., Zhang, Q., Song, Y., Fu, X., Li, Y., et al. (2019). Internalization and toxicity: A preliminary study of effects of nanoplastic particles on human lung epithelial cell. *Science of the Total Environment*, *694*, 133794.

Yee, M. S.-L., Hii, L.-W., Looi, C. K., Lim, W.-M., Wong, S.-F., Kok, Y.-Y., et al. (2021). Impact of microplastics and Nanoplastics on human health. *Nanomaterials*, *11*(2). https://doi.org/10.3390/nano11020496.

Yu, P., Liu, Z., Wu, D., Chen, M., Lv, W., & Zhao, Y. (2018). Accumulation of polystyrene microplastics in juvenile Eriocheir sinensis and oxidative stress effects in the liver. *Aquatic Toxicology*, *200*, 28–36.

Zazouli, M., Nejati, H., Hashempour, Y., Dehbandi, R., Nam, V. T., & Fakhri, Y. (2022). Occurrence of microplastics (MPs) in the gastrointestinal tract of fishes: A global systematic review and meta-analysis and meta-regression. *Science of the Total Environment*, *815*, 152743. https://doi.org/10.1016/j.scitotenv.2021.152743.

Zhang, D., Cui, Y., Zhou, H., Jin, C., Yu, X., Xu, Y., et al. (2020). Microplastic pollution in water, sediment, and fish from artificial reefs around the Ma'an archipelago, Shengsi, China. *Science of the Total Environment*, *703*, 134768. https://doi.org/10.1016/j.scitotenv.2019.134768.

Zhang, H., Wang, J., Zhou, B., Zhou, Y., Dai, Z., Zhou, Q., et al. (2018). Enhanced adsorption of oxytetracycline to weathered microplastic polystyrene: Kinetics, isotherms and influencing factors. *Environmental Pollution*, *243*, 1550–1557.

Zhang, Q., Xu, E. G., Li, J., Chen, Q., Ma, L., Zeng, E. Y., et al. (2020). A review of microplastics in table salt, drinking water, and air: Direct human exposure. *Environmental Science & Technology*, *54*(7), 3740–3751.

Zheng, H., Wang, J., Wei, X., Chang, L., & Liu, S. (2021). Proinflammatory properties and lipid disturbance of polystyrene microplastics in the livers of mice with acute colitis. *Science of the Total Environment*, *750*, 143085. https://doi.org/10.1016/j.scitotenv.2020.143085.

CHAPTER TEN

Remediation plan of nano/microplastic toxicity in food

Vandana Chaudhary[a], Neha Thakur[b], Suman Chaudhary[c], and Sneh Punia Bangar[d,*]

[a]Department of Dairy Technology, College of Dairy Science and Technology, Lala Lajpat Rai University of Veterinary and Animal Sciences, Hisar, Haryana, India
[b]Department of Livestock Products Technology, Lala Lajpat Rai University of Veterinary and Animal Sciences, Hisar, Haryana, India
[c]Department of Veterinary Physiology and Biochemistry, Lala Lajpat Rai University of Veterinary and Animal Sciences, Hisar, Haryana, India
[d]Department of Food, Nutrition and Packaging Sciences, Clemson University, Clemson, SC, United States
*Corresponding author: e-mail address: snehpunia69@gmail.com

Contents

Advances in Food and Nutrition Research, Volume 103
ISSN 1043-4526
https://doi.org/10.1016/bs.afnr.2022.07.004

Abstract

Microplastic pollution is causing a stir globally due to its persistent and ubiquitous nature. The scientific collaboration is diligently working on improved, effective, sustainable, and cleaner measures to control the nano/microplastic load in the environment especially wrecking the aquatic habitat. This chapter discusses the challenges encountered in nano/microplastic control and improved technologies like density separation, continuous flow centrifugation, oil extraction protocol, electrostatic separation to extract and quantify the same. Although it is still in the early stages of research, biobased control measures, like meal worms and microbes to degrade microplastics in the environment have been proven effective. Besides the control measures, practical alternatives to microplastics can be developed like core-shell powder, mineral powder, and biobased food packaging systems like edible films and coatings developed using various nanotechnological tools. Lastly, the existing and ideal stage of global regulations is compared, and key research areas are pinpointed. This holistic coverage would enable manufacturers and consumers to reconsider their production and purchase decisions for sustainable development goals.

1. Introduction

Growing concern and research on plastic pollution have engaged the scientists working on the environmental health and safety of nanomaterials. Governments and organizations worldwide have begun to improve plastic recycling and management policies; however, improper plastic disposal is still a global trend resulting in unregulated release to the environment (Barría, Brandts, Tort, Oliveira, & Teles, 2020). Plastics are affected by many factors in the environment, including UV radiation, the mechanical force of water, and biological metabolism, resulting in weathering and fragmentation, leading to smaller plastic particles, nanoplastics (NPs) and microplastics (MPs) (Liu et al., 2021). In the past decade, the production of plastics including polycarbonate (PC), polystyrene (PS), polyethylene (PE), poly(ethylene terephthalate) (PET), and poly(vinyl chloride) (PVC) have increased vigorously, and consequently, nanoplastics and microplastics. Being very small in size, nanoplastics (NPs) that is 1 nm to 1 μm and microplastics (MPs) that is 1 μm to 5 mm remain marine, terrestrial, and freshwater ecosystems (Wang et al., 2021).

Moreover, terrestrial and aquatic biota interact with plastics, spreading concern about negative ecological impacts. Plastics ingested by the organisms can bioaccumulate and reach human beings through trophic transfer in the food chain. Exposure of humans to MPs and NPs may cause oxidative stress, inflammation, immune dysfunction, neurotoxicity, neoplasia, metabolic changes, and energy homeostasis (Prata, da Costa, Lopes, Duarte, & Rocha-Santos, 2020). Considering the environmental burden of plastics, several countries have banned single-use plastics; however, there is no regulatory requirement to prevent plastic contamination in food, mostly due to a lack of analytical techniques for quantification of these materials, data on adverse effects on human health, and contamination mitigation strategies. Today plastic is ubiquitous and human exposure to NPs and MPs through food, water, and the air is inevitable. Therefore, it is, necessary to evaluate the impact of NPs and MPs in food and on human health since nonbiological ultrafine particles can be considered neither inherently safe nor hazardous.

The present chapter highlights the impact of micro and nanosized plastic pollutants on the ecosystem. In addition, a detailed understanding of the control measures for accumulated nano/microplastics in food is brought out. Finally, biobased alternatives to reduce nano/microplastics in food and global regulations and legislature are discussed for future remedial aspects.

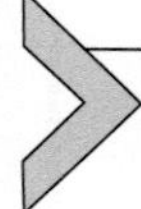

2. Challenges in nanoplastic/microplastic (NP/MPs) remediation in food

Due to the presence of NP/MPs all over the globe and in all kinds of food chains, these are becoming the most challenging classes of micropollutants. Various forms and sizes of NP/MPs, like primary and secondary NP/MPs, make the situation more complex and resistant to biodegradation. This variety further limits the efficiency of sampling, detection, and bioremediation techniques to counter the NP/MPs (Enfrin et al., 2021). The different challenges (Fig. 1) that come in the way to combat with NP/MPs are discussed in this section.

2.1 Actual number of NP/MPs are much more than expected

Nanoplastics/microplastics (NP/MPs) pollution is becoming more widely recognized as a serious environmental problem; yet, the large-scale pattern of MPs in agriculture soils, as well as the environmental consequences, are unknown (Hu et al., 2022). NP/MPs can be formed from the fragmentation of a variety of items including plastics generated from agricultural practices, fishing gear, plastics bottles, taps, lids, bags used in the household, the plastic used for packaging of food items, straws, cigarette butts, industrial plastic, microbeads used in the cosmetic industry as well as other debris coming from the weathering of all of these (Enfrin et al., 2021; Gallo et al., 2018). MPs can be found in all marine parts, occurring on the surface of the sea, beaches,

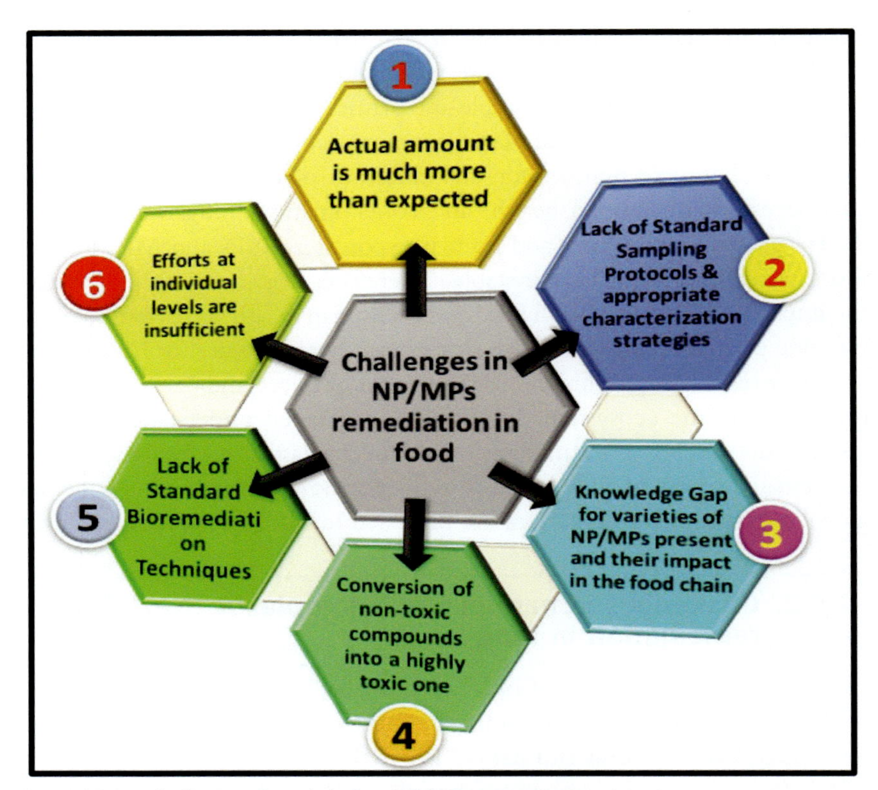

Fig. 1 Major challenges faced during NP/MPs remediation.

seabed, sediments, and in water columns. Thus, the estimation of its quantity by merely observing open ocean will be a wrong prediction or underestimation, as it represents only a fraction of the total input. More than twice the amount of plastic litter in the open ocean is found at the seabed, with half washed up in sea beaches and the other half floating on or under the surface (Andrady, 2011; Gallo et al., 2018). Many academics and industries around the world are interested in learning more about NP/MPs in the environment. The accuracy, efficiency, reliability, robustness, and repeatability of the method(s) utilized to isolate microplastics from ambient media are critical to the success of these studies. However, as microplastics have been discovered in more complex media, several various methods for determining the levels of these pollutants have been devised. Recovery rates are rarely utilized to confirm microplastic procedures, according to a meta-analysis. The most investigated media is sediment, while saline solutions are the most commonly utilized reagent. With the exception of water, all reagents recovered more than 80% of the spiked microplastics. Microplastics could be underestimated by about 14% based on typical recoveries. To assist the standardization, the quality of recovery studies should be improved (Way, Hudson, Williams, & Langley, 2022).

2.1.1 NP/MPs amount is increasing at an exponential rate

The production of plastics has been increasing exponentially since its beginning in the 1950s. Due to mass plastic utilization and manufacture, global production of plastic reached approximately 311 million tons in 2014. It will reach near around 1800 million tons by 2050. In addition, the actual amount of plastics leaks to the oceans on a global scale is unknown. Still, quantitative estimations of all sources, input loads, and originating sectors suggest that it can be around eight million tons annually (Gallo et al., 2018). It has been estimated that the oceans may already have more than 150 million tons, of which about 250,000 tons, broken down into five trillion pieces of plastic be floating at the oceans' surface. The total estimated amount of global plastic in the ocean will be nearly double to 250 million tons by 2025 (Eriksen et al., 2014). Thus, the production of plastics all over the world is following a clear-cut exponential trend. Besides, the microplastics spread at sea-bottom are far more widespread than previously anticipated, with accumulation patterns same as the increasing production (Avio, Gorbi, & Regoli, 2017; Gallo et al., 2018; GESAMP, 2016). Soil is still a neglected area with insufficient NP/MP estimation. There are currently insufficient microplastics monitoring data in soil systems, and standardized analytical methodologies for soil microplastics are absent as well. Microplastics have the potential to alter soil physicochemical qualities and biota, necessitating further research into their occurrence and effects in soils (Wang, Ge, Yu, & Li, 2020). Moreover, to change the quality and properties of plastic items, emerging compounds such as phthalates, bisphenol A (BPA), and polybrominated diphenyl ethers (PBDEs) are frequently utilized. However, as plastics degrade after being released into the environment, the usage of these compounds may make the ensuing plastic products more toxic to flora, wildlife, and humans (Rai, Mehrotra, Priya, Gnansounou, & Sharma, 2021).

2.1.2 Rate of accumulation in different tissues through the food chain is very fast

Generally, plastics persist as it is in the environment for up to hundreds of years in marine conditions, but the fragmentation process fastens the breakage of plastic into NP/MPs in a very short period, further facilitating uptake by marine biota all through the food chain (Gallo et al., 2018). A study reported that polyethylene and polystyrene were found in different tissues, i.e., gill, gills of quagga mussel, digestive gland, intestines, gonad, and muscle of mussels. In addition, the feeding pattern was found independent of the MPs type in mussel (Pedersen et al., 2020; Von Moos, Burkhardt-Holm, & Köhler, 2012; Wang et al., 2020; Wei, Chao-Yang, Rong-Rong, Yan-Yu, & Ai-Li, 2021). Moreover, consumption of seafood contaminated with NP/MPs is one way of accumulation in the human food chain (Revel, Châtel, & Mouneyrac, 2018). It was assumed that only Europeans consumed ~11,000 MPs in the form of bivalves every year. Thus NP/MPs are potentially hazardous to humans (Van Cauwenberghe & Janssen, 2014; Wei et al., 2021).

Furthermore, both the polyethylene and polystyrene accumulation reached the maximum level in different tissues only at a short span of 48 h of exposure in the liver of zebrafish, brine shrimp, and mussels (Lu, Zhang, Deng, Jiang, & Zhao, 2016; Wang, Mao, Zhang, Ding, & Sun, 2019; Wei et al., 2021). Polystyrene also found to regulate the antioxidant enzyme activities in diverse aquatic organisms, including red tilapia liver, zebrafish gut, crab liver, and hepatopancreas, thus confirming disturbance among these organisms (Ding, Zhang, Razanajatovo, Zou, & Zhu, 2018; Qiao et al., 2019; Wang et al., 2021; Yu et al., 2018). Thus, there is a high demand to determine the effect of NP/MPs on the size, shape, and nutritional quality of aquatic organisms to carry out their bioremediation (Wei et al., 2021). If strong preventive measures are not taken immediately, the worst environmental impacts will occur even in a short time. The volume of plastic waste grows in lockstep with demand and population growth in cities. The agroecosystems are endangered by poor postconsumer plastic and microplastic waste management. Microplastic contaminates the agroecosystem and affects the food chain. Possible solutions can be sustainable agroecosystems for safer food production that are available through nonconventional methods. Proper laws attempt can help to decrease further threats of plastic in the food chain and to human health (Okeke et al., 2022).

2.2 Lack of standard sampling protocols and appropriate characterization strategies

It is estimated that about 80% of ocean plastic arise due to improper collection schemes, waste management techniques, and collection infrastructures of municipalities of many countries, while the remaining 20% comes due to careless littering and leakage from the waste management system itself (Andrady, 2017; Gallo et al., 2018). The key challenge is to develop standardized new technologies and methodologies for better extraction, quantification, and characterization of NP/MPs in terms of their different sizes, chemistry, number distributions, levels of surface contamination, and properties, as all these further affects their bioremediation process (Allen et al., 2019; Enfrin et al., 2021; Gasperi et al., 2018; Godoy, Martín-Lara, Calero, & Blázquez, 2019). Secondary microplastics form as a result of bulk plastic fragmentation owing to environmental degradation, while primary microplastics are created specifically for that purpose. The great majority of plastics in the environment are of secondary origin, with primary plastics accounting for between 15% and 31% of total plastics in the environment (Boucher & Friot, 2017).

As NPs/MPs buoyancy and other behavioral changes can be affected due to alteration of density, surface, and bulk properties, which can further affect the bioremediation process in the food chain (Brandon, Goldstein, & Ohman, 2016; Morris, 2016). Unluckily, there are no appropriate analytical methods to detect and quantify nanoplastics in the food chain (Lusher, Hollman, & Mendoza, 2017). Therefore, appropriate characterization techniques with sound base knowledge are required to carry out their bioremediation (Allen et al., 2019). The characterization of microplastics is also challenging due to the different nature of the water systems in which they are present (Fu, Min, Jiang, Li, & Zhang, 2020). Before being released into the environment, plastics may undergo a variety of physical, chemical, and biological modifications, some of which result in the development of nano- or microplastics. Hydrolysis, photooxidation, chemical oxidation, natural organic matter adsorption/attachment, as well as flocculant aggregation are all examples of these processes. Plastic particles may be subjected to photodegradation, hydrolysis, chemical oxidation, biodegradation, mechanical stress, and other processes after being released into the natural environment. Plastics will be exposed to one or more of these weathering pathways at some point during their existence, either simultaneously or sequentially; nevertheless, most microplastics research focuses on just a few of these processes when simulating environmentally relevant systems. Weathering studies must examine the processes that occur both before and after microplastics are released into the environment in order to replicate microplastics that are reflective of those found in the environment (Alimi et al., 2022).

2.3 Lack of standard bioremediation techniques

Microbial remediation is considered a greener option among the various NPs/MPs remediation processes. Enzymatic processes, substrates and cosubstrates concentration, temperature, pH, oxidative stress, and other biotic and abiotic variables influence microbial breakdown of plastics. As a result, it is critical to understand the key mechanisms of the microorganism to use plastic particles as their sole carbon source for growth and development. Starting with the biological and toxicological implications of NPs/MPs, this review critically evaluated the function of various microorganisms and their enzymatic mechanisms involved in biodegradation of NPs/MPs in wastewater (WW) stream, municipal sludge, municipal solid waste (MSW), and composting. Adoption of enzymatic remediation methods should also be explored (Zhou et al., 2022).

To perform bioremediation, there must be effective microbial strains that can degrade the NPs/MPs at the appropriate time period, as there is too much to combat. Although, bacterial-induced degradation of NPs/MPs is still very uncommon, it has to formulate strategies for in situ (by adding microorganisms at the same site of their presence) bioremediation of them to save money, labor and time. In addition, there are not proper culturing techniques and production at the commercial level of efficient strain that already present or tested for bioremediation of NPs/MPs. Therefore, considering the importance of bioengineered solutions, it requires extensive research and development to make them suitable for large-scale applications (Barceló & Pico, 2019; Enfrin et al., 2021).

2.4 Knowledge gap for varieties of NP/MPs present and their impact on the food chain

There is still a lot of confusion about the actual threat posed to aquatic biota due to the ingestion of NPs/MPs and other hazardous chemicals or contaminants associated with them. In some studies, no physical or very less harmful effect have been observed. In contrast, in some others, even a very low amount of endocrine disrupter released from MPs showed potential harm to the aquatic ecosystem, biodiversity, and availability of food (Avio et al., 2017; Bakir, Rowland, & Thompson, 2014; Gallo et al., 2018). For example, in mussels, *Mytilus galloprovincialis*, bioaccumulation of polyaromatic hydrocarbons were recorded in different tissues after their exposure to NPs/MPs. Same way, adsorption of different pollutants like nonylphenol and phenanthrene and additive chemicals (Triclosan and PBDE-47) were observed in tidal flatworms (lugworms) when they were exposed to NP/MPs (Avio et al., 2015; Gallo et al., 2018; GESAMP, 2016).

It is important to note that there is still a huge knowledge gap regarding the ecotoxic effects of long-term exposures of complex mixtures and their relevant concentrations (Gallo et al., 2018). Some compounds, such as alkylphenols, have enough potential to disturb proper male reproductive development leading to feminization in fish and the complete opposite sex in mollusks. In addition, tin-containing plastic stabilizers prompt imposex conditions in gastropods while provoking immunological disorders in fishes (Bergman et al., 2013; Khalid, Aqeel, Noman, Khan, & Akhter, 2021).

The proper bioremediation process of NPs/MPs first needs deep research to understand their abundance, distribution, and hazardous consequences on natural populations of terrestrial and aquatic environments. There are very limited information except for a few laboratory studies regarding the combined toxic effects of NPs/MPs with heavy metals on marine organisms. Therefore, a multidisciplinary approach is needed to characterize risks associated with the absorption of heavy metals on the surface of NPs/MPs, which have the potential to persist for ages in the environment and pose a great threat to lives by accumulating in their bodies (Abbasi et al., 2020; Khalid, Aqeel, et al., 2021; Khalid, Rizvi, et al., 2021).

2.5 Conversion of nontoxic compounds into a highly toxic one

Chemicals of rising concern, which may represent a serious risk to humans and entire ecosystems, are receiving a great scientific attention nowadays. Among these, NP/MP residues are a highly studied group of compounds. NPs/MPs have been frequently proved to be present in the environment and some of them can be hazardous to organisms and accumulate in their tissues. Owing to their enormous specific surface area, microplastics are more prone to attract other chemicals in nature, preventing decomposition. Some of these compounds are toxic and hazardous, and they can be transported through the food chain with microplastics to harm other organisms (Liu, Xu, Ye, & Zhang, 2022).

During the bioremediation process, sometimes microbes transform less toxic compounds into highly toxic or recalcitrant compounds. For example, inorganic compounds of heavy metal mercury, which are poorly hazardous, are converted into methyl mercury by the action of anaerobic bacteria living in water bodies. Methyl mercury is highly toxic and resistant to degradation comparatively (Paranjape & Hall, 2017). Hence, it is necessary to study the bioremediation process in detail at every step.

2.6 Efforts at individual levels are insufficient

As discussed above, there is still high demand to carry out dedicated research to fill the knowledge gaps regarding the effects of NPs/MPs litter in the marine environment, the food chain, and human health (Wagner, Scherer, Alvarez-Muñoz, Brennholt, & Bourrain, 2014). The solution to this alarming situation of NPs/MPs accumulation in the food chain cannot be attained by individual efforts. The effectiveness of various regulatory and policy interventions imposed or proposed on producers or end customers is poorly understood. A comforting option for regulating microplastics pollution is a bottom-up hybrid regulatory approach that includes price-based, right-base legislation and behavioral frameworks based on best practices in microplastic waste management (Deme et al., 2022). That would be a very narrow approach to combat the huge variety of NPs/MPs compounds omnipresent. Hence, it is necessary to combine actions by all, i.e., individuals, families, the scientific, industry, policy makers, trade and civil society, or the general public, to reduce the amount of NPs/MPs flow and the harmful chemicals they release into the aquatic-terrestrial and aerial environment. All have to work together by coming on the same side to save their own existence by saving our environment and earth (Gallo et al., 2018; Heidbreder, Bablok, Drews, & Menzel, 2019).

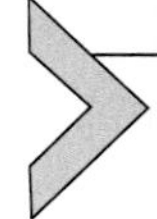

3. Control measures for accumulated nano/microplastics in food

It is a challenging task to control NPs/MPs in food due to multiple factors discussed already. Still, the environmentalists and food technologists suggest certain counteractive measures that can dent the exponential rise of NPs/MPs in food, which range from several micrometers to as fine as few nanometers (Gallo et al., 2018). These measures include using extraction techniques and refining them to suit special needs for NPs/MPs. In addition, bioremediation means like NP/MPs filtering organisms (e.g., mussels) (Barceló & Pico, 2019), digesting organisms (e.g., meal worms) can be explored. A better NP/MP extraction system, identification, and characterization would easily capture the particles and develop a suitable weathering plan for them (Enfrin et al., 2021).

3.1 Extraction and quantification techniques

Control measures largely depend upon the detection techniques being followed. Usually, the tests are run at the testing facility for either one type of plastic or a single fragment, ignoring a vast majority of harmful NPs/MPs originating from food. Additionally, the lack of consistency in the lab and field findings concerning concentrations of NPs/MPs in samples makes the remediation measures difficult. Hence, refined detection and characterization tools become of utmost importance to control the ongoing problem of NPs/MPs toxicity from food packaging due to leaching into the environment and ultimately accumulating in the human body. Thus, the quantification methods are discussed in detail in this subsection of the chapter.

Estimating NPs/MPs load in the environment includes a cascade of steps, i.e., sample collection, isolation and purification. Postseparation, identification, and quantification are made as well. To analyze NPs/MPs density separation using chemicals or enzymes to remove biological debris helps in precise estimation (Strungaru, Jijie, Nicoara, Plavan, & Faggio, 2019). The existing techniques are a sum of visual methods based on morphological characteristics for identification and analytical methods for quantification.

The simplest method of extraction of microplastics is *visual identification*. Visual identification is also known as optical sorting and can be done with the help of the human eye or by a dissection microscope for finer particles. It is based on sorting larger particles depending upon their physical features like shape, size, and density, which is a simple and commonly practiced technique. Additionally, nanoparticles can be optically selected and sorted based on their quantum mechanical features. The optical forces are a reflection of nanoparticles' quantum mechanical properties as well as their optical qualities. Light interacts with nanomaterials, causing not only an energy transfer from photons to electron quantum mechanical motion, but also a momentum transfer between them. Optical forces are generated by changes in photon momentum, which drive the macroscopic mechanical motion of nanoparticles (Fujiwara, Yamauchi, Wada, Ishihara, & Sasaki, 2021). However, separation of finer microplastics can become tricky, and hence, a more precise method is needed.

A solution to this problem is a reliable method based on varying densities of the substances known as *density separation*. In this method, a liquid of known and intermediate density is used to float or sink particles lower or higher in densities than it. Most commonly, a brine solution of known density is used as the liquid to separate sediments. To separate high-density polymers, the brine density should be increased to avoid sinking toxic plastics into the brine, which later might be discarded into the environment if not separated. Advanced density separation methods with separation of high and low-density polymers in a single step are also available (Quinn, Murphy, & Ewins, 2017). Once the extraction protocol is complete, the floating fraction containing different NPs/MPs is separated from the intermediate solution using various procedures like centrifugation or elutriation

(Bellasi, Binda, Pozzi, Boldrocchi, & Bettinetti, 2021). Due to their form, shape, and large surface, density separation using water can recover some types of plastics (PE, PP) from soil samples or fibers from sediments. Since it can obtain slightly greater densities with a higher extraction effectiveness for bulkier polymers, NaCl has been the most popular salt for density separation (high-density polyethylene). Superb recovery rates of up to 99% can be achieved with the right separating solution that is capable of distinguishing the heavier polymers. Separation may also necessitate a single or numerous washings of the sediment, depending on the separation solution chosen (Kundu et al., 2021). Additional characteristics such as cost, sample density, NP/MP size (since this separation is size dependent), etc. are considered when choosing a separation solution (Picó & Barceló, 2021).

Another innovative technique is *oil extraction* that utilizes the oleophilic characteristics of the NPs/MPs. The sample is mixed with oil in a shaker to ensure the mixing of oil and plastic particles. Then this mixture is subjected to funnel extraction and filtering. The filtered mass containing oil and plastic particles is treated chemically to remove oil, and the remaining particles proceed for spectroscopic analysis for quantification. The main advantage of this method is that, unlike density separation, it is not influenced by the organic matter present in the floating fraction. Oils with a short processing time and low cost per sample like canola oil are ideal for this purpose (Crew, Gregory-Eaves, & Ricciardi, 2020). For seven polymers, this approach showed a 90–100% recovery ratio, showing that it is more efficient than density separation in a salt solution. Salt solution separation is more complicated, time-consuming, and expensive than oil extraction. However, oil interferes with Fourier Transform Infrared Spectroscopy (FTIR) during identification, necessitating a wash with 90% ethanol following extraction (Junhao et al., 2021).

Magnetic extraction developed by Grbic et al. (2019), magnetizes polymers and removes them magnetically, taking advantage of their hydrophobic surface. Fe nanoparticles that have been hydrophobized make it easier for them to stick to NPs/MPs. The efficiency of this approach is proportional to the surface area to volume ratio of the particles; for smaller particles, more Fe nanoparticles can bind per unit mass of plastic. Thus, magnetic extraction can help recover tiny NPs/MPs and complement existing approaches.

Depolymerization followed by quantification is another tool to separate NPs/MPs from the sample or food source. Using this technique, PET (poly(ethylene terephthalate) NPs/MPs originating from food can be detected accurately and precisely. The principle of this method is alkaline

PET depolymerization along with phase transfer catalysis. It is followed by oxidation and fractionations. These steps serve two major functions, which are firstly, to eliminate background noise created by interfering species during the estimation and secondly, to preconcentrate the terephthalic acid (TPA) monomer. This TPA is ultimately quantified by HPLC (Castelvetro et al., 2020). This is a novel procedure since it is the first to use a method that allows for the quantification of total mass of PET and polyamide (mainly nylon 6 and nylon 6,6) in complex matrices with high selectivity, accuracy, and sensitivity, as opposed to the traditional method of mechanical separation by flotation and identification by microspectroscopies (Cerasa, Teodori, & Pietrelli, 2021).

Electrostatic properties of plastic particles in food can be exploited for their better extraction from the sample. This is known as *electrostatic precipitation* or electrocoagulation and is a comparatively simple technique with the added advantage of short processing time. Conventional electrostatic precipitators can be modified to suit the needs of NP/MP separation from a given sample by reducing the sample mass (Felsing et al., 2018). Additionally, holistic protocols like using Fenton's oxidation to remove biological material, followed by centrifugation of sample to separate microplastics, dyeing (Nile red), and image processing for quantification, can be useful (Grause, Kuniyasu, Chien, & Inoue, 2021). Electrocoagulation differs from coagulation in that metal ions and hydroxyl ions are generated in situ by electrodes in electrocoagulation. NPs/MPs can be removed from secondary effluents using a mix of coagulation and electrocoagulation with membrane filtering (Estahbanati, Kiendrebeogo, Mostafazadeh, Drogui, & Tyagi, 2021).

3.2 Filter organisms and artificial mussels

Commercial shellfish from coastal regions are a source of public concern due to rising anthropogenic inputs into aquatic ecosystems. As a result, heavy metals play an important role in everything from mussel watch monitoring to human health risk assessment. There has been an influx of documented studies using bivalves for pollution studies in the past, present, and predicted future (Yap et al., 2021).

Bio-indicator organisms in the environment work on the principle of bioaccumulation. These organisms' intake or ingest the NPs/MPs at a rate greater than their rate of elimination. Common indicator organisms include bivalve mollusks, gastropod fish, and amphibians. The uptake of nano/microplastics by these organisms in the aquatic environment depends upon multiple abiotic factors like water salinity, conductivity, pH, temperature, and biotic factors like age, body size, and reproductive status of the bio-accumulating organism. Biomonitoring is an interesting tool for the surveillance of detrimental nano/microplastics in aquatic habitats. The bioindicators commonly include zooplankton, bivalve mollusks, algae, insect, gastropods, macrophytes, fish, and amphibians. They help evaluate actual aquatic metal pollution, bioremediation, toxicology prediction, and research on toxicological mechanisms. Apart from the organisms mentioned above, filter creatures like bivalve mollusks have been thoroughly studied and established as microplastic pollution studies indicators. Filter organisms are the most susceptible to nano/microplastics accumulation in the aquatic environment. Microplastics occur in the soft tissues, faces, and pseudo feces of these organisms. They can play an important role in the bioremediation of NPs/MPs by acting as environmental sentinels (Staichak et al., 2021). The current difficulties are to improve the precision of bivalves as bio-monitors in biomonitoring studies, whether in coastal or freshwater settings. The tunicate *Ciona robusta*, for example, is a useful bioindicator for microplastic and nanoplastic contamination in the ocean because of its ability to filter huge amounts of water and accumulate particles (Valsesia, Jeremie, et al., 2021).

Biological indicators often attract flak of certain communities due to ethical issues, and hence, synthetic indicators offer a long-term solution. In monitoring programs, passive sampling is becoming more widely acknowledged as a method for determining dissolved contaminant concentrations (Morrone, Cappelletti, Tatone, Astoviza, & Colombo, 2021). Artificial mussels (AM) function as synthetic indicators and they are used as a device for passive sampling. AM consists of an impervious Perspex tube whose ends are closed with a semipermeable membrane. Inside this tube are Chelex-100 beads that trap metals (Labuschagne et al., 2020). These AMs can be modified to trap the NPs/MPs of interest. They can play an important role in the surveillance, monitoring, and regulation of nano/microplastics in aquatic, marine, and freshwater habitats.

3.3 Biodegradation

Due to the highly polymerized chemical structure and very durable chemical qualities of plastic trash, it takes about 10–1000 years to degrade in the natural environment, resulting in vast accumulation of plastic waste and a major environmental burden (Luo et al., 2021). The larval forms of certain beetles can digest certain forms of plastic, especially styrofoam (polystyrene), and ultimately reduce it to carbon dioxide. Examples of the insects include Darkling beetle (*Tenebrio molitor, Tenebrio obscures, Zophobas morio*), Lesser wax moth (*Achroia grisella*), Indian meal moth (*Plodia interpunctella*), and Rice moth larvae (*Corcyra cephalonica*) (Liaqat, Hussain, Malik, Aslam, & Mumtaz, 2020). Research on plastics biodegradation with insects has become a popular subject as a way to combat the problem of plastic waste accumulation in the natural environment. Mealworms (*T. molitor*) have been shown to breakdown polystyrene (PS) into lower molecular weight chemicals in recent investigations (Luo et al., 2021). Besides, the polymers degraded by them include polystyrene, polyethylene, LDPE, polyester polyurethane, polyurethane, polystyrene, polyvinylchloride, and polylactic acid. It is noteworthy that the gut microbiota of these insects is the reason behind the biodegradation of plastics. Related insect groups with similar gut microflora can be further studied for a breakthrough in plastic degradation by microbes to better understand the degradation pathways, including enzymes involved. This would help establish synthetic protocols at the industrial level to minimize the toxic effects of NPs/MPs being formed in the environment due to unchecked degradation of commercial polymers. Most studied are the mealworms, larval form of darkling beetles. However, recent studies have shown that the plastic digesting ability of these insects is limited and plastic forms like PE, PVC remain largely undigested by them (Huang, Li, & Lim, 2021). After being digested by the larvae, the thermal stability of plastics deteriorates, and the ionic strengths of their developed characteristic gases drop (Zhu et al., 2021). The potential of microorganisms to degrade petroleum-based plastics like polyethylene, polypropylene and polystyrene can be exploited commercially to develop cost-effective remediation technologies (Wu, Yang, & Criddle, 2017). Studies with germ-free species (i.e., those without a microbiome) are desirable to validate plastic breakdown by macroinvertebrates, as the physiological homeostasis of organisms like *T. molitor* is effected by associated changes in digestive enzyme expression by axenic cultures (Lear et al., 2021).

Another interesting field of the research in this regard is the engineering or manipulation of plastic-degrading enzymes (Zurier & Goddard, 2021). Extracellular and intracellular enzymes are two categories that have been used to classify plastic-degrading enzymes, as well as other enzymes involved in the biological degradation of polymers. Extracellular enzymes, however, are the most studied of the two groups, as they have a wide spectrum of reactivity, ranging from oxidative to hydrolytic functionality. Microbial lactases, peroxidases, lipases, esterases, and cutinases have been found to operate similarly to these varied categories of enzymes (Amobonye, Bhagwat,

Singh, & Pillai, 2021). The enzymes involved in the four steps of plastic breakdown, namely, bio-deterioration, bio-fragmentation, assimilation, and mineralization, need to be understood and explored for their commercial use for effluent treatment. The culturing and characterization of the microbes producing these enzymes is difficult and involves many technicalities. The field of protein engineering for in situ NP/MP degradation is still in its early stage and needs further research. Microbial technology is the safest technology for remediation of NPs/MPs due to the unique makeup of the degrading enzymes (Enfrin et al., 2021). The information on diverse microorganisms with plastic-degrading potentials that has been available so far has been based on pure culture isolates. This clearly shows that the great diversity of microorganisms found in many natural settings has not been fully utilized, with no yeast species discovered as plastic bio-degraders, for example. The use of metagenomics, which allows for the discovery of both culturable and unculturable bacteria, will aid in the identification of microbes and biocatalysts that have the ability to degrade plastics (Amobonye et al., 2021). Future research should focus on isolating and purifying keystone species for plastic degradation, as well as simulating gut conditions with synthetic bacterial communities.

3.4 Bans by government

Regulations by the government can help close the toxicological pathway of nano/microplastics in the environment before it opens. Control of environmental damage can be ensured by taking a proactive stance by the policy makers (Tagg & Labrenz, 2018). Containment, separation, and mitigation are the three pillars of the possible NP/MP control. Containment would include proper disposal, preferably land filling with minimal microplastic leaching. Separation and treatment of sludge would allow reduction at the source and hinder the entry of toxic NP/MP fragments into the environment. Mitigation would include preventive measures to reduce the content of NPs/MPs in circulation. It can be enforced by stringent laws pertaining to littering and substantially funded programs pertaining to raising awareness (Zurier & Goddard, 2021).

3.5 Improved utilization of NP/MPs

Improved utilization of the nano/microplastics in food can be done by advanced waste collection systems, enhanced recyclability of plastics, better traceability of foods, advocating responsible consumer behavior as well as use of biobased plastics discussed later in this chapter (Mwanza, 2021).

Improved recyclability of food packaging is an effective remediation measure. It can be attained by merging the knowledge and findings of advanced research in the field of depolymerization chemistry, separation sciences, food packaging biotechnology, complete economic assessments, as well as characterization sciences (Ragauskas et al., 2021). Green chemistry, solid waste management, and the circular economy are all disciplines of study that, when combined with legislative action, can help to lessen the environmental impact of plastics. As environmental scientists focus on the effects of nanoplastics, it is critical to emphasize the importance of enhancing the plastic life cycle by completely describing the depth and breadth of negative consequences of the plastic pollution disaster, all the way down to the nanoscale (Allan et al., 2021). Public policies or legislation can have a substantial impact on the shift to biodegradable products or climate-friendly behavior. A powerful national-level regulatory regime can influence consumer attitudes, perceptions, and behavior (Adeyanju et al., 2021).

3.6 Improved characterization and recent advances

Newer, nondestructive, and cleaner technologies like dark field hyperspectral microscopy can distinguish between NPs/MPs facilitating improved detection and quantification and even biodistribution in the body. Nanotoxicology of chemically different NPs/MPs can also be effectively studied (Nigamatzyanova & Fakhrullin, 2021). There are two types of detection systems of the NPs/MPs, i.e., particle and mass-based. Recently, the latter has been gaining great attention and include infrared (IR) spectroscopy, near IR spectroscopy, Raman spectroscopy, coherent anti-Stokes Raman scattering (CARS), and stimulated Raman scattering (SRS). Although a combination of mass and particle-based methods paints a better picture, scanning electron microscopy/energy-dispersive X-ray spectroscopy (SEM/EDX) strengthens the findings and helps in better characterization (Ivleva, 2021). Valsesia, Quarato, et al. (2021) developed a method for extracting NPs/MPs from salt-water mussel tissues and analyzing it using Raman spectroscopy. The method combines an enzyme digestion/filtering step to remove the biological matrix with a detection procedure that employs a micro-machined surface with well-defined submicron depths and diameters. The self-assembly of suspended nanoparticles into the cavities is driven by capillary forces in a drying droplet of analyte solution, leaving the individual particles isolated from each other over the surface. When using confocal Raman microscopy to analyze the resulting array, it is possible to explore the individual submicron pNP trapped in the cavities structure on a size-selective basis.

In order to reduce the nano/microplastics in food, one needs to look for alternatives that are harmless and easily degraded in the natural environment, without causing a problem of residues. The following section and subsections (Sections 4.1, 4.2 and 4.3) aim at suggesting alternatives to initially reduce and ultimately eliminate synthetic polymers in the food packaging industry. Hence, while discussing bioremediation, it is of utmost importance to learn about biobased alternatives to beat plastic pollution as prevention is always better than cure.

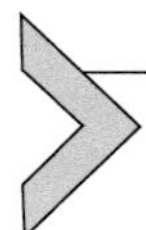

4. Biobased alternatives to reduce nano/microplastics in food

A new generation of packaging, bioplastic packaging, is picking up traction worldwide due to its environmental friendliness and renewable nature. Bioplastic is a term used to epitomize human-made or -processed organic macromolecules sustainable packaging materials contrived from biobased (agricultural and marine) sources and are biodegradable (Abral et al., 2019; European Bioplastics, 2020). Bioplastics are not only acting as a substitute to traditional food packaging but are also opening new avenues for a plethora of new sets of properties (Murariu & Dubois, 2016). These found their main applications in food coating, packaging, encapsulation, and the formation of edible films (Fabra, López-Rubio, & Lagaron, 2016). Generally, biobased plastics are classified into three categories: synthetic biobased plastic polymers orchestrated from bioderived monomers, microbial biobased plastics, and natural plastics directly extracted from biobased raw materials. Fig. 2 shows a schematic categorization of biobased plastics along with examples from each category.

4.1 Synthetic biobased plastic polymers

4.1.1 Polylactic acid (PLA)

Polylactic acid (PLA) is a green, novel, nontoxic, biodegradable, biobased, innovative thermoplastic packaging material (Murariu & Dubois, 2016). This polymer with enhanced strength, transparency, and high elastic modulus can be derived from renewable sources like starch and/or sugar (Carrasco, Pagès, Gámez-Pérez, Santana, & Maspoch, 2010). Its glass transition lies between 50 and 80 °C, and its crystalline melting temperature varies from 130 and 180 °C. PLA exhibits a unique viable, biodegradable substitute to plastics (Mallegni, Phuong, Coltelli, Cinelli, & Lazzeri, 2018).

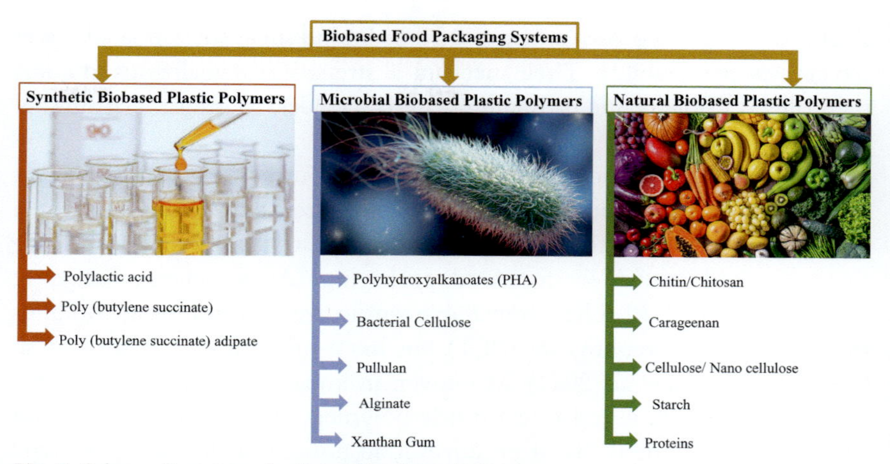

Fig. 2 Schematic categorization of biobased plastics.

PLA is an aliphatic polyester with lactic acid as its basic constitutional unit. Lactic acid is obtained from the fermentation of carbohydrate sources like corn, wheat, potato, or agricultural wastes such as whey, molasses, which can produce lactic acid. However, corn is the most preferred biomaterial as it provides superior quality feedstock for the process of fermentation, thereby resulting in the generation of immaculate, pure lactic acid (Ncube, Ude, Ogunmuyiwa, Zulkifli, & Beas, 2020). It can be produced from lactic acid either by polycondensation reaction or through ring-opening polymerization of lactide monomer.

Generally, bio-decomposition of PLA takes place in approximately 6–24 months, depending on the environmental conditions. To reduce the bio-decomposition time, da Silva, Menezes, Montagna, Lemes, & Passador, (2019) incorporated kraft lignin at different concentrations in PLA composites. They deduced that PLA-lignin composites with 10% lignin were able to reduce bio-decomposition time significantly. Moreover, these composites exhibited comparable tensile strength to that of pure PLA packaging. In an experiment by Gordobil, Delucis, Egüés, and Labidi (2015), it was indicated that lignin imbibition in PLA composites leads to escalation in thermal stability but was responsible for reducing the degree of crystallinity because lignin hampers maneuverability of PLA chains in the course of crystallization. As per the fact that the degradation initiates during the amorphous phase and expedites to the crystalline phase, it may be held responsible for the bio-decomposition of the PLA-lignin composite. PLA multilayer films with gelatin supplemented with an extract from almond shells substantiated lower oxygen gas penetrability. Their mechanical strength was analogous to commercially available plastic packaging material (Valdés, Martínez, Garrigos, & Jimenez, 2021).

PLA fibers display low odor retention and outstanding water-resistant characteristics. In addition, due to its fat and oil resistance and good aroma barrier characteristics, these are popular raw materials for manufacturing thermoformed containers for food packaging (Auras, Harte, Selke, & Hernandez, 2003). PLA has comparable tensile strength and elastic modulus to polyethylene terephthalate (PET) but has a much lower elongation at break (Tamburini et al., 2021). Moreover, its impact strength is comparable to that of polystyrene (relatively brittle polymer). Another drawback with PLA food packaging is that it produces loud noise, which is not perceived as a desirable property (Diaz, Pao, & Kim, 2016). Zych et al. (2021) found that plasticization of PLA with epoxidized soybean oil methyl ester achieved elongation at a break of nearly 800%. Further to this, these films recorded significantly lower noise compared to the packaging material of pure PLA.

4.1.2 Polybutylene succinate and polybutylene succinate adipate

Polybutylene succinate (PBS) is a biodegradable aliphatic polyester derived from the polycondensation reaction of succinic acid and 1,4-butanediol (Hirotsu, Toshiyuki, Masuda, & Nayakama, 2000). PBS has various advantages, including heat proof and well-balanced mechanical properties, which are useful in a variety of applications (Zeng et al., 2009). PBS melts at a temperature of 115 °C, lower than that of PLA. Because of its lower melting temperature, it takes lesser processing time and very well blends with other packaging materials. PBS's biodegradability is an alluring aspect that can be used for single-use packaging. It is also available in direct food contact grades (Rudnik, 2019). Due to superior fat transfer resistance at elevated temperatures, PBS is a reasonable choice against petroleum-based polymers and perfluorinated chemicals (Thurber & Curtzwiler, 2020). A blend of PBS, glycerol, starch from tapioca and empty fruit bunch fiber enhanced the flexibility and tenacity. This may be attributed to the effective load transfer mechanism by the incorporated fibers. However, a bit higher loss of mass was observed for initial stages of thermal decomposition (Ayu et al., 2020). An active packaging film was developed by incorporating quercetin in PBS matrix. A remarkable effect on color, ultraviolet blocking and opacity was noted although there was a marginal decline in mechanical characteristics. In addition, the films disclosed extraordinary free radical eradication and antibacterial pursuit (Łopusiewicz, Zdanowicz, Macieja, Kowalczyk, & Bartkowiak, 2021). A blend of PBS and polybutylene adipate terephthalate with diversified proportions was prepared by de Matos Costa et al. (2020). The blended films with 25% PBS revealed significant values of elastic modulus and deformation at break that is 135 MPa and 390%, respectively. These displayed a nice combination of satisfactory barrier attributes, compostability and mechanical properties and can be used successfully in place of petro based packaging. Polybutylene succinate adipate (PBSA) is synthesized by adding adipic acid to source materials during PBS synthesis. PBS has a higher crystallinity and is better suited for molding, whereas PBSA has a lower crystallinity and is better suited for film applications. In an experiment by Vytejckova et al. (2017), it was demonstrated that even though the mechanical, physical and chemical attributes of generally used polyamide or polyethylene and innovative PBS or PBSA based packaging materials were not comparable, then also PBS based films could be used without for packing poultry products without compromising on the storage life and quality.

4.2 Microbial biobased plastic polymers

4.2.1 Polyhydroxyalkanoates (PHA)

Polyhydroxyalkanoates (PHA) are biobased, biodegradable, compostable biopolyesters produced by fermentation of excess carbon-based feedstocks by bacteria like *Bacillus* sp., and *Cupriavidus necator* (Sen & Baidurah, 2021). PHAs can be generated in three ways using various carbon sources: Pathway I is primarily utilized by organisms like *C. necator* and *Bacillus* sp. that produce polyhydroxy butyrate (PHB), whereas pathways II and III are found in *Pseudomonas* sp. that produces mcl-PHA (Saratale et al., 2020). PHAs are distinguished based on the number of monomeric carbons, with short-chain length including 3–6 carbons, medium-chain length encompassing 7–16 carbons, and long-chain length with more than 16 carbons. The type of organism and the carbon feedstock utilized for growth decides the monomer composition, structure of macromolecule, and the physical and chemical properties of the PHA polymer. Approximately 150 PHA monomers have been identified so far. Depending on the kind of application, the properties of PHA polymer can be customized by incorporating different polymers into the polymer chain. Besides, the variability in chain length, many PHA with different functional groups such as halogens and aromatic groups have been successfully fabricated (Tortajada, da Silva, & Prieto, 2013).

The first PHA discovered was polyhydroxy butyrate. PHB and its copolymer with polyhydroxy valerate are perhaps the most commonly used PHA packaging resins. PHB displays extraordinary mechanical properties that are only rivaled by polypropylene and polyethylene. Its superior mechanical properties, high elastic modulus, and tensile strength with superb moisture and gas barrier properties mark its suitability for food packaging (Lim, You, Li, & Li, 2017). Despite some desirable attributes, its practical industrial application is limited because of innate aging due to secondary crystallization leading to brittleness with time (Chen, 2009), development of big spherulites due to sluggish crystallization results in high fracturability (El-Hadi, Schnabel, Straube, Muller, & Henning, 2002), thermally instable (Ariffin, Nishida, Shirai, & Hassan, 2008) and increased cost of production (Wang, Yin, & Chen, 2014).

4.2.2 Bacterial cellulose

One of the sustainable, renewable sources of bioplastic is bacterial cellulose. It is a linear and unbranched microbial polymer generated as an exopolysaccharide by certain bacteria like *Acetobacter*, *Gluconacetobacter*, *Sarcina*, *Agrobacterium*, etc. Even though its molecular formula resembles that of plant cellulose, it lacks pectin, lignin, and hemicellulose, which are present as the main constituent in plant cellulose. The absence of these complex carbohydrates makes its isolation and purification process simpler and low energy task compared to plant-derived cellulose, which includes the usage of toxic chemicals (Huang et al., 2014). It is made up of ultrafine nanofibrils that lead to the development of a three-dimensional mesh-like structure that is further stabilized by inter and intramolecular hydrogen bonding (Tsouko et al., 2015). Bacterial cellulose illustrates superior characteristics comparing to plant-derived cellulose in terms of surpassing fiber content, degree of crystallinity, and tensile strength (Shah, Ul-Islam, Khattak, & Park, 2013). Scientists have made tremendous efforts to develop highly functional bacterial cellulose composites with tailor-made characteristics (Padrão et al., 2016).

It was deduced that bacterial cellulose functionalization could be achieved by in situ and ex situ processes (Cacicedo et al., 2016). In situ approach is the most commonly used strategy for biopolymer composite synthesis, which involves adding reinforcing materials such as agar, sodium alginate, starch, montmorillonite carboxymethylcellulose, etc. initially to the culture medium. The supremacy of this process lies in its simplicity. In addition, the added compounds become an integral part of the 3D fibril structure, conferring stability and favorable characteristics to the composite.

A major constraint of this process is that these added compounds may be insoluble in culture media and even inhibit the growth of bacteria (Lin et al., 2016). Ex situ modification strategy, however, follows the bacterial cellulose manufacturing process. The bioactive materials are impregnated into a porous, nanofibrillar bacterial cellulose matrix in this postmanufacturing process. The precedence of this process is the ability to use bioactive compounds and preserve the native structure of bacterial cellulose. One major drawback is that only nanoparticles can enter the bacterial cellulose pores (Andriani, Apriyana, & Karina, 2020). Several permutations and combinations are possible to meet the needs of the food packaging industry due to the chemical reactivity of bacterial cellulose due to hydroxyl groups.

In combination with silver nanoparticles, bacterial cellulose, when dispersed in chitosan produced nanocomposite films, leads to improve tensile and barrier characteristics (Salari, Khiabani, Mokarram, Ghanbarzadeh, & Kafil, 2018). Augmented optical, mechanical, and biodegradability could be achieved by combining bacterial cellulose with carboxymethylcellulose (Bandyopadhyay, Saha, Brodnjak, & Saha, 2018). In conclusion, bacterial cellulose biobased packaging is currently one of the most dynamically developing trends in the food industry, with positive effects on the natural environment, human health, and the quality of stored food products. However, many challenges remain for low-cost commercial production of BNC-based packaging materials, such as low yield of known bacterial nanocellulose strains and relatively high operating costs (e.g., expensive culture media, bioactive agents), particularly when compared to synthetic alternatives.

4.2.3 Pullulan

Pullulan is a water-soluble, edible, nonmutagenic, nonionic, commercially available exopolysaccharide derived from the fermentation medium resembling yeast *Aureobasidium pullulans*. As a packaging material, pullulan can be fabricated into odorless, thin, tasteless, and transparent packaging material (Kraśniewska, Pobiega, & Gniewosz, 2019). It presents a linear, unbranched polymer composed of maltotriose monomeric units bonded by α-(1-6)-glycosidic bonds. Its distinctive properties like pliability, elasticity, and solubility are governed by a single-linkage pattern (Chen, Wu, & Pan, 2012; Tong, Xiao, & Lim, 2013); in contrast, the hydroxyl groups are obligated for its barrier properties (Cazzolino, Campanella, Türe, Olsson, & Farris, 2016). Besides, its stability in aqueous solutions over a wide pH range, it has low viscosity compared to other polysaccharides, and exhibits good oxygen barrier properties in films and coatings (Singh & Saini, 2008). Because of its high cost, its use in food packaging applications is hampered. As a result, pullulan is mingled with other biopolymers for cost reduction and improving its material properties.

The evolution of pullulan and apple composite films had better tensile strength and elasticity at P value <0.05 when compared to pure pullulan films (Luís, Ramos, & Domingues, 2021). Pullulan and whey protein isolate biocomposite films containing nano silicon dioxide at different concentrations developed by Hassannia-Kolaee, Khodaiyan, Pourahmad, and Shahabi-Ghahfarrokhi (2016) were reported to have enhanced tensile strength and declined elongation at break values. Besides, an increment in water solubility, absorbability of water, and moisture content were illustrated with an increase in nano silicon dioxide concentration that may be attributed to the high tenacity of the biocomposite matrix. Pullulan alkyl esters were synthesized with varying degrees of substitution and carboxylic anhydrides. Films made from pullulan esters demonstrated maximal barrier efficiency against oxygen and moisture, proving their suitability for extending the shelf life of packed food products (Niu, Shao, Chen, & Sun, 2019). Bio-nanocomposite packaging developed by intermixing pullulan with cellulose nanofibers confirmed improved tenacity by 60% and thermal stability. Water vapor and oxygen transmission rates diminished by 32% and 38%, respectively, with the addition of cellulose nanofibers (Zhou et al., 2021). Its proven safety record as a nature-friendly and biocompatible biopolymer has gained widespread regulatory recognition. According to the United States Food and Drug Administration, pullulan falls in the "generally recognized as safe (GRAS)" category (FDA, 2002).

4.2.4 Alginate

Alginates are structural polysaccharides obtained from brown algae (*Phaeophyceae*) and some bacteria like *Pseudomonas* and *Azotobacter*. Alginate is composed of a monomeric unit consisting of (1,4)-linked-D-mannuronic acid (M) and -L-guluronic acid (G) residues (Puscaselu, Gutt, & Amariei, 2020). The various combinations of M and G blocks result in at least 200 different alginates (Andriamanantoanina & Rinaudo, 2010). Alginates are reported to have low toxicity, compatibility with living organisms, and environment-friendly, superior film-forming ability but are highly hydrophilic, leading to amplification in water vapor transmission rate (Venkatesan et al., 2015). To overcome this shortcoming, cellulose or montmorillonite nanoparticles are integrated into their matrix (Jost, Kobsik, Schmid, & Noller, 2014). In addition to this, alginates are tasteless and odorless (Kontominas, 2020). Various other biocompatible compounds are added or incorporated to improve the characteristics of alginate-based packaging.

The examples include addition of microfibrillated cellulose and calcium chloride for good mechanical properties, addition of calcium chloride for declining water vapor transmission rate and addition of glycerol or sorbitol for enhancing flexibility (Theagarajan, Dutta, Moses, & Anandha Ramakrishnan, 2019). A composite film of alginate, silver and nanocrystals of cellulose as fillers was developed by Yadav, Liu, and Chiu (2019). On comparison with native alginate films, the tensile modulus of composite films rose by 39–57% and water vapor permeability dropped by 17–36%. These films depicted better UV and water barrier characteristics.

4.2.5 Xanthan gum

Xanthan Gum (XG) is a high molecular weight, extracellular polysaccharide synthesized by fermentation of carbohydrates by bacteria *Xanthomonas campestris*. The basic monomeric unit of XG is D-glucose units with a trisaccharide side chain. Two mannose units of the side chain are differentiated by guluronic acid. Molecule of XG when dissolved in polar solvents like water flaunts considerable charged groups, which further is linked to the ionic strength of the solution. By raising the ionic strength of solution, branching chains of XG can be tucked together with the main chain resulting in more coherent structure that have soared glass transition temperature (Ahmad, Mustafa, & Che Man, 2014). When in contact with water, xanthan gum displays remarkable endurance against a wide range of pH variations, acid, and alkalis by the toughening and insulating effect developed by anionic trisaccharide side chains (Balasubramanian, Kim, & Lee, 2018). Because of its superb properties, such as nontoxicity and nonsensitizing effect, XG is considered as a key natural and industrial biopolymer. As a result, the USFDA has given its acceptance as a food additive in the food sector without any limits. It was also authorized as a safe food additive and so registered in the Code of Federal Regulations as a stabilizer and emulsifier (CFR). Xanthan gum when combined with starch revamped their tensile strength but was not able to improve its water vapor permeation (Sapper, Talens, & Chiralt, 2019). Particularly, XG applications are not only limited to the food packaging or processing, but also includes treatment of water, prevention of corrosion, agriculture sector, medical field, etc. However, constraints like as reduced surface area, inferior physical properties, heat stability, and microbial activity can make it unsuitable for some applications. As a result, many endeavors have been made to alter XG by employing diverse transformation procedures in order to improve its mechanical, physical and chemical properties (Abu Elella et al., 2021).

4.3 Natural biobased plastic polymers

4.3.1 Chitin/chitosan

Chitin is the second most abundant natural polymer after cellulose. Chitin, is derived from exoskeletons of crustaceans, has emerged as a promising alternative to petroleum-based packaging materials in the food packaging industry (Barikani, Oliaei, Seddiqi, & Honarkar, 2014). Chitin composition is identical to cellulose except for an acetamide group on the alpha carbon atom instead of the secondary hydroxyl group in the cellulose molecule. It is utilized for the generation of chitosan by alkali deacetylation (Kumar, Ye, Dobretsov, & Dutta, 2019).

Chitosan is a soluble form of chitin and it is commonly applied to the packaging industry due to its biodegradability, nontoxicity, biocompatibility, low cost, and abundant natural availability (Kausar, 2017). Properties of chitosan are mainly dependent on acetylation, which may vary from 0% to 70%. The existence of the nonpolar acetyl group imparts hydrophobic characteristic property to chitosan (Nilsen-Nygaard, Strand, Varum, Draget, & Nordgard, 2015). Chitosan-based packaging films have been shown to exhibit quite good mechanical properties and are also less permeable to gases. However, natural polymers display a higher affinity for moisture, thereby illustrating increased water vapor permeability (Elsabee & Abdou, 2013). Numerous strategies have been employed to improve the characteristics and properties of pristine polymers by amalgamating bioactive substances or as blends with other natural biopolymers. Muthulakshmi et al., (2021) concluded that pores present in the composite of chitosan and microbial-derived extracellular polymeric substances allow a continuous exchange of gases. Still, they were able to minimize the moisture transfer, improving its applicability for food packaging. The moisture barrier, mechanical and optical properties of chitosan and rice starch film improved by ultrasonic treatment. It was concluded that the formation of cross-linkages by rice starch prevented the inside seepage of water in composite films. In addition, tensile strength and elongation at break also improved (Brodnjak & Todorova, 2018).

Bioplastics deduced from a combination of chitosan, montmorillonite, and ginger essential oil exhibited quite good oxygen barrier properties, thereby retarding oxidation in foods rich in unsaturated fats (Souza et al., 2019). Woranuch and Yoksan (2013) demonstrated that the inclusion of eugenol loaded chitosan nanoparticles in thermoplastic flour improved moisture barrier properties and flaunted superior antioxidant activity. Collagen has piqued the interest of researchers as a potential replacement for synthetic polymers. The developed composite films with chitosan unveiled superb thermal stability, compatibility, and adhesion (Ahmad, Nirmal, Danish, Chuprom, & Jafarzedeh, 2016). Increased tensile strength, ultraviolet barrier properties, and decreased water vapor permeability were reported in biodegradable chitosan and gelatin biocomposite packaging films (Ahmed & Ikram, 2016). When blended with different protein/carbohydrates, essential oils, etc., it can be assumed that chitosan offers a variety of preprogrammed characteristics imparting desirable attributes to the biobased packaging.

4.3.2 Carrageenan

Carrageenan is high molecular weight biopolymers extracted from the cell walls of the *Rhodophyceae* family of seaweeds. These are water-soluble, extremely flexible, linear sulfated galactan polysaccharides with spiral helical structures and are also present in the cavities of the cellulose network in the plants (Abdou & Sorour, 2014). These spiral helical confirmations are responsible for the formation of different types of gels at room temperature. It usually occurs in two forms that are native and degraded. Carrageenan is commercially used as a food additive in the food processing industry as stabilizers, gelling agents, thickeners, etc. Still, owing to its inherent properties, it is acknowledged as the base material for the production of biobased packaging material (Martiny et al., 2020). In addition, bio-nano composite films of carrageenan have elevated biodegradability index, which can mitigate environmental impacts (Aga et al., 2021).

The packaging material devised from carrageenan has good gas barrier properties but showcases poor water barrier characteristics, restraining their use in food packaging (Alves et al., 2011). Hence, in an attempt to optimize the barrier properties of the carrageenan matrix, it is often intermingled with other biopolymers. A composite film of kappa-carrageenan and polyvinyl alcohol positively altered the water vapor transmissibility, tensile strength, bursting ability, and water solubility of the film (Irianto, Agusman, Fransiska, & Musfira, 2019). Water vapor permeability of carrageenan films impregnated with olive leaves extract was significantly reduced by 54%; moreover, the films exhibited more flexibility (Martiny et al., 2020). In agreement with the previous study, a steady refinement was noticed in mechanical properties, and water vapor permeability of *Ipomoea batatas* and kappa-carrageenan blended films (Bharti et al., 2021).

4.3.3 Cellulose/nano cellulose

Cellulose is an organic, sustainable, structural polysaccharide composed of D-glucose monomers joined together with β-1,4 glycoside bonds. It is the most abundant biopolymer on earth. Cellulose can be derived from multiple bio-sources like wood, cotton, hemp, agricultural by-products, etc. (Bharimalla, Deshmukh, Vigneshwaran, Patil, & Prasad, 2016). The elementary organization of cellulose is fabricated from microsized string-like microfibers that are further made up of nanosized microfibrils (Yu, Ji, Bourg, Bilgen, & Meredith, 2020). Succinctly aside from the manufacturing of paper, cellulose without alteration has very few applications in the food packaging industry. Although the enzymatic, chemical, or mechanical modification of cellulose leads to carboxymethyl cellulose, methyl cellulose, hydroxypropyl cellulose which can be judiciously used as a coating material for food packaging applications. Being nontoxic and having exceptional strength to weight ratio marks it a preferred choice for food packaging material (Stenstad, Andresen, Tanem, & Stenius, 2007). Cellophane, a regenerated and bioresorbable form of cellulose, has been preferred as a food packaging material for a long time due to its superior impediment capability against fats, bacteria, and gas (Petersen et al., 1999). Cellulose-based film fabricated from delignified banana stem fibers by Ai et al. (2021) demonstrated higher gas and moisture permeability than commercially available polyethylene. The enhanced permeability helped in promoting the ethylene release, thereby outstretching the storage period of mangoes.

To make the most of benefits from cellulose, its strands can be grouped together to develop highly organized structures, which can then be segregated as nanoparticles termed as nanocellulose (Foster et al., 2018). Nanocellulose express a number of unique attributes, including low density, greater flexibility, and resilience while remaining relatively inert (Lavoine & Bergström, 2017) as well as thermal properties (Gan, Sam, Abdullah, & Omar, 2020). It was also recommended by Gan et al. (2020) that the incorporation of nanosize filler in nanocellulose-reinforced composites gives them remarkable capabilities, making them suitable candidates to reinstate traditional synthetic polymer composites. Nanocellulose is considered as a futuristic material for food packaging due to its admirable rigidity equivalent to that of polyethylene terephthalate lower oxygen portability comparable to that of ethylene vinyl alcohol. Regardless of better mechanical and gas permeability characteristics, nanocellulose has a very high affinity for water. Conducive to enhancing the effectiveness of the aforementioned polymers to a level appropriate for food packaging applications development of nanocomposites is considered an effective alternative (Silva, Dourado, Gama, & Poças, 2020). Several research studies have indicated that nano-structured cellulose fibers if used as a strengthening material in composite packaging, a considerable improvement in gas permeation property, thermal stability, and biodegradability may be observed (Abdul Khalil et al., 2016). Ghaderi, Mousavi, Yousefi, and Labbafi (2014) illustrated that nanocomposites formed from cellulose nanofibers extracted from the sugarcane bagasse with polylactic acid ameliorated the water vapor permeability. Pan, Li, and Tao (2020) substantiated that fish gelatin film, when reinforced with microcrystalline cellulose, had tenacity and elasticity values higher than those of pure films while the elongation at break was lower.

4.3.4 Starch

Starches are widely available polysaccharides and one of the most affordable groups of biodegradable polymers. These are also referred to as hydrocolloid biopolymers. Biopolymers are made from various starches, including rice, potato, corn, cassava, tapioca, and others (Thakur et al., 2018). Starch-based packaging is promising due to desirable attributes like flexibility, transparency, environment friendly, and low cost (Suput, Lazić, Popović, & Hromiš, 2015). Due to the firm configuration of polysaccharide molecules, they block the diffusion of oxygen and carbon dioxide gases. However, these biomolecules are prone to water transmission through films; therefore, they are coupled with lipids or other biopolymers for counteracting this limitation. Composite biofilms contrived with cassava, and pinhão thermoplastic starch, compostable polyester polybutylene adipate co-terephthalate, green tea, and rosemary extracts helped improve water vapor permeability (Muller, Carpiné, Yamashita, & Waszczynskyj, 2020).

4.3.5 Proteins

Proteins used for film-forming substances are derived from renewable sources and can easily be degraded compared to plastic analogs. Proteins are macromolecules having precise amino acid sequences joined by amide linkage and molecular arrangement, which can be degraded by proteases (Clarinval & Halleux, 2005). They are frequently used as film-forming substances. Proteins benefit from their amphiphilic nature as well as electrostatic charge and denaturation properties (Han, 2014). Various changes can be brought about in the secondary, tertiary, and quaternary structure of proteins to suit the needs of film-forming substances. These variations can be done using heat, irradiation, chemical, mechanical treatment, pressure, and enzymatic applications. The most commonly used proteins in edible film and coating formulations include proteins derived from milk (casein and whey protein), plant sources (soy protein, corn zein), and wheat gluten, pea protein, rice bran protein collagen, egg albumin from eggs, myofibrillar protein of fish and keratin, etc. (Ramos, Fernandes, Silva, Pintado, & Malcata, 2012).

5. Global regulations and legislature

Over 33% of all trash produced is not collected and ends up in the environment, including rivers, lakes, and oceans. Misgivings about the poisonous composition of plastic pollution and its effects on the environment, wildlife, and human health are mounting. Global plastic manufacturing is estimated to be around 300 million tons per year, with less than 10% recycled (Mihai et al., 2022). Considering the persistent nature, long transport range, globally presence, hazardous effect on the environment and food chain, and potential to cause a threat to humans, NPs/MPs have been given place or special focus by many regulatory authorities of the world in their meting bodies like United Nations Environment Program (UNEP), Stockholm Convention on persistent organic pollutants (POPs), the Basel Convention on hazardous wastes and Conference of the Parties (COP), etc. being held time to time (Gallo et al., 2018).

Nano/microplastics pollution is an unavoidable situation, as it will ultimately affect us all via food chains, and further accumulation is not sustainable for the earth. However, people's opinion also compels the leaders/governments to make regulations for NPs/MPs. For example, the European Union recently made a strategy for plastics and NPs/MPs in a Circular Economy, which awaked the people for drawbacks of the linear economy, including single-use plastics. Though some people consider NPs/MPs as a good plastic source, they need to be aware of threats of these as well (Mitrano & Wohlleben, 2020).

The involvement of people to save the environment by controlling plastic pollution traces back to the late 1960s and early 1970s. Nowadays, media is also playing an important link that plays a role in bridging the gap between science and the public and can raise awareness. One such example of the role played by media is the Blue Planet II show by BBC, in which they showed plastic litter in surroundings and talked about the risk of plastic. It has motivated the people to save the environment, and the public also demanded strict regulations/action plans to control it; further compelling leaders to make legislation (Mitrano & Wohlleben, 2020). Using cooperation through preexisting regional seas programs, in addition to these partially applicable and diversified universal regulatory measures, could be an effective technique for preventing or reducing the consequences of toxins entering the marine environment (Stöfen-O'Brien, Naji, Brooks, Jambeck, & Khan, 2022). Many regulations have been made by international, regional, national, and local bodies, which is summarized in Table 1.

Table 1 Different regulatory framework and their contributions toward control of plastics (NPs/MPs).

Regulatory framework	Contributions	Reference
International initiatives		
UNEP (United Nations Environment Program)	A report was given on marine plastic debris and MPs by UNEP; also, UN General Assembly adopted resolution no. A /RES/70/1 "Transforming Our World: The 2030 Agenda for Sustainable Development" Clean Seas #campaign was started 14 Different sea programs were started to control plastic pollutions Shared seas approach to reduce marine pollution	Brennholt, Heß, & Reifferscheid, 2018; Gago, Booth, Tiller, Maes, & Larreta, 2020
The United Nations Convention on the Law of the Sea (UNCLOS) 1982	Come into force in 1994 and is the highest authority on the global level to reduce marine pollution, commonly called the "Constitution for the Oceans"	Gagain, 2012; Gago et al., 2020
International Convention for the Prevention of Pollution from Ships 1978 (MARPOL), adopted by the International Maritime Organization (IMO)	It bans ships from releasing all forms of plastic waste into the sea/oceans, but it has a challenge, as more than 80% of waste comes from land, not from ships	Vince & Hardesty, 2018
National Oceanic and Atmospheric Administration (NOAA)	Planned to reduce its NPs/MPs pollution and its associated environmental impacts by 2025	Gago et al., 2020
Stockholm Convention on persistent organic pollutants (POPs)	Took into account the risks of additives in microplastics with compounds having endocrine disruptor properties Listing of new POPs was done	Gallo et al., 2018
Basel Convention on hazardous wastes	It acknowledges plastic marine plastic as a global environmental issue and health concern	Gallo et al., 2018

Continued

Table 1 Different regulatory framework and their contributions toward control of plastics (NPs/MPs).—cont'd

Regulatory framework	Contributions	Reference
The Group of 7 (G7)	It consists of the USA, Germany, Italy, Canada, France, Japan, and the UK. It works in different areas like climate change, nuclear safety, sustainable development, resource efficiency, and marine pollution	G7, 2015; Brennholt et al., 2018
World Economic Forum (WEF)	It gives an industry agenda on "New Plastic Economy": Rethinking the Future of Plastics	Brennholt et al., 2018
World Bank	It started Pollution Management and Environmental Health (PMEH) program Is financing many campaigns for reducing pollution and health improvement	PMEH, 2016
Regional initiatives		
European Union (EU)	Registration, Evaluation, Authorization, and Restriction of Chemicals (REACH) was proposed Dispersion of microplastics in environment banned by 2022 Marine Strategy Framework Directive (MSFD) Goal to achieve Good Environmental Status (GES) of the European marine environment by 2020	European Parliament, 2019; da Costa, Mouneyrac, Costa, Duarte, & Rocha-Santos, 2020
European Water Framework Directive (WFD)	It was established in October 2000 for the betterment of the aquatic environment	Brennholt et al., 2018
The Packaging and Packaging Waste Directive	The main objectives of this are to ensure smooth trade along with prevention of environment to provide a high level of protection	EU, 1994; Brennholt et al., 2018
Plastic Bags Directive	Such kind of drive was come into force for the first time, which considered the management of packaging and waste generated from its use	EU, 2015

Waste Framework Directive	Water Protection and Wastewater Treatment Directives were given under this regulation Also, other campaigns were started to promote recovery and recycling of waste targets	EU (2008, 2015)
Circular Economy Package	It was decided under this drive to shift systemically from a linear to a circular economy model from 2015 onward to reduce the waste generated. European Commission adopted this Package with primarily five priority sectors along with other plastics to achieve sustainable development	Eriksen et al., 2014
Industrial Emissions Directive (IED)	It focuses on preventing, control and reducing the impact of industrial emissions on all kinds of the environment, i.e., aerial, aquatic, and terrestrial, ensuring a high	EU, 2010; Brennholt et al., 2018
Green Paper on a European Strategy on Plastic Waste in the Environment	This initiative was taken to launch a wide reflection on all probable responses to the public policy on plastic wastes; as the same were not addressed in the EU waste legislations This Green Paper is the first of its kind systematic approach to NP/MPs in the environment, referring to the problem of MPs and their fate at the EU level	European Commission, 2013
Initiatives at national levels		
USA (Microbead Free Waters Act)	The use of microbeads in cosmetics was banned along with other regulations to control MPs	Brennholt et al., 2018; Mitrano & Wohlleben, 2020
Australia	Litter Regulations 1993 Protection of the Environment Operations Act 1997	Brennholt et al., 2018

Continued

Table 1 Different regulatory framework and their contributions toward control of plastics (NPs/MPs).—cont'd

Regulatory framework	Contributions	Reference
United Kingdom (UK)	Environmental Protection Act (EPA) to save environment in 1990 Clean Neighborhoods and Environment Act (CNEA) was formulated in 2005	Brennholt et al., 2018
China	Strict rules were given regarding the import and manufacture of nonlisted new substances in IECSC Also, make a plan to forbid MPs manufacturing after December 31, 2020 and also for sale by December 31, 2022	Mitrano & Wohlleben, 2020
South Korea	Cosmetics Act was passed to stop using scrubbing beads in the cosmetics industry Also taken into consideration all other MPs issues for the polymer which are responsible for environmental pollution by putting a lot of waste	Mitrano & Wohlleben, 2020
India	Plastic Waste Management Amendment Rules (PWM), 2021, Ministry of Environment, Forest, and Climate Change (MoEF & CC) which prohibits identified single-use plastic items India Plastic Challenge—Hackathon 2021 India Plastic Pact (IPP) Un-Plastic Collective (UPC) by UN-Environment Program—India GloLitter Partnerships Project to prevent and reduce marine plastic litter from the maritime transport and fisheries sectors was launched by the International Maritime Organization (IMO) and the Food and Agriculture Organization of the United Nations (FAO) to assist developing countries	Indian Plastic Challenge Hackthone, 2021; Indian Plastic Pact, 2021; UPC, 2021; Go Litter Partnership, 2021

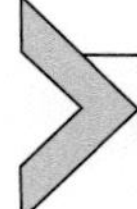

6. Common steps taken by these regulatory instruments

All regulatory bodies functioning, whether at the local, regional, or international level, handle NP/MPs pollution problem through one or more locations of intervention, which, broadly, may be classified as:

(1) Preventive—by focusing on 4Rs (Reduce, Rejuvenate, Reuse, and Recycle): The focus should be on using the 4Rs principle to conserve the environment from plastic waste disposal;
(2) Removal—by waste fragments monitoring and clean-up initiatives;
(3) Mitigation—litter disposal and development of discharge regulations;
(4) Educational—by organizing awareness campaigns and monetary/incentive attitudes (Chen, 2015; da Costa et al., 2020)
(5) Circular Economy—with a focus on resource efficiency and recovery. The circular economy is the need of the hour as it is restorative and regenerative by design. This means materials constantly flow around a closed-loop systems, rather than being used once and then discarded. In the case of plastic, this means simultaneously keeping the value of plastics in the economy without leakage into the natural environment (da Costa et al., 2020).

7. Conclusion

Biobased plastics can revolutionize the food packaging industry and have the exhilarating ability to transform the way people live their lives through mindful sustainability. However, the biobased packaging is still not at par with the traditional polymers because of high production cost, low endurance, and substantiality. Hence, the need of the hour is to work diligently and continuously make strides in developing biocompatible biobased packaging that complies with industry and safety requirements, is cost-effective, environment friendly, and can vie with traditional packaging while also helping to mitigate the nuisance of nano/microplastic in the environment. The world can look forward to biodegradable polymers revolutionizing the way products are made. Consequently, biobased-driven innovations in packaging will pay to a sustainable environment, economy, and society.

References

Abbasi, S., Moore, F., Keshavarzi, B., Hopke, P. K., Naidu, R., Rahman, M. M., et al. (2020). PET-microplastics as a vector for heavy metals in a simulated plant rhizosphere zone. *Science of the Total Environment*, *744*, 140984.

Abdou, E. S., & Sorour, M. A. (2014). Preparation and characterization of starch/carrageenan edible films. *International Food Research Journal*, *21*(1), 189–193.

Abdul Khalil, H. P. S., Davoudpour, Y., Saurabh, C. K., Hossain, M. S., Adnan, A. S., Dungani, R., et al. (2016). A review on nanocellulosic fibres as new material for sustainable packaging: Process and applications. *Renewable and Sustainable Energy Reviews*, *64*, 823–836.

Abral, H., Basri, A., Muhammad, F., Fernando, Y., Hafizulhaq, F., Mahardika, M., et al. (2019). A simple method for improving the properties of the sago starch films prepared by using ultrasonication treatment. *Food Hydrocolloids*, *93*, 276–283.

Abu Elella, M. H., Goda, E. S., Gab-Allah, M. A., Hong, S. E., Pandit, B., Lee, S., et al. (2021). Xanthan gum-derived materials for applications in environment and eco-friendly materials: A review. *Journal of Environmental Chemical Engineering*, *9*(1), 104702.

Adeyanju, G. C., Augustine, T. M., Volkmann, S., Oyebamiji, U. A., Ran, S., Osobajo, O. A., et al. (2021). Effectiveness of intervention on behaviour change against use of non-biodegradable plastic bags: A systematic review. *Discover Sustainability*, *2*(1), 1–15.

Aga, M. B., Dar, A. H., Nayik, G. A., Panesar, P. S., Allai, F., Khan, S. A., et al. (2021). Recent insights into carrageenan-based bio-nanocomposite polymers in food applications: A review. *International Journal of Biological Macromolecules*, *192*, 197–209.

Ahmad, N. H., Mustafa, S., & Che Man, Y. B. (2014). Microbial polysaccharides and their modification approaches: A review. *International Journal of Food Properties*, *18*(2), 332–347.

Ahmad, M., Nirmal, N. P., Danish, M., Chuprom, J., & Jafarzedeh, S. (2016). Characterisation of composite films fabricated from collagen/chitosan and collagen/soy protein isolate for food packaging applications. *RSC Advances*, *6*, 82191–82204.

Ahmed, S., & Ikram, S. (2016). Chitosan and gelatin based biodegradable packaging films with UV-light protection. *Journal of Photochemistry and Photobiology B: Biology*, *163*, 115–124.

Ai, B., Zheng, L., Li, W., Zheng, X., Yang, Y., Xiao, D., et al. (2021). Biodegradable cellulose film prepared from banana pseudo-stem using an ionic liquid for mango preservation. *Frontiers in Plant Science*, *12*, 625878.

Alimi, O. S., Claveau-Mallet, D., Kurusu, R. S., Lapointe, M., Bayen, S., & Tufenkji, N. (2022). Weathering pathways and protocols for environmentally relevant microplastics and nanoplastics: What are we missing? *Journal of Hazardous Materials*, *423*, 126955.

Allan, J., Belz, S., Hoeveler, A., Hugas, M., Okuda, H., Patri, A., et al. (2021). Regulatory landscape of nanotechnology and nanoplastics from a global perspective. *Regulatory Toxicology and Pharmacology*, *122*, 104885.

Allen, S., Allen, D., Phoenix, V. R., Le Roux, G., Jiménez, P. D., Simonneau, A., et al. (2019). Atmospheric transport and deposition of microplastics in a remote mountain catchment. *Nature Geoscience*, *12*(5), 339–344.

Alves, V. D., Castelló, R., Ferreira, A. R., Costa, N., Fonseca, I. M., & Coelhoso, I. M. (2011). Barrier properties of carrageenan/pectin biodegradable composite films. *Procedia Food Science*, *1*, 240–245.

Amobonye, A., Bhagwat, P., Singh, S., & Pillai, S. (2021). Plastic biodegradation: Frontline microbes and their enzymes. *Science of the Total Environment*, *759*, 143536.

Andrady, A. L. (2011). Microplastics in the marine environment. *Marine Pollution Bulletin*, *62*(8), 1596–1605.

Andrady, A. L. (2017). The plastic in microplastics: A review. *Marine Pollution Bulletin*, *119*(1), 12–22.

Andriamanantoanina, H., & Rinaudo, M. (2010). Relationship between the molecular structure of alginates and their gelation in acidic conditions. *Polymer International, 59*, 1531–1541.

Andriani, D., Apriyana, A. Y., & Karina, M. (2020). The optimization of bacterial cellulose production and its applications: A review. *Cellulose, 9*, 1–20.

Ariffin, H., Nishida, H., Shirai, Y., & Hassan, M. A. (2008). Determination of multiple thermal degradation mechanisms of poly(3-hydroxybutyrate). *Polymer Degradation and Stability, 93*(8), 1433–1439.

Auras, R. A., Harte, B., Selke, S., & Hernandez, R. (2003). Mechanical, physical, and barrier properties of poly(Lactide) films. *Journal of Plastic Film and Sheeting, 19*, 123–135.

Avio, C. G., Gorbi, S., Milan, M., Benedetti, M., Fattorini, D., d'Errico, G., et al. (2015). Pollutants bioavailability and toxicological risk from microplastics to marine mussels. *Environmental Pollution, 198*, 211–222.

Avio, C. G., Gorbi, S., & Regoli, F. (2017). Plastics and microplastics in the oceans: From emerging pollutants to emerged threat. *Marine Environmental Research, 128*, 2–11.

Ayu, R. S., Khalina, A., Harmaen, A. S., Zaman, K., Isma, T., Liu, Q., et al. (2020). Characterization study of empty fruit bunch (EFB) fibers reinforcement in poly(butylene) succinate (PBS)/starch/glycerol composite sheet. *Polymers, 12*, 1571.

Bakir, A., Rowland, S. J., & Thompson, R. C. (2014). Enhanced desorption of persistent organic pollutants from microplastics under simulated physiological conditions. *Environmental Pollution, 185*, 16–23.

Balasubramanian, R., Kim, S. S., & Lee, J. (2018). Novel synergistic transparent k -carrageenan/xanthan gum/gellan gum hydrogel film: Mechanical, thermal and water barrier properties. *International Journal of Biological Macromolecules, 118*, 561–568.

Bandyopadhyay, S, Saha, N, Brodnjak, U, V, & Saha, P. (2018). Bacterial cellulose based greener packaging material: A bioadhesive polymeric film. *Materials Research Express, 5*(11), 115405.

Barceló, D., & Pico, Y. (2019). Microplastics in the global aquatic environment: Analysis, effects, remediation and policy solutions. *Journal of Environmental Chemical Engineering*, 7(5), 103421.

Barikani, M., Oliaei, E., Seddiqi, H., & Honarkar, H. (2014). Preparation and application of chitin and its derivatives: A review. *Iranian Polymer Journal, 23*(4), 307–326.

Barría, C., Brandts, I., Tort, L., Oliveira, M., & Teles, M. (2020). Effect of nanoplastics on fish health and performance: A review. *Marine Pollution Bulletin, 151*, 110791.

Bellasi, A., Binda, G., Pozzi, A., Boldrocchi, G., & Bettinetti, R. (2021). The extraction of microplastics from sediments: An overview of existing methods and the proposal of a new and green alternative. *Chemosphere, 130357*.

Bergman, Å., Heindel, J. J., Jobling, S., Kidd, K., Zoeller, T. R., & World Health Organization. (2013). *State of the science of endocrine disrupting chemicals 2012*. World Health Organization.

Bharimalla, A. K., Deshmukh, S. P., Vigneshwaran, N., Patil, P. G., & Prasad, V. (2016). Nanocellulose-polymer composites for applications in food packaging: Current status, future prospects and challenges. *Polymer-Plastics Technology and Engineering, 56*(8), 805–823.

Bharti, S. K., Pathak, V., Arya, A., Alam, T., Rajkumar, V., & Verma, A. K. (2021). Packaging potential of *Ipomoea batatas* and κ-carrageenan biobased composite edible film: Its rheological, physicomechanical, barrier and optical characterization. *Journal of Food Processing and Preservation, 45*, e15153.

Boucher, J., & Friot, D. (2017). *Primary microplastics in the oceans: A global evaluation of sources. vol. 10*. Gland, Switzerland: IUCN.

Brandon, J., Goldstein, M., & Ohman, M. D. (2016). Long-term aging and degradation of microplastic particles: Comparing in situ oceanic and experimental weathering patterns. *Marine Pollution Bulletin, 110*(1), 299–308.

Brennholt, N., Heß, M., & Reifferscheid, G. (2018). Freshwater microplastics: Challenges for regulation and management. In *Freshwater microplastics* (pp. 239–272). Cham: Springer.
Brodnjak, U., & Todorova, D. (2018). *Chitosan and rice starch films as packaging materials* (pp. 275–280). https://doi.org/10.24867/GRID-2018-p34.
Cacicedo, M. L., Castro, M. C., Servetas, I., Bosnea, L., Boura, K., Tsafrakidou, P., et al. (2016). Progress in bacterial cellulose matrices for biotechnological applications. *Bioresource Technology, 213*, 172–180.
Carrasco, F., Pagès, P., Gámez-Pérez, J., Santana, O. O., & Maspoch, M. L. (2010). Processing of poly(lactic acid): Characterization of chemical structure, thermal stability and mechanical properties. *Polymer Degradation and Stability, 95*, 116–125.
Castelvetro, V., Corti, A., Bianchi, S., Ceccarini, A., Manariti, A., & Vinciguerra, V. (2020). Quantification of poly (ethylene terephthalate) micro-and nanoparticle contaminants in marine sediments and other environmental matrices. *Journal of Hazardous Materials, 385*, 121517.
Cazzolino, C. A., Campanella, G., Türe, H., Olsson, R. T., & Farris, S. (2016). Microfibrillated cellulose and borax as mechanical, O2 -barrier, and surface modulating agents of pullulan biocomposite coating on BOPR. *Carbohydrate Polymers, 143*, 179–187.
Cerasa, M., Teodori, S., & Pietrelli, L. (2021). Searching Nanoplastics: From sampling to sample processing. *Polymers, 13*(21), 3658.
Chen, G. Q. (2009). A microbial polyhydroxyalkanoates (PHA) based bio- and materials industry. *Chemical Society Reviews, 38*, 2434–2446.
Chen, C. L. (2015). Regulation and management of marine litter. In *Marine anthropogenic litter* (pp. 395–428). Cham: Springer.
Chen, J., Wu, S., & Pan, S. (2012). Optimization of medium for pullulan production using a novel strain of *Aureobasidium pullulans* isolated from sea mud through response surface. *Carbohydrate Polymers, 87*, 771–774.
Clarinval, A. M., & Halleux, J. (2005). Classification of biodegradable polymers. In *Biodegradable polymers for industrial applications* (pp. 3–31). Cambridge, UK: Woodhead Publishing.
Crew, A., Gregory-Eaves, I., & Ricciardi, A. (2020). Distribution, abundance, and diversity of microplastics in the upper St. Lawrence River. *Environmental Pollution, 260*, 113994.
da Costa, J. P., Mouneyrac, C., Costa, M., Duarte, A. C., & Rocha-Santos, T. (2020). The role of legislation, regulatory initiatives and guidelines on the control of plastic pollution. *Frontiers in Environmental Science, 8*, 104.
da Silva, T. F., Menezes, F., Montagna, L. S., Lemes, A. P., & Passador, F. R. (2019). Effect of lignin as accelerator of the biodegradation process of poly (lactic acid)/lignin composites. *Materials Science and Engineering: B, 251*, 114441.
de Matos Costa, A. R., Crocitti, A., Hecker de Carvalho, L., Carroccio, S. C., Cerruti, P., & Santagata, G. (2020). Properties of biodegradable films based on poly(butylene succinate) (pbs) and poly(butylene adipate-*co*-terephthalate) (PBAT) blends. *Polymers, 12*, 2317.
Deme, G. G., Ewusi-Mensah, D., Olagbaju, O. A., Okeke, E. S., Okoye, C. O., Odii, E. C., et al. (2022). Macro problems from microplastics: Toward a sustainable policy framework for managing microplastic waste in Africa. *Science of the Total Environment, 804*, 150170.
Diaz, C, A, Pao, H, Y, & Kim, S. (2016). Film Performance of poly (lactic acid) blends for packaging applications. *Journal of Applied Packaging Research, 8*(3), 4.
Ding, J., Zhang, S., Razanajatovo, R. M., Zou, H., & Zhu, W. (2018). Accumulation, tissue distribution, and biochemical effects of polystyrene microplastics in the freshwater fish red tilapia (*Oreochromis niloticus*). *Environmental Pollution, 238*, 1–9.
El-Hadi, A., Schnabel, R., Straube, E., Muller, E., & Henning, S. (2002). Correlation between degree of crystallinity, morphology, glass temperature, mechanical properties and biodegradation of poly (3-hydroxyalkanoate) PHAs and their blends. *Polymer Testing, 21*, 665–674.

Elsabee, M. Z., & Abdou, E. S. (2013). Chitosan based edible films and coatings: A review. *Materials Science and Engineering: C*, *33*(4), 1819–1841.
Enfrin, M., Hachemi, C., Hodgson, P. D., Jegatheesan, V., Vrouwenvelder, J., Callahan, D. L., et al. (2021). Nano/micro plastics—Challenges on quantification and remediation: A review. *Journal of Water Process Engineering*, *42*, 102128.
Eriksen, M., Lebreton, L. C., Carson, H. S., Thiel, M., Moore, C. J., Borerro, J. C., et al. (2014). Plastic pollution in the world's oceans: More than 5 trillion plastic pieces weighing over 250,000 tons afloat at sea. *PLoS One*, *9*(12), e111913.
Estahbanati, M. K., Kiendrebeogo, M., Mostafazadeh, A. K., Drogui, P., & Tyagi, R. D. (2021). Treatment processes for microplastics and nanoplastics in waters: State-of-the-art review. *Marine Pollution Bulletin*, *168*, 112374.
EU. (1994). *European Parliament and Council Directive 94/62/EC of 20 December 1994 on packaging and packaging waste*. http://eurlex.europa.eu. Retrieved 12 Nov 2016.
EU. (2008). *Directive 2008/98/EC of the European Parliament and of the Council of 19 November 2008 on waste and repealing certain Directives*. http://eur-lex.europa.eu.
EU. (2010). *Directive 2010/75/EU of the European Parliament and of the Council of 24 November 2010 on industrial emissions (integrated pollution prevention and control)*. http://eur-lex.europa.eu.
EU. (2015). *Directive (EU) 2015/720 of the European Parliament and of the Council of 29 April 2015 amending Directive 94/62/EC as regards reducing the consumption of lightweight plastic carrier bags*. http://eurlex.europa.eu.
European Bioplastics. (2020). *Bioplastics*. Berlin, Germany: Author.
European Commission. (2013). *Green Paper on a European Strategy on Plastic Waste in the Environment*. http://eur-lex.europa.eu/legal-content/EN/TXT/?uri¼CELEX:52013DC0123.
European Parliament. (2019). Directive (EU) 2019/904 of the European Parliament and of the council of 5 June 2019 on the reduction of the impact of certain plastic products on the environment. *Official Journal of the European Union*, *155*, 1–19.
Fabra, M. J., López-Rubio, A., & Lagaron, J. M. (2016). Use of the electrohydrodynamic process to develop active/bioactive bilayer films for food packaging applications. *Food Hydrocolloids*, *55*, 11–18.
FDA. (2002). *Agency response letter: GRAS notice no. GRN 000099 [pullulan]*. College Park, Maryland: US Food and Drug Administration (US FDA), Center for Food Safety and Applied Nutrition (CFSAN), Office of Food Additive Safety.
Felsing, S., Kochleus, C., Buchinger, S., Brennholt, N., Stock, F., & Reifferscheid, G. (2018). A new approach in separating microplastics from environmental samples based on their electrostatic behavior. *Environmental Pollution*, *234*, 20–28.
Foster, E. J., Moon, R. J., Agarwal, U. P., Bortner, M. J., Bras, J., Camarero-Espinosa, S., et al. (2018). Current characterization methods for cellulose nanomaterials. *Chemical Society Reviews*, *47*(8), 2609–2679.
Fu, W., Min, J., Jiang, W., Li, Y., & Zhang, W. (2020). Separation, characterization and identification of microplastics and nanoplastics in the environment. *Science of the Total Environment*, *721*, 137561.
Fujiwara, H., Yamauchi, K., Wada, T., Ishihara, H., & Sasaki, K. (2021). Optical selection and sorting of nanoparticles according to quantum mechanical properties. *Science Advances*, 7(3), eabd9551.
G7. (2015). *Declaration, L. Annex to the Leaders' Declaration G7 Summit 7–8 June 2015*.
Gagain, M. (2012). Climate change, sea level rise, and artificial islands: Saving the Maldives' statehood and maritime claims through the constitution of the oceans. *Colorado Journal of International Environmental Law & Policy*, *23*, 77.
Gago, J., Booth, A. M., Tiller, R., Maes, T., & Larreta, J. (2020). Microplastics pollution and regulation. In T. Rocha-Santos (Ed.), *Handbook of Microplastics in the Environment* (pp. 1–27). Springer.

Gallo, F., Fossi, C., Weber, R., Santillo, D., Sousa, J., Ingram, I., et al. (2018). Marine litter plastics and microplastics and their toxic chemicals components: The need for urgent preventive measures. *Environmental Sciences Europe*, *30*(1), 1–14.
Gan, P., Sam, S., Abdullah, M. F., & Omar, M. F. (2020). Thermal properties of nanocellulose-reinforced composites: A review. *Journal of Applied Polymer Science*, *137*, 48544.
Gasperi, J., Wright, S. L., Dris, R., Collard, F., Mandin, C., Guerrouache, M., et al. (2018). Microplastics in air: Are we breathing it in? *Current Opinion in Environmental Science & Health*, *1*, 1–5.
GESAMP. (2016). Sources, fate and effects of microplastics in the marine environment: part two of a global assessment. In P. J. Kershaw, & C. M. Rochman (Eds.), *Vol. 93. IMO/FAO/UNESCO-IOC/UNIDO/WMO/IAEA/UN/UNEP/UNDP* (p. 220). Joint Group of Experts on the Scientific Aspects of Marine Environmental Protection Rep. Stud. GESAMP.
Ghaderi, M., Mousavi, M., Yousefi, H., & Labbafi, M. (2014). All-cellulose nanocomposite film made from bagasse cellulose nanofibers for food packaging application. *Carbohydrate Polymers*, *104*, 59–65.
Go Litter Partnership. (2021). https://www.imo.org/en/OurWork/PartnershipsProjects/Pages/GloLitter-Partnerships-Project-.aspx.
Godoy, V., Martín-Lara, M. A., Calero, M., & Blázquez, G. (2019). Physical-chemical characterization of microplastics present in some exfoliating products from Spain. *Marine Pollution Bulletin*, *139*, 91–99.
Gordobil, O., Delucis, R., Egüés, I., & Labidi, J. (2015). Kraft lignin as filler in PLA to improve ductility and thermal properties. *Industrial Crops and Products*, *72*, 46–53.
Grause, G., Kuniyasu, Y., Chien, M. F., & Inoue, C. (2021). Separation of microplastic from soil by centrifugation and its application to agricultural soil. *Chemosphere*, *132654*.
Grbic, J., Nguyen, B., Guo, E., You, J. B., Sinton, D., & Rochman, C. M. (2019). Magnetic extraction of microplastics from environmental samples. *Environmental Science & Technology Letters*, *6*(2), 68–72.
Han, J. (2014). Edible films and coatings: A review. *Innovations in Food Packaging*, 213–255.
Hassannia-Kolaee, M., Khodaiyan, F., Pourahmad, R., & Shahabi-Ghahfarrokhi, I. (2016). Development of eco-friendly bionanocomposite: Whey protein isolate/pullulan films with nano-SiO2. *International Journal of Biological Macromolecules*, *86*, 139–144. https://www.imo.org/en/OurWork/PartnershipsProjects/Pages/GloLitter-Partnerships-Project-.aspx.
Heidbreder, L. M., Bablok, I., Drews, S., & Menzel, C. (2019). Tackling the plastic problem: A review on perceptions, behaviors, and interventions. *Science of the Total Environment*, *668*, 1077–1093.
Hirotsu, T., Toshiyuki, T., Masuda, T., & Nayakama, K. (2000). Plasma surface treatments and biodegradation of poly (butylene succinate) sheets. *Journal of Applied Polymer Science*, *78*(5), 1121–1129.
Hu, J., He, D., Zhang, X., Li, X., Chen, Y., Wei, G., et al. (2022). National-scale distribution of micro (meso) plastics in farmland soils across China: Implications for environmental impacts. *Journal of Hazardous Materials*, *424*, 127283.
Huang, X., Li, Y., & Lim, K. Y. T. (2021). Investigating the plastic decomposing ability of Tenebrio molitor using carbon dioxide sensors. *APD Trove*, *4*, 1–11.
Huang, Y., Zhu, C., Yang, J., Nie, Y., Chen, C., & Sun, D. (2014). Recent advances in bacterial cellulose. *Cellulose*, *21*, 1–30.
Indian Plastic Challenge Hackthone. (2021). https://pib.gov.in/PressReleaseIframePage.aspx?PRID=1725458.
Indian Plastic Pact. (2021). https://www.cii.in/PressreleasesDetail.aspx?enc=SISYEhhuG40HN4iTne/gqhwseTpLnTJigim5UwrxUpQ=https://www.indiaplasticspact.org/.

Irianto, H. E., Agusman, Fransiska, D., & Musfira. (2019). Mechanical and barrier properties of composite films based on kappa-carrageenan-polyvinyl alcohol. *International Journal of Engineering and Advanced Technology*, *8*, 489–499.

Ivleva, N. P. (2021). Chemical analysis of microplastics and nanoplastics: Challenges, advanced methods, and perspectives. *Chemical Reviews*, *121*(19), 11886–11936.

Jost, V., Kobsik, K., Schmid, M., & Noller, K. (2014). Influence of plasticizer on the barrier, mechanical and grease resistance properties of alginate cast films. *Carbohydrate Polymers*, *110*, 309–319.

Junhao, C., Xining, Z., Xiaodong, G., Li, Z., Qi, H., & Siddique, K. H. (2021). Extraction and identification methods of microplastics and nanoplastics in agricultural soil: A review. *Journal of Environmental Management*, *294*, 112997.

Kausar, A. (2017). Scientific potential of chitosan blending with different polymeric materials: A review. *Journal of Plastic Film and Sheeting*, *33*(4), 384–412.

Khalid, N., Aqeel, M., Noman, A., Khan, S. M., & Akhter, N. (2021). Interactions and effects of microplastics with heavy metals in aquatic and terrestrial environments. *Environmental Pollution*, *290*(12), 118104.

Khalid, N., Rizvi, Z. F., Yousaf, N., Khan, S. M., Noman, A., et al. (2021). Rising metals concentration in the environment: A response to effluents of leather industries in Sialkot. *Bulletin of Environmental Contamination and Toxicology*, *106*(3), 493–500.

Kontominas, M. G. (2020). Use of alginates as food packaging materials. *Food*, *9*, 1440.

Kraśniewska, K., Pobiega, K., & Gniewosz, M. (2019). Pullulan– Biopolymer with potential for use as food packaging. *International Journal of Food Engineering*, *15*(9).

Kumar, S., Ye, F., Dobretsov, S., & Dutta, J. (2019). Chitosan nanocomposite coatings for food, paints, and water treatment applications. *Applied Sciences*, *9*(12), 2409.

Kundu, A., Shetti, N. P., Basu, S., Reddy, K. R., Nadagouda, M. N., & Aminabhavi, T. M. (2021). Identification and removal of micro-and nano-plastics: Efficient and cost-effective methods. *Chemical Engineering Journal, 129816*.

Labuschagne, M., Wepener, V., Nachev, M., Zimmermann, S., Sures, B., & Smit, N. J. (2020). The application of artificial mussels in conjunction with transplanted bivalves to assess elemental exposure in a platinum mining area. *Water*, *12*(1), 32.

Lavoine, N., & Bergström, L. (2017). Nanocellulose-based foams and aerogels: Processing, properties, and applications. *Journal of Materials Chemistry, A*, *5*, 16105–16117.

Lear, G., Kingsbury, J. M., Franchini, S., Gambarini, V., Maday, S. D. M., Wallbank, J. A., et al. (2021). Plastics and the microbiome: Impacts and solutions. *Environmental Microbiome*, *16*(1), 1–19.

Liaqat, S., Hussain, M., Malik, M. F., Aslam, A., & Mumtaz, K. (2020). Microbial ecology: A new perspective of plastic degradation. *Pure and Applied Biology*, *9*(4), 2138–2150.

Lim, J., You, M., Li, J., & Li, Z. (2017). Emerging bone tissue engineering via Polyhydroxyalkanoate (PHA)-based scaffolds. *Materials Science and Engineering: C*, *79*, 917–929.

Lin, S.-P., Liu, C.-T., Hsu, K.-D., Hung, Y.-T., Shih, T.-Y., & Cheng, K. C. (2016). Production of bacterial cellulose with various additives in a PCS rotating disk bioreactor and its material property analysis. *Cellulose*, *23*, 367–377.

Liu, Q., Chen, Z., Chen, Y., Yang, F., Yao, W., & Xie, Y. (2021). Microplastics and Nanoplastics: Emerging contaminants in food. *Journal of Agricultural and Food Chemistry*, *69*(36), 10450–10468.

Liu, L., Xu, M., Ye, Y., & Zhang, B. (2022). On the degradation of (micro) plastics: Degradation methods, influencing factors, environmental impacts. *Science of the Total Environment*, *806*, 151312.

Łopusiewicz, Ł., Zdanowicz, M., Macieja, S., Kowalczyk, K., & Bartkowiak, A. (2021). Development and characterization of bioactive poly(butylene-succinate) films modified with quercetin for food packaging applications. *Polymers*, *13*, 1798.

Lu, Y., Zhang, Y., Deng, Y., Jiang, W., Zhao, Y., et al. (2016). Uptake and accumulation of polystyrene microplastics in zebrafish (Danio rerio) and toxic effects in liver. *Environmental Science & Technology, 50(7)*, 4054–4060.

Luís, Â., Ramos, A., & Domingues, F. (2021). Pullulan–Apple FiberBiocomposite films: Optical, mechanical, barrier, antioxidant and antibacterial properties. *Polymers, 13*, 870.

Luo, L., Wang, Y., Guo, H., Yang, Y., Qi, N., Zhao, X., et al. (2021). Biodegradation of foam plastics by Zophobas atratus larvae (Coleoptera: Tenebrionidae) associated with changes of gut digestive enzymes activities and microbiome. *Chemosphere, 282*, 131006.

Lusher, A., Hollman, P., & Mendoza, J. (2017). *Microplastics in fisheries and aquaculture: status of knowledge on their occurrence and implications for aquatic organisms and food safety*. FAO.

Mallegni, N., Phuong, T. V., Coltelli, M.-B., Cinelli, P., & Lazzeri, A. (2018). Poly (lactic acid) (PLA) based tear resistant and biodegradable flexible films by blown film extrusion. *Materials, 11*, 148.

Martiny, T. R., Pacheco, B. S., Pereira, C. M., Mansilla, A., Astorga-España, M. S., Dotto, G. L., et al. (2020). A novel biodegradable film based on κ-carrageenan activated with olive leaves extract. *Food Science & Nutrition, 8*(7), 3147–3156.

Mihai, F. C., Gündoğdu, S., Markley, L. A., Olivelli, A., Khan, F. R., Gwinnett, C., et al. (2022). Plastic pollution, waste management issues, and circular economy opportunities in rural communities. *Sustainability, 14*(1), 20.

Mitrano, D, M, & Wohlleben, W. (2020). Microplastic regulation should be more precise to incentivize both innovation and environmental safety. *Nature Communications, 11*(1), 1–12.

Morris, B. A. (2016). *The science and technology of flexible packaging: multilayer films from resin and process to end use* (2nd). Elsevier.

Morrone, M., Cappelletti, N. E., Tatone, L. M., Astoviza, M. J., & Colombo, J. C. (2021). The use of biomimetic tools for water quality monitoring: Passive samplers versus sentinel organisms. *Environmental Monitoring and Assessment, 193*(3), 1–13.

Muller, P. S., Carpiné, D., Yamashita, F., & Waszczynskyj, N. (2020). Influence of pinhão starch and natural extracts on the performance of thermoplastic cassava starch/PBAT extruded blown films as a technological approach for bio-based packaging material. *Journal of Food Science, 85*, 2832–2842.

Murariu, M., & Dubois, P. (2016). PLA composites: From production to properties. *Advanced Drug Delivery Reviews, 107*, 17–46.

Muthulakshmi, L., Annaraj, J., Ramakrishna, S., Ranjan, S., Dasgupta, N., Mavinkere, R. S., et al. (2021). A sustainable solution for enhanced food packaging via a science-based composite blend of natural-sourced chitosan and microbial extracellular polymeric substances. *Journal of Food Processing and Preservation, 45*(1), e15031.

Mwanza, B. G. (2021). Microplastics pollution: Integrated approaches and solutions. In *AIJR Abstracts* (pp. 11–13).

Ncube, L. K., Ude, A. U., Ogunmuyiwa, E. N., Zulkifli, R., & Beas, I. N. (2020). Environmental impact of food packaging materials: A review of contemporary development from conventional plastics to Polylactic acid based materials. *Materials, 13*(21), 4994.

Nigamatzyanova, L., & Fakhrullin, R. (2021). Dark-field hyperspectral microscopy for label-free microplastics and nanoplastics detection and identification in vivo: A Caenorhabditis elegans study. *Environmental Pollution, 271*, 116337.

Nilsen-Nygaard, J., Strand, S., Varum, K., Draget, K., & Nordgard, C. T. (2015). Chitosan: Gels and interfacial properties. *Polymers, 7*, 552–579.

Niu, B., Shao, P., Chen, H., & Sun, P. (2019). Structural and physiochemical characterization of novel hydrophobic packaging films based on pullulan derivatives for fruits preservation. *Carbohydrate Polymers, 208*, 276–284.

Okeke, E. S., Okoye, C. O., Atakpa, E. O., Ita, R. E., Nyaruaba, R., Mgbechidinma, C. L., et al. (2022). Microplastics in agroecosystems-impacts on ecosystem functions and food chain. *Resources, Conservation and Recycling*, *177*, 105961.

Padrão, J., Gonçalves, S., Silva, J. P., Sencadas, V., Lanceros-Méndez, S., Pinheiro, A. C., et al. (2016). Bacterial cellulose-lactoferrin as an antimicrobial edible packaging. *Food Hydrocolloid*, *58*, 126–140.

Pan, L., Li, P., & Tao, Y. (2020). Preparation and properties of microcrystalline cellulose/fish gelatin composite film. *Materials*, *13*, 4370.

Paranjape, A. R., & Hall, B. D. (2017). Recent advances in the study of mercury methylation in aquatic systems. *Facets*, *21(1)*, 85–119.

Pedersen, A. F., Gopalakrishnan, K., Boegehold, A. G., Peraino, N. J., Westrick, J. A., & Kashian, D. R. (2020). Microplastic ingestion by quagga mussels, Dreissena bugensis, and its effects on physiological processes. *Environmental Pollution*, *260*, 113964.

Petersen, K., Nielsen, P. V., Bertelsen, G., Lawther, M., Olsen, M. B., Nilsson, N. H., et al. (1999). Potential of biobased materials for food packaging. *Trends in Food Science Technology*, *10*, 52–68.

Picó, Y., & Barceló, D. (2021). Analysis of micro and nanoplastics: How green are the methodologies used? *Current opinion in green and sustainable. Chemistry*, *31*, 100503.

PMEH. (2016). Pollution Management and Environmental Health (PMEH) program. World Bank Group. *Annual Report.*

Prata, J. C., da Costa, J. P., Lopes, I., Duarte, A. C., & Rocha-Santos, T. (2020). Environmental exposure to microplastics: An overview on possible human health effects. *Science of the Total Environment*, *702*, 134455.

Puscaselu, R. G., Gutt, G., & Amariei, S. (2020). The use of edible films based on sodium alginate in meat product packaging: An eco-friendly alternative to conventional plastic materials. *Coatings*, *10*, 166.

Qiao, R., Sheng, C., Lu, Y., Zhang, Y., Ren, H., & Lemos, B. (2019). Microplastics induce intestinal inflammation, oxidative stress, and disorders of metabolome and microbiome in zebrafish. *Science of the Total Environment*, *662*, 246–253.

Quinn, B., Murphy, F., & Ewins, C. (2017). Validation of density separation for the rapid recovery of microplastics from sediment. *Analytical Methods*, *9*(9), 1491–1498.

Ragauskas, A. J., Huber, G. W., Wang, J., Guss, A., O'Neill, H. M., Lin, C. S. K., et al. (2021). New technologies are needed to improve the recycling and upcycling of waste plastics. *A Journal of Chemistry and Sustainability, Energy and Materials*, *14*(19), 3982–3984.

Rai, P., Mehrotra, S., Priya, S., Gnansounou, E., & Sharma, S. K. (2021). Recent advances in the sustainable design and applications of biodegradable polymers. *Bioresource Technology*, *325*, 124739.

Ramos, Ó. L., Fernandes, J. C., Silva, S. I., Pintado, M. E., & Malcata, F. X. (2012). Edible films and coatings from whey proteins: A review on formulation, and on mechanical and bioactive properties. *Critical Reviews in Food Science and Nutrition*, *52*(6), 533–552.

Revel, M., Châtel, A., & Mouneyrac, C. (2018). Micro (nano) plastics: A threat to human health? *Current Opinion in Environmental Science & Health*, *1*, 17–23.

Rudnik, E. (2019). *Compostable polymer materials* (p. 410). Warsaw, Poland: Newne.

Salari, M., Khiabani, M. S., Mokarram, R. R., Ghanbarzadeh, B., & Kafil, H. S. (2018). Development and evaluation of chitosan based active nanocomposite films containing bacterial cellulose nanocrystals and silver nanoparticles. *Food Hydrocolloids*, *84*, 414–423.

Sapper, M., Talens, P., & Chiralt, A. (2019). Improving functional properties of cassava starch-based films by incorporating xanthan, gellan, or pullulan gums. *International Journal of Polymer Science*, *2019*, 5367164.

Saratale, R. G., Cho, S. K., Ghodake, G. S., Shin, H. S., Saratale, G. D., Park, Y., et al. (2020). Utilization of noxious weed water hyacinth biomass as a potential feedstock for biopolymers production: A novel approach. *Polymers*, *12*, 1704.

Sen, K. Y., & Baidurah, S. (2021). Renewable biomass feedstocks for production of sustainable biodegradable polymer. *Current Opinion in Green and Sustainable Chemistry*, *27*, 1–6.
Shah, N., Ul-Islam, M., Khattak, W. A., & Park, J. K. (2013). Overview of bacterial cellulose composites: A multipurpose advanced material. *Carbohydrate Polymers*, *98*, 1585–1598.
Silva, F. A. G. S., Dourado, F., Gama, M., & Poças, F. (2020). Nanocellulose bio-based composites for food packaging. *Nanomaterials*, *10*, 2041.
Singh, R. S., & Saini, G. K. (2008). Production, purification and characterization of pullulan from a novel strain of *Aureobasidium pullulans* FB-1. *Journal of Biotechnology*, *136*, S506–S507.
Souza, V. G. L., Pires, J. R. A., Rodrigues, C., Rodrigues, P. F., Lopes, A., Silva, R. J., et al. (2019). Physical and morphological characterization of chitosan/montmorillonite films incorporated with ginger essential oil. *Coatings*, *9*(11), 700.
Staichak, G., Ferreira-Jr, A. L., Silva, A. C. M., Girard, P., Callil, C. T., & Christo, S. W. (2021). Bivalves with potential for monitoring microplastics in South America. *Case Studies in Chemical and Environmental Engineering*, *4*, 100119.
Stenstad, P., Andresen, M., Tanem, B. S., & Stenius, P. (2007). Chemical surface modifications of microfibrillated cellulose. *Cellulose*, *15*, 35–45.
Stöfen-O'Brien, A., Naji, A., Brooks, A. L., Jambeck, J. R., & Khan, F. R. (2022). Marine plastic debris in the Arabian/Persian Gulf: Challenges, opportunities and recommendations from a transdisciplinary perspective. *Marine Policy*, *136*, 104909.
Strungaru, S. A., Jijie, R., Nicoara, M., Plavan, G., & Faggio, C. (2019). Micro-(nano) plastics in freshwater ecosystems: Abundance, toxicological impact and quantification methodology. *TrAC Trends in Analytical Chemistry*, *110*, 116–128.
Suput, D. Z., Lazić, V. L., Popović, S. Z., & Hromiš, N. M. (2015). Edible films and coatings: Sources, properties and application. *Food and Feed Research*, *42*(1), 11–22.
Tagg, A. S., & Labrenz, M. (2018). Closing microplastic pathways before they open: A model approach. *Environmental Science and Technology*, *52*(6), 3340–3341.
Tamburini, E., Costa, S., Summa, D., Battistella, L., Fano, E. A., & Castaldelli, G. (2021). Plastic (PET) vs bioplastic (PLA) or refillable aluminium bottles—What is the most sustainable choice for drinking water? A life-cycle (LCA) analysis. *Environmental Research*, *196*, 110974.
Thakur, S., Chaudhary, J., Sharma, B., Verma, A., Tamulevicius, S., & Thakur, V. K. (2018). Sustainability of bioplastics: Opportunities and challenges. *Current Opinion in Green and Sustainable Chemistry*, *13*, 68–75.
Theagarajan, R., Dutta, S., Moses, J. A., & Anandha Ramakrishnan, C. (2019). In A. A. Shakeel (Ed.), *Alginates for food packaging applications* (pp. 207–232). Beverly, MA, USA: Scrivener Publishing LLC.
Thurber, H., & Curtzwiler, G. W. (2020). Suitability of poly(butylene succinate) as a coating for paperboard convenience food packaging. *International Journal of Biobased Plastics*, *2*(1), 1–12.
Tong, Q., Xiao, Q., & Lim, L. T. (2013). Effects of glycerol, sorbitol, xylitol and fructose plasticisers on mechanical and moisture barrier properties of pullulan-alginate-carboxymethylcellulose blend films. *International Journal of Food Science and Technology*, *48*, 870–878.
Tortajada, M., da Silva, L. F., & Prieto, M. A. (2013). Second-generation functionalized medium-chain-length polyhydroxyalkanoates: The gateway to high-value bioplastic applications. *International microbiology: the official journal of the Spanish Society for Microbiology*, *16*(1), 1–15.
Tsouko, E., Kourmentza, C., Ladakis, D., Kopsahelis, N., Mandala, I., Papanikolaou, S., et al. (2015). Bacterial cellulose production from industrial waste and by-product streams. *International Journal of Molecular Sciences*, *16*(7), 14832–14849.

UPC. (2021). *Un-Plastic Collective*. https://www.cii.in/PressreleasesDetail.aspx?enc=pkWBYAQGvSE3mnpRnfOorC2LJaFUXez3Zjb80e6kTn8=https://www.wwfindia.org/about_wwf/making_businesses_sustainable/un_plastic_collective/.

Valdés, A., Martínez, C., Garrigos, M. C., & Jimenez, A. (2021). Multilayer films based on poly(lactic acid)/Gelatin supplemented with cellulose nanocrystals and antioxidant extract from almond Shell by-product and its application on Hass avocado preservation. *Polymers*, *13*, 3615.

Valsesia, A., Jeremie, P., Jessica, P., Dora, M., Rita, M., Daniela, M., et al. (2021). Detection, counting and characterization of nanoplastics in marine bioindicators: A proof of principle study. *Microplastics and Nanoplastics*, *1*(1), 1–13.

Valsesia, A., Quarato, M., Ponti, J., Fumagalli, F., Gilliland, D., & Colpo, P. (2021). Combining microcavity size selection with Raman microscopy for the characterization of nanoplastics in complex matrices. *Scientific Reports*, *11*(1), 1–12.

Van Cauwenberghe, L., & Janssen, C. R. (2014). Microplastics in bivalves cultured for human consumption. *Environmental Pollution*, *193*, 65–70.

Venkatesan, J., Lowe, B., Anil, S., Manivasagan, P., Al Kheraif, A. A., Kang, K. H., et al. (2015). Seaweed polysaccharides and their potential biomedical applications. *Starch*, *67*, 381–390.

Vince, J, & Hardesty, B, D. (2018). Governance solutions to the tragedy of the commons that marine plastics have become. *Frontiers in Marine Science*, *5*, 214.

Von Moos, N., Burkhardt-Holm, P., & Köhler, A. (2012). Uptake and effects of microplastics on cells and tissue of the blue mussel Mytilus edulis L. after an experimental exposure. *Environmental Science & Technology*, *46*(20), 11327–11335.

Vytejckova, S., Vápenka, L., Hradecký, J., Dobiáš, J., Hajšlová, J., Loriot, C., et al. (2017). Testing of polybutylene succinate based films for poultry meat packaging. *Polymer Testing*, *60*, 357–364.

Wang, W., Ge, J., Yu, X., & Li, H. (2020). Environmental fate and impacts of microplastics in soil ecosystems: Progress and perspective. *Science of the Total Environment*, *708*, 134841.

Wagner, M., Scherer, C., Alvarez-Muñoz, D., Brennholt, N., Bourrain, X., et al. (2014). Microplastics in freshwater ecosystems: What we know and what we need to know. *Environmental Sciences Europe*, *26*(1), 1–19.

Wang, L., Wu, W. M., Bolan, N. S., Tsang, D. C., Li, Y., Qin, M., et al. (2021). Environmental fate, toxicity and risk management strategies of nanoplastics in the environment: Current status and future perspectives. *Journal of Hazardous Materials*, *401*, 123415.

Wang, Y., Mao, Z., Zhang, M., Ding, G., Sun, J., et al. (2019). The uptake and elimination of polystyrene microplastics by the brine shrimp, Artemia parthenogenetica, and its impact on its feeding behavior and intestinal histology. *Chemosphere, 234*, *234*, 123–131.

Wang, Y., Yin, J., & Chen, G. Q. (2014). Polyhydroxyalkanoates, challenges and opportunities. *Current Opinion in Biotechnology*, *30*, 59–65.

Way, C., Hudson, M. D., Williams, I. D., & Langley, G. J. (2022). Evidence of underestimation in microplastic research: A meta-analysis of recovery rate studies. *Science of the Total Environment*, *805*, 150227.

Wei, Q., Chao-Yang, Hu, Rong-Rong, Zhang, Yan-Yu, Gu, Ai-Li, Sun, et al. (2021). Comparative evaluation of high-density polyethylene and polystyrene microplastics pollutants: Uptake, elimination and effects in mussel. *Marine Environmental Research*, *169*, 105329.

Woranuch, S., & Yoksan, R. (2013). Eugenol-loaded chitosan nanoparticles: II. Application in bio-based plastics for active packaging. *Carbohydrate Polymers*, *96*, 586–592.

Wu, W. M., Yang, J., & Criddle, C. S. (2017). Microplastics pollution and reduction strategies. *Frontiers of Environmental Science & Engineering*, *11*(1), 1–4.

Yadav, M., Liu, Y.-K., & Chiu, F.-C. (2019). Fabrication of cellulose nanocrystal/silver/alginate bionanocomposite films with enhanced mechanical and barrier properties for food packaging application. *Nanomaterials, 9*, 1523.
Yap, C. K., Sharifinia, M., Cheng, W. H., Al-Shami, S. A., Wong, K. W., & Al-Mutairi, K. A. (2021). A commentary on the use of bivalve Mollusks in monitoring metal pollution levels. *International Journal of Environmental Research and Public Health, 18*(7), 3386.
Yu, P., Liu, Z., Wu, D., Chen, M., Lv, W., & Zhao, Y. (2018). Accumulation of polystyrene microplastics in juvenile Eriocheir sinensis and oxidative stress effects in the liver. *Aquatic Toxicology, 200*, 28–36.
Yu, Z., Ji, Y., Bourg, V., Bilgen, M., & Meredith, J. C. (2020). Chitin- and cellulose-based sustainable barrier materials: A review. *Emergent Materials, 3*(6), 919–936.
Zeng, J. B., Li, Y. D., Zhu, Q. Y., Yang, K. K., Wang, X. L., & Wang, Y. Z. (2009). A novel biodegradable multiblock poly(ester urethane) containing poly(L-lactic acid) and poly(butylene succinate) blocks. *Polymer, 50*, 1178–1186.
Zhou, W., He, Y., Liu, F., Liao, L., Huang, X., Li, R., et al. (2021). Carboxymethyl chitosan-pullulan edible films enriched with galangal essential oil: Characterization and application in mango preservation. *Carbohydrate Polymers, 256*, 117579.
Zhou, Y., Kumar, M., Sarsaiya, S., Sirohi, R., Awasthi, S. K., Sindhu, R., et al. (2022). Challenges and opportunities in bioremediation of micro-nano plastics: A review. *Science of the Total Environment, 802*, 149823.
Zhu, P., Pan, X., Li, X., Liu, X., Liu, Q., Zhou, J., et al. (2021). Biodegradation of plastics from waste electrical and electronic equipment by greater wax moth larvae (Galleria mellonella). *Journal of Cleaner Production, 310*, 127346.
Zurier, H. S., & Goddard, J. M. (2021). Biodegradation of microplastics in food and agriculture. *Current Opinion in Food Science, 37*, 37–44.
Zych, A., Perotto, G., Trojanowska, D., Tedeschi, G., Bertolacci, L., Francini, N., et al. (2021). Super tough polylactic acid plasticized with epoxidized soybean oil methyl ester for flexible food packaging. *ACS Applied Polymer Materials, 3*(10), 5087–5095.

Printed in the United States
by Baker & Taylor Publisher Services